Environmental Statistics

with S-PLUS

APPLIED ENVIRONMENTAL STATISTICS

Steven P. Millard, *Series Editor*

Environmental Statistics with S-PLUS
Steven P. Millard and Nagaraj K. Neerchal

FORTHCOMING TITLES

Groundwater Monitoring and Regulations
Charles Davis

Statistical Tools for Environmental Quality
Michael E. Ginevan and Douglas E. Splitstone

Environmental Statistics

with S-PLUS

Steven P. Millard
Nagaraj K. Neerchal

CRC Press
Taylor & Francis Group
Boca Raton London New York

CRC Press is an imprint of the
Taylor & Francis Group, an **informa** business

CRC Press
Taylor & Francis Group
6000 Broken Sound Parkway NW, Suite 300
Boca Raton, FL 33487-2742

First issued in paperback 2019

© 2001 by Taylor & Francis Group, LLC
CRC Press is an imprint of Taylor & Francis Group, an Informa business

No claim to original U.S. Government works

ISBN-13: 978-0-8493-7168-4 (hbk)
ISBN-13: 978-0-367-39814-9 (pbk)
Library of Congress Card Number 00-058565

Library of Congress Cataloging-in-Publication Data

Millard, Steven P.
   Environmental statistics with S-Plus / Steven P. Millard, Nagaraj K. Neerchal.
      p.   cm.-- (CRC applied environmental statistics series)
   Includes bibliographical references and index.
   ISBN 0-0893-7168-6 (alk. paper)
   1. Environmental sciences--Statistical methods--Data processing. 2. S-Plus. I.
Neerchal, Nagaraj K. II. Title. III. Series

   GE45.S73 M55 2000
   363.7′007′27--dc21   00-058565

Visit the Taylor & Francis Web site at
http://www.taylorandfrancis.com

and the CRC Press Web site at
http://www.crcpress.com

PREFACE

The environmental movement of the 1960s and 1970s resulted in the creation of several laws aimed at protecting the environment, and in the creation of Federal, state, and local government agencies charged with enforcing these laws. Most of these laws mandate monitoring or assessment of the physical environment, which means someone has to collect, analyze, and explain environmental data. Numerous excellent journal articles, guidance documents, and books have been published to explain various aspects of applying statistical methods to environmental data analysis. Only a very few books attempt to provide a comprehensive treatment of environmental statistics in general, and this book is an addition to that category.

This book is a survey of statistical methods you can use to collect and analyze environmental data. It explains *what* these methods are, *how* to use them, and *where* you can find references to them. It provides insight into what to think about *before* you collect environmental data, how to collect environmental data (via various random sampling schemes), and also how to make sense of it *after* you have it. Several data sets are used to illustrate concepts and methods, and they are available both with software and on the CRC Press Web so that the reader may reproduce the examples. The appendix includes an extensive list of references.

This book grew out of the authors' experiences as teachers, consultants, and software developers. It is intended as both a reference book for environmental scientists, engineers, and regulators who need to collect or make sense of environmental data, and as a textbook for graduate and advanced undergraduate students in an applied statistics or environmental science course. Readers should have a basic knowledge of probability and statistics, but those with more advanced training will find lots of useful information as well.

A unique and powerful feature of this book is its integration with the commercially available software package S-PLUS, a popular and versatile statistics and graphics package. S-PLUS has several add-on modules useful for environmental data analysis, including ENVIRONMENTALSTATS for S-PLUS, S+SPATIALSTATS, and S-PLUS for ArcView GIS. Throughout this book, when a data set is used to explain a statistical method, the commands for and results from the software are provided. Using the software in conjunction with this text will increase the understanding and immediacy of the methods.

This book follows a more or less sequential progression from elementary ideas about sampling and looking at data to more advanced methods of estimation and testing as applied to environmental data. Chapter 1 provides an introduction and overview, Chapter 2 reviews the Data Quality Objectives (DQO) and Data Quality Assessment (DQA) process necessary in the design and implementation of any environmental study, and Chapter 3 talks about exploratory data analysis (EDA). Chapter 4 explains the idea of a population, sample, random variable, and probability distribution. Chapter 5 details various methods for estimating characteristics of a population (probability distribution) based on a sample (data). Chapter 6 discusses prediction intervals, tolerance intervals, and control charts, which have been used in the manufacturing industry for a long time and have been proposed as good methods to use in groundwater monitoring. Chapter 7 reviews the basic ideas in hypothesis testing, including balancing the two possible errors a decision maker can make (e.g., declaring a site contaminated when it really is not, or declaring a site not contaminated when it really is). This chapter also illustrates tests for goodness-of-fit and outliers, classical and nonparametric methods for comparing one, two, or several groups (e.g., background vs. potentially contaminated sites), and the multiple comparisons problem. Chapter 8 returns to the DQO process of Chapter 2 and illustrates how to determine required sample sizes based on the statistical theory presented in Chapters 6 and 7. Chapter 9 discusses linear models, including correlation, simple regression, testing for trend, and multiple regression. This chapter also explains the idea of calibration and how this relates to measuring chemical concentrations and determining various limits associated with the chemical measurement process (i.e., decision limit, detection limit, and quantitation limit). Chapter 10 continues the ideas on calibration discussed in Chapter 9 by explaining how to handle environmental data that contain "less-than-detection-limit" results. Chapter 11 examines methods for dealing with data collected over time that may be serially correlated. Chapter 12 considers how to handle data collected over space that may be spatially correlated. Finally, Chapter 13 discusses the immense field of risk assessment, which usually involves both "hard" data and expert judgment.

TYPOGRAPHIC CONVENTIONS

Throughout this book, we use the following typographic conventions:

- The **bold font** is used for section headings, figure and table titles, equation numbers, and what you click on dialog boxes.
- The *italic font* and ***bold italic font*** are used for emphasis.
- The `courier font` is used to display commands that you type into the S-PLUS Command or Script Window, and the names of variables and functions within S-PLUS (S-PLUS objects).
- The *`italic courier font`* is used for mathematical notation (e.g., variable names, function definitions, etc.).

A NOTE ABOUT S-PLUS AND GRAPHICS

Throughout this book we assume the reader has access to S-PLUS and ENVIRONMENTALSTATS for S-PLUS, knows how to start S-PLUS, and knows how to load the ENVIRONMENTALSTATS for S-PLUS module. Also, in Chapter 12 where we deal with spatial statistics, we assume the reader has access to and is running S+SPATIALSTATS, ArcView GIS, and S-PLUS for ArcView GIS.

At the time this book was being written, the current version of S-PLUS for Windows was S-PLUS 2000 Release 2, and the current version of S-PLUS for UNIX was Version 5.1. The current version of ENVIRONMENTALSTATS for S-PLUS was Version 1.1 Release 2, but Version 2.0 (which includes pull-down menus for the Windows version of S-PLUS) was in Beta Release and should be available by the time this book is published. Throughout this book, we have included examples demonstrating how to use S-PLUS and ENVIRONMENTALSTATS for S-PLUS to create the figures and analyses shown in this book. We assume the reader is using one of the above-mentioned versions of S-PLUS or a later version.

The current UNIX version of S-PLUS only works at the command line, so if you use this version of S-PLUS you can safely ignore sections that begin with the heading Menu (the next UNIX version of S-PLUS, however, will include pull-down menus). If you use the Standard Edition of S-PLUS for Windows, you can only use the pull-down menus and toolbars; you do not have access to the command line, so you can safely ignore sections that begin with the heading Command. If you use the Professional Edition of S-PLUS for Windows, you can use both the pull-down menus and toolbars and the command line, so you can apply the information listed under both headings.

Many of the examples of using the command line under a Command heading use the `attach` function to attach a data frame to your search list. This is done in order to be able to reference the columns of the data frame explicitly without having to use subscript operators. Please be aware that it is possible you may have a data object in your working directory with the same name as the column of the data frame that is being used, in which case your data object will "mask" the column of the data frame. For example, the data frame `epa.94b.tccb.df` contains a column named `Area`. If you already have a data object named `Area` in your working directory, then any examples that involve attaching `epa.94b.tccb.df` and using the column `Area` will not work correctly. In these cases, you must change the name of your data object or use the `$` or `[` operator to direct S-PLUS to the correct data set (e.g., `epa.94b.tccb.df$Area`).

In S-PLUS and ENVIRONMENTALSTATS for S-PLUS it is very easy to produce color plots. In fact, most of the built-in plotting functions produce color plots by default. In this book, however, all of the plots are black and white or grayscale due to the high cost of color printing. The steps for producing color plots are still included in the examples in this book, but the pictures in the book will be in black and white, whereas in many cases the pictures on your computer screen will be in color.

All of the graphs you can create with ENVIRONMENTALSTATS for S-PLUS use traditional S-PLUS graphics and, as such, they are not editable in the Windows version of S-PLUS. If you use the Windows version of S-PLUS, you can convert traditional plots to editable plots by right-clicking on the data part of the graph and choosing Convert to Objects from the context menu.

ABOUT THE AUTHORS

Steven P. Millard was born in Williamsburg, Virginia and raised in Arlington, Virginia. He received a bachelor's degree in mathematics from Pomona College in Claremont, California in 1980, and a Ph.D. in biostatistics from the University of Washington, Seattle in 1995. He has taught at the University of California at Santa Barbara and Saint Martin's College in Lacy, Washington. He ran a consulting unit at the University of Washington and also worked for CH2M Hill on the second Love Canal Habitability Study. From 1990 to 1993 Dr. Millard ran the training program in S-PLUS at Statistical Sciences, Inc. (now part of MathSoft, Inc.). From 1993 through the end of the century, he was a private statistical consultant through his company Probability, Statistics & Information (PSI), working on projects ranging from analyzing water quality data from the Everglades National Park to developing software for automated housing appraisal. Currently, Dr. Millard is the Manager of Consulting Services for the Data Analysis Products Division of MathSoft, Inc. He is the creator of ENVIRONMENTALSTATS for S-PLUS, an add-on module to S-PLUS for environmental data analysis.

Nagaraj K. Neerchal was born in a west coast village in southern India. He received his bachelor's and master's degrees in statistics, in 1981 and 1982, respectively, from the Indian Statistical Institute in Calcutta. He received a Ph.D. in statistics from Iowa State University, Ames in 1986. That same year he joined the Department of Mathematics and Statistics at the University of Maryland Baltimore County, where he is currently a Professor of Statistics. He served as the interim chair of the department 1999 – 2000. Dr. Neerchal's main areas of research interest are time series analysis and methods of analyzing correlated categorical data. He has worked with various environmental applications involving pollution monitoring, sampling, and data quality issues related to policy making. Dr. Neerchal has done extensive consulting for various government agencies and private organizations on statistical problems. He was a Senior Research Scientist at Pacific Northwest National Laboratories, Richland, Washington.

ACKNOWLEDGMENTS

This book is the result of knowledge we have acquired over the course of our careers. Several people were instrumental in imparting this knowledge to us. Steve Millard would like to thank Don Bentley (Mathematics Department, Pomona College), Dennis Lettenmaier (Department of Civil Engineering, University of Washington), and Peter Guttorp (Department of Statistics, University of Washington) for their guidance in his statistical education, and all of the people at MathSoft and on S-news who have responded to his many questions about S-PLUS. Nagaraj Neerchal gratefully acknowledges the continued support of Drs. Phil Ross, Barry Nussbaum, Ron Shafer, and Pepi Lacayo and the constant encouragement of Professor Bimal Sinha.

Several people have helped us prepare this book. We thank Robert Stern at Chapman & Hall/CRC Press for all of his time and help as the managing editor, and Chris Andreasen, our copy editor, for her careful review of the manuscript. We are also grateful to Dr. Richard O. Gilbert of Battelle for his reviews and suggestions, and his help in general as a colleague. We are grateful to Charles Davis, Henry Kahn, and Bruce Peterson for reviewing drafts of some of the chapters. A cooperative agreement between the University of Maryland Baltimore County and the U.S. Environmental Protection Agency supported some of Nagaraj Neerchal's work that was incorporated into this book.

Data used in the examples of Chapters 11 and 12 were from Nagaraj Neerchal's collaborations with Dr. Susan Brunenmeister of the Chesapeake Bay Program and Dr. Steve Weisberg of Versar Inc. (now with the Southern California Water Research Project). We thank them for their collaboration.

Finally, we would like to thank our families, Stacy Selke and Chris Millard, and Chetana, Harsha, and Siri Neerchal, for their continued love and support of our endeavors.

TABLE OF CONTENTS

1   INTRODUCTION

The environmental movement of the 1960s and 1970s resulted in the creation of several laws aimed at protecting the environment, and in the creation of Federal, state, and local government agencies charged with enforcing these laws. In the U.S., laws such as the Clean Air Act, the Clean Water Act, the Resource Conservation and Recovery Act, and the Comprehensive Emergency Response and Civil Liability Act mandate some sort of monitoring or comparison to ensure the integrity of the environment. Once you start talking about monitoring a process over time, or comparing observations from two or more sites, you have entered the world of numbers and statistics. In fact, more and more environmental regulations are mandating the use of statistical techniques, and several excellent books, guidance documents, and journal articles have been published to explain how to apply various statistical methods to environmental data analysis (e.g., Berthoux and Brown, 1994; Gibbons, 1994; Gilbert, 1987; Helsel and Hirsch, 1992; McBean and Rovers, 1998; Ott, 1995; Piegorsch and Bailer, 1997; ASTM, 1996; USEPA, 1989a,b,c; 1990; 1991a,b,c; 1992a,b,c,d; 1994a,b,c; 1995a,b,c; 1996a,b; 1997a,b). Only a very few books attempt to provide a comprehensive treatment of environmental statistics in general, and even these omit some important topics.

This explosion of regulations and mandated statistical analysis has resulted in at least four major problems.

- Mandated procedures or those suggested in guidance documents are not always appropriate, or may be misused (e.g., Millard, 1987a; Davis, 1994; Gibbons, 1994).
- Statistical methods developed in other fields of research need to be adapted to environmental data analysis, and there is a need for innovative methods in environmental data analysis.
- The backgrounds of people who need to analyze environmental data vary widely, from someone who took a statistics course decades ago to someone with a Ph.D. doing high-level research.
- There is no single software package with a comprehensive treatment of environmental statistics.

This book is an attempt to solve some of these problems. It is a survey of statistical methods you can use to collect and analyze environmental data. It explains *what* these methods are, *how* to use them, and *where* you can find references to them. It provides insight into what to think about *before* you collect environmental data, how to collect environmental data (via various

1

random sampling schemes), and also and how to make sense of it *after* you have it. Several data sets are used to illustrate concepts and methods, and are available in ENVIRONMENTALSTATS for S-PLUS (see below) and/or on the CRC Press Web site at www.crcpress.com so that the reader may reproduce the examples. You will also find a list of relevant URLs on this Web site. The appendices of this book include an extensive list of references and an index.

A unique and powerful feature of this book is its integration with the commercially available software package S-PLUS, a popular and powerful statistics and graphics package. S-PLUS has several add-on modules useful for environmental data analysis, including ENVIRONMENTALSTATS for S-PLUS, S+SPATIALSTATS, and S-PLUS for ArcView GIS. Throughout this book, when a data set is used to explain a statistical method, the commands for and results from the software are provided. Using the software in conjunction with this text will increase the understanding and immediacy of the methods.

INTENDED AUDIENCE

This book grew out of the authors' experience as teachers, consultants, and software developers. It is intended as both a reference book for environmental scientists, engineers, and regulators who need to collect or make sense of environmental data, and as a textbook for graduate and advanced undergraduate students in an applied statistics or environmental science course. Readers should have a basic knowledge of probability and statistics, but those with more advanced training will find lots of useful information as well. Readers will find that topics are introduced at an elementary level, but the theory behind the methods is explained as well, and all pertinent equations are included. Each topic is illustrated with examples.

ENVIRONMENTAL SCIENCE, REGULATIONS, AND STATISTICS

As a brief introduction to some of the problems involved in environmental statistics, this section discusses three examples where environmental science, regulations, and statistics intersect. Each of these examples illustrates several issues that need to be considered in sampling design and statistical analysis. We will discuss many of these issues both in general terms and in detail throughout this book.

Groundwater Monitoring at Hazardous and Solid Waste Sites

The Resource Conservation and Recovery Act (RCRA) requires that groundwater near hazardous waste sites and municipal solid waste sites be monitored to ensure that chemicals from the site are not leaking into the groundwater (40 CFR Parts 264 and 265; 40 CFR Part 258). So how do you design a program to monitor groundwater?

Several Federal and state guidance documents have been published addressing the design and statistical issues associated with complying with RCRA regulations (e.g., USEPA, 1989b; 1991c; 1992b,c). The current practice at most sites is to start in a phase called *detection monitoring* in which groundwater is sampled at *upgradient* and *downgradient* wells a number of times each year. The groundwater at the upgradient wells is supposed to represent the quality of *background* groundwater that has not been affected by leakage from the site. Figure 1.1 is a simple schematic for the physical setup of such a monitoring program. More detailed figures can be found in Sara (1994, Chapters 9 to 11) and Gibbons (1995, p. 186).

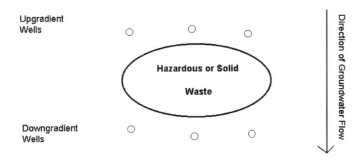

Figure 1.1   Simple schematic of an aerial view of a groundwater monitoring system

During detection monitoring, the groundwater samples are analyzed for *indicator parameters* such as pH, conductance, total organic carbon, and total organic halides. For each indicator parameter and each downgradient well, the value of the parameter at the downgradient well is compared to the value at the upgradient well(s). If the value of the parameter is deemed to be "above background" then, depending on the permit, the owner/operator of the site may be required to take more samples, or the site may enter a second phase of monitoring called *assessment monitoring* or *compliance monitoring*.

In assessment or compliance monitoring, the owner/operator of the site is required to start analyzing the groundwater samples for other chemical parameters, such as the concentrations of specific chemicals. The concentrations of chemicals in the groundwater from downgradient wells are compared to fixed concentration limits (Ground Water Protection Standards or GWPS) such as a Maximum Contaminant Level (MCL) or Alternative Concentration Limit (ACL). For any specified chemical, if the concentration is "above" the GWPS, the site enters a third phase called ***corrective action monitoring***.

There are several basic scientific design and statistical issues involved in monitoring groundwater, including:

- How do you determine what constitutes an upgradient well and what constitutes a downgradient well? Can you be sure the gradient will stay the same over time?
- What chemicals are contained in the site? Are the mandated indicator parameters for detection monitoring good indicators of leakage of these particular chemicals?
- During assessment monitoring, are you required to test for chemicals that are not contained in the site? If so, why?
- For detection monitoring, how do you determine whether an indicator parameter at a downgradient well is "above" background? For each indicator parameter, what kind of increase in value is important to detect and how soon?
- Is there "significant" spatial and/or temporal variability in any of the indicator parameters or chemical concentrations in the upgradient area? If so, how do you account for this when comparing upgradient and downgradient wells? Is it possible to use intrawell comparisons instead of comparing downgradient wells with upgradient wells?
- What other sources of random variation are present? Is there a lot of variability between samples taken on the same day? Is there a lot of variability in the chemical measurement process? Are different laboratories being used, and if so, is there a lot of variability between labs?
- For assessment monitoring, what is the basis of each GWPS? How do you tell whether chemical concentrations at downgradient wells are "above" the GWPS?
- How do you account for the possibility of false alarms, which involve increased monitoring costs, and the possibility of missing contamination, which involves a potential threat to public health?

Soil Cleanup at Superfund Sites

The Comprehensive Emergency Response and Civil Liability Act (CERCLA), also known as "Superfund," requires a remedial investigation/feasibility study (RI/FS) at each site on the National Priorities List (NPL) to determine the extent of contamination and the risks posed to human health and the environment. The U.S. Environmental Protection Agency (USEPA) has developed several guidance documents that discuss design and analysis issues for various stages of this process (USEPA, 1987a,b; 1989a,c; 1991b; 1992b,d; 1994b; 1996b,c).

The guidance documents **Soil Screening Guidance: User's Guide** (USEPA, 1996b) and **Soil Screening Guidance: Technical Background Document** (USEPA, 1996c) discuss the use of Soil Screening Levels (SSLs) at Superfund sites that may have future residential land use to determine whether soil in a particular area requires further investigation and/or remediation, or if it can be left alone (or at least does not require any further attention under CERCLA). The guidance suggests stratifying the site into areas that are contaminated, areas unlikely to be contaminated, and areas that may be contaminated. Within each stratum, the guidance suggests dividing the area into exposure areas (EAs) that are up to a half acre in size, and taking soil samples within each EA. For each EA, the concentration of a particular chemical of concern is compared with the SSL. If the concentration is "greater" then the SSL, then the EA requires further investigation.

This soil screening guidance involves several basic scientific design and statistical issues, including:

- How do you know what chemicals you are looking for?
- How do you know the boundary of the area to look at? How do you determine which areas are contaminated, which are not, and which might be?
- What is the basis for an SSL for a particular chemical?
- How do you determine whether the chemical concentration in the soil is "greater" than the SSL?
- What are the sources of random variation in the data (e.g., field variability, collector variability, within lab variability, between lab variability, etc.), and how do you account for them?
- How do you account for the possibility of false alarms (saying the chemical concentration is greater than the SSL when it is not), which involves unnecessary costs for further investigation, and the possibility of missing contamination (saying the chemical concentration is less than the SSL when in fact it not), which involves a potential threat to public health?

Monitoring Air Quality

The Clean Air Act is the comprehensive Federal law that regulates air emissions from area, stationary, and mobile sources. This law authorizes the USEPA to establish National Ambient Air Quality Standards (NAAQS), which are national targets for acceptable concentrations of specific pollutants in the air.

There are two kinds of standards: primary and secondary. Primary standards set limits to protect public health, including the health of "sensitive" populations such as asthmatics, children, and the elderly. Secondary standards set limits to protect public welfare, including effects on soils, water, crops, vegetation, buildings, property, animals, wildlife, weather, visibility, transportation, and other economic values, as well as personal comfort and well-being.

The USEPA has set national air quality standards for seven principal air pollutants, called *criteria air pollutants*: carbon monoxide (CO), lead (Pb), nitrogen dioxide (NO_2), volatile organic compounds (VOCs), ozone (O_3), particulate matter (PM-10), and sulfur dioxide (SO_2). As an example, the primary standard for ozone is based on comparing a specific limit to the daily maximum ozone measurement among the network of stations monitoring a specific area. The monitors at each station record ozone concentrations continuously in time, so there are several possible ways to define the "daily maximum ozone concentration." Between 1978 and 1997, the daily maximum ozone concentration was based on averaging the concentrations for each hour to produce 24 observations per station per day, and the maximum daily concentration at a station was defined to be the maximum of these 24 values. The daily maximum value (over all of the monitoring stations) could exceed 0.12 parts per million (ppm) only three times or fewer within a 3-year period.

The proposed new standard for ozone divides the 24-hour day into three 8-hour blocks, concentrations are averaged within each 8-hour block to produce three observations per station per day, and the maximum daily concentration at a station is defined to be the maximum of these three values. The standard looks at the fourth highest daily maximum concentration over all stations within a single year. If the 3-year average of these annual fourth highest daily maximum concentrations is less than 0.08 ppm, then the standard is met.

The scientific basis for national ambient air quality standards depends on knowledge about the effects of air pollutants on health and the environment, which involves clinical, epidemiological, and field studies, all of which depend on scientific design and statistical analysis. Also, USEPA monitors trends in air quality over time, which involves time series analysis. Nychka et al. (1998) present several different types of statistical analyses of air quality data.

OVERVIEW

This section explains the background and layout of this book.

What is Environmental Statistics?

Environmental statistics is simply the application of statistical methods to problems concerning the environment. Examples of activities that require the use of environmental statistics include:

- Monitoring air or water quality.
- Monitoring groundwater quality near a hazardous or solid waste site.
- Using risk assessment to determine whether an area with potentially contaminated soil needs to be cleaned up, and, if so, how much.
- Assessing whether a previously contaminated area has been cleaned up according to some specified criterion.
- Using hydrological data to predict the occurrences of floods.

The term "environmental statistics" must also include work done in various branches of ecology, such as animal population dynamics and general ecological modeling, as well as other fields, such as geology, chemistry, epidemiology, oceanography, and atmospheric modeling. This book concentrates on statistical methods to analyze chemical concentrations and physical parameters, usually in the context of mandated environmental monitoring.

Environmental statistics is a special field of statistics. Probability and statistics deal with situations in which the outcome is not certain. They are built upon the concepts of a *population* and a *sample* from the population.

Probability deals with predicting the characteristics of the sample, given that you know the characteristics of the population (e.g., what is the probability of picking an ace out of a deck of 52 well-shuffled standard playing cards?). *Statistics* deals with inferring the characteristics of the population, given information from one or more samples from the population (e.g., after 100 times of randomly choosing a card from a deck of 20 unknown playing cards and then replacing the card in the deck, no ace has appeared; therefore the deck probably does not contain any aces).

The field of environmental statistics is relatively young and employs several statistical methods that have been developed in other fields of statistics, such as sampling design, exploratory data analysis, basic estimation and hypothesis testing, quality control, multiple comparisons, survival analysis, and Monte Carlo simulation. Nonetheless, special problems have motivated innovative research, and both traditional and new journals now report on statistical methods that have been developed in the context of environmental monitoring. (See Appendix A: References for a comprehensive list of journal articles, guidance documents, and general textbooks that deal with environmental statistics and related topics.)

Where Do the Data Come from?

If a law says that you have to monitor the environment because your factory is emitting chemicals into the air or water, or because you run a hazardous waste site and you have to make sure nothing is seeping into the groundwater, or because you want to develop a shopping mall on a piece of real estate that used to be occupied by a plant that put creosote on railroad ties, then you have to figure out how to collect and analyze the data. But before you collect the data, you have to figure out what the question is that you are trying to answer.

The **Data Quality Objectives (DQO) Process** is a formal method for deciding what the question is, what data need to be collected to answer the question, how the data will be collected, and how a decision will be made based on the data. It is the first and most important step of any environmental study. The DQO process is discussed in Chapter 2, along with various methods of random sampling.

The actual collection and laboratory analysis of a physical sample from the environment that eventually leads to a reported number (e.g., concentration of trichloroethylene) involves several steps (Clark and Whitfield, 1994; Ward et al., 1990, p. 19):

1. Sample collection
2. Sample handling
3. Transportation
4. Sample receipt and storage at laboratory
5. Sample work up
6. Sample analysis
7. Data entry
8. Data manipulation
9. Data reporting

Figure 1.2, reproduced from Clark and Whitfield (1994, p. 1065), illustrates these steps, along with sources of variability that cause the measured concentration to deviate from the true concentration. Sources of variability are discussed in more detail in Chapter 2.

Once the physical samples are collected, they must be analyzed in the laboratory to produce measures of chemical concentrations and/or physical parameters. This is a whole topic in itself. An environmental chemist does not just scoop a piece of dirt out of a collection vial, dissolve it in water or chemicals in a test tube, stick the tube in a machine, and record the concentrations of all the chemicals that are present. In fact, the process of measuring chemical concentrations in soil, water, or air has its own set of DQO steps! Chapter 9 discusses some of the aspects of chemometrics and the important topic of machine calibration and detection limit.

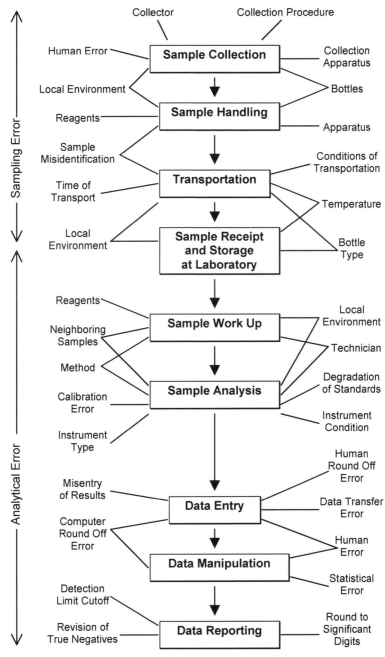

Figure 1.2 The steps involved in producing environmental data, and their associated sources of variability. (From Clark and Whitfield, 1994. *Water Resources Bulletin* **30**(6), 1063–1079. With permission of American Water Resources Association, Herndon, VA.)

How Do You Analyze the Data Once You Have It?

When most people think of environmental statistics, they think of the data analysis part. The design of the sampling program and the DQO process are much more important, however. In fact, one of the steps in the DQO process is to specify how you will analyze the data once you have it.

Chapters 3 to 9 discuss various ways of analyzing environmental data, starting with exploratory data analysis (EDA) in Chapter 3. Chapter 4 explains the idea of a population, sample, random variable, and probability distribution, and gives examples of each of these in the context of a real data set. Chapter 5 discusses various methods for estimating characteristics of a population (probability distribution) based on a sample (data). Chapter 6 discusses prediction intervals, tolerance intervals, and control charts, which have been proposed as good methods to use in groundwater monitoring. Chapter 7 reviews the basic ideas in hypothesis testing, including balancing the two possible errors a decision-maker can make (e.g., declaring a site contaminated when it really is not, or declaring a site not contaminated when it really is). This chapter also illustrates tests for goodness-of-fit and outliers, classical and nonparametric methods for comparing one, two, or several groups (e.g., background vs. potentially contaminated sites), and the multiple comparisons problem. Chapter 8 returns to the DQO process of Chapter 2 and illustrates how to determine required sample sizes based on the statistical theory presented in Chapters 6 and 7. Chapter 9 discusses linear models, including correlation, simple regression, testing for trend, and multiple regression. In addition, this chapter explains the idea of calibration and how this relates to measuring chemical concentrations and determining various limits associated with the process (i.e., decision limit, detection limit, quantitation limit, etc.).

Special Topics

Chapter 10 continues the ideas on calibration discussed in Chapter 9 by explaining how to handle environmental data that contain "less-than-detection-limit" results. Chapter 11 examines methods for dealing with data collected over time that may be serially correlated. Chapter 12 considers how to handle data collected over space that may be spatially correlated. Finally, Chapter 13 discusses the immense field of risk assessment, which usually involves both "hard" data and expert judgment.

DATA SETS AND CASE STUDIES

Throughout this book, we use several data sets to illustrate statistical concepts and methods of analyzing environmental data. Many of these data sets are taken from regulatory guidance documents, but a few of them are larger data sets from the real world of environmental monitoring.

SOFTWARE

As mentioned earlier, throughout this book we use the software package S-PLUS and some of its add-on modules to display computations and graphs. S-PLUS is a popular and powerful software program for statistical and graphical analysis. You can produce results either by using drop-down menus and toolbars, or by typing commands in a command window. ENVIRONMENTALSTATS for S-PLUS is an add-on module to S-PLUS that provides several graphical and statistical methods that are used extensively in environmental statistics. S+SPATIALSTATS is an add-on module for statistical analysis of spatial data and includes kriging. S-PLUS for ArcView GIS is an add-on module that lets you link the statistical and graphical tools in S-PLUS with the mapping and visual tools of ArcView. In this book, when a data set is used to explain a statistical method, the commands for and results from the software are provided. Information on the software providers is listed below.

- **S-PLUS, ENVIRONMENTALSTATS FOR S-PLUS, S+SPATIALSTATS, and S-PLUS for ArcView GIS**

    MathSoft, Inc.
    Data Analysis Products Division
    1700 Westlake Ave N, Suite 500
    Seattle, WA 98109 U.S.A.
    800-569-0123
    mktg@splus.mathsoft.com
    www.splus.mathsoft.com

    MathSoft International
    Knightway House, Park Street
    Bagshot, Surrey GU19 5AQ
    United Kingdom
    +44 1276 475350
    splus@mathsoft.co.uk
    www.splus.mathsoft.com

- **ArcView GIS**

    Environmental Systems Research Institute, Inc. (ESRI)
    380 New York Street
    Redlands, CA 92373-8100 U.S.A.
    909-793-2853
    www.esri.com

SUMMARY

- Several laws mandate some sort of monitoring or comparison to ensure the integrity of the environment.
- This explosion of regulations and mandated statistical analysis has resulted in several problems, including inappropriate or misused statistical procedures, a need for more research, a wide variety of backgrounds in the people who need to use environmental statistics, and a lack of comprehensive software. This book is an attempt to address these problems.
- Probability deals with predicting the characteristics of the sample, given that you know the characteristics of the population.
- Statistics deals with inferring the characteristics of the population, given information from one or more samples from the population.
- Environmental statistics is the application of statistical methods to problems concerning the environment, such as monitoring air or water quality, or comparing chemical concentrations at two sites.
- This book discusses the importance of determining the question that needs to be answered, planning the design of an environmental study, how to look at data once you have collected it, and how to use statistical methods to help in the decision-making process.
- S-PLUS and its add-on modules, including ENVIRONMENTALSTATS for S-PLUS, are useful tools for the analysis of environmental data.

EXERCISES

1.1. Compile a partial list of Federal, state, and local agencies that deal with the environment. Two good places to start are the government section of the telephone book, and national agency sites on the World Wide Web. (The URL of the U.S. Environmental Protection Agency is www.epa.gov. Also, www.probstatinfo.com provides links to agencies that deal with the environment.)

1.2. Compile a partial list of national, state, and local regulations that require some sort of environmental monitoring.

1.3. Look in the Yellow Pages (published or on the Web) under the heading "Environmental." What kinds of listings are there?

2 DESIGNING A SAMPLING PROGRAM, PART I

The DQO Process

The first and most important step of any environmental study is to define the objectives of the study and to design the sampling program. This chapter discusses the basic scientific method and the Data Quality Objectives (DQO) process, which is a formal mechanism for implementing the scientific method and identifying important information that needs to be known in order to make a decision based on the outcome of the study (e.g., clean up the site or leave it alone). One of the major steps in the DQO process involves deciding how you will sample the environment in order to produce the information you need to make a decision. We therefore also discuss several sampling methods in this chapter as well. Finally, we present a real-world case study to illustrate the DQO process.

THE BASIC SCIENTIFIC METHOD

Any scientific study, whether it involves monitoring pollutants in the environment, determining the efficacy of a new drug, or attempting to improve the precision of airplane parts, can eventually be boiled down to trying to determine a cause and effect relationship. Over the centuries, science has developed a set of rules to follow to try to rationally determine cause and effect. These rules can be summarized as follows:

1. Form a hypothesis about the relationship between the supposed cause and the observed effect (e.g., the presence of a pollutant in a river has decreased the population of a particular species of fish).
2. Perform an experiment in which one set of subjects (e.g., fish, people, petri dishes, etc.) is exposed to the cause, and another set of subjects experiences exactly the same conditions as the first set of subjects, except they are not exposed to the cause. The group of subjects exposed to the cause is termed the *experimental group* or *exposed group*, and the other group is termed the *control group*. All subjects must be similar to one another and be randomly assigned to the experimental or control group.
3. Record and analyze the results of the experiment.
4. Revise the hypothesis based on the results. Repeat Steps 2 to 4.

The scientific method recognizes the fact that our environment is constantly changing and that any of these changes may create an observed effect which may or may not be related to some cause we have hypothesized. The only way to determine whether a specific cause creates a specific effect is through careful experimentation that matches the experimental group with the control group on every possible condition except for allowing the experimental group to be exposed to the cause. In reality, this is often extremely difficult or impossible to achieve.

Observational Experiments

Often in environmental studies, the experiment is not actually controlled by scientists doing the investigation, but rather it is an *observational experiment* (often called an *epidemiological study*) in which the experimental group is a group of people, organisms, aquifers, etc. that has been exposed to a cause (e.g., a pollutant) by virtue of physical location or other factors. A control group is then selected based upon the characteristics of the experimental group. An observational experiment can be very useful for initially identifying possible causes and effects, but suffers from the major drawback that the experimental group is self-selected, rather than randomly assigned, and therefore there might be something peculiar about this group that caused the observed effect, rather than the hypothesized cause. The tobacco industry used this argument for years to claim that there is no "proof" that smoking causes cancer, since all studies on smoking and cancer in humans are observational experiments (smoking was not randomly assigned to one group and no smoking to another). On the other hand, if many, many observational experiments result in the same conclusions, then this is very good evidence of a direct cause and effect.

The Necessity of a Good Sampling Design

No amount of advanced, cutting-edge statistical theory and techniques can rescue a study that has produced poor quality data, not enough data, or data irrelevant to the question it was meant to answer. From the very start of an environmental study, there must be a constant dialog between the data producers (field and lab personnel, data coders, etc.), the data users (scientists and statisticians), and the ultimate decision maker (the person or organization for whom the study was instigated in the first place). All persons involved in the study must have a clear understanding of the study objectives and the limitations associated with the chosen physical sampling and analytical (measurement) techniques before anyone can make any sense of the resulting data.

The DQO Process is the Scientific Method

The DQO process is really just a formalization of sampling design and the scientific method. We will explain the DQO process in detail, but first we need to understand the concepts of population, sample, random sampling, and hypothesis test.

WHAT IS A POPULATION AND WHAT IS A SAMPLE?

In everyday language, the word "population" refers to all the people or organisms contained within a specific country, area, region, etc. When we talk about the population of the United States, we usually mean something like "the total number of people who currently reside in the U.S."

In the field of statistics, however, the term *population* is defined operationally by the question we ask: it is the entire collection of measurements about which we want to make a statement (Zar, 1999, p. 16; Berthoux and Brown, 1994, p. 7; Gilbert, 1987, Chapter 2).

For example, if the question is "What does the concentration of dissolved oxygen look like in this stream?", the question must be further refined until a suitable population can be defined: "What is the average concentration of dissolved oxygen in a particular section of a stream at a depth of 0.5 m over a particular 3-day period?" In this case, the population is the set of all possible measurements of dissolved oxygen in that section of the stream at 0.5 m within that time period. The section of the stream, the time period, the method of taking water samples, and the method of measuring dissolved oxygen all define the population.

A *sample* is defined as some subset of a population (Zar, 1999, p. 17; Berthoux and Brown, 1994, p. 7; Gilbert, 1987, Chapter 2). If the sample contains all the elements of the population, it is called a *census*. Usually, a population is too large to take a census, so a portion of the population is sampled. The statistical definition of the word sample (a selection of individual population members) should not be confused with the more common meaning of a *physical sample* of soil (e.g., 10g of soil), water (e.g., 5ml of water), air (e.g., 20 cc of air), etc.

RANDOM VS. JUDGMENT SAMPLING

Judgment sampling involves subjective selection of the population units by an individual or group of individuals (Gilbert, 1987, Chapter 3). For example, the number of samples and sampling locations might be determined based on expert opinion or historical information. Sometimes, public opinion might play a role and samples need to be collected from areas known to be highly polluted. The uncertainty inherent in the results of a judgment sample cannot be quantified and statistical methods cannot be applied to judgment samples. Judgment sampling does *not* refer to using prior information and the knowledge of experts to define the area of concern, define the population, or plan the study. Gilbert (1987, p. 19) also describes "haphazard" sampling, which is a kind of judgment sampling with the attitude that "any sample will do" and can lead to "convenience" sampling, in which samples are taken in convenient places at convenient times.

Probability sampling or ***random sampling*** involves using a random mechanism to select samples from the population (Gilbert, 1987, Chapter 3). All statistical methods used to quantify uncertainty assume some form of random sampling has been used to obtain a sample. At the simplest level, a simple random sample is used in which each member of the population has an equal chance of being chosen, and the selection of any member of the population does not influence the selection of any other member. Other probability sampling methods include stratified random sampling, composite sampling, and ranked set sampling. We will discuss these methods in detail later in this chapter.

THE HYPOTHESIS TESTING FRAMEWORK

Every decision you make is based on a set of assumptions. Usually, you try to make the "best" decision given what you know. The simplest decision involves only two possible choices, with the choice you make depending only on whether you believe one specific condition is true or false. For example, when you get up in the morning and leave your home, you have to decide whether to wear a jacket or not. You may make your choice based on whether you believe it will rain that day. Table 2.1 displays the framework for your decision-making process.

	What Happens	
Your Decision	It Does Not Rain	It Rains
Wear Jacket	Mistake I	Correct Decision
Leave Jacket	Correct Decision	Mistake II

Table 2.1   Hypothesis testing framework for wearing a jacket

If you wear your jacket and it rains, you made a "correct" decision, and if you leave your jacket at home and it does not rain then you also made a "correct" decision. On the other hand, if you wear your jacket and it does not rain, then you made a "mistake" in some sense because you did not need to wear your jacket. Also, if you leave your jacket at home and it does rain, then you also made a "mistake." Of course, you may view the "cost" of making mistake II as greater than the cost of making mistake I.

You can think of the above example as an illustration of a hypothesis test. We will say the null hypothesis is that it will not rain today. You then gather information from the television, Internet, radio, newspaper, or personal observation and make a decision based on this information. If the null hypothesis is true and you decide not to leave your jacket at home, then you made the "correct" decision. If, on the other hand, the null hypothesis is false and you decide to leave your jacket at home, you have made a mistake.

Decisions regarding the environment can also be put into the hypothesis testing framework. Table 2.2 illustrates this framework in the context of deciding whether contamination is present in the environment. In this case, the null hypothesis is that no contamination is present.

	Reality	
Your Decision	No Contamination	Contamination
Contamination	Mistake: Type I Error (Probability = α)	Correct Decision (Probability = $1-\beta$)
No Contamination	Correct Decision	Mistake: Type II Error (Probability = β)

Table 2.2   Hypothesis testing framework for deciding on the presence of contamination in the environment when the null hypothesis is "no contamination"

Statisticians call the two kinds of mistakes you can make a *Type I error* and a *Type II error*. Of course, in the real world, once you make a decision, you take an action (e.g., clean up the site or do nothing), and you hope to find out eventually whether the decision you made was the correct decision.

For a specific decision rule, the probability of making a Type I error is usually denoted with the Greek letter α (alpha). This probability is also called the *false positive rate*. The probability of making a Type II error is usually denoted with the Greek letter β (beta). This probability is also called the *false negative rate*. The probability $1-\beta$ denotes the probability of correctly deciding there is contamination when in fact it is present. This probability is called the *power* of the decision rule. We will talk more about the hypothesis testing framework in Chapters 6 and 7, and we will talk about power in Chapter 8.

COMMON MISTAKES IN ENVIRONMENTAL STUDIES

The most common mistakes that occur in environmental studies include the following:

- **Lack of Samples from Proper Control Populations**. If one of the objectives of an environmental study is to determine the effects of a pollutant on some specified population, then the sampling design must include samples from a proper control population. This is a basic tenet of the scientific method. If control populations were not sampled, there is no way to know whether the observed effect was really due to the hypothesized cause, or whether it would have occurred anyway.
- **Using Judgment Sampling to Obtain Samples.** When judgment sampling is used to obtain samples, there is no way to quantify the precision and bias of any type of estimate computed from these samples.
- **Failing to Randomize over Potentially Influential Factors.** An enormous number of factors can influence the final measure associated with a single sampling unit, including the person doing the sampling, the device used to collect the sample, the weather and field conditions when the sample was collected, the method used to analyze the sample, the laboratory to which the sample was sent, etc. A good sampling design controls for as many potentially influencing factors as possible, and randomizes over the factors that cannot be controlled. For example, if there are four persons who collect data in the field, and two laboratories are used to analyze the results, you would not send all the samples collected by persons 1 and 2 to laboratory 1 and all the samples collected by persons 3 and 4 to laboratory 2, but rather send samples collected by each person to each of the laboratories.
- **Collecting Too Few Samples to Have a High Degree of Confidence in the Results.** The ultimate goal of an environmental study is to answer one or more basic questions. These questions should be stated in terms of hypotheses that can be tested using statistical procedures. In this case, you can determine the probability of rejecting the null hypothesis when in fact it is true (a Type I error), and the probability of not rejecting the null hypothesis when if fact it is false (a Type II error). Usually, the Type I error is set in advance, and the probability of correctly rejecting the null hypothesis when in fact it is false (the power) is calculated for various sample sizes. Too often, this step of determining power and sample size is neglected, resulting in a study from which no conclusions can be drawn with any great degree of confidence.

Following the DQO process will keep you from committing these common mistakes.

THE DATA QUALITY OBJECTIVES PROCESS

The Data Quality Objectives (DQO) process is a systematic planning tool based on the scientific method that has been developed by the U.S. Environmental Protection Agency (USEPA, 1994a). The DQO process provides an easy-to-follow, step-by-step approach to decision-making in the face of uncertainty. Each step focuses on a specific aspect of the decision-making process. Data Quality Objectives are the qualitative and quantitative statements that:

- Clarify the study objective.
- Define the most appropriate type of data to collect.
- Determine the most appropriate conditions under which to collect the data.
- Specify acceptable levels of decision errors that will be used as the basis for establishing the quantity and quality of data needed to support the decision.

Once the DQOs are specified, it is the responsibility of the team to determine the most cost-effective sampling design that meets the DQOs. A sampling design or sampling plan is a set of instructions to use to scientifically investigate the study objective and come up with a quantifiable answer. We will discuss several commonly used sampling designs later in this chapter, but it is important to note that the actual method of analysis/estimation is part of the design, and various available choices must be considered to choose the most resource-effective combination of the method of sampling and the method of analysis/estimation.

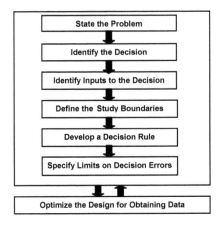

Figure 2.1   The Data Quality Objectives (DQO) Process

Figure 2.1 displays a flowchart of the DQO process. The DQO process is described in detail in USEPA (1994a). The DQO process is a series of seven planning steps, as outlined below.

1. **State the Problem.** Clearly and concisely define the problem so that the focus of the study will be unambiguous. Activities include:

   - Identify members of the planning team.
   - Identify the primary decision maker and each team member's role and responsibilities.
   - Develop a concise description of the problem by describing what is causing the problem and reviewing prior studies and existing information to gain a better understanding of the problem.
   - Specify available resources and relevant deadlines for the study.

2. **Identify the Decision.** Define the decision statement that the study will attempt to resolve. Activities include:

   - Identify the principal study question. For example: Is the concentration of 1,2,3,4-tetrachlorobenzen (TcCB) in the soil at a specific site "significantly" above "background" levels?
   - Define possible (alternative) actions to take based on the answer to the principal study question. For example, if the concentration of TcCB is significantly above background level, then require remediation; otherwise, do nothing.
   - Combine the principal study question and the alternative actions into a decision statement.
   - Organize multiple decisions.

3. **Identify Inputs to the Decision.** Identify the information that needs to be obtained and the measurements that need to be taken to resolve the decision statement. Activities include:

   - Identify the information required to resolve the decision statement. For example, what is the distribution of TcCB concentrations at the site of concern, and what is the distribution of TcCB concentrations for a "background" site?
   - Determine the sources for each item of information. Sources may include previous studies, scientific literature, expert opinion, and new data collections.
   - Identify the information that is needed to establish the action level. For example, will the "background" level of TcCB be established based on sampling a nearby "control" site or based on a regulatory standard?

- Confirm that appropriate analytical methods exist to provide the necessary data. Develop a list of potentially appropriate measurement methods, noting the method detection limit and limit of quantitation for each method.

4. **Define the Study Boundaries.** Define the spatial and temporal boundaries that are covered by the decision statement. A clear connection should be made between the study boundaries and the statement of the problem so that the decisions are relevant to the problem stated in Step 1. Steps 3 and 4 involve defining the population(s) of interest, how it (they) will be sampled, and how the physical samples will be measured. Activities for Step 4 include:

- Specify the characteristics that define the population of interest.
- Define the geographic area within which all decisions apply.
- When appropriate, divide the population into relatively homogeneous strata.
- Determine the timeframe within which all decisions apply.
- Determine when to collect the data.
- Define the scale of decision making. For example, will only one decision be made for the whole site, or will the site be divided into smaller sub-areas and a separate decision made for each sub-area?
- Identify any practical constraints on data collection.

5. **Develop a Decision Rule.** Define the statistical parameter of interest, specify the action level, and integrate the previous DQO outputs into a single statement that describes the logical basis for choosing among alternative actions. Activities include:

- Specify one or more statistical parameter(s) that characterize(s) the population and that are most relevant to the decision statement. For example, the average concentration of TcCB, the median concentration of TcCB, and the 95th percentile of the concentration of TcCB.
- Specify the action level(s) for the study.
- Combine the outputs from the previous DQO steps into one or more "If ...then..." decision rules that define the conditions that would cause the decision maker to choose among alternative actions. For example, "If the average concentration of TcCB in a sub-area is greater than the background average concentration then remediate the sub-area."

6. **Specify Tolerable Limits on Decision Errors.** Define the decision maker's tolerable decision error rates based on a consideration of the consequences of making an incorrect decision. Since decisions are

made based on a sample rather than a census, and there is uncertainty in the measurements, incorrect decisions can be made. The probabilities of making incorrect decisions are referred to as decision error rates. Activities for this step include:

- Determine the full possible range of the parameter of interest.
- Identify the decision errors and formulate the null hypothesis.
- Specify a range of parameter values where the consequences of decision errors are relatively minor (gray region).
- Assign probability values to points above and below the action level that reflect the tolerable probability for the occurrence of decision errors.

Example of a Decision Performance Goal Diagram

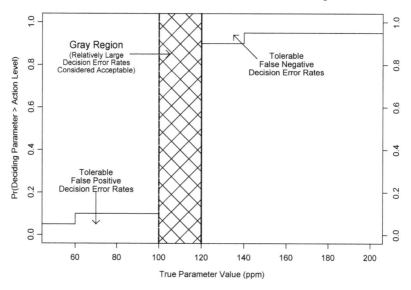

Figure 2.2   Decision performance goal diagram for the null hypothesis that the parameter is less than the action level of 100 ppm

Figure 2.2, similar to Figure 6-2 of USEPA (1994a, p. 36), illustrates a decision performance goal diagram in which the full range of the parameter is assumed to be between about 50 and 200 ppm. The null hypothesis is that the true value of the parameter is less than the action level of 100 ppm. The tolerable false positive decision error rate is 5% as long as the true value is less than 60 ppm. If the true value is between 60 and 100 ppm, then the tolerable false positive rate is 10%. If the true value is between 100 and 120 ppm, this is a gray region where the consequences of a decision error are relatively minor. If the true value is between 120 and 140 ppm, the tolerable false negative decision error rate is 10%, and if the true value is bigger than 140 ppm the tolerable false negative decision error rate is 5%.

7. **Optimize the Design.** Evaluate information from the previous steps and generate alternative sampling designs. Choose the most resource-efficient design that meets all DQOs. Activities include:

- Review the DQO outputs and existing environmental data.
- Develop several possible sampling designs.
- Formulate the mathematical expressions required to solve the design problems for each design alternative. These expressions include the relationship between sample size and decision error rates, and the relationship between sample size and cost of the design.
- Select the optimal sample size that satisfies the DQOs for each possible sampling design.
- Select the most resource-effective sampling design that satisfies all of the DQOs.
- Document the operational details and theoretical assumptions of the selected sampling design in the Sampling and Analysis Plan.

Example of a Power Curve

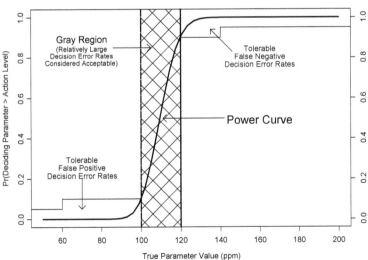

Figure 2.3   Example of a power curve for the null hypothesis that the parameter is less than the action level of 100 ppm

Figure 2.3, similar to Figure 7-1 of USEPA (1994a, p. 40) illustrates a power curve for a particular sampling design drawn on top of the decision performance goal diagram shown in Figure 2.2. Here you can see that this particular sampling design satisfies the performance goals.

Steps 3 and 4 should include developing a *quality assurance project plan (QAPP)* to document *quality assurance (QA)* and *quality control (QC)* and insure the integrity of the final results (USEPA, 1998b). A good QAPP covers instructions for sample collection in the field, handling, laboratory analysis and reporting, data coding, statistical analyses, and reports. Embedded in these instructions is the *chain of custody procedures* for documenting who has custody of the samples and the current conditions of the samples from the point of collection in the field to the analysis at the laboratory. Chain of custody procedures are used to ensure that samples are not lost, tampered with, or improperly stored or handled. See Keith (1991) and USEPA (1998b) for more information.

Step 7 requires information from previous studies and/or a pilot study to quantify the amount of variability that is typical in samples. Once this information is available, you can estimate the required sample size based on the statistical method of analysis that will be used (see Chapter 8). Even if information on sample variability is available from previous studies, however, it is almost always advisable to conduct a pilot study in order to "fine tune" the QA/QC sampling plan and the overall sampling design. Sample size requirements should take into account that a certain proportion of the samples will be unusable due to loss, mislabeling, mishandling, or some other factor that keeps the samples from meeting the specified QA/QC standards.

The DQO process is iterative. During the design optimization step (Step 7), you may discover that there are not enough funds and/or staff available to answer the question, given the required decision error rates specified in Step 6 and the required sample sizes determined in Step 7. In this case, you may need to adjust the budget of the study or the decision error rates, or investigate alternative methods of sampling that may yield smaller variability. Once a sampling design is chosen, you should develop a written protocol for implementing the sampling design and QAPP programs. See Gilbert (1987), Keith (1991), and USEPA (1989a; 1992b; 1994a,b; 1996) for more information on sampling design.

SOURCES OF VARIABILITY AND INDEPENDENCE

Figure 1.2 in Chapter 1 displays several sources of variability in measurements from environmental studies. Some potential sources of variability include:

- Natural variability over space.
- Natural variability over time (including seasonal and year-to-year fluctuations).
- Field sampling variability (sample collection, sample handling, and transportation).
- Within laboratory variability (day-to-day, machine-to-machine, technician-to-technician, etc.).
- Between laboratory variability.

When you are designing a sampling program, it is very important to be aware of these different sources of variability and to know how much they contribute to the overall variability of a measure. For example, RCRA regulations for groundwater monitoring at hazardous and solid waste sites require a minimum of only one upgradient well (see Figure 1.1 in Chapter 1). If there is substantial natural spatial variability in the concentration of some chemical of concern, then we may not be able to tell whether a difference in concentrations between the upgradient well and a downgradient well is due to actual contamination showing up at the downgradient well or simply due to natural spatial variability.

Independent vs. Dependent Observations

A key concept in statistical design and analysis is the idea of *independent observations.* Here is a non-technical definition: observations are independent of one another if knowing something about one observation does not help you predict the value of another observation. As an example of independent observations, suppose you have a standard deck of 52 playing cards, which you shuffle before picking a single card. You look at the card, put it back in the deck, shuffle the deck again, and pick another card. Knowing that the card you picked the first time was the 10 of diamonds will not help you predict what the next card will be when you pick from the deck again. The values of the two different cards that you pick are independent. On the other hand, if you had kept the 10 of diamonds after you picked it and not returned it to the deck, then you know that the next card you pick cannot be the 10 of diamonds. In this case, the two observations are dependent.

The idea of independence is closely linked to the idea of accounting for sources of variability. Suppose we are monitoring the concentration of arsenic in groundwater around a hazardous waste site and the flow of groundwater is extremely slow. If we sample say biweekly or monthly, then the temporal variability at a monitoring well will be small, so that observations taken close together in time will resemble one another more than observations taken further apart in time. Thus, if we combine monthly observations over 2 years, these 24 observations would not be considered to be independent of each other; they would exhibit temporal correlation (see Chapter 11).

We would have to adjust the sampling frequency to something like quarterly to try to produce observations that act like they are independent.

Similarly, if there is no appreciable natural spatial variability of arsenic concentrations over the area we are monitoring, then the combined observations from an upgradient well and a downgradient well could be considered to be independent (assuming there is no leakage from the waste site). On the other hand, if there is substantial spatial variability (but no leakage from the waste site), and the concentrations at the downgradient well tend to be larger than the concentrations at the upgradient well, then the combined observations from the upgradient well and the downgradient well would not be considered independent; knowing which well an observation comes from (upgradient or downgradient) helps us predict the value of the observation. If we know the average concentrations at the upgradient and downgradient wells (e.g., 5 ppb at the upgradient well and 10 ppb at the downgradient well), we can subtract these averages from the observations to produce "residual values." The combined residual values would be considered independent, because knowing whether the residual value came from the upgradient or downgradient well does not help us predict the actual value of the residual value.

Most standard statistical methods assume the observations are independent. This is a reasonable assumption as long as we are careful about accounting for potential sources of variability and randomizing over potentially influential factors (e.g., making sure each laboratory analyzes samples collected by each different collector). Time series analysis, discussed in Chapter 11, and spatial statistics, discussed in Chapter 12, are special ways of accounting for variability over time and space. A very common method of accounting for physical sources of variability in the field is stratified random sampling, which is discussed in the next section.

METHODS OF RANDOM SAMPLING

This section describes four methods of random sampling: simple random sampling, systematic sampling, stratified random sampling, composite sampling, and ranked set sampling. Other methods of random sampling include two-stage and multi-stage random sampling, double random sampling, sequential random sampling, and adaptive random sampling. See Gilbert (1987), Keith (1991, 1996), Thompson (1992), and Cochran (1977) for more information.

In our discussions, we will explain the various methods of random sampling in English. To compare the methods of sampling, however, we need to use equations and talk about concepts like the population mean, sample mean, and variance of the sample mean. We will discuss these concepts in detail in Chapters 3 and 4. For now, you may want to skip the equations and return to them after you have read these later chapters.

Simple Random Sampling

Simple random sampling is, true to its name, the simplest type of sampling design. Recall that a sample is a subset of the population. A simple random sample (SRS) is obtained by choosing the subset in such a way that every individual in the population has an equal chance of being selected, and the selection of one particular individual has no effect on the probability of selecting another individual. The word **random** denotes the use of a probabilistic mechanism to ensure that all units have an equal probability of being selected, rather than the colloquial meaning "haphazard."

In order to realize an SRS from a population, a complete listing of the units in the population may be needed. These units are referred to as **sampling units**. For example, suppose we are interested in determining the concentration of a chemical in the soil at a Superfund site. One way to define the population is as the set of all concentrations from all possible physical samples of the soil (down to some specified depth). In the DQO process, we may further refine our definition of the population to mean the set of all concentrations from all possible physical samples taken on a grid that overlays the site. The grid points are then the sampling units, and an SRS will consist of a subset of these grid points in which each grid point has an equal probability of being included in the sample. Figure 2.4 illustrates the idea of simple random sampling for a grid with $N = 160$ points for which $n = 16$ points were selected at random.

Example of SRS with N=160 and n=16

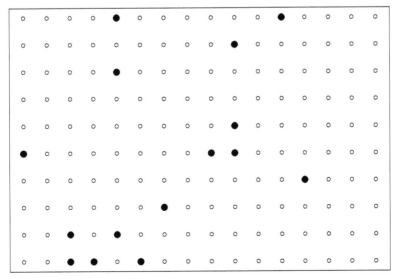

Figure 2.4   Illustration of simple random sampling on a grid

In Chapters 3 and 4 we will discuss the concepts of the population mean, population variance, sample mean, and sample variance. One way to compare sampling methods is based on the variance of the sample mean (how much wiggle it has). For simple random sampling, the population mean or average, denoted by μ, is estimated by the sample mean:

$$\hat{\mu}_{SRS} = \bar{x} = \frac{1}{n} \sum_{i=1}^{n} x_i \qquad (2.1)$$

where n denotes the number of sampling units selected in the SRS and x_i denotes the value of the measurement on the ith sampling unit that was selected. The variance of the sample mean is given by:

$$Var\left(\hat{\mu}_{SRS}\right) = Var\left(\bar{x}\right) = \sigma_{\bar{x}}^2 = \frac{\sigma^2}{n}\left(1 - \frac{n}{N}\right) \qquad (2.2)$$

where σ^2 denotes the population variance and N denotes the number of sampling units in the population (Cochran, 1977, p. 23; Gilbert, 1987, p. 28). Note that as the sample size n increases, the variance of the sample mean decreases. For a finite population, if we take a census so that $n = N$, the sample mean is the same as the population mean and the variance of the sample mean is 0. The quantity n/N in Equation (2.2) is called the **finite population correction factor**. For an infinite population where $N = \infty$, the finite population correction factor is 0 so that the variance of the sample mean depends only on the sample size n.

$$Var\left(\hat{\mu}_{SRS}\right) = Var\left(\bar{x}\right) = \frac{\sigma^2}{n} \qquad (2.3)$$

Often in environmental studies, the finite population correction factor is set to 0 because the size of the population N is much, much larger than the sample size n.

Determining Grid Size

A common question is: "If I decide to overlay the area with a grid and define my population by the set of all possible points on the grid, how fine should the grid be? Should the points be spaced by 10 feet, one foot, half of a foot, or some other distance?" For simple random sampling, the answer is:

Make the grid of possible points to sample from as fine as possible, since you would like to extrapolate your results to the whole area. When you overlay the area with a grid, you are in effect taking a systematic sample of all possible sampling points (see the next section). You therefore may have introduced bias, because even if you sample from all of the points on the grid (instead of taking a random sample of points), the mean based on sampling all of the grid points may not equal the true population mean. As you make the grid finer, however, the mean of all grid points will get close to the true population mean.

Systematic Sampling

Systematic sampling involves choosing a random starting point, and then sampling in a systematic way, for example, along a line every 2 feet, or on a square or triangular grid. Figure 2.5 illustrates systematic sampling on a triangular grid. In this figure, the coordinates of the sampling point in the lower left-hand corner were generated by random numbers; once this coordinate was determined, the other sampling coordinates were completely specified.

Example of Systematic Sample on a Triangular Grid

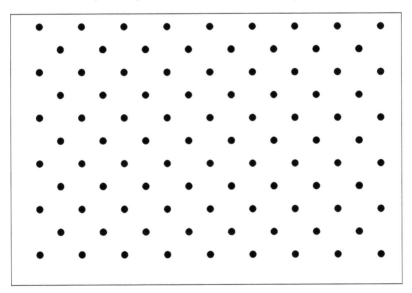

Figure 2.5   Illustration of systematic sampling with a triangular grid

Systematic sampling is useful when you are trying to uncover "hot spots" of highly contaminated areas. Gilbert (1987, Chapter 10) discusses how to determine grid size when using square, rectangular, or triangular grids to search for hot spots. Elipgrid-PC is software for determining grid spacing to detect hotspots with a specified probability. It was originally developed at Oak Ridge National Laboratory and is available at the following URL: http://etd.pnl.gov:2080/DQO/software/elipgrid.html. Visual Sample Plan (VSP; see http://terrassa.pnl.gov:2080/DQO/software/vsp) is an updated version of Elipgrid-PC that also includes sample size calculations for simple random sampling.

If you are trying to estimate a mean or a total, systematic sampling works as well as or better than simple random sampling if there are no trends or natural strata in the population you are sampling. In general, however, it is difficult to obtain reliable estimates of variability with systematic sampling, and if some kind of natural trend or pattern is present, systematic sampling may yield biased estimates (Gilbert, 1987, Chapter 10).

Stratified Random Sampling

Another type of sampling design is called stratified random sampling. In stratified random sampling, the population (site or process) is divided into two or more non-overlapping strata, sampling units are defined within each stratum, then separate simple random or systematic samples are chosen within each stratum. Population members within a stratum are thought to be in some way more similar to one another than to population members in a different stratum. If this is in fact the case, then the sample mean based on stratified random sampling is less variable than the sample mean based on simple random sampling. Figure 2.6 illustrates the idea of stratified random sampling.

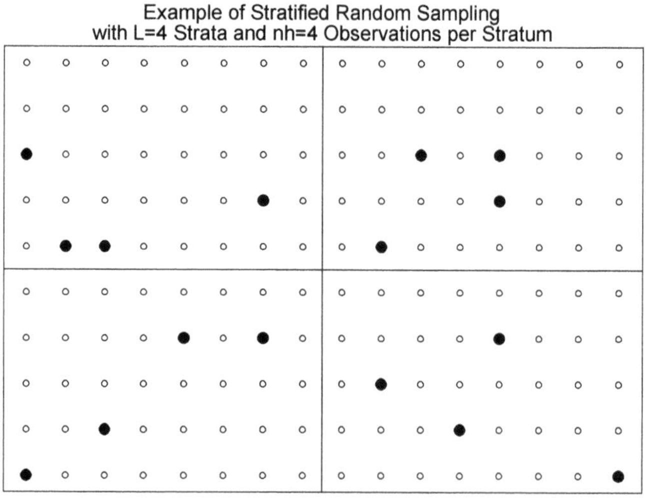

Figure 2.6   Illustration of stratified random sampling

It is important to note that the formulas for estimators such as the sample mean or proportion and the associated confidence intervals (see Chapter 5) need to be modified to be applicable to data from stratified random sampling. For stratified random sampling, the population mean of the hth stratum is estimated by choosing n_h sampling units and computing the sample mean:

$$\hat{\mu}_h = \bar{x}_h = \frac{1}{n_h} \sum_{i=1}^{n_h} x_{hi} \qquad (2.4)$$

where x_{hi} denotes the value of the measurement on the ith sampling unit that was selected in stratum h. The population mean over all strata is estimated by a weighted average of the sample means from each stratum:

$$\hat{\mu}_{Stratified} = \sum_{h=1}^{L} W_h \hat{\mu}_h = \sum_{h=1}^{L} W_h \bar{x}_h \qquad (2.5)$$

where

$$W_h = \frac{N_h}{N} \qquad (2.6)$$

$$N = \sum_{h=1}^{L} N_h \qquad (2.7)$$

L denotes the total number of strata, N_h denotes the number of sampling units in the hth stratum, and N denotes the total number of sampling units (Cochran, 1977, p. 91; Gilbert, 1987, p. 46). If we let n denote the total sample size, then

$$n = \sum_{h=1}^{L} n_h \qquad (2.8)$$

The estimator of the population mean given in Equation (2.5) is *not* the same as the simple sample mean taken over all observations in all the strata unless $n_h/N_h = n/N$ for all of the L strata, that is, the sampling fraction is the same in all of the strata, which is called ***proportional allocation***.

The variance of the stratified sample mean is given by:

$$Var\left(\hat{\mu}_{Stratified}\right) = \sum_{h=1}^{L} W_h^2 \, Var\left(\bar{x}_h\right)$$

(2.9)

$$= \sum_{h=1}^{L} W_h^2 \, \frac{\sigma_h^2}{n_h}\left(1 - \frac{n_n}{N_h}\right)$$

where σ_h^2 denotes the variance of the hth stratum (Cochran, 1977, p. 92; Gilbert, 1987, p. 47). Cochran (1977, pp. 96–99) and Gilbert (1987, pp. 50–52) show how to minimize this variance if you know the variability within each stratum and the cost of taking a sample within each stratum.

When Should You Use Stratified Random Sampling?

The main advantage of the stratified random sampling design is its ability to provide a greater coverage of the population. By choosing sampling units from all strata, you are avoiding the possibility that all of the sampled units may come from the same or adjacent geographical areas and certain parts of the population are not represented. Stratified sampling is beneficial if the population you are sampling is fairly heterogeneous and you have a good idea of how to divide the population up into strata that are fairly homogeneous.

Composite Sampling

So far in our discussions of simple and stratified random sampling, we have inherently assumed that the physical samples are measured only once (e.g., a sample of groundwater is pumped from a monitoring well and the concentration of arsenic in that sample is measured). Sometimes, you may want to take two samples very close together in space or time, sometimes you may want to split a physical sample into two or more subsamples and take a measure on each, or sometimes you may want the laboratory to perform two or more analyses on a single physical sample. Each of these is an example of a ***replicate***, so you must be very explicit about what you mean when you use this term. Replicates are used in QA/QC studies to estimate within-sample variability. If the within-sample variability is fairly large (e.g., of the same order as between-sample variability), and the cost of analysis is fairly small compared to the cost of sample collection, you may want to specify two or more replicates in the sampling design.

Often, however, for environmental studies the cost of collecting a sample is relatively small compared to the cost of analyzing it. For instance, in sampling from a site which has highly radioactive material, the entire crew needs to rehearse the sample collection procedure to minimize the exposure and the time spent at the site for sampling. Once the sampling crew is in the field taking soil samples, however, the additional cost of collecting more soil samples is relatively small compared to the cost of analyzing the soil samples for radioactivity. In such situations, cost-effective sampling designs can be achieved by composite sampling (Lovison et al., 1994; USEPA, 1995a).

Composite samples are obtained by physically mixing the material obtained from two or more sampling units. Then measurements are obtained by analyzing subsamples (replicates) from each composite sample. Figure 2.7 illustrates composite sampling in which a composite sample is created by mixing together three grab samples (sampling units). A total of four composite samples are created from 12 grabs. Two subsamples (replicates) are taken from each of the four composite samples and measured, yielding a total of eight measurements.

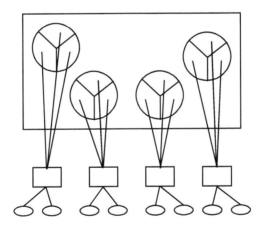

Figure 2.7   Illustration of composite sampling in which three grabs are mixed together to form a composite sample, then two subsamples are taken from the composite sample

Compositing simply represents a physical rather than mathematical mechanism of averaging the measurements from individual sampling units that make up the composite. When the cost of collecting samples is small, a large number of samples may be collected to ensure a large coverage of the population.

Compositing is a cost-effective means of estimating population means. Estimating the uncertainty and obtaining confidence intervals are not always possible with measurements from composite samples since information regarding the extremes is lost when grab samples are composited. Compositing can also be used to efficiently identify "hot-spots" if the sampled material does not deteriorate in storage. In this case, the individual units are strategically composited in groups and the "hot" samples are identified based on measurements from the composite (Sengupta and Neerchal, 1994). It is also possible to apply compositing with stratification or other types of designs as a screening mechanism.

Composite Sampling vs. Simple Random Sampling

Suppose we take n grab samples and divide them evenly into g groups of size h (so $h = n/g$), then composite the grabs within each group to produce g composite samples. We then take r subsamples (replicates) from each composite sample for a total of gr measurements. In Figure 2.7, we have $n = 12$, $g = 4$, $h = 3$, and $r = 2$.

Let x_{ij} denote the measurement from the jth grab sample in the ith group ($i = 1,2,...,g$; $j = 1,2,...,h$). For simple random sampling, our estimate of the population mean is the sample mean based on the n grab samples:

$$\bar{x} = \frac{1}{gh} \sum_{i=1}^{g} \sum_{j=1}^{h} x_{ij} \qquad (2.10)$$

This is an unbiased estimator of the population mean μ, and the variance of this estimator is given by:

$$Var\left(\bar{x}\right) = \frac{\sigma^2}{gh} \qquad (2.11)$$

where σ^2 denotes the population variance of the grab samples:

$$\sigma^2 = Var\left(x_{ij}\right) \qquad (2.12)$$

Let y_{ik} denote the kth subsample (replicate) taken from the ith composite sample (group), where $i = 1,2,...,g$ and $k = 1,2,...,r$. The average of all of the subsamples is:

$$\bar{y} = \frac{1}{gr} \sum_{i=1}^{g} \sum_{k=1}^{r} y_{ik} \qquad (2.13)$$

The ith composite sample is formed by physically mixing the h grab samples within the ith group, so this is a mechanical way to produce a weighted average of the h grab samples within that group. We can therefore write y_{ik} as:

$$y_{ik} = \sum_{j=1}^{h} w_{ijk} x_{ij} \qquad (2.14)$$

where $0 \le w_{ij} \le 1$ and

$$\sum_{j=1}^{h} w_{ijk} = 1 \qquad (2.15)$$

The weights w_{ijk} represent the contribution of the jth grab sample to the ith composite. These weights are assumed to be random with an expected value of $1/h$ and a variance of σ_w^2. In the case of perfect mixing for the ith composite, each weight is exactly equal to $1/h$, the variance of the weights is 0 ($\sigma_w^2 = 0$), and the values of all of the r subsamples are exactly the same as the average of the h grab samples comprising the ith composite group (assuming no measurement error). It is important to note that subsample measurements coming from the same composite (y_{i1}, $y_{i2},...,y_{ir}$) are correlated because they are based on the same set of h grab samples. However, measurements from two different composites are uncorrelated because different composites are based on distinct sets of grab samples. It can be shown that the average of all the subsamples given in Equation (2.13) is an unbiased estimator of the population mean with variance given by:

$$Var\left(\bar{y}\right) = \frac{\sigma^2}{gh} + \frac{h}{gr} \sigma^2 \sigma_w^2 \qquad (2.16)$$

How do you know when it is a good idea to use composite sampling? One way to decide is by comparing the variance of the estimator of the popu-

lation mean under simple random sampling with the variance of the estimator under composite sampling. Comparing Equations (2.11) and (2.16), we
see that the variance of the estimator based on compositing can never be
smaller than the variance of the estimator based on measuring each of the individual grab samples.

Let us consider the case now where you have n grab samples but your
budget only allows you to measure r of the n grab samples, where $r < n$. Is
it better to measure r of the n grab samples and compute the mean based on
these observations, or is it better to combine all n grab samples into one
composite, take r subsamples, and compute the mean based on these observations? In the first case, the variance of the sample mean is given by:

$$Var\left(\bar{x}\right) = \frac{\sigma^2}{r} \qquad (2.17)$$

In the second case, we have $g = 1$ and $h = n$, so based on Equation (2.16) the
variance of the mean of all of the subsamples is given by:

$$Var\left(\bar{y}\right) = \frac{\sigma^2}{n} + \frac{n}{r}\sigma^2\sigma_w^2 \qquad (2.18)$$

The ratio of these two variances is given by:

$$\frac{Var\left(\bar{y}\right)}{Var\left(\bar{x}\right)} = \frac{r}{n} + n\sigma_w^2 \qquad (2.19)$$

Composite sampling will be better than measuring r of the n grab samples
when this ratio is less than 1, which will happen if σ_w^2 is sufficiently small
(i.e., the grabs are well mixed in the composite).

Designing a sampling plan that may include compositing samples requires you to determine the number of grab samples (n), the number of composite groups (g) or the number of grab samples per group (h), and the number of subsamples per composite (r). Thorough mixing reduces σ_w^2 and
makes composite sampling more competitive with simple random sampling.

Using Composite Sampling to Detect "Hot Spots"

As mentioned earlier, information regarding the extremes is lost when grab samples are composited. If, however, the grabs can be stored without any physical deterioration and are available for measurement at a later time, then we can use compositing to efficiently determine which, if any, grab samples exceed a given threshold without necessarily having to measure each individual grab sample. We shall give a simple example and refer the reader to Sengupta and Neerchal (1994) for more details.

Suppose we have $n = 7$ grab samples, identified by $1,2,...7$, and we want to identify any grab sample that exceeds some concentration level L, a level that is deemed to be of concern (e.g., a soil screening level as described in Chapter 1). We can use the following procedure.

1. Make one composite consisting of all seven grabs. If this composite measurement does not exceed $L/7$, then declare all grabs to be "safe." If the measurement exceeds $L/7$, conclude that at least one of the grabs is contaminated and go to Step 2. For this example, we will assume that if the composite measure is greater than $L/7$, it is also less than $2(L/7)$, so that we can conclude that at most one of the grabs is "contaminated."

2. Form 4 composites as follows:

Composite	Constituent Grabs
A	1, 3, 5, 7
B	2, 3, 6, 7
C	4, 5, 6, 7

    Measure composites A, B, and C and compare each measurement to $L/4$.

3. Note that each of A, B, and C could either be above $L/4$ (denoted by a +) or below $L/4$ (denoted by a -). The table below lists the eight possible outcomes and which grab this outcome indicates as being contaminated.

A	B	C	Contaminated Grab
+	-	-	1
-	+	-	2
+	+	-	3
-	-	+	4
+	-	+	5
-	+	+	6
+	+	+	7
-	-	-	None

As an example, suppose that the grab measurements are 1, 3, 7, 29, 4, 9 and 2 ppm. For this method, we do not measure each grab sample but instead make one composite of all the seven grabs. Assuming perfect mixing ($\sigma_w = 0$), the composite measurement is simply the average of the grab measurements, which in this case is 7.86 ppm. Now suppose that a sample is considered "hot" if it exceeds $L = 28$ ppm, so we will compare the concentration of the composite sample to $L/7 = 28/7 = 4$ ppm. Since the composite measurement is between 4 ppm and 8 ppm, we conclude that there is at most one "hot" grab. Thus we form three more composite as in Step 2. The resulting measurements are shown in Table 2.3.

Composite	Constituent Grabs	Measurement (ppm)
A	1,3,5,7	3.5
B	2,3,6,7	5.25
C	4,5,6,7	11

Table 2.3   Example of using compositing to determine which grab samples are "hot"

Comparing these measurements to 7 ppm, we look up (-,-,+) in the table shown in Step 3 above and conclude that grab 4 is possibly "hot." One more measurement on grab 4 will reveal that it is in fact contaminated. Note that we made a total of five measurements instead of seven. For the procedure illustrated here we assume perfect mixing and no measurement errors. Sengupta and Neerchal (1994) show that for a number of situations this procedure leads to savings over testing every grab sample.

Ranked Set Sampling

Ranked Set Sampling (RSS) was first introduced by McIntyre (1952) and is described in Patil et al. (1994b), Johnson et al. (1996), and USEPA (1995b). In RSS, a large number of sampling units are selected from the population and only a subset of those are actually measured. Ranking based on an auxiliary variable is used to determine which of the selected sampling units are measured, hence its name. RSS is worth considering if

- The cost of measurements (lab analyses) is far greater than the cost of collecting the samples.
- An auxiliary characteristic is available that is highly correlated with the main characteristic of interest and is also inexpensive to measure.

We will first describe the procedure to obtain a ranked set sample and then discuss its advantages and disadvantages.

Steps to Create a Ranked Set Sample

In this section we will illustrate how to create a ranked set sample of size $m = 3$ objects, and then explain how to extend this to general values of m.

1. First collect three simple random samples, each of size 3. Let S_{ij} denote the jth sampling unit within the ith SRS ($i = 1,2,3$; $j = 1,2,3$).

SRS	Sampling Units		
1	S_{11}	S_{12}	S_{13}
2	S_{21}	S_{22}	S_{23}
3	S_{31}	S_{32}	S_{33}

2. Next, order the sampling units within each SRS in increasing order according to the values for the auxiliary variable. Let $S_{i(j)}$ denote the jth ordered (ranked) sampling unit within the ith SRS ($i = 1,2,3$; $j = 1,2,3$).

SRS	Sampling Units		
1	$S_{1(1)}$	$S_{1(2)}$	$S_{1(3)}$
2	$S_{2(1)}$	$S_{2(2)}$	$S_{2(3)}$
3	$S_{3(1)}$	$S_{3(2)}$	$S_{3(3)}$

3. Create the ranked set sample by selecting the jth ranked sampling unit for the jth SRS ($j = 1,2,3$). That is, the ranked set sample is $S_{1(1)}$, $S_{2(2)}$, and $S_{3(3)}$ (the diagonal in the table from the upper left to the lower right). These sampling units are shown in boldface in the table above. The actual variable of interest is measured only on these units, and we will denote these measurements by $x_{1(1)}$, $x_{2(2)}$, and $x_{3(3)}$.

In general, a RSS of size m requires taking m simple random samples each of size m (a total of m^2 sampling units).

Ranking a large number of sampling units is difficult and prone to error, especially if the ranking has to be done visually. It is therefore recommended that the size of each SRS that needs to ranked should be no larger than three to five sampling units. You can increase the sample size of your ranked set sample by replicating the process r times. For example, if you want a total of $n = 15$ sampling units in your ranked set sample, simply create $r = 5$ ranked set samples, each of size $m = 3$. This will require a total of $5 \times 3^2 = 45$ sampling units, as opposed to $15^2 = 225$ sampling units.

The Auxiliary Variable

As we discussed earlier in the context of composite sampling, often the laboratory analyses are much more expensive than the cost of collecting samples (per unit). The auxiliary characteristic used for ranking, on the other hand, should be inexpensive to measure. Often, the ranking can be done visually. For example, if we are interested in estimating the average biomass volume in a forest, a ranking of small, medium, and large can be done visually, even though measuring the biomass volume will involve time-intensive analyses. In a contaminated site, visible soil characteristics may provide a means of ranking the units while a soil sample will have to sent to the lab to actually obtain measurements.

Ranked Set Sampling vs. Simple Random Sampling: Estimating the Mean

The sample mean based on RSS is an unbiased estimator of the population mean μ. If we create r ranked set samples, each of size m, to produce a ranked set sample of size $n = rm$, then the variance of the sample mean based on RSS is

$$
Var\left(\bar{x}_{RSS}\right) = \frac{\sigma^2}{n}\left[1 - \frac{1}{m\sigma^2}\sum_{i=1}^{m}\left(\mu_{(i)} - \mu\right)^2\right] \qquad (2.20)
$$

where $\mu_{(i)}$ denotes the expected value of the ith order statistic from a random sample of size m (Patil et al., 1994b). Comparing Equation (2.20) with Equation (2.3), we see that the sample mean based on RSS has a smaller variance than the sample mean based on a SRS of the same size. Thus, fewer samples are needed to achieve the same precision, leading to a savings in sampling costs. It is worth emphasizing, however, that to obtain unbiased estimates of variances you need two or more cycles (see Stokes, 1980 and Bose and Neerchal, 1998). Mode et al. (1999) discuss when ranked set sampling is cost effective based on considering the cost of measuring the auxiliary variable, the cost of ranking, and the cost of measuring the variable of interest.

Equation (2.20) assumes there is perfect correlation between the auxiliary variable and the variable of interest, and that there are no errors in ranking based on the auxiliary variable. Nevertheless, the variance of the sample mean based on RSS is always less than or equal to the variance of the sample mean based on SRS. If there is no correlation between the auxiliary variable and the variable of interest, then RSS will simply produce the same results as SRS.

Ranked Set Samples Are More Regularly Spaced Than Simple Random Samples

We conclude our discussion of RSS by illustrating a very important property of the RSS design with a simple example. Suppose we want to estimate the average weight of a herd of elephants. Furthermore, assume this simple herd of elephants has one calf for every mother elephant and has no father elephants at all. We will consider two options:

- **Simple Random Sampling.** Pick two elephants at random, weigh them, and use their average weight as an estimate of the average weight for the entire herd. Note that this procedure gives an unbiased estimate.
- **Ranked Set Sampling.** Pick two elephants randomly in the morning, pick the smaller of the two, and weigh it. Pick two elephants randomly in the afternoon, pick the larger of the two, and weigh it. Use the average weight of the two elephants chosen in this way as an estimate of the average weight for the entire herd. It can be shown that this estimate is also an unbiased estimate of the average weight of the herd.

Table 2.4 shows the results of SRS with $n = 2$ elephants, where c denotes a calf and M denotes a mother. For RSS, Table 2.5 shows the possible results for the morning sample (smaller of the two is chosen), Table 2.6 shows the possible results for the afternoon sample (larger of the two is chosen), and Table 2.7 shows the final results.

		2nd Elephant	
		c	M
1st Elephant	c	cc	cM
	M	cM	MM

Table 2.4   Possible results of SRS from the elephant herd with sample size $n = 2$

		2nd Elephant	
		c	M
1st Elephant	c	c	c
	M	c	M

Table 2.5   Possible results of the morning SRS in which two elephants are selected and then the smaller elephant is weighed

2nd Elephant

	c	M
c	c	M
M	M	M

1st Elephant

Table 2.6   Possible results of the afternoon SRS in which two elephants are selected and then the larger elephant is weighed

Afternoon Elephant

	c	M	M	M
c	cc	cM	cM	cM
c	cc	cM	cM	cM
c	cc	cM	cM	cM
M	cM	MM	MM	MM

Morning Elephant

Table 2.7   Possible results of RSS for the elephant herd with final sample size
$n = 2$

For SRS, the probability of getting two mothers in the sample (resulting in an overestimate of the mean) or two calves in the sample (resulting in an underestimate of the mean) is $2/4 = 50\%$. For RSS, the probability of getting two mothers in the sample or two calves in the sample is only $6/16 = 37.5\%$. Clearly RSS is superior to SRS in the sense that RSS gives a "representative" sample more often. It can be shown theoretically that RSS gives the more representative samples more often in general for any herd of elephants regardless of the proportion of calves in the herd (Lacayo and Neerchal, 1996).

The above example captures the salient feature of RSS of producing less variable and more "representative" samples. This is the reason why the sample mean from a RSS has a smaller variance than the sample mean from an SRS of the same size. See Patil et al. (1994b) and the references therein for a number of theoretical results comparing RSS to SRS for various standard estimation problems.

CASE STUDY

We end this chapter with a case study that illustrates using the DQO process to systematically develop a sampling plan to achieve a specific objective. For our case study, we are interested in determining the percentage of employees of a facility who have knowledge of specific critical facts related to emergency preparedness. Although the case study is based on a real project in which one of the authors was involved, all references to the actual facility and people have been removed. The case study is presented in terms of the seven steps of the DQO process. Neptune et al. (1990) present a case study of using the DQO process to design a remedial investigation/feasibility study (RI/FS) at a Superfund site.

1. State the Problem

A large facility deals with hazardous material on an everyday basis. The facility is required by a Federal law to impart Emergency Preparedness Training (EPT) to its employees every year. The facility has about 5,000 employees housed in approximately 20 buildings. In the past, the facility has provided training by requiring all employees to attend a training session. Assuming a conservative estimate of about 45 minutes of employee time per training session (including travel time and not counting the time of the training staff), the total time spent on EPT adds up to approximately $5,000 \times 0.75 = 3,750$ employee hours. At an average billing rate of $80 per hour per employee, the cost of EPT coming out of overhead and research programs is about $3,750 \times \$80 = \$300,000$ per year. In the past, there has been no follow-up to verify that the employees have retained the material presented at the training session.

To save money and still comply with the law, the facility wants to change the training procedure as follows. Instead of each employee attending the formal training session, each employee will receive a list of emergency preparedness (EP) facts the general staff needs to know to be prepared for an emergency. A description of these EP facts is given below, listed in their order of importance (highest to lowest).

1. Emergency telephone number.
2. Location of the nearest fire alarm pull box.
3. Location of the Building Staging Area.
4. Building alarms and the corresponding required actions by staff members.

The facility wants to verify that the EP facts are retained by the employees during the year so that they can recall them should an emergency arise. The facility will ensure that a high percentage of employees in each facility know all of the above facts by actually testing a certain number of occupants of each building. Complete EPT will be required for all occupants of a building for which a prescribed proportion of the randomly selected employees fail to demonstrate the knowledge of these facts.

2. Identify the Decision

It is important that staff members who spend off-hours in the facility on a regular basis know each EP fact listed above without exception. However, it may be sufficient if only a high percentage of the remaining staff members know these facts. For example, it may be sufficient if 70% or more of the occupants of each building are familiar with the EP facts. We need to verify that a certain large proportion of the occupants of each building are familiar with the EP facts 1 through 4. For each building, if we decide that the proportion of occupants who know the EP facts is too small, then we will send everyone in that building to formal EP training.

3. Identify Inputs to the Decision

A complete listing of the staff and the respective locations within the facility is available. Training manuals containing the EP facts can be provided to each staff member and they can be interviewed either in person or by telephone to verify that they have knowledge of all four of the EP facts.

4. Define the Study Boundaries

Since the retention of EP information may not last very long, it is important to realize that this investigation needs to be repeated every year. A face-to-face interview or some honor code may need to be imposed to ensure that each staff member is in fact providing the verification based on memory rather than using notes.

5. Develop a Decision Rule

The number of employees who know the EP facts will be verified for each building. For each building, if 70% or more of the staff members who work in that building know the EP facts, we will declare the whole building is prepared for an emergency. If less than 70% of the staff members in a building know the EP facts, then everyone in the building will be asked to go through the formal EP training session.

6. Specify Acceptable Limits for the Decision Errors

Interviewing every single staff member to determine the true percentage who know EP facts 1 to 4 would be almost as time consuming as simply sending all of the employees to the EPT itself. Therefore, cost considerations dictate taking a representative sample of staff in each building and interviewing each person in the sample to estimate the percentage of staff in each building who know the EP facts. For each building, a random sample of its occupants will be chosen (the recommended sample sizes are given in the tables below). Each selected staff member will be interviewed to determine whether he/she knows facts 1 to 4 of the EP checklist. If the number of DUDs (Doesn't Understand Directions) in the sample does not exceed t (t is given in tables below), then the building can be declared to be emergency prepared. Otherwise, the building is considered unprepared for an emergency and all staff in the building will be required to go through EPT.

We assume that the verification is error-free for all the staff members included in the sample. The decision, however, is based on information from only a subset of the occupants of the building. Therefore, we must recognize that there are risks involved with our decision. Table 2.8 summarizes the two kinds of decision errors that can be made and the consequences of each.

Decision	True Percentage of DUDs	
	$\leq 30\%$	$> 30\%$
Train Occupants	Decision Error I; Waste of Money	Correct Decision
Building is Emergency Prepared	Correct Decision	Decision Error II; Compliance Issue

Table 2.8   Decision errors and their consequences for the EPT case study

Risks involved with the two kinds error shown in the table above are competing with each other. That is, when the probability of making Decision Error I goes down, the probability of making Decision Error II goes up, and vice-versa. The approach taken in developing a statistical decision rule is to hold the probability of making the most critical error to a predetermined low level and to do the best we can with lowering the probability of making the other error. In this example, it is clear that controlling the probability of making the error leading to compliance issues (Decision Error II) should be a higher priority than controlling the probability of making the error leading to a waste of money (Decision Error I).

In a series of meetings with the managers involved with emergency preparedness training and legal issues of the facility, the following levels of decision errors were determined to be acceptable: the probability of deciding that a building is emergency prepared should not exceed 1% when in fact 30% or more of its occupants are DUDs; also, when no more than 10% of the occupants of a building are DUDs, the probability that the building will be declared emergency prepared is at least 90%. So the required Decision Error II rate is no more than 1%, and the required Decision Error I rate is no more than 10%. Note that the decision error rates must be specified relative to the true percentage of DUDs in a building. Figure 2.8 illustrates these performance goals for the sampling design (see USEPA, 1994a).

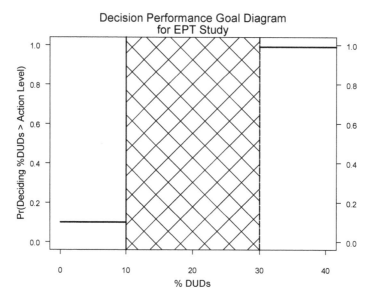

Figure 2.8   Decision performance goal diagram for the emergency preparedness training study with the null hypothesis that the %DUDs is less than the action level of 30%

7. Optimize the Design

A representative sample from each building will be taken using simple random sampling without replacement. The probability calculation related to this method of sampling is based on the hypergeometric distribution (explained in detail in Chapter 4). Table 2.9 lists the number of employees residing in 3 of the 20 buildings, along with the required sample size for each building and the maximum number of DUDs that can be present in the sample without classifying the building as being unprepared for an emergency. The probabilities of making the two types of decision errors are also given in the table.

Building	# of Occupants	Sample Size	Maximum # DUDs (t)	Pr(Error I) (in %)	Pr(Error II) (in %)
A	160	41	6	7.8	0.9
D	247	46	7	6.7	1
F	342	47	7	7.5	0.9

Table 2.9   Sampling plans with probability of Decision Error I ≤ 10% and probability of Decision Error II ≤ 1%.

For example, Building A has 160 occupants. A random sample of 41 occupants from the building must be interviewed and if no more than 6 of those interviewed do not know EP facts 1 to 4, then the building is declared emergency prepared. The probability of facing a compliance issue by failing to identify Building A as unprepared for building emergencies when in fact 30% or more of its occupants are DUDs is 0.9% (probability of Error II). On the other hand, the probability of wrongly identifying Building A as unprepared for emergencies and wasting money on training when in fact no more than 10% of the occupants are DUDs is 7.8% (probability of Error I).

Figure 2.9 displays the power curve for the sampling design for Building A drawn on top of the decision performance goal diagram shown in Figure 2.8. Here you can see that this particular sampling design satisfies the performance goals.

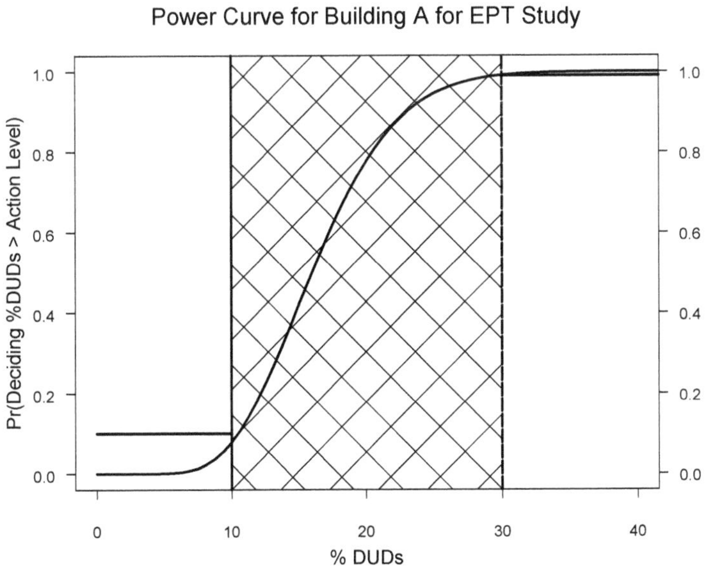

Power Curve for Building A for EPT Study

Figure 2.9   Power curve for Building A with the null hypothesis that the %DUDs is less than the action level of 30%, using a sample size of $n = 41$ and maximum number of DUDs of $t = 6$

To explore how the required sample size is affected by our specification of the Decision Error II rate, we create Table 2.10, which is similar to Table 2.9 except that we allow the Decision Error II rate to be 5% instead of 1%. Comparing the required sample sizes in these two tables, you can see that about 10 to 15 more people per building have to be interviewed to reduce the Decision Error II rate from 5% to 1%.

Building	# of Occupants	Sample Size	Maximum # DUDs (t)	Pr(Error I) (in %)	Pr(Error II) (in %)
A	160	31	5	6	4
D	247	32	5	8	4
F	342	32	5	8	4

Table 2.10   Sampling plans with probability of Decision Error I ≤ 10% and probability of Decision Error II ≤ 5%.

In the discussion above and in the recommended sampling plans, we have set the highest acceptable percentage of DUDs in a building at 30%. If this percentage is lowered, then a larger sample size is required to ensure the same decision error rates of 1% or 5% for Decision Error II and 10% for Decision Error I. Table 2.11 gives the required sample sizes for the case when the passing percentage is "no more than 20% DUDS," the Decision Error II rate is 1%, and the Decision Error I rate is 10%. Table 2.12 shows the same thing, except that the Decision Error II rate is allowed to be 5% instead of 1%. The required sample size for each building in Table 2.11 is nearly two times the corresponding sample size given in Table 2.9.

Building	# of Occupants	Sample Size	Maximum # DUDs (t)	Pr(Error I) (in %)	Pr(Error II) (in %)
A	160	87	11	7	1
D	247	104	13	10	1
F	342	119	15	8	0.9

Table 2.11   Sampling plans with probability of Decision Error I ≤ 10% and probability of Decision Error II ≤ 1%, with maximum percentage of DUDs set to 20%

Building	# of Occupants	Sample Size	Maximum # DUDs (t)	Pr(Error I) (in %)	Pr(Error II) (in %)
A	160	68	9	8	5
D	247	83	11	9	4
F	342	84	11	10	5

Table 2.12   Sampling plans with probability of Decision Error I ≤ 10% and probability of Decision Error II ≤ 5%, with maximum percentage of DUDs set to 20%

Figure 2.10 illustrates how the sampling rate (percentage of occupants selected to be interviewed) increases as the maximum allowed percentage of DUDs decreases from 50% to 10% for Buildings D and F with 247 and 342 occupants, respectively. The figure shows a slow increase followed by a rapid increase. We see that the curves start an uphill climb at around 30%.

This documents and, we believe, also justifies the choice of 30% as the highest acceptable level of DUDs.

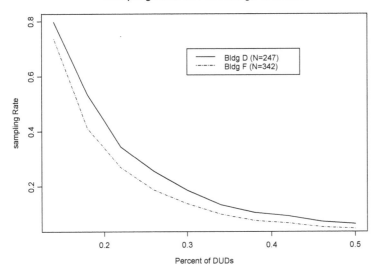

Figure 2.10   Sampling rate vs. maximum allowed percentage of DUDs for Buildings D and F

SUMMARY

- The first and most important step of any environmental study is to define the objectives of the study and design the sampling program.
- The *scientific method* recognizes the fact that our environment is constantly changing and that any of these changes may create an observed effect which may or may not be related to some cause we have hypothesized.
- The basic scientific method consists of forming a hypothesis, performing an experiment using an experimental (exposed) group and a control group, analyzing the results of the experiment, and revising the hypothesis.
- The *Data Quality Objectives (DQO) process* is a formalization of the scientific method that is a systematic way to create a legitimate and effective sampling and analysis plan.
- The term *population* is defined operationally by the question we ask: it is the entire collection of measurements about which we want to make a statement.

- A *sample* is defined as some subset of a population. The statistical definition of the word sample (a selection of individual population members) should not be confused with the more common meaning of a *physical sample* of soil (e.g., 10g of soil), water (e.g., 5ml of water), air (e.g., 20 cc of air), etc.
- *Probability sampling* or *random sampling* involves using a random mechanism to select samples from the population. All statistical methods used to quantify uncertainty assume some form of random sampling has been used to obtain a sample.
- Decisions involving the environment can often be put into the *hypothesis testing framework*. You choose a null and alternative hypothesis, then decide which one is probably true based on the information you have. You then make a decision based on your belief.
- Common mistakes in environmental studies include lack of samples from proper control populations, using judgment sampling instead of random sampling, failing to randomize over potentially influential factors, and collecting too few samples to have a high degree of confidence in the results.
- The DQO process consists of seven steps: state the problem, identify the decision, identify inputs to the decision, define the study boundaries, develop a decision rule, specify acceptable limits on decision errors, and optimize the design.
- The DQO process should include developing a *quality assurance project plan (QAPP)* to ensure the integrity of the data collected for the study.
- Optimizing a sampling plan requires knowledge of the various sources of variability. It is usually a good idea to perform a pilot study to estimate the magnitudes of the sources of variability, and to "fine-tune" the QA/QC procedures.
- This chapter describes four methods of random sampling: simple random sampling (SRS), stratified random sampling, composite sampling, and ranked set sampling (RSS). Other methods of random sampling include two-stage and multi-stage random sampling, double random sampling, sequential random sampling, and adaptive random sampling.

EXERCISES

2.1. Concrete pipes that have been used to transport crude oil from the field to a refinery are passing in the proximity of a town. There are complaints that the pipes have leaked. A judgment sampling plan (sample the joints in the pipeline) is being recommended as a preliminary step in investigating the complaints. How would you proceed? Discuss the pros and cons of judgment sampling in this context.

2.2. Soil excavated from a contaminated site has been placed in 200-gallon barrels and stored at a temporary storage facility. Barrels containing very high contamination are to be disposed of at a permanent disposal site (burial ground), and the rest of the barrels will be allowed to stay in the temporary facility. The analysis costs are very high and therefore disposal by batches of five barrels is being contemplated. Explain how you would use compositing in this project to save analysis costs.

2.3. Suppose for the elephant example discussed in the section on ranked set sampling that the proportion of calves is 25% and the proportion of mothers is 75%. Compute the probability that the ranked set sample consists of only calves. Compute the probability of obtaining such a sample under SRS and show that RSS gives rise to such extreme samples less often.

2.4. In the elephant example we obtained a ranked set sample of size $n = 2$. Extend the example to an RSS of size $n = 3$ and show that RSS is superior to SRS in that it gives rise to extreme samples less often.

2.5. For the case study discussed at the end of this chapter, the number of DUDs in a sample selected from a specific building is a hypergeometric random variable. Use the S-PLUS menu or the S-PLUS function `phyper` to verify the decision error rates shown in Table 2.9. (See Chapter 4 for an explanation of the hypergeometric distribution.)

3   LOOKING AT DATA

What is Going On?

Once you have a collection of observations from your environmental study, you should thoroughly examine the data in as many ways as possible and relevant. When the first widely available commercial statistical software packages came out in the 1960s, the emphasis was on statistical summaries of data, such as means, standard deviations, and measures of skew and kurtosis. It is still true that "a picture is worth a thousand words," and no amount of summary or descriptive statistics can replace a good graph to explain your data. John Tukey coined the acronym **EDA**, which stands for *Exploratory Data Analysis*. Helsel and Hirsch (1992, Chapters 1, 2, and 16) and USEPA (1996) give a good overview of statistical and graphical methods for exploring environmental data. Cleveland (1993, 1994) and Chambers et al. (1983) are excellent general references for methods of graphing data. This chapter discusses the use of summary statistics and graphs to describe and look at environmental data.

SUMMARY STATISTICS

Summary statistics (also called *descriptive statistics*) are numbers that you can use to summarize the information contained in a collection of observations. Summary statistics are also called *sample statistics* because they are statistics computed from a sample; they do not describe the whole population.

One way to classify summary or descriptive statistics is by what they measure: location (central tendency), spread (variability), skew (long-tail in one direction), kurtosis (peakedness), etc. Another way to classify summary statistics is by how they behave when unusually extreme observations are present: sensitive vs. robust. Table 3.1 summarizes several kinds of descriptive statistics based on these two classification schemes. In this section we will give an example of computing summary statistics, and then discuss their formulas and what they measure.

Summary Statistics for TcCB Concentrations

The guidance document USEPA (1994b, pp. 6.22–6.25) contains measures of 1,2,3,4-Tetrachlorobenzene (TcCB) concentrations (in parts per billion, usually abbreviated ppb) from soil samples at a "Reference" site and a

"Cleanup" area. The Cleanup area was previously contaminated and we are interested in determining whether the cleanup process has brought the level

Statistic	What It Measures / How It Is Computed	Robust to Extreme Values?
Mean	Center of distribution Sum of observations divided by sample size Where the histogram balances	No
Trimmed Mean	Center of distribution Trim off extreme observations and compute mean Where the trimmed histogram balances	Somewhat, depends on amount of trim
Median	Center of the distribution Middle value or mean of middle values Half of observations are less and half are greater	Very
Geometric Mean	Center of distribution Exponentiated mean of log-transformed observations Estimates true median for a lognormal distribution	Yes
Variance	Spread of distribution Average of squared distances from the mean	No
Standard Deviation	Spread of distribution Square root of variance In same units as original observations	No
Range	Spread of distribution Maximum minus minimum	No
Interquartile Range	Spread of distribution 75th percentile minus 25th percentile Range of middle 50% of data	Yes
Median Absolute Deviation	Spread of distribution Median of distances from the median	Yes
Geometric Standard Deviation	Spread of distribution Exponentiated standard deviation of log-transformed observations	No
Coefficient of Variation	Spread of distribution/Center of distribution Standard deviation divided by mean Sometimes multiplied by 100 and expressed as a percentage	No
Skew	How the distribution leans (left, right, or centered) Average of cubed distances from the mean	No
Kurtosis	Peakedness of the distribution Average of quartic distances from the mean, then subtract 3	No

Table 3.1   A description of commonly used summary statistics

of TcCB back down to what you would find in soil typical of that particular geographic region. The data are shown in Table 3.2.

Area	Observed TcCB (ppb)						
Reference	0.22	0.23	0.26	0.27	0.28	0.28	0.29
	0.33	0.34	0.35	0.38	0.39	0.39	0.42
	0.42	0.43	0.45	0.46	0.48	0.50	0.50
	0.51	0.52	0.54	0.56	0.56	0.57	0.57
	0.60	0.62	0.63	0.67	0.69	0.72	0.74
	0.76	0.79	0.81	0.82	0.84	0.89	1.11
	1.13	1.14	1.14	1.20	1.33		
Cleanup	ND	0.09	0.09	0.12	0.12	0.14	0.16
	0.17	0.17	0.17	0.18	0.19	0.20	0.20
	0.21	0.21	0.22	0.22	0.22	0.23	0.24
	0.25	0.25	0.25	0.25	0.26	0.28	0.28
	0.29	0.31	0.33	0.33	0.33	0.34	0.37
	0.38	0.39	0.40	0.43	0.43	0.47	0.48
	0.48	0.49	0.51	0.51	0.54	0.60	0.61
	0.62	0.75	0.82	0.85	0.92	0.94	1.05
	1.10	1.10	1.19	1.22	1.33	1.39	1.39
	1.52	1.53	1.73	2.35	2.46	2.59	2.61
	3.06	3.29	5.56	6.61	18.40	51.97	168.64

Table 3.2   TcCB concentrations from USEPA (1994b, pp. 6.22–6.25)

There are 47 observations from the Reference site and 77 in the Cleanup area. Note that in Table 3.2, there is one observation in the Cleanup area coded as "ND," which stands for nondetect. This means that the concentration of TcCB for this soil sample (if any was present at all) was so small that the procedure used to quantify TcCB concentrations could not reliably measure the true concentration. (We will talk more about why observations are sometimes coded as nondetects in Chapters 9 and 10.) For the purposes of this example, we will assume the nondetect observation is less than the smallest observed value, which is 0.09 ppb, but we will set it to the assumed detection limit of 0.09. (In Chapter 10, we talk extensively about statistical methods for handling data sets containing nondetects.)

Figure 3.1 displays two histograms (see the next section, Graphs for a Single Variable), one for the Reference area TcCB concentrations, and one for the Cleanup area TcCB concentrations. Note that these histograms do not share the same x-axis. Also note that in the histogram for the Cleanup area data, the bars for the three largest observations (18.40, 51.97, and 168.64) do not show up because of the scale of the y-axis. Figure 3.2 displays the same two histograms, but on the (natural) logarithmic scale so that the two histograms can share the same x-axis.

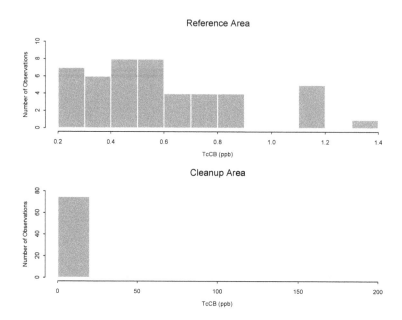

Figure 3.1   Histograms of TcCB concentrations in Reference and Cleanup areas

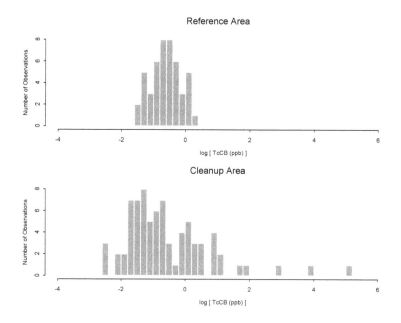

Figure 3.2   Histograms of log-transformed TcCB concentrations in Reference and Cleanup areas

Here are the summary statistics for the original TcCB data.

	Cleanup	Reference
Sample Size:	77	47
# Missing:	0	0
Mean:	3.915	0.5985
Median:	0.43	0.54
10% Trimmed Mean:	0.6846	0.5728
Geometric Mean:	0.5784	0.5382
Skew:	7.566	0.8729
Kurtosis:	61.6	2.993
Min:	0.09	0.22
Max:	168.6	1.33
Range:	168.5	1.11
1st Quartile:	0.23	0.39
3rd Quartile:	1.1	0.75
Standard Deviation:	20.02	0.2836
Geometric Standard Deviation:	3.898	1.597
Interquartile Range:	0.87	0.36
Median Absolute Deviation:	0.3558	0.2669
Coefficient of Variation:	5.112	0.4739

Here are the summary statistics for the log-transformed TcCB data.

	Cleanup	Reference
Sample Size:	77	47
# Missing:	0	0
Mean:	-0.5474	-0.6196
Median:	-0.844	-0.6162
10% Trimmed Mean:	-0.711	-0.6207
Skew:	1.663	0.0307
Kurtosis:	6.889	2.262
Min:	-2.408	-1.514
Max:	5.128	0.2852
Range:	7.536	1.799
1st Quartile:	-1.47	-0.9416
3rd Quartile:	0.09531	-0.2878
Standard Deviation:	1.36	0.468
Interquartile Range:	1.565	0.6538
Median Absolute Deviation:	1.063	0.4825
Coefficient of Variation:	-2.485	-0.7553

Here are a few things to briefly note about the data in Table 3.2 and the summary statistics above:

- Most of the observations in the Cleanup area are similar to or smaller than the observations in the Reference area.
- The Cleanup area data contain several observations that are one, two, and even three orders of magnitude larger than the rest of the observations (this is why we plotted the histograms in Figure 3.2 on a logarithmic scale).

- The mean or average TcCB concentration in the Cleanup area is much larger than in the Reference area (about 4 ppb vs. 0.6 ppb).
- The median TcCB concentration is smaller for the Cleanup area than the Reference area (0.4 ppb vs. 0.5 ppb).
- The standard deviation and coefficient of variation are both two orders of magnitude larger for the Cleanup area.
- The skew of the Cleanup area is an order of magnitude larger than the skew of the Reference area.

All of these characteristics of the Cleanup area data indicate that probably the TcCB contamination has been cleaned up in most of the area, but there are a few places within the Cleanup area with residual contamination. These particular places might be called "hot spots."

In ENVIRONMENTALSTATS for S-PLUS, the data in Table 3.2 are stored in the data frame `epa.94b.tccb.df` (see the help file Datasets: USEPA (1994b)). Here are the steps for using ENVIRONMENTALSTATS for S-PLUS to produce summary statistics for the TcCB data by area.

Menu

To produce summary statistics for the original TcCB data using the ENVIRONMENTALSTATS pull-down menu, follow these steps.

1. Open the Object Explorer, and click on the **Find S-PLUS Objects** button (the binoculars icon).
2. In the Pattern box, type **epa.94b.tccb.df**, then click **OK**.
3. Highlight the shortcut **epa.94b.tccb.df** in the Object column of the Object Explorer.
4. On the S-PLUS menu bar, make the following menu choices: **EnvironmentalStats>EDA>Summary Statistics**. This will bring up the Full Summary Statistics dialog box.
5. In the Data Set box, make sure **epa.94b.tccb.df** is selected.
6. In the Variable(s) box, choose **TcCB**.
7. In the Grouping Variables box, select **Area**.
8. Click **OK** or **Apply**.

To produce summary statistics for the log-transformed TcCB data using the ENVIRONMENTALSTATS pull-down menu, it will simplify things if we first make a new data frame called `new.epa.94b.tccb.df` to contain the original data and the log-transformed TcCB observations. To create the data frame `new.epa.94b.tccb.df`, follow these steps.

1. Highlight the shortcut **epa.94b.tccb.df** in the Object column of the Object Explorer.
2. On the S-PLUS menu bar, make the following menu choices: **Data>Transform**. This will bring up the Transform dialog box.

3. In the Target Column box, type **log.TcCB**.
4. In the Variable box, choose **TcCB**.
5. In the Function box choose **log**.
6. Click on the **Add** button, then click **OK**. At this point, you will get a warning message telling you that you have created a new copy of the data frame epa.94b.tccb.df that masks the original copy. Close the message window. Also, the modified data frame pops up in a data window. Close the data window.
7. In the left-hand column of the Object Explorer, click on the **Data** folder. In the right-hand column of the Object Explorer, right-click on **epa.94b.tccb.df** and choose **Properties**. In the Name box rename this data frame to **new.epa.94b.tccb.df** and click **OK**.

To produce summary statistics for the log-transformed TcCB data using the ENVIRONMENTALSTATS pull-down menu, follow these steps.

1. In the Object Explorer, highlight the shortcut **new.epa.94b.tccb.df** in the Object column (right-hand column).
2. On the S-PLUS menu bar, make the following menu choices: **EnvironmentalStats>EDA>Summary Statistics**. This will bring up the Full Summary Statistics dialog box.
3. In the Data Set box, make sure **new.epa.94b.tccb.df** is selected.
4. In the Variable(s) box, choose **log.TcCB**.
5. In the Grouping Variables box, select **Area**.
6. Click on the **Statistics** tab and deselect (uncheck) **Geometric Mean** and **Geometric Standard Deviation**. Click **OK** or **Apply**.

Command

To produce summary statistics for the original TcCB data using the S-PLUS Command or Script Window, type these commands.

```
attach(epa.94b.tccb.df)
full.summary(split(TcCB, Area))
```

To produce summary statistics for the log-transformed TcCB data, type this command.

```
full.summary(split(log(TcCB), Area))
detach()
```

Formulas for Summary Statistics

The formulas for various kinds of summary statistics are shown below. In all of these formulas and throughout this book, we will denote the n observations by:

$$x_1, x_2, \ldots, x_n$$

and denote the observations ordered from smallest to largest by:

$$x_{(1)}, x_{(2)}, \ldots, x_{(n)}$$

That is, $x_{(1)}$ denotes the smallest value and $x_{(n)}$ denotes the largest value.

Measures of Location (Central Tendency)

Equations (3.1) to (3.4) below show the formulas for four measures of location or central tendency. Often in environmental statistics, we are very interested in the central tendency of a group of measures, either because we want to compare the central tendency to some standard, or because we would like to compare the central tendency of data from one area with the central tendency of data from another area.

Mean

The *mean* (sometimes called *average*) is simply the sum of the observations divided by the sample size.

$$\bar{x} = \frac{1}{n} \sum_{i=1}^{n} x_i \tag{3.1}$$

It indicates approximately where the histogram balances (it is not necessarily exactly where the histogram balances because of the subjectivity of choosing the histogram classes). For example, looking at Figure 3.1 and the summary statistics above, we see that the histogram for the Reference area TcCB data balances at about 0.6 ppb, whereas the histogram for the Cleanup area balances at about 4 ppb. These differences in means between the two areas also demonstrate that the mean is sensitive to extreme values.

Trimmed Mean

The *trimmed mean* involves first trimming off a certain percentage of the smallest and largest observations, and then taking the mean of what is left (Helsel and Hirsch, 1992, p. 7; Hoaglin et al., 1983, pp. 306–311). In the formula below, α is some number between 0 and 0.5 that denotes the trimming fraction, and $[y]$ denotes the largest integer less than or equal to y.

$$\bar{x}_{trimmed} = \frac{1}{n - 2[\alpha n]} \sum_{i=[\alpha n]+1}^{n-[\alpha n]} x_{(i)} \qquad (3.2)$$

Because we purposely trim off extreme observations, the trimmed mean is not as sensitive to extreme values as the mean. For the TcCB data, you can see that the mean for the Cleanup area is about 4 ppb, but the 10% trimmed mean is only about 0.7 ppb (very close to the mean for the Reference area).

Median

The *median* is simply the 50% trimmed mean (Helsel and Hirsch, 1992, pp. 5–6; Hoaglin et al., 1983, p. 308). That is, if there is an odd number of observations, the median is the middle value, and if there is an even number of observations, the median is the mean of the two middle values.

$$Median = \begin{cases} x_{\left(\frac{n+1}{2}\right)} & \text{if } n \text{ is odd} \\[2em] \dfrac{x_{\left(\frac{n}{2}\right)} + x_{\left(\frac{n}{2}+1\right)}}{2} & \text{if } n \text{ is even} \end{cases} \qquad (3.3)$$

Half of the observations lie below the median and half of them lie above the median.

The median is very robust to extreme values. For example, although the mean TcCB concentration in the Cleanup area is much larger than in the Reference area (about 4 ppb vs. 0.6 ppb), the median TcCB concentration is smaller for the Cleanup area than the Reference area (0.4 ppb vs. 0.5 ppb). You could take almost half of your data and keep increasing it and the median would stay the same.

Geometric Mean

The *geometric mean* is often used to describe positive-valued data. It is the exponentiated mean of the log-transformed observations (Helsel and Hirsch, 1992, p. 6). That is, you take the logarithms of the original observations, compute the mean of these transformed observations, then exponentiate this mean:

$$\overline{x}_g = \exp\left[\frac{1}{n}\sum_{i=1}^{n}\log\left(x_i\right)\right] \qquad (3.4)$$

The geometric mean estimates the true **median** for a lognormal distribution (see Chapter 4 for an explanation of the lognormal distribution). For example, note that for the Reference area TcCB data, both the median and geometric mean are 0.54 ppb. The geometric mean is always less than or equal to the sample mean, with equality only if all the observations are the same value (Zar, 1999, p. 28).

Just as the median is robust to extreme observations, so is the geometric mean. The geometric mean for the Cleanup area data is about the same as for the Reference area (0.6 vs. 0.5 ppb).

Measures of Spread (Variability)

Equations (3.5) to (3.16) below show the formulas for seven measures of spread or variability. The spread of a distribution lets us know how well we can characterize it. If the spread is relatively small, then the sample mean or median is a fairly "representative" observation, whereas if the spread is large, then several observations could be much smaller or much larger than the sample mean or median. When we are comparing chemical concentrations from two or more areas, it is also useful to see whether the variability in the data is about the same for all of the areas. If it is not, this may indicate that something unusual is going on. For example, looking at the log-transformed TcCB data in Figure 3.2 and the associated summary statistics, we can see that the central tendency (mean, median, etc.) of the Reference and Cleanup area is about the same, but the spread is much larger in the Cleanup area.

Range

Probably the simplest measure of spread is the *range* of the data; the distance between the largest and smallest value.

$$Range = x_{(n)} - x_{(1)} \qquad (3.5)$$

The range quickly gives you an idea about the differences in the orders of magnitude of the observations. For the Reference area TcCB data, the range is about 1 ppb, whereas it is about 169 ppb for the Cleanup area. Obviously, the range is very sensitive to extreme observations.

Interquartile Range

The *interquartile range* (often abbreviated *IQR*) is a modified form of the range. The 25th, 50th, and 75th percentiles of the data are also called the quartiles of the data, so the interquartile range is the distance between the 75th percentile of the data and the 25th percentile (Chambers et al., 1983, p. 21; Helsel and Hirsch, 1992, p. 8; Hoaglin et al., 1983, pp. 38, 59).

$$IQR = x_{0.75} - x_{0.25} \qquad (3.6)$$

In the above equation, x_p denotes the $p100^{th}$ percentile of the data. In Chapter 5 we will talk about how to compute the percentiles for a set of observations. For now all you need to know is that for the $p100^{th}$ percentile, about $p100\%$ of the observations are less than this number and about $(1-p)100\%$ of the observations are greater than this number.

Unlike the range, the interquartile range is not affected by a few extreme observations; it measures only the range of the middle 50% of the data. For the TcCB data the IQR is about 0.4 ppb for the Reference area and 0.9 ppb for the Cleanup area.

Variance

The *variance* is the mean or average of the squared distances between each observation and the mean.

$$s_{mm}^2 = \frac{1}{n} \sum_{i=1}^{n} (x_i - \bar{x})^2 \qquad (3.7)$$

The sample variance estimates the population variance (see Chapter 5). The formula above is called the method of moments estimator (see Chapter 5), but the more commonly used formula for the sample variance is the unbiased estimator.

$$s^2 = \frac{1}{n-1} \sum_{i=1}^{n} (x_i - \bar{x})^2 \qquad (3.8)$$

Equation (3.8) is the one used by default in ENVIRONMENTALSTATS for S-PLUS and S-PLUS.

For reasons related to estimation (see Chapter 5) and hypothesis testing (see Chapter 7), the variance and standard deviation (see below) are the two most commonly used statistics to quantify the spread of a set of observations. Unlike the interquartile range, the variance is very sensitive to extreme values, even more so than the mean, because it involves squaring the distances between the observations and the mean. For the TcCB data, the variance for the Reference area is about 0.08 ppb^2 whereas for the Cleanup area it is about 400 ppb^2.

Standard Deviation

The **standard deviation** is simply the square root of the variance. The formula based on the method of moments estimator of variance is:

$$s_{mm} = \sqrt{\frac{1}{n} \sum_{i=1}^{n} (x_i - \bar{x})^2} \tag{3.9}$$

but the most commonly used formula for the sample standard deviation is based on the unbiased estimator of variance:

$$s = \sqrt{\frac{1}{n-1} \sum_{i=1}^{n} (x_i - \bar{x})^2} \tag{3.10}$$

Equation (3.10) is the one used for the summary statistics for the TcCB data and is used by default in ENVIRONMENTALSTATS for S-PLUS and S-PLUS.

The standard deviation is often preferred to the variance as a measure of spread because it maintains the original units of the data. Just like the variance, the standard deviation is sensitive to extreme values. For the TcCB data, the standard deviation for the Reference area is about 0.3 ppb whereas for the Cleanup area it is about 20 ppb.

Geometric Standard Deviation

The **geometric standard deviation** is sometimes used to describe positive-valued data (Leidel et al., 1977). It is the exponentiated standard deviation of the log-transformed observations.

$$s_g = e^{s_y} \tag{3.11}$$

where

$$s_y = \sqrt{\frac{1}{n-1} \sum_{i=1}^{n} (y_i - \bar{y})^2} \qquad (3.12)$$

$$y_i = \log(x_i) \qquad (3.13)$$

Unlike the sample geometric mean, the sample geometric standard deviation does not estimate any population parameter that is usually used to characterize the lognormal distribution.

Median Absolute Deviation

The **median absolute deviation** (often abbreviated **MAD**) is the median of the distances between each observation and the median (Helsel and Hirsch, 1992, pp. 8–9; Hoaglin et al., 1983, pp. 220, 346, 365–368).

$$MAD = Median\left(|x_1 - m|, |x_2 - m|, \dots, |x_n - m| \right) \qquad (3.14)$$

where

$$m = Median\left(x_1, x_2, \dots, x_n\right) \qquad (3.15)$$

Unlike the variance and standard deviation, the median absolute deviation is unaffected by a few extreme observations. For the TcCB data, the MAD is 0.27 ppb for the Reference area and 0.36 ppb for the Cleanup area. You could take almost half of your data and keep increasing it and the MAD would stay the same.

Coefficient of Variation

The **coefficient of variation** (sometimes denoted CV) is simply the ratio of the standard deviation to the mean.

$$CV = \frac{s}{\bar{x}} \qquad (3.16)$$

The coefficient of variation is a unitless measure of how spread out the distribution is relative to the size of the mean. It is usually used to characterize positive, right-skewed distributions such as the lognormal distribution (see Chapter 4). It is sometimes multiplied by 100 and expressed as a percentage (Zar, 1999, p. 40). Like the mean and standard deviation, the coefficient of variation is sensitive to extreme values. For the TcCB data, the CV is 0.5 for the Reference area and ten times as large for the Cleanup area.

Measures of Deviation from a Symmetric or Bell-Shaped Distribution

Equations (3.17) to (3.19) below show the formulas for two statistics that are used to measure deviation from a symmetric or bell-shaped histogram: the skew and kurtosis. A bell-shaped (and thus symmetric) histogram is a good indication that the data may be modeled with a normal (Gaussian) distribution (see Chapter 4). Many statistical hypothesis tests assume the data follow a normal distribution (see Chapter 7). In the days before it was easy to create plots and perform goodness-of-fit tests with computer software, the skew and kurtosis were often reported. Nowadays, they are not so widely reported, although they are still used to fit distributions in the system of Pearson curves (Johnson et al., 1994, pp. 15–25).

Skew

The **skew** or **coefficient of skewness** is based on the mean or average of the cubed distances between each observation and the mean. The average of the cubed distances is divided by the cube of the standard deviation to produce a unitless measure.

$$Skew_{mm} = \frac{\frac{1}{n} \sum_{i=1}^{n} (x_i - \bar{x})^3}{s_{mm}^3} \qquad (3.17)$$

The formula above uses method of moments estimators. This is the formula that is used by default to compute the skew in ENVIRONMENTALSTATS for S-PLUS. Another formula is sometimes used based on unbiased estimators.

$$Skew = \frac{\frac{n}{(n-1)(n-2)} \sum_{i=1}^{n} (x_i - \bar{x})^3}{s^3} \qquad (3.18)$$

The skew measures how the observations are distributed about the mean. If the histogram is fairly symmetric, then the skew is 0 or close to 0. If there are a few or several large values to the right of the mean (greater than the mean) but not to the left of the mean, the skew is positive and the histogram is said to be *right skewed* or *positively skewed*. If there are a few or several small values to the left of the mean (less than the mean) but not to the right of the mean, the skew is negative and the histogram is said to be *left skewed* or *negatively skewed*.

Because environmental data usually involve measures of chemical concentrations, and concentrations cannot fall below 0, environmental data often tend to be positively skewed (see Figure 3.1). For the log-transformed TcCB data shown in Figure 3.2 , the Reference area has a skew of about 0.03 since this histogram is close to being symmetric, but the Cleanup area has a skew of about 1.7.

Kurtosis

The *kurtosis* or *coefficient of kurtosis* is based on the average of the distances between each observation and the mean raised to the 4th power. The average of these distances raised to the 4th power is divided by the square of the standard deviation to produce a unitless measure. The formula below uses method of moments estimators. This is the formula that is used by default to compute the kurtosis in ENVIRONMENTALSTATS for S-PLUS.

$$Kurtosis_{mm} = \frac{\frac{1}{n} \sum_{i=1}^{n} \left(x_i - \bar{x}\right)^4}{s_{mm}^4} \qquad (3.19)$$

The kurtosis measures how peaked the histogram is relative to an idealized bell-shaped histogram. This idealized bell-shaped histogram is based on the normal (Gaussian) distribution (see Chapter 4), which has a kurtosis of 3. If the histogram has too many observations in the tails compared to the idealized histogram then the kurtosis is larger than 3. If the histogram has short tails and most of the observations are tightly clustered around the mean, then the kurtosis is less than 3. For the log-transformed TcCB data shown in Figure 3.2, the kurtosis for the Reference area is about 2, whereas it is about 7 for the Cleanup area.

GRAPHS FOR A SINGLE VARIABLE

One of the main strengths of S-PLUS and its add-on modules ENVIRONMENTALSTATS for S-PLUS and S+SPATIALSTATS is the great variety of graphs you can easily produce. As Helsel and Hirsch (1992, p. 17) point out, the two main purposes of graphs are to help you visually explore the data, and also to help you illustrate important points about the data when presenting your results to others. This main section discusses graphs you can use when you are concentrating on a single variable (Table 3.3). The next main section discusses graphs you can create when you are interested in looking at the relationship between two or more variables.

Plot Type	Data Characteristics	Plot Characteristics
Dot Plot, Bar Chart	Numeric data where each value has a label or is associated with a category	Displays each value or percentage and its label
Pie Chart	Numeric data where each value has a label or is associated with a category	Displays percentages
Strip Plot, Histogram, Density Plot, Boxplot	Numeric data	Displays the distribution of the values
Quantile Plot or Empirical CDF Plot	Numeric data	Displays the quantiles of the values
Probability Plot or Q-Q Plot, Tukey Mean-Difference Q-Q Plot	Numeric data	Compares the distribution of the data with some specified probability distribution, or compares the distributions of two data sets

Table 3.3   A description of plots for a single variable

Dot Plots

A *dot plot* (also called a *dot chart*) lets you display quantitative data for which each measurement has a distinct label (Cleveland, 1994, p. 150). You can also use a dot plot to display percentages of observations in each category based on a categorical variable.

Table 3.4 shows the median of daily maxima ozone concentration (ppb) for June to August, 1974, for some cities in the northeastern United States. Figure 3.3 shows a dot plot of these data. In Figure 3.3, you can immediately see that Boston and Springfield have lower ozone levels (about 35 ppb) than the cluster of cities with ozone levels above 50 ppb. You can also see that Waltham, with 42 ppb, is between Boston and Springfield and the other cities.

Cleveland (1994, pp. 152, 244–247) states that our perception of the data is enhanced when the data are ordered from largest to smallest. (An exception is when the labels have a natural ordering, in which case the dot plot should be arranged with that ordering.) Figure 3.4 shows the dot plot with the values ordered by ozone concentration.

City	Ozone Concentration (ppb)
Boston	36
Cambridge	54
Fitchburg	65
Glens Falls	64
Lowell	63
Medford	52
Pittsfield	65
Quincy	62
Rensselaer	60
Schenectady	56
Springfield	34
Waltham	42
Worcester	66

Table 3.4   Median of daily maxima ozone concentrations (ppb) for June to August, 1974 for some cities in the northeastern United States. (From Cleveland et al. 1975.)

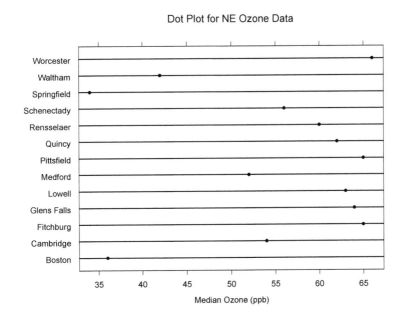

Figure 3.3   Dot plot of ozone data from northeastern U.S. cities

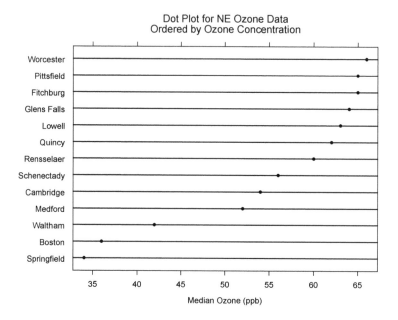

Figure 3.4   Dot plot of ozone data from northeastern U.S. cities, ordered by ozone concentration

In S-PLUS, the data in Table 3.4 are stored in the vectors `ozone.city` and `ozone.median` (see the S-PLUS help file for ozone). Also, the vector `ozone.quartile` contains the 75th percentile of the daily maxima ozone concentrations, and the list `ozone.xy` contains (negative) longitudes and (positive) latitudes of each city. These data contain information on 41 cities in the northeastern U.S. For simplicity, we have used data only for the 13 cities with latitudes greater than or equal to 42 degrees north.

Menu

To produce the unordered dot plot shown in Figure 3.3 using the S-PLUS pull-down menu, it will simplify things if we first make a new data frame called `ozone.NE` to contain the information. To create `ozone.NE`, follow these steps.

1. On the S-PLUS menu bar, make the following menu choices: **Data>Select Data**. This will bring up the Select Data dialog box. Under Source, click on the **New Data** button. Under New Data, in the Name box, type **ozone.NE**. Click **OK**.

2. You will now see a data window with the title ozone.NE. Right-click on the first column and choose **Insert Column**. This will bring up the Insert Columns dialog box.

3. In the Name(s) box, type **Median**. In the Fill Expression box, type **ozone.median**. Click **OK**.
4. Repeat Steps 2 and 3, except right-click on the second column, and type **Quartile** in the Name(s) box, and **ozone.quartile** in the Fill Expression box.
5. Repeat Steps 2 and 3, except right-click on the third column, and type **City** in the Name(s) box, and **ozone.city** in the Fill Expression box. Also, choose **factor** under the Column Type drop-down list.
6. Repeat Steps 2 and 3, except right-click on the fourth column, and type **Longitude** in the Name(s) box, and **ozone.xy$x** in the Fill Expression box.
7. Repeat Steps 2 and 3, except right-click on the fifth column, and type **Latitude** in the Name(s) box, and **ozone.xy$y** in the Fill Expression box.
8. Left-click on the **Exit** button (X in upper right-hand corner).

To produce the unordered dot plot shown in Figure 3.3 using the S-PLUS pull-down menus or toolbars, follow these steps.

1. Highlight the **Data** folder in the left-hand column of the Object Explorer.
2. On the S-PLUS menu bar, make the following menu choices: **Graph>2D Plot**. This will bring up the Insert Graph dialog box. Under the Axes Type column, **Linear** should be highlighted. Under the Plot Type column, select **Dot Plot** and click **OK**. (Alternatively, left-click on the **2D Plots** button, then left-click on the **Dot** button.)
3. The Line/Scatter Plot dialog box should appear. In the Data Set box, select **ozone.NE**.
4. In the x Columns box, choose **Median**.
5. In the y Columns box, choose **City**.
6. In the Subset Rows with box, type **Latitude >= 42**.
7. Click **OK**.

See the S-PLUS documentation for information on how to modify the *x*- and *y*-axis labels and add a main title.

To produce the ordered dot plot shown in Figure 3.4 using the S-PLUS pull-down menus, it will simplify things if we first add a new variable to the data frame `ozone.NE`.

1. On the S-PLUS menu bar, make the following menu choices: **Data>Transform**. This will bring up the Transform dialog box.
2. In the Data Set box, choose **ozone.NE**.
3. In the Target Column box, type **City.Ordered**.
4. In the Expression box, type **reorder.factor(City, Median)**.
5. Click **OK**. Close the ozone.NE data window.

To produce the ordered dot plot shown in Figure 3.4 using the S-Plus pull-down menus or toolbars, follow these steps.

1. Highlight the **Data** folder in the Object Explorer.
2. On the S-Plus menu bar, make the following menu choices: **Graph>2D Plot**. This will bring up the Insert Graph dialog box. Under the Axes Type column, **Linear** should be highlighted. Under the Plot Type column, select **Dot Plot** and click **OK**. (Alternatively, left-click on the **2D Plots** button, then left-click on the **Dot** button.)
3. The Line/Scatter Plot dialog box should appear. In the Data Set box, select **ozone.NE**.
4. In the x Columns box, choose **Median**.
5. In the y Columns box, choose **City.Ordered**.
6. In the Subset Rows with box, type **Latitude >= 42**.
7. Click **OK**.

See the S-Plus documentation for information on how to modify the *x*- and *y*-axis labels and add a main title.

Command

To produce the data frame `ozone.NE` using the S-Plus Command or Script Window, type this command.

```
ozone.NE <- data.frame(Median = ozone.median,
   Quartile = ozone.quartile, City = ozone.city,
   Longitude = ozone.xy$x, Latitude = ozone.xy$y)
```

To produce the unordered dot plot shown in Figure 3.3, type this command.

```
dotplot(City ~ Median, data = ozone.NE,
   subset = Latitude >= 42,
   xlab = "Median Ozone (ppb)",
   main = list(cex = 1.25,
     "Dot Plot for NE Ozone Data"))
```

To add the new variable `City.Ordered` to the data frame `ozone.NE`, type this command.

```
ozone.NE <- cbind(ozone.NE, City.Ordered =
   reorder.factor(ozone.NE$City, ozone.NE$Median))
```

To produce the ordered dot plot shown in Figure 3.4 using the Command or Script Window, type this command.

```
dotplot(City.Ordered ~ Median, data = ozone.NE,
   subset = Latitude >= 42,
   xlab = "Median Ozone (ppb)",
```

```
main = list(cex = 1.25, paste(
  "Dot Plot for NE Ozone Data",
  "Ordered by Ozone Concentration", sep = "\n")))
```

Bar Charts

Just like a dot plot, a ***bar chart*** (also called a ***bar plot***) also lets you display quantitative data for which each measurement has a distinct label. You can also use a bar chart to display percentages of observations in each category based on a categorical variable. The problem with bar charts is that the length of the bar is an imposing visual aspect and for bar plots with a baseline other than 0 the lengths of the bars do not represent meaningful information (Cleveland, 1994, p. 152).

Figure 3.5 is the bar chart version of the dot plot shown in Figure 3.4. The baseline for this figure is about 32.5 ppb, so the length of each bar is encoding the median ozone for that city minus 32.5, which does not really mean anything. Figure 3.6 shows the bar chart with zero as the baseline.

Huff and Geis (1954, pp. 61–62) argue that you should always include zero in the scale of your graph and not doing so is dishonest. Cleveland (1994, pp. 92–93) argues, however, that for graphical communications in science and technology you should expect the viewer to read the tick mark labels, and the need to include zero should not outweigh the need to display important variation in the data. The large difference between Springfield and Boston vs. the other cities is lost in Figure 3.6.

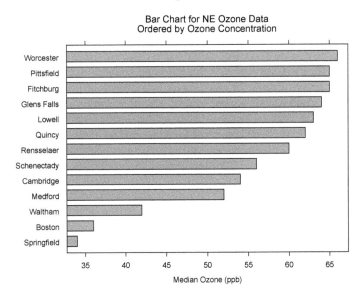

Figure 3.5   Bar chart of ozone data, ordered by ozone concentration

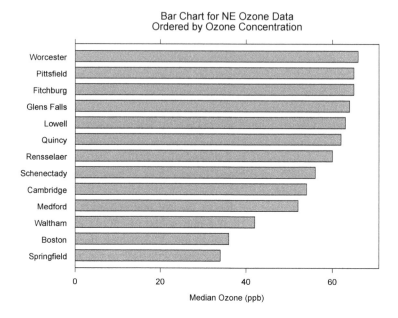

Figure 3.6   Bar chart of ozone data, ordered by ozone concentration, with a zero baseline

Menu

To produce the ordered bar chart shown in Figure 3.5 using the S-Plus pull-down menus or toolbars, follow these steps.

1. Highlight the **Data** folder in the left-hand column of the Object Explorer.
2. On the S-Plus menu bar, make the following menu choices: **Graph>2D Plot**. This will bring up the Insert Graph dialog box. Under the Axes Type column, **Linear** should be highlighted. Under the Plot Type column, select **Bar, Horiz.** and click **OK**. (Alternatively, left-click on the **2D Plots** button, then left-click on the **Horiz Bar** button.)
3. The Line/Scatter Plot dialog box should appear. In the Data Set box, select **ozone.NE**.
4. In the x Columns box, choose **Median**.
5. In the y Columns box, choose **City.Ordered**.
6. In the Subset Rows with box, type **Latitude >= 42**.
7. Click on the **Position** tab.
8. In the Bar Base box, make sure **Data min** is selected.
9. Click **OK**.

To produce the bar chart in Figure 3.6 with the scale starting at zero, repeat the above steps, but in Step **8** make sure **Zero** is selected in the Bar Base box.

Command

To produce the ordered bar chart shown in Figure 3.5 using the S-PLUS Command or Script Window, type this command.

```
barchart(City.Ordered ~ Median, data = ozone.NE,
   subset = Latitude >= 42,
   xlab = "Median Ozone (ppb)",
   main = list(cex = 1.25, paste(
      "Bar Chart for NE Ozone Data",
      "Ordered by Ozone Concentration", sep = "\n")))
```

To produce the bar chart in Figure 3.6 with the scale starting at zero, type this command.

```
barchart(City.Ordered ~ Median, data = ozone.NE,
   subset = Latitude >= 42,
   xlim = c(0, max(ozone.NE$Median[
      ozone.NE$Latitude >= 42]) + 5),
   scales = list(x = list(axs = "i")),
   xlab = "Median Ozone (ppb)",
   main = list(cex = 1.25, paste(
      "Bar Chart for NE Ozone Data",
      "Ordered by Ozone Concentration", sep = "\n")))
```

Pie Charts

A *pie chart* is used to display and compare percentages. Experiments in graphical perception, however, have shown that they convey information far less reliably than dot plots or bar charts (Cleveland, 1994, pp. 262–264).

Silver Concentration (µg/L)											
0.8	0.1	2	0.2	<5	<1	<2	<25	2.7	<1	1.2	3.2
<20	<10	<5	<0.1	<10	<1	2	<5	560	<0.2	<20	<1
1	10	<10	<5	<0.5	1.4	<0.2	<6	1	<10	0.1	5	2
1	<1	4.4	90	<20	<0.3	<2.5	<10	0.7	<1	1.5	<1
<0.2	2	<0.2	<1	<1	<1	<0.1					

Table 3.5   Silver concentrations from an inter-laboratory comparison. (From Helsel and Cohn, 1988.)

Table 3.5 shows silver concentrations that were used in an inter-laboratory comparison (Helsel and Cohn, 1988). Out of 56 observations, 34

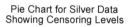
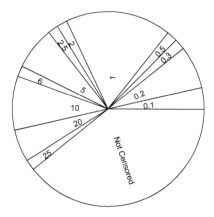

Pie Chart for Silver Data
Showing Censoring Levels

Figure 3.7   Pie chart showing percentage of observations at each of the censoring levels for the silver data

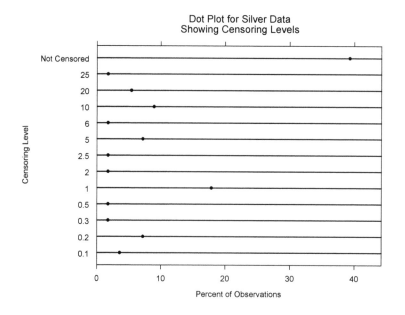

Dot Plot for Silver Data
Showing Censoring Levels

Figure 3.8   Dot plot for same data as in Figure 3.7

were reported as <DL, where DL denotes the detection limit. (Observations reported as less than a detection limit are also called **censored observations**.) The observations in Table 3.5 contain 12 distinct detection limits. (We will talk more about the meaning of detection limits in Chapters 9 and 10. For now all you need to know is that if an observation is reported as <DL, this is usually interpreted to mean that for the specific laboratory, physical sample, and instrument set up, the concentration of the chemical can only be quantified as being less than this number, that is, anywhere between 0 and DL.)

Table 3.6 shows the number and percentage of observations at each censoring level for the data in Table 3.5. Figure 3.7 shows a pie chart of these percentages, and Figure 3.8 shows the corresponding dot plot.

Censoring Level	Number Observations	Percent Observations
0.1	2	4
0.2	4	7
0.3	1	2
0.5	1	2
1	10	18
2	1	2
2.5	1	2
5	4	7
6	1	2
10	5	9
20	3	5
25	1	2
Not Censored	22	39

Table 3.6   Number and percentage of observations recorded at each censoring level for the data in **Table 3.5**

In ENVIRONMENTALSTATS for S-PLUS, the data in Table 3.5 are stored in the data frame helsel.cohn.88.silver.df (see the help file for helsel.cohn.88.silver.df). This data frame contains the variables Ag.orig (a character vector with the observations as they were originally coded and as shown in Table 3.5), Ag (a numeric vector with the silver concentrations, where censored values are coded to the censoring level), and Censored (a logical vector indicating whether the observation was censored).

Menu

To produce the pie chart and dot plot shown in Figure 3.7 and Figure 3.8 using the S-PLUS pull-down menu, it will simplify things if we first add a variable to the data frame helsel.cohn.88.silver.df that indicates either the censoring level or that the observation was not censored, then make a new data frame called `silver.table` to contain the information shown in Table 3.6. To add the censoring level variable to the data frame `helsel.cohn.88.silver.df`, follow these steps.

1. Open the Object Explorer, and click on the **Find S-PLUS Objects** button (the binoculars icon).
2. In the Pattern box, type **helsel.cohn.88.silver.df**, then click **OK**.
3. Highlight the shortcut **helsel.cohn.88.silver.df** in the Object column of the Object Explorer.
4. On the S-PLUS menu bar, make the following menu choices: **Data>Transform**. This will bring up the Transform dialog box.
5. In the Target Column box, type **Censoring.Level**. In the Variable box, choose **Censored**. In the Function box, choose **ifelse**. Click on the **Add** button. Modify the Expression box so that the expression is as follows: **ifelse(Censored, Ag, "Not Censored")**. Click **OK**. At this point, you will get a warning message telling you that you have created a new copy of the data frame that masks the original copy. Close the message window. Also, the modified data frame pops up in a data window. Close the data window.
6. In the left column of the Object Explorer, click on the **Data** folder.
7. In the Object column of the Object Explorer, right-click on the shortcut **helsel.cohn.88.silver.df** and choose **Properties**.
8. In the Name box, rename this data frame to **new.helsel.cohn.88.silver.df**, then click **OK**.

To create `silver.table`, follow these steps.

1. Make sure **new.helsel.cohn.88.silver.df** is highlighted in the right-hand column of the Object Explorer.
2. On the S-PLUS menu bar, make the following menu choices **Data>Tablulate**. This brings up the Tabulate dialog box.
3. In the Data Set box, **new.helsel.cohn.88.silver.df** should be selected. In the Variables box, left-click on **Censoring.Level**. In the Save In box, type **silver.table**. Click **OK**.
4. At this point, you can close the Report window and the data window with the title silver.table (left-click on the X button in the upper left-hand corner).
5. In the Object Explorer, click on the **Data** folder in the left-hand column, and in the right-hand column click on the shortcut **silver.table**.
6. On the S-PLUS menu bar, make the following menu choices: **Data>Transform**. This brings up the Transform dialog box.
7. In the Data Set box, select **silver.table**. In the Target Column box, type **Percent**. In the Expression box, type **100*Count/sum(Count)**. Click **Apply**.

8. In the Target Column box, type **Numeric.Censoring.Level**. In the Expression box, type **as.numeric(as.character(Censoring.Level))**. Click **Apply**. (At this point, you will get a warning message telling you "1 missing value generated coercing from character to numeric".)

9. In the Target Column box, type **Ordered.Censoring.Level**. In the Expression box, type **reorder.factor(Censoring.Level, Numeric.Censoring.Level)**. Click **OK**.

10. On the S-PLUS menu bar, make the following menu choices: **Data>Sort**. This brings up the Sort Columns dialog box.

11. Under the From group, in the Data Set box select **silver.table**, in the Columns box select **<ALL>**, and in the Sort By Columns box, select **Numeric.Censoring.Level**. Under the To group, in the Data Set box, select **silver.table**, and in the Columns box, select **<ALL>**. Click **OK**.

To produce the pie chart shown in Figure 3.7 using the S-PLUS pull-down menus or toolbars, follow these steps.

1. Highlight the shortcut **silver.table** in the Object column of the Object Explorer.

2. On the S-PLUS menu bar, make the following menu choices: **Graph>2D Plot**. This will bring up the Insert Graph dialog box. Under the Axes Type column, click on **Pie**. Under the Plot Type column, make sure **Pie Chart** is selected, and click **OK**. (Alternatively, left-click on the **2D Plots** button, then left-click on the **Pie** button.)

3. The Pie Chart dialog box should appear. The Data Set box should display **silver.table**.

4. In the x Columns box, choose **Count**.

5. Click on the **Labels** tab.

6. Under the Label 1 group, in the Type box, select **Column**. In the Data Set box, **silver.table** should be selected. In the Column box, select **Censoring.Level**.

7. Click **OK**.

To produce the dot plot shown in Figure 3.8 using the S-PLUS pull-down menus or toolbars, follow these steps.

1. Highlight the shortcut **silver.table** in the Object column of the Object Explorer.

2. On the S-PLUS menu bar, make the following menu choices: **Graph>2D Plot**. This will bring up the Insert Graph dialog box. Under the Axes Type column, **Linear** should be highlighted. Under

the Plot Type column, select **Dot Plot** and click **OK**. (Alternatively, left-click on the **2D Plots** button, then left-click on the **Dot** button.)
3. The Line/Scatter Plot dialog box should appear. The Data Set box should display **silver.table**.
4. In the x Columns box, choose **Percent**.
5. In the y Columns box, choose **Ordered.Censoring.Level**.
6. Click **OK**.

Command

To produce a table of counts for each censoring level using the S-PLUS Command or Script Window, type these commands.

```
Censoring.Level <-
    as.character(helsel.cohn.88.silver.df$Ag)
Censoring.Level[!helsel.cohn.88.silver.df$Censored] <-
    "Not Censored"
silver.table <- table(Censoring.Level)
```

To transform these results into a data frame and create the same variables we created under the Menu section, type these commands.

```
Censoring.Level <- factor(names(silver.table))
Count <- as.vector(silver.table)
Percent <- 100 * Count / sum(Count)
Numeric.Censoring.Level <-
    as.numeric(as.character(Censoring.Level))
Ordered.Censoring.Level <- reorder.factor(
    Censoring.Level, Numeric.Censoring.Level)
silver.table <- data.frame(Censoring.Level, Count,
    Percent, Numeric.Censoring.Level,
    Ordered.Censoring.Level)
silver.table <-
    silver.table[order(Numeric.Censoring.Level),]
```

To produce the pie chart shown in Figure 3.7 using the Command or Script Window, type these commands.

```
piechart(~ Count, names = as.character(
    silver.table$Censoring.Level), data = silver.table,
    main = list(paste("Pie Chart for Silver Data",
    "Showing Censoring Levels", sep = "\n"), cex=1.25))
```

To produce the dot plot shown in Figure 3.8, type these commands.

```
dotplot(Ordered.Censoring.Level ~ Percent,
   data = silver.table,
   xlim = c(0, max(silver.table$Percent) + 5),
   scales = list(x = list(axs = "i")),
   xlab = "Percent of Observations",
   ylab = "Censoring Level",
   main = list(paste("Dot Plot for Silver Data",
   "Showing Censoring Levels", sep = "\n"), cex=1.25))
```

Strip Plots

A *strip plot*, also called a *one-dimensional scatterplot*, is simply a plot of each observation showing the value of the observation (Chambers et al., 1983, p. 19; Cleveland, 1994, p. 133). Strip plots are used to look at the distribution of one data set, or to compare the distributions of two or more data sets.

Figure 3.9 shows strip plots of the log-transformed TcCB data for the Reference and Cleanup areas (see Table 3.2). The strip plots show that most of the observations for the Cleanup area are comparable to (or even smaller than) the observations for the Reference area, but, as we found out from looking at the summary statistics for these data, there are a few very large "outliers" in the Cleanup area. This may indicate a few "hot spots" in the cleanup area that were missed during the remediation process.

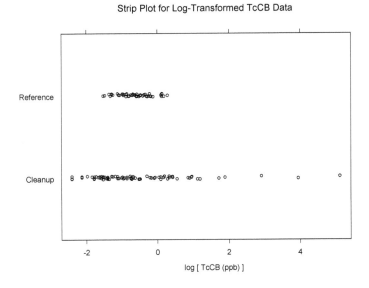

Figure 3.9   Strip plots of log-transformed TcCB concentrations in the Reference and Cleanup areas

Menu

To produce the strip plots shown in Figure 3.9 using the S-Plus pull-down menus or toolbars, follow these steps.

1. In the Object Explorer, highlight the **Data** folder in the left-hand column. In the right-hand (Object) column, highlight the shortcut **new.epa.94b.tccb.df** for the version of this data frame that you created earlier.

2. On the S-Plus menu bar, make the following menu choices: **Graph>2D Plot**. This will bring up the Insert Graph dialog box. Under the Axes Type column, **Linear** should be highlighted. Under the Plot Type column, select **Scatter Plot** and click **OK**. (Alternatively, left-click on the **2D Plots** button, then left-click on the **Scatter** button.)

3. The Line/Scatter Plot dialog box should appear. The Data Set box should display **new.epa.94b.tccb.df**.

4. In the x Columns box, choose **log.TcCB**.

5. In the y Columns box, choose **Area**.

6. Click on the **Symbol** tab.

7. In the Jitter Symbols box, select **Y only**. Click **OK**.

Command

To produce the strip plot shown in Figure 3.9 using the S-Plus Command or Script Window, type this command.

```
stripplot(Area ~ log(TcCB), data = epa.94b.tccb.df,
   jitter = T, xlab = "log [ TcCB (ppb) ]",
   main = list(cex = 1.25,
   "Strip Plot for Log-Transformed TcCB Data"))
```

Histograms

Unlike a strip plot, a *histogram* does not usually display every observation in a data set, but rather summarizes the distribution of the data by placing observations into intervals (also called classes or bins) and counting the number of observations in each interval (Chambers et al., 1983, pp. 24–26; Cleveland, 1993, pp. 6–8; Cleveland, 1994, pp. 134–136; Helsel and Hirsch, 1992, p. 19). The y-axis for a histogram usually displays one of four possible scales:

*   **Number (Counts)**. The height of the bar indicates how many observations fall into that interval.
*   **Percent of Total**. The height of the bar indicates the percentage of observations (out of the total number of observations) that fall into that interval.

- **Fraction of Total** (called **Frequency** in S-PLUS). The height of the bar indicates the same thing as for Percent of Total except it is expressed as a fraction between 0 and 1 instead of a percent.
- **Density**. The height of the bar times the width of the interval indicates the fraction of observations (out of the total number of observations) that fall into that interval.

It is usually a good idea to use a scale other than raw counts if you are comparing two data sets with different sample sizes (Cleveland, 1994, p. 134).

Figure 3.10 displays the Percent of Total histograms for the log-transformed TcCB data for the Reference and Cleanup areas (see Table 3.2 and Figure 3.2). As we saw with the strip plots in Figure 3.9, the histograms indicate that most of the observations for the Cleanup area are comparable to (or even smaller than) the observations for the Reference area, but there are a few very large "outliers" in the Cleanup area.

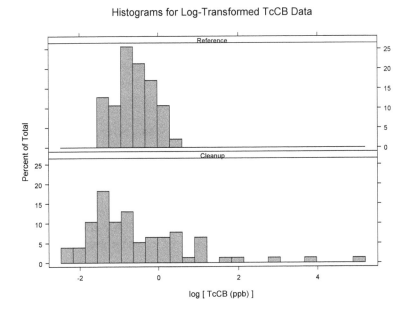

Histograms for Log-Transformed TcCB Data

Figure 3.10   Percent of Total histograms of the log-transformed TcCB data for the Reference and Cleanup areas

A histogram is a useful way of displaying the distribution of a data set, but as Helsel and Hirsch (1992, p. 19) point out, the appearance of a histogram depends upon how you decide to divide the data into intervals. As the interval widths get smaller and smaller, the histogram get more ragged and approaches (starts to look like) a strip plot, and as the interval widths get larger and larger, the histogram gets less informative and approaches a plot with one interval containing all of the data. Most of the time, you should choose the interval width so that you do not lose too much accuracy, but there is no fixed rule for what to choose (Cleveland, 1994, p. 135).

Menu

To produce the histograms shown in Figure 3.10 using the S-Plus pull-down menus or toolbars, follow these steps.

1. In the Object Explorer, highlight the **Data** folder in the left-hand column. In the right-hand (Object) column, left-double-click the shortcut **new.epa.94b.tccb.df** for the version of this data frame that you created earlier. This will bring up a data window.
2. On the S-Plus menu bar, make the following menu choices: **Graph>2D Plot**. This will bring up the Insert Graph dialog box. Under the Axes Type column, **Linear** should be highlighted. Under the Plot Type column, select **Histogram** and click **OK**. (Alternatively, left-click on the **2D** Plots button, then left-click on the **Histogram** button.)
3. The Histogram/Density dialog box should appear. Under the Data Columns group, in the Data Set box, select **new.epa.94b.tccb.df**.
4. Under the Data Columns group, in the x Columns box, choose **log.TcCB**.
5. Click on the **Options** tab.
6. Under Histogram Specs, in the Output Type box, select **Percent**.
7. Click on the **Histogram Bars** tab.
8. Make sure the **Draw Bars** and **Draw Histograms** boxes are checked. Click **OK**. A histogram of the data for both areas is displayed in a graphsheet.
9. **Click** on the data window to bring it forward. **Left click** on the top of the Area column to highlight that column, then **left-click** in a cell of the column, then **drag** the column to the top of the graphsheet and **drop** it.

Command

To produce the histograms shown in Figure 3.10 using the S-Plus Command or Script Window, type this command.

```
histogram(~ log(TcCB) | Area, data = epa.94b.tccb.df,
   nint = 25, layout = c(1,2),
   xlab = "log [ TcCB (ppb) ]",
   main = list(cex = 1.25,
     "Histograms for Log-Transformed TcCB Data"))
```

Density Plots

If we could take larger and larger sample sizes from the population and make the intervals (bins) of our histogram narrower and narrower, we would eventually end up with a picture that resembles the true probability distribution or density of the underlying population (we will talk more about probability distributions and probability densities in Chapter 4). A *density plot* is an estimate of the true underlying probability distribution, based on a sample of observations. With a density plot, the area underneath the curve between two points on the *x*-axis is an estimate of the true probability that a value will fall into that interval.

The basic algorithm to construct a density plot is as follows: for each *x*-value, a window (local neighborhood) is constructed about that value, the number of observations within the window are counted, and these counts are then scaled to produce an estimate of the probability density at that *x*-value. Depending on the algorithm, observations farther away from the *x*-value may be down-weighted compared to observations close to the *x*-value. See Chambers et al. (1983, pp. 32–36) and Silverman (1986) for details.

Figure 3.11 shows a density plot overlaying a histogram based on the log-transformed Reference area TcCB data. Compare this plot to the strip plot and histogram shown in Figure 3.9 and Figure 3.10, respectively. In the strip plot (Figure 3.9), we can see a gap in the data around 0 on the *x*-axis, but this gap is hidden in the histogram (Figure 3.10). In the density plot (Figure 3.11), the gap shows up again in the density line.

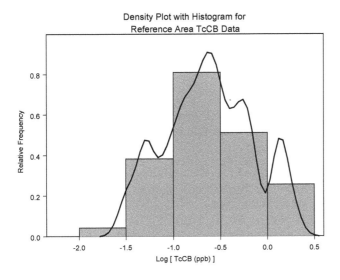

Figure 3.11   Density plot and histogram of log-transformed Reference area TcCB data

Menu

To produce the density plot and histogram shown in Figure 3.11 using the S-PLUS pull-down menus or toolbars, follow these steps.

1. In the Object Explorer, highlight the **Data** folder in the left-hand column. In the right-hand (Object) column, highlight the shortcut **new.epa.94b.tccb.df** for the version of this data frame that you created earlier.

2. On the S-PLUS menu bar, make the following menu choices: **Graph>2D Plot**. This will bring up the Insert Graph dialog box. Under the Axes Type column, **Linear** should be highlighted. Under the Plot Type column, select **Histogram with Density Line** and click **OK**. (Alternatively, left-click on the **2D Plots** button, then left-click on the **Histogram Density** button.)

3. The Histogram/Density dialog box should appear. Under the Data Columns group, the Data Set box should display **new.epa.94b.tccb.df**.

4. In the x Columns box, choose **log.TcCB**.

5. In the Subset Rows with box, type **Area == "Reference"**.

6. Click on the **Options** tab.

7. Under the Density Options group, in the Window Width box, type **0.35**.

8. Click on the **Histogram Bars** tab.

9. Make sure the **Draw Bars** and **Draw Histograms** boxes are checked.

10. Click on the **Density Line** tab.

11. In the Weight box, select **2**.

12. Click **OK**.

Command

To produce the density plot shown in Figure 3.11 (along with a strip plot, not a histogram) using the S-PLUS Command or Script Window, type this command.

```
densityplot(~ log(TcCB), data = epa.94b.tccb.df,
   subset = Area == "Reference", width = 0.35,
   xlab = "log [ TcCB (ppb) ]",
   main = list(cex = 1.25, paste(
     "Density Plot With Strip Plot for",
     "Reference Area TcCB Data", sep = "\n")))
```

Boxplots

A *boxplot* (sometimes called a *box-and-whisker plot*) is a simple graphical display invented by John Tukey that summarizes the distribution of a set
of observations (Chambers et al., 1983, pp. 21–24; Cleveland, 1993,
pp. 25–27; Cleveland, 1994, pp. 139–143; Helsel and Hirsch, 1992,
pp. 24–26). A standard boxplot consists of

- A "box" defined by the 25th and 75th percentiles.
- A line or point on the box at the median (50th percentile).
- A line (whisker) drawn from the 25th percentile to the smallest observation within one "step" of the 25th percentile (the lower adjacent
  value).
- A line (whisker) drawn from the 75th percentile to the largest observation within one "step" of the 75th percentile (the upper adjacent
  value).
- Observations over one step away from the box ends plotted with a
  symbol (e.g., asterisk) or line.

Illustration of a Boxplot

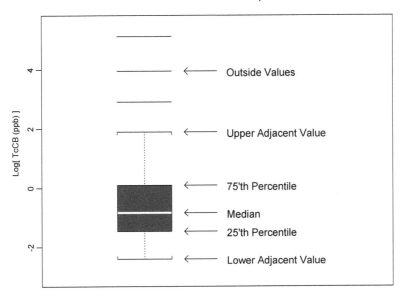

Figure 3.12   A boxplot based on the log-transformed TcCB data for the Cleanup area

Figure 3.12 illustrates a boxplot using the log-transformed TcCB soil data
for the Cleanup area. A "step" is conventionally defined as 1.5 times the interquartile range. (Recall that the interquartile range is the difference between the 75th and 25th percentiles.) Given this algorithm for constructing

boxplots, you would expect to see at least one "outside value" only about 0.7% of the time if you were always looking at data from a normal distribution.

A boxplot quickly allows you to see:

- The center of the data (the median).
- The spread (variability) of the data.
- The skewness of the data (by looking at the relative lengths of the box halves, and the whiskers).
- The presence of any "unusual" (outside) values.

Outside values are not necessarily outliers or bad observations! Indeed, since most environmental data is right skewed, you may see lots of boxplots with outside values when you plot variables on their original scales.

Boxplots are very useful for comparing the distributions of two data sets. Figure 3.13 displays side-by-side boxplots for the log-transformed TcCB data for the Reference and Cleanup areas (see Table 3.2). As we saw with the strip plots in Figure 3.9 and the histograms in Figure 3.10, the boxplots indicate that most of the observations for the Cleanup area are comparable to (or even smaller than) the observations for the Reference area, but there are a few very large "outliers" in the Cleanup area.

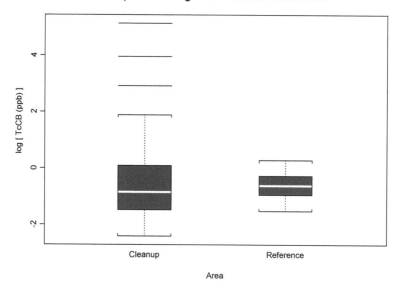

Figure 3.13   Boxplots of the log-transformed TcCB data for the Reference and Cleanup areas

Menu

To produce the boxplots shown in Figure 3.13 using the S-PLUS pull-down menus or toolbars, follow these steps.

1. In the Object Explorer, highlight the **Data** folder in the left-hand column. In the right-hand (Object) column, highlight the shortcut **new.epa.94b.tccb.df** for the version of this data frame that you created earlier.
2. On the S-PLUS menu bar, make the following menu choices: **Graph>2D Plot**. This will bring up the Insert Graph dialog box. Under the Axes Type column, **Linear** should be highlighted. Under the Plot Type column, select **Boxplot** and click **OK**. (Alternatively, left-click on the **2D** Plots button, then left-click on the **Box** button.)
3. The Boxplot dialog box should appear. Under the Data Columns group, the Data Set box should display **new.epa.94b.tccb.df**.
4. In the x Columns box, choose **Area**.
5. In the y Columns box, choose **log.TcCB**.
6. Click **OK**.

Command

To produce the boxplots shown in Figure 3.13 using the S-PLUS Command or Script Window, type these commands:

```
attach(new.epa.94b.tccb.df)
boxplot(split(log(TcCB), Area), xlab = "Area",
   ylab = "log [ TcCB (ppb) ]",
   main = "Boxplots for Log-Transformed TcCB Data")
detach()
```

or this command (which produces horizontal boxplots):

```
bwplot(Area ~ log(TcCB), data = epa.94b.tccb.df,
   xlab = "log [ TcCB (ppb) ]",
   main = list(cex = 1.25,
   "Boxplots for Log-Transformed TcCB Data"))
```

Quantile Plots or Empirical CDF Plots

Loosely speaking, the pth *quantile* of a population is the (a) number such that a fraction p of the population is less than or equal to this number. The pth quantile is the same as the $100p^{th}$ percentile; for example, the 0.5 quantile is the same as the 50th percentile. For a population, a plot of the quantiles on the x-axis vs. the percentage or fraction of the population less than or equal to that number on the y-axis is called a *cumulative distribution function plot* or *cdf plot* (we will talk more about cumulative distribution functions in Chapter 4). The *y*-axis is usually labeled as the *cumulative probability* or *cumulative frequency*.

When we have a sample of data from some population, we usually do not know what percentiles our observations correspond to because we do not know the true population percentiles, so we use the sample data to estimate them. A *quantile plot* (also called an *empirical cumulative distribution function plot* or *empirical cdf plot*) plots the ordered data (sorted from smallest to largest) on the *x*-axis vs. the estimated cumulative probabilities on the *y*-axis (Chambers et al., 1983, pp. 11–19; Cleveland, 1993, pp. 17–20; Cleveland, 1994, pp. 136–139; Helsel and Hirsch, 1992, pp. 21–24). (Sometimes the *x*- and *y*-axes are reversed.) The specific formulas that are used to estimate the cumulative probabilities are discussed later in the sub-section Computing Plotting Positions.

Figures 3.14 and 3.15 show quantile plots for the Reference area TcCB data. Based on these plots, you can easily pick out the median as about 0.55 ppb and the quartiles as about 0.4 ppb and 0.75 ppb (compare these numbers to the ones listed on page 57). You can also see that the quantile plot quickly rises, then pretty much levels off after about 0.8 ppb, which indicates that the data are skewed to the right (see the histogram for the Reference area data in Figure 3.1). Helsel and Hirsch (1992, p. 22) note that quantile plots, unlike histograms, do not require you to figure out how to divide the data into classes, and, unlike boxplots, all of the data are displayed in the graph.

You can also use quantile plots to compare the empirical cdf of a data set with the cdf from some specified theoretical distribution, or with the empirical cdf of another data set. Figure 3.16 shows the empirical cdf for the Reference area TcCB data compared to the cdf of a lognormal distribution (the parameters of this distribution were estimated from the data; we will talk more about the lognormal distribution in Chapter 4). We see that the lognormal distribution appears to fit these data quite well. Usually, however, quantile-quantile (Q-Q) plots, not empirical cdf plots, are used to visually assess the goodness of fit of a theoretical distribution to a data set (see the next section Probability Plots or Quantile-Quantile (Q-Q) Plots).

Just as with histograms and boxplots, you can also use empirical cdf plots to compare data sets. Figure 3.17 compares the empirical cdf for the Reference area with the empirical cdf for the cleanup area for the log-transformed TcCB data. As we saw with the strip plots (Figure 3.9), histograms (Figure 3.10), and boxplots (Figure 3.13), the Cleanup area has quite a few extreme values compared to the Reference area.

Quantile Plot for Reference Area TcCB Data

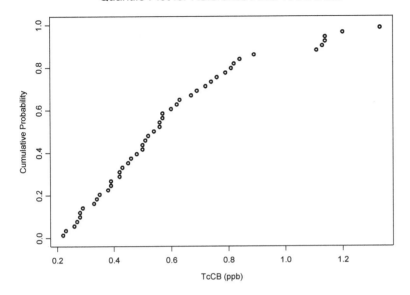

Figure 3.14   Quantile plot of Reference area TcCB data (points)

Quantile Plot for Reference Area TcCB Data

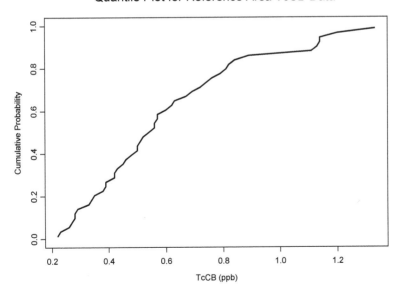

Figure 3.15   Quantile plot of Reference area TcCB data (line)

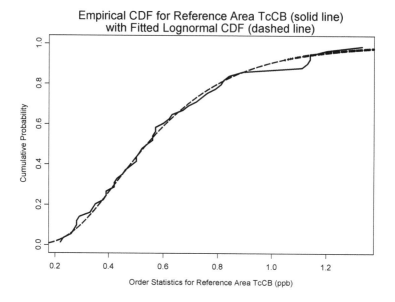

Figure 3.16   Empirical cdf of Reference area TcCB data compared to a lognormal cdf

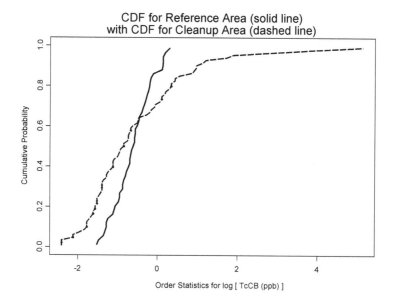

Figure 3.17   Quantile plots comparing log-transformed TcCB data at the Reference and Cleanup areas

Menu

To produce the quantile plot of the Reference area TcCB data shown in Figure 3.14 using the ENVIRONMENTALSTATS for S-PLUS pull-down menu, follow these steps.

1. In the Object Explorer, highlight the **Data** folder in the left-hand column. In the right-hand (Object) column, highlight the shortcut **new.epa.94b.tccb.df** for the version of this data frame that you created earlier.

2. On the S-PLUS menu bar, make the following menu choices: **EnvironmentalStats>EDA>CDF Plot>Empirical CDF**. This will bring up the Plot Empirical CDF dialog box.

3. The Data Set box should display **new.epa.94b.tccb.df**.

4. In the Variable(s) box, choose **TcCB**.

5. In the Subset Rows with box, type **Area=="Reference"**.

6. Click on the **Plot Options** tab.

7. For Plot Type, choose **Points**.

8. Click **OK** or **Apply**.

To produce the quantile plot of the Reference area TcCB data shown in Figure 3.15, repeat the above steps, but omit Steps 6 and 7.

To produce the quantile plot shown in Figure 3.16 comparing the Reference area TcCB data to a lognormal distribution, follow these steps.

1. On the S-PLUS menu bar, make the following menu choices: **EnvironmentalStats>EDA>CDF Plot>Compare Two CDFs**. This will bring up the Compare Two CDFs dialog box.

2. The Data Set box should display **new.epa.94b.tccb.df**.

3. In the x Variable box, choose **TcCB**.

4. In the Subset Rows with box, type **Area=="Reference"**.

5. Under the Distribution Information Group, make sure that the **Estimate Parameters** box is checked.

6. In the Distribution box, select **Lognormal**.

7. Click **OK** or **Apply**.

To produce the quantile plot shown in Figure 3.17 comparing the log-transformed Reference area TcCB data to the log-transformed Cleanup area TcCB data, follow these steps.

1. On the S-PLUS menu bar, make the following menu choices: **EnvironmentalStats>EDA>CDF Plot>Compare Two CDFs**. This will bring up the Compare Two CDFs dialog box.

2. For the Compare Data to choice, select **Other Data**.

3. The Data Set box should display **new.epa.94b.tccb.df**.

4. In the Variable 1 box, choose **log.TcCB**.

5. In the Variable 2 box, choose **Area**.

6. Click on the **Variable 2 is a Grouping Variable** box to check it.
7. Click **OK** or **Apply**.

Command

The commands below show you how to produce the quantile plots shown in Figures 3.14 – 3.17 using the S-PLUS Command or Script Window. Note that the function `ecdfplot` is part of ENVIRONMENTALSTATS for S-PLUS, and that the S-PLUS function `cdf.compare` has been modified in ENVIRONMENTALSTATS for S-PLUS. All of the commands assume the data frame `epa.94b.tccb.df` has been attached to your search list via the following command:

```
attach(epa.94b.tccb.df)
```

To produce the quantile plot of the Reference area TcCB data shown in Figure 3.14, type the following command.

```
ecdfplot(TcCB[Area=="Reference"], type = "p",
  xlab = "TcCB (ppb)",
  main = paste("Quantile Plot for",
    "Reference Area TcCB Data"))
```

To produce the quantile plot of the Reference area TcCB data shown in Figure 3.15, type the following command.

```
ecdfplot(TcCB[Area=="Reference"], xlab = "TcCB (ppb)",
  main = paste("Quantile Plot for",
    "Reference Area TcCB Data"))
```

To produce the quantile plot shown in Figure 3.16 comparing the Reference area TcCB data to a lognormal distribution, type the following command.

```
cdf.compare(TcCB[Area=="Reference"], dist = "lnorm",
  xlab = paste("Order Statistics for",
    "Reference Area TcCB (ppb)"),
  main = paste("Empirical CDF for ",
    "Reference Area TcCB (solid line)\n",
    "with Fitted Lognormal CDF (dashed line)",
    sep = ""))
```

To produce the quantile plot shown in Figure 3.17 comparing the log-transformed Reference area TcCB data to the log-transformed Cleanup area TcCB data, type the following command.

```
cdf.compare(log(TcCB[Area=="Reference"]),
  log(TcCB[Area == "Cleanup"]),
  xlab = "Order Statistics for log [ TcCB (ppb) ]",
```

```
main = paste("CDF for Reference Area (solid line)",
    "with CDF for Cleanup Area (dashed line)",
    sep = "\n"))
```

To detach the data frame epa.94b.tccb.df from your search list, type the following command:

```
detach(epa.94b.tccb.df)
```

Computing Plotting Positions

In a quantile plot, the estimated cumulative probabilities are often called *plotting positions*, since they determine the positions relative to the y-axis. The estimated cumulative probabilities are usually computed by one of the following formulas.

$$\hat{p}_i = \frac{\#\left[x_j \leq x_{(i)}\right]}{n} \tag{3.20}$$

$$\hat{p}_i = \frac{i}{n} \tag{3.21}$$

$$\hat{p}_i = \frac{i - a}{n - 2a + 1} \tag{3.22}$$

where $\hat{p}_i$ denotes the estimated cumulative probability associated with the ith order statistic $x_{(i)}$, $i=1, 2, ..., n$, and a is some constant between 0 and 1 (Cleveland, 1993, p. 18; D'Agostino, 1986a, pp. 8, 25).

The estimator in Equation (3.20) says that for the ith ordered observation, the proportion of the population less than or equal to this observation is estimated by the proportion of observations in the sample less than or equal to this observation. This estimator is sometimes called the *empirical probabilities estimator* and is intuitively appealing. The estimator in Equation (3.21) is equivalent to the estimator in Equation (3.20) if there are no ties in the data. The disadvantage of these two estimators is that they imply the largest observed value is the maximum possible value of the distribution (the 100th percentile). This may be satisfactory if we can model the population with a finite discrete distribution, but it is usually not satisfactory if we want to model the population with a continuous distribution.

When the underlying distribution is assumed to be continuous, Equation (3.22) with some specified value of the constant a is usually used to compute estimated cumulative probabilities. Based on certain principles from statistical theory, certain values of the constant a make sense for specific underlying distributions. These details are discussed in the ENVIRONMENTALSTATS for S-PLUS help files under Glossary: Empirical Cumulative Distribution Function (ecdf), and Glossary: Probability Plot or Quantile-Quantile (Q-Q) Plot.

Table 3.7, adapted from Helsel and Hirsch (1992, p. 23) and Stedinger et al. (1993), displays commonly used plotting positions based on Equation (3.22). In practice, the Blom or Hazen plotting positions are probably the most frequently used (e.g., Chambers et al., 1983, p. 12; Cleveland, 1993, p. 18), but as Helsel and Hirsch (1992, pp. 23–24) point out, hydrologists usually use Weibull plotting positions for flow-duration and flood-frequency curves. In fact, to simplify things, Helsel and Hirsch (1992, p. 24) use Cunnane plotting positions throughout their book because they are very close to Blom plotting positions and because some hydrologists use them instead of Weibull plotting positions. For moderate and large sample sizes, there will be very little difference between the plotting positions listed in Table 3.7.

Name	a	Distribution Often Used With	References/Examples
Weibull	0	Weibull, Uniform	Weibull (1939), Stedinger et al. (1993)
Median	0.3175	Several	Filliben (1975), Vogel (1986)
Bloom	0.375	Normal and Others	Blom (1958), Looney and Gulledge (1985)
Cunnane	0.4	Several	Cunnane (1978), Chowdhury et al. (1991)
Gringorten	0.44	Gumbel	Gringorton (1963), Vogel (1986)
Hazen	0.5	Several	Hazen (1914), Chambers et al. (1983), Cleveland (1993)

Table 3.7   Plotting position formulas based on Equation (3.22) (after Helsel and Hirsch, 1992, p. 23, and Stedinger et al., 1993)

For any general value x such that

$$x_{(i)} \leq x \leq x_{(i+1)}$$

(3.23)

the estimated cumulative probability is usually defined as follows for discrete distributions:

$$\hat{p}\left(x\right) = \hat{p}_i \tag{3.24}$$

and as follows for continuous distributions:

$$\hat{p}\left(x\right) = \left(1 - r\right)\hat{p}_i + r\,\hat{p}_{i+1} \tag{3.25}$$

$$r = \frac{x - x_{(i)}}{x_{(i+1)} - x_{(i)}} \tag{3.26}$$

Equation (3.24) says that the quantile plot stays flat until it hits a value on the x-axis corresponding to one of the order statistics, then it makes a jump, so it is a step function. Equations (3.25) and (3.26) use linear interpolation to compute the value of the estimated cumulative probability at a particular value of x; in this case the quantile plot looks like lines connecting the points, just as in Figure 3.15.

Probability Plots or Quantile-Quantile (Q-Q) Plots

A *probability plot* or *quantile-quantile (Q-Q)* plot is a graphical display invented by Wilk and Gnanadesikan (1968) to compare a data set to a particular probability distribution or to compare it to another data set. The idea is that if two population distributions are exactly the same, then they have the same quantiles (percentiles), so a plot of the quantiles for the first distribution vs. the quantiles for the second distribution will fall on the 0-1 line (i.e., the straight line y=x with intercept 0 and slope 1). If the two distributions have the same shape and spread but different locations, then the plot of the quantiles will fall on the line y=a+x (parallel to the 0-1 line) where a denotes the difference in locations. If the distributions have different locations and differ by a multiplicative constant b, then the plot of the quantiles will fall on the line y=a+bx (D'Agostino, 1986a, p. 25; Helsel and Hirsch, 1986, p. 42). Various kinds of differences between distributions will yield various kinds of deviations from a straight line.

Comparing Data to a Theoretical Probability Distribution

In practice, we take samples from a population, so a Q-Q plot uses empirical quantiles (i.e., the order statistics), and the corresponding estimated cumulative probabilities (plotting positions) are computed using one of the Equations (3.20) to (3.22) above (Chambers et al., 1983, pp. 48–57; Cleveland, 1993, pp. 21–25, 28–32; Cleveland, 1994, pp. 143–149; Helsel and Hirsch, 1992, pp. 40–45). To compare a data set to a theoretical probability distribution, the empirical quantiles (order statistics) are plotted on the *y*-axis and the corresponding quantiles from the assumed probability distribution are plotted on the *x*-axis. The quantiles of the assumed probability distribution are computed based on the estimated cumulative probabilities associated with the empirical quantiles of the data set.

As an example, let us compare the Reference area TcCB data to a normal (Gaussian) distribution. Table 3.8 below displays the order statistics, Blom plotting positions, and corresponding quantiles from a standard normal distribution for the first three and last three order statistics of the Reference area TcCB data.

Order Statistic	Plotting Position	Normal Quantile
0.22	0.013	-2.2
0.23	0.034	-1.8
0.26	0.056	-1.6
...	...	...
1.14	0.944	1.6
1.20	0.966	1.8
1.33	0.987	2.2

Table 3.8   Order statistics, Blom plotting positions, and standard normal quantiles for the Reference area TcCB data

Figure 3.18 shows the normal Q-Q plot for these data, along with a fitted regression line (see Chapter 9). In this figure you can see that the points do not tend to fall on the line, but rather seem to make a U shape. This indicates that the Reference area data are skewed to the right relative to a symmetrical, bell-shaped normal distribution (see Figure 3.1). Figure 3.19 shows the normal Q-Q plot for the log-transformed Reference area data. Here you can see the points do tend to fall on the line, indicating that a log-normal distribution may be a good model for these data. Compare this figure to Figure 3.16.

An interesting feature of normal Q-Q plots is that if a sample of data comes from a normal distribution, and the data are plotted against quantiles of a standard normal distribution (with mean 0 and variance 1), then the intercept of the fitted line estimates the mean of the population, and the slope of the fitted line estimates the standard deviation (Nelson, 1982, p. 113;

Normal Q-Q Plot for Reference Area TcCB Data

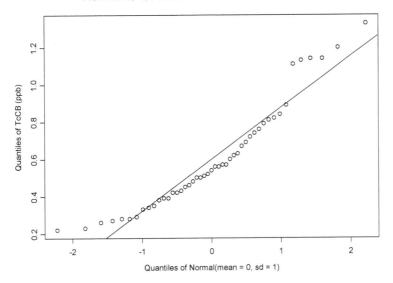

Figure 3.18   Normal Q-Q plot for Reference area TcCB data

Normal Q-Q Plot for Log-Transformed Reference Area TcCB Data

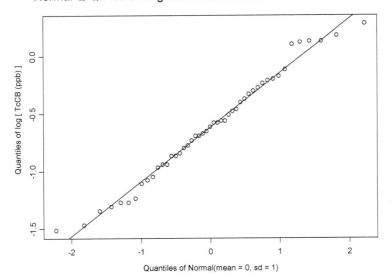

Figure 3.19   Normal Q-Q plot for the log-transformed Reference area TcCB data

Cleveland, 1993, p. 31). For the fitted line in Figure 3.19, we can eyeball the intercept at about –0.6 and the slope at about 0.5. These numbers are comparable to the mean and standard deviation estimated in the usual way shown on page 57.

Comparing Two Data Sets

You can use a Q-Q plot to determine whether two data sets come from the same parent distribution. The empirical quantiles of the first data set are plotted on the y-axis vs. the empirical quantiles of the second data set on the x-axis. If the two data sets have different sample sizes, then the empirical quantiles of the smaller data set are used, and the corresponding quantiles for the larger data set are computed based on the plotting positions of the smaller data set and linear interpolation (Chambers et al., 1983, pp. 54–55; Cleveland, 1993, p. 21; Cleveland, 1994, p. 144; Helsel and Hirsch, 1992, pp. 43–45). Specifically, let

$$Y_{(1)}, \ Y_{(2)}, \ \cdots, \ Y_{(m)}$$

denote the m ordered observations in the first data set, let

$$X_{(1)}, \ X_{(2)}, \ \cdots, \ X_{(n)}$$

denote the n ordered observation in the second data set, assume both samples come from continuous distributions, and assume $m < n$. Let

$$\hat{p}_1, \ \hat{p}_2, \ \cdots, \ \hat{p}_m$$

denote the plotting positions corresponding to the m empirical quantiles from the first data set and let

$$\hat{p}_1^*, \ \hat{p}_2^*, \ \cdots, \ \hat{p}_n^*$$

denote the plotting positions corresponding the n empirical quantiles from the second data set (see Equation (3.22)). Then the empirical quantile $y_{(i)}$ is plotted against

$$x^{*}_{(i)} = (1 - r)\, x_{(j)} + r\, x_{(j+1)} \qquad (3.27)$$

where

$$r = \frac{\hat{p}_i - \hat{p}^{*}_j}{\hat{p}^{*}_{j+1} - \hat{p}^{*}_j} \qquad (3.28)$$

$$\hat{p}^{*}_j \le \hat{p}_i \le \hat{p}^{*}_{j+1} \qquad (3.29)$$

(Note: The S-PLUS function qqplot and the S-PLUS menu selection **Graph>2D Plot>Quantile-Quantile Plot** do not compute the quantiles for the larger data set in this way.)

As an example, let us compare the log-transformed Reference and Cleanup area TcCB data. Recall that the Reference area has 47 observations and the Cleanup area has 77 observations. Table 3.9 below displays the order statistics and Blom plotting positions for the first three and last three order statistics of the log-transformed Reference area data. Note that the first column in this table is the log of the first column in Table 3.8, and the second column in this table is the same as the second column in Table 3.8. The third column in Table 3.9 displays the corresponding quantiles for the log-transformed Cleanup area data.

Quantiles for Reference Area	Plotting Position	Quantiles for Cleanup Area
−1.51	0.013	−2.41
−1.47	0.034	−2.40
−1.35	0.056	−2.12
...	...	...
0.13	0.944	1.77
0.18	0.966	2.88
0.29	0.987	4.66

Table 3.9   Empirical quantiles and plotting positions for the log-transformed Reference and Cleanup area TcCB data

Figure 3.20 shows the Q-Q plot comparing the Reference and Cleanup areas for the log-transformed TcCB data, along with the 0-1 line. In this figure you can see the points do not tend to fall on the 0-1 line, but instead tend to fall along two different lines, both with a steeper slope than 1. Q-Q plots that exhibit this kind of pattern indicate that one of the samples (the Cleanup area data in this case) probably comes from a "mixture" distribution: some of the observations come from a distribution similar in shape and scale to the Reference area distribution, and some of the observations come from a distribution that is shifted to the right and more spread out relative to the Reference area distribution because of residual contamination (see Figures 3.2, 3.9, 3.10, 3.13, and 3.17).

Q-Q Plot Comparing Cleanup and Reference Area TcCB Data

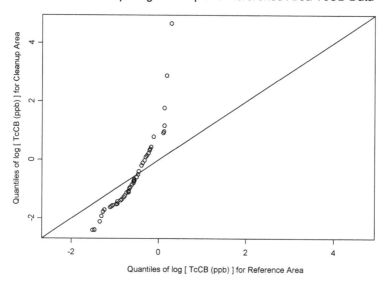

Quantiles of log [TcCB (ppb)] for Reference Area

Figure 3.20   Q-Q plot comparing log-transformed Cleanup and Reference area TcCB data

Assessing Departures from Linearity

Because the points on a Q-Q plot represent a sample from a population, even if the two underlying distributions are in fact the same (or differ only by an additive and/or multiplicative constant), the points will never all lie exactly on a straight line. Also, the extreme points have more variability than the points toward the center, and Q-Q plots based on small sample sizes are more variable than Q-Q plots based on large sample sizes. It is therefore important to have a good idea of what a "typical" Q-Q plot might look like for the kind of data, distributions, and sample sizes you are looking at. In ENVIRONMENTALSTATS for S-PLUS you can use the menu selection **EnvironmentalStats>EDA>Q-Q Plot>Q-Q Plot Gestalt** or the command line function `qqplot.gestalt` to create several "typical" Q-Q plots for any specified distribution and sample size.

Table 3.10 lists common patterns of departures from linearity that you might see in a Q-Q plot (D'Agostino, 1986a, p. 47; Helsel and Hirsch, 1992, pp. 30–33). A U-shaped Q-Q plot indicates that the underlying distribution for the observations on the *y*-axis is skewed to the right relative to the underlying distribution for the observations on the *x*-axis (e.g., Figure 3.18). An upside-down-U-shaped Q-Q plot indicates the *y*-axis distribution is skewed-left relative to the *x*-axis distribution. An S-shaped Q-Q plot indicates the *y*-axis distribution has shorter tails than the *x*-axis distribution. Conversely, a plot that is bent down on the left, linear in the middle, and bent up on the right indicates that the *y*-axis distribution has longer tails than the *x*-axis distribution. A Q-Q plot in which the points tend to fall on two different lines indicates one of the data sets probably comes from a mixture distribution (e.g., Figure 3.20). A Q-Q plot in which most of the points fall pretty much along a line but one or a few of the extreme points are way above (or below) the rest of the points indicates these few points are "outliers."

Pattern	Cause
U shape	*y*-axis distribution right skewed relative to *x*-axis distribution
Upside-down U shape	*x*-axis distribution right skewed relative to *y*-axis distribution
S shape	*x*-axis distribution has heavy tails compared to *y*-axis distribution
Bent down on left, linear in middle, and bent up on right	*y*-axis distribution has heavy tails compared to *x*-axis distribution
Two separate lines	*y*-axis distribution is a mixture of two different distributions
Most points linear but one or a few way above (below) the rest	One or more outliers

Table 3.10   Common patterns of deviation from linearity for Q-Q plots and their causes

Tukey Mean-Difference Q-Q Plots

Instead of adding a fitted regression line to a Q-Q plot, an even better way to assess deviation from linearity is to use a Tukey mean-difference Q-Q plot, also called an m-d plot (Cleveland, 1993, pp. 22–23). This is a plot of the differences between the quantiles on the *y*-axis vs. the average of the quantiles on the *x*-axis. If the two sets of quantiles come from the same parent distribution, then the points in an m-d plot should fall roughly along the horizontal line $y=0$. If one set of quantiles come from the same distribution with a shift in median, then the points in this plot should fall along a horizontal line above or below the line $y=0$. If the parent distributions of the quantiles differ in scale, then the points on this plot will fall at an angle.

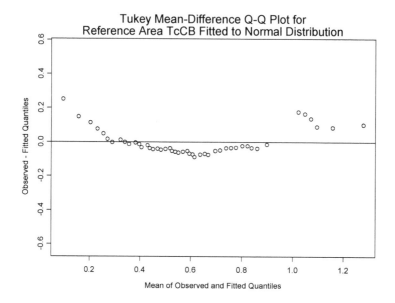

Figure 3.21   Tukey mean-difference Q-Q plot for Reference area TcCB data fitted to a normal distribution

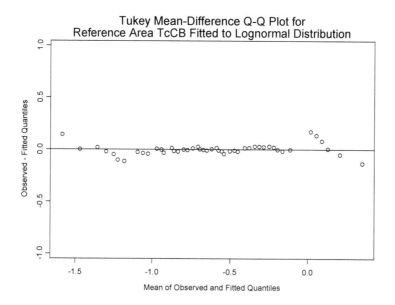

Figure 3.22   Tukey mean-difference Q-Q plot for Reference area TcCB data fitted to a lognormal distribution

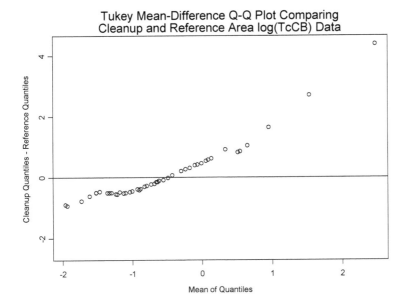

Figure 3.23   Tukey mean-difference Q-Q plot comparing log-transformed cleanup
and reference area TcCB data

A Tukey mean-difference Q-Q plot enhances our perception of how the points in the Q-Q plot deviate from a straight line, because it is easier to judge deviations from a horizontal line than from a line with a non-zero slope. Figures 3.21 to 3.23 are the Tukey mean-difference analogues of the Q-Q plots shown in Figures 3.18 to 3.20.

Menu

This section shows you how to use the ENVIRONMENTALSTATS for S-PLUS pull-down menu to create Figures 3.18-3.23. To create the normal Q-Q plot for the Reference area TcCB data shown in Figure 3.18, follow these steps.

1. In the Object Explorer, highlight the **Data** folder in the left-hand column. In the right-hand (Object) column, highlight the shortcut **new.epa.94b.tccb.df** for the version of this data frame that you created earlier.
2. On the S-PLUS menu bar, make the following menu choices: **EnvironmentalStats>EDA>Q-Q Plot>Q-Q Plot**. This will bring up the Q-Q Plot dialog box.
3. The Data Set box should display **new.epa.94b.tccb.df**.
4. In the x Variable box, choose **TcCB**.

5. In the Subset Rows with box, type **Area=="Reference"**.
6. In the Distribution box, make sure **Normal** is selected.
7. Click on the **Plotting** tab.
8. Click on the **Add a Line** box to select this option.
9. Click **OK** or **Apply**.

To create the normal Q-Q plot for the log-transformed Reference area data shown in Figure 3.19, follow the same steps as above, except in Step 6 make sure **Lognormal** is selected in the Distribution box. To create the Q-Q plot comparing the Reference and Cleanup areas for the log-transformed TcCB data shown in Figure 3.20, follow these steps.

1. On the S-PLUS menu bar, make the following menu choices: **EnvironmentalStats>EDA>Q-Q Plot>Q-Q Plot**. This will bring up the Q-Q Plot dialog box.
2. Set the Compare Data to button to **Other Data**.
3. The Data Set box should display **new.epa.94b.tccb.df**.
4. In the Variable 1 box, choose **log.TcCB**.
5. In the Variable 2 box, choose **Area**.
6. Click on the **Variable 2 is a Grouping Variable** box to select this option.
7. Click on the **Plotting** tab.
8. Click on the **Equal Axes** box to select this option.
9. Click on the **Add a Line** box to select this option.
10. For Q-Q Line Type button, select **0-1**.
11. Click **OK** or **Apply**.

Note that this plot has the Reference area quantiles on the *y*-axis and the Cleanup area quantiles on the *x*-axis.

To create the Tukey mean-difference Q-Q plot for the Reference area TcCB data fitted to a normal distribution shown in Figure 3.21, follow these steps.

1. On the S-PLUS menu bar, make the following menu choices: **EnvironmentalStats>EDA>Q-Q Plot>Q-Q Plot**. This will bring up the Q-Q Plot dialog box.
2. The Data Set box should display **new.epa.94b.tccb.df**.
3. In the x Variable box, choose **TcCB**.
4. In the Subset Rows with box, type **Area=="Reference"**.
5. Click on the **Estimate Parameters** box to select this option.
6. In the Distribution box, make sure **Normal** is selected.
7. Click on the **Plotting** tab.
8. For the Plot Type button, select **Tukey M-D**.
9. Click on the **Add a Line** box to select this option.
10. Click **OK** or **Apply**.

To create the Tukey mean-difference Q-Q plot for the Reference area data fitted to a lognormal distribution shown in Figure 3.22, follow the same steps as above, except in Step 6 make sure **Lognormal** is selected in the Distribution box. To create the Tukey mean-difference Q-Q plot comparing the Reference and Cleanup areas for the log-transformed TcCB data shown in Figure 3.23, follow these steps.

1. On the S-PLUS menu bar, make the following menu choices: **EnvironmentalStats>EDA>Q-Q Plot>Q-Q Plot**. This will bring up the Q-Q Plot dialog box.
2. Set the Compare Data to button to **Other Data**.
3. The Data Set box should display **new.epa.94b.tccb.df**.
4. In the Variable 1 box, choose **log.TcCB**.
5. In the Variable 2 box, choose **Area**.
6. Click on the **Variable 2 is a Grouping Variable** box to select this option.
7. Click on the **Plotting** tab.
8. For the Plot Type button, choose **Tukey M-D**.
9. Click on the **Add a Line** box to select this option.
10. Click **OK** or **Apply**.

Note that this plot is the mirror image (on the y-axis) of Figure 3.23 because it is subtracting the Cleanup area quantiles from the Reference area quantiles instead of vice-versa.

Command

The commands below show you how to produce the Q-Q plots and Tukey mean-difference Q-Q plots shown in Figures 3.18 to 3.23 using the S-PLUS Command or Script Window. Note that the S-PLUS function qqplot has been modified in ENVIRONMENTALSTATS for S-PLUS. All of the commands assume the data frame epa.94b.tccb.df has been attached to your search list via the following command:

```
attach(epa.94b.tccb.df)
```

To produce the Q-Q plot of the Reference area TcCB data fitted to a normal distribution shown in Figure 3.18, type the following command.

```
qqplot(TcCB[Area=="Reference"], add.line = T,
    ylab = "Quantiles of TcCB (ppb)",
    main = paste("Normal Q-Q Plot for",
      "Reference Area TcCB Data"))
```

To produce the Q-Q plot of the Reference area TcCB data fitted to a lognormal distribution shown in Figure 3.19, type the following command.

```
qqplot(TcCB[Area=="Reference"], dist = "lnorm",
   add.line = T,
   ylab = "Quantiles of log [ TcCB (ppb) ]",
   main = paste("Normal Q-Q Plot for",
     "Log-Transformed Reference Area TcCB Data"))
```

To create the Q-Q plot comparing the Reference and Cleanup areas for the log-transformed TcCB data shown in Figure 3.20, type the following command.

```
qqplot(log(TcCB[Area=="Reference"]),
   log(TcCB[Area == "Cleanup"]),
   plot.pos.con = 0.375, equal.axes = T, add.line = T,
   qq.line.type = "0-1",
   xlab = paste("Quantiles of log [ TcCB (ppb) ]",
     "for Reference Area"),
   ylab = paste("Quantiles of log [ TcCB (ppb) ]",
     "for Cleanup Area"),
   main = paste("Q-Q Plot Comparing Cleanup and",
     "Reference Area TcCB Data"))
```

To create the Tukey mean-difference Q-Q plot for the Reference area TcCB data fitted to a normal distribution shown in Figure 3.21, type the following command.

```
qqplot(TcCB[Area=="Reference"], plot.type = "Tukey",
   estimate.params = T, add.line = T,
   main = paste("Tukey Mean-Difference Q-Q Plot for",
     "\nReference Area TcCB Fitted to ",
     "Normal Distribution", sep = ""))
```

To create the Tukey mean-difference Q-Q plot for the Reference area data fitted to a lognormal distribution shown in Figure 3.22, type the following command.

```
qqplot(TcCB[Area=="Reference"], dist = "lnorm",
   plot.type = "Tukey", estimate.params = T,
   add.line = T,
   main = paste("Tukey Mean-Difference Q-Q Plot for",
     "\nReference Area TcCB Fitted to ",
     "Lognormal Distribution", sep = ""))
```

To create the Tukey mean-difference Q-Q plot comparing the Reference and Cleanup areas for the log-transformed TcCB data shown in Figure 3.23, type this command.

```
qqplot(log(TcCB[Area=="Reference"]),
   log(TcCB[Area == "Cleanup"]), plot.type = "Tukey",
   plot.pos.con = 0.375, add.line = T,
   ylab = "Cleanup Quantiles - Reference Quantiles",
   main = paste("Tukey Mean-Difference Q-Q Plot ",
     "Comparing\nCleanup and Reference Area ",
     "log(TcCB) Data", sep = ""))
```

To detach the data frame `epa.94b.tccb.df` from your search list, type the following command:

```
detach(epa.94b.tccb.df)
```

Box-Cox Data Transformations and Q-Q Plots

Two common assumptions for several standard parametric hypothesis tests are:

1. The observations come from a normal distribution.
2. If several groups are involved, the variances are the same among all groups.

(See Chapter 7.) A standard linear regression model (see Chapter 9) makes the above assumptions, and also assumes a linear relationship between the response variable and the predictor variable or variables.

Often, especially with environmental data, the above assumptions do not hold because the original data are skewed and/or they follow a distribution that is not really shaped like a normal distribution. It is sometimes possible, however, to transform the original data so that the transformed observations in fact come from a normal distribution or close to a normal distribution. The transformation may also induce homogeneity of variance and a linear relationship between the response and predictor variable(s) (if this is relevant).

Sometimes, theoretical considerations indicate an appropriate transformation. For example, count data often follow a Poisson distribution (see Chapter 4), and it can be shown that taking the square root of observations from a Poisson distribution tends to make these data look more bell-shaped (Johnson et al., 1992, p. 163; Johnson and Wichern, 1992, p. 164; Zar, 1999, pp. 275–278). A common example in the environmental field is that chemical concentration data often appear to come from a lognormal distribution or some other positively skewed distribution. In this case, taking the logarithm of the observations often appears to yield normally distributed data. Usually, a data transformation is chosen based on knowledge of the process generating the data, as well as graphical tools such as quantile-quantile plots and histograms.

Although data analysts knew about using data transformations for several years, Box and Cox (1964) presented a formalized method for deciding on a data transformation. Given a random variable X from some distribution with only positive values, the Box-Cox family of power transformations is defined as:

$$Y = \begin{cases} \dfrac{\left(X^\lambda - 1\right)}{\lambda} & , \quad \lambda \neq 0 \\[2em] \log(X) & , \quad \lambda = 0 \end{cases} \qquad (3.30)$$

where λ (lambda) denotes the power of the transformation and Y is assumed to come from a normal distribution. This transformation is continuous in λ. Note that this transformation also preserves ordering. That is, if

$$X_1 < X_2$$

then

$$Y_1 < Y_2$$

Box and Cox (1964) proposed choosing the appropriate value of λ based on maximizing the likelihood function (see Chapter 5). Note that for non-zero values of λ, instead of using the formula of Box and Cox in Equation (3.30), you may simply use the power transformation

$$Y = X^\lambda \qquad (3.31)$$

since these two equations differ only by a scale difference and origin shift, and the essential character of the transformed distribution remains unchanged (Draper and Smith, 1981, p. 225).

The value $\lambda=1$ corresponds to no transformation. Values of λ less than 1 shrink large values of X, and are therefore useful for transforming positively skewed (right-skewed) data. Values of λ larger than 1 inflate large values of X, and are therefore useful for transforming negatively skewed (left-skewed) data (Helsel and Hirsch, 1992, pp. 13–14; Johnson and Wichern, 1992, p. 165). Commonly used values of λ include 0 (log transformation), 0.5 (square-root transformation), −1 (reciprocal), and −0.5 (reciprocal root).

Transformations are not "tricks" used by the data analyst to hide what is going on, but rather useful tools for understanding and dealing with data (Berthouex and Brown, 1994, p. 64). Hoaglin (1988) discusses "hidden" transformations that are used everyday, such as the pH scale for measuring acidity. It is often recommend that when dealing with several similar data sets, it is best to find a common transformation that works reasonably well for all the data sets, rather than using slightly different transformations for each data set (Helsel and Hirsch, 1992, p. 14; Shumway et al., 1989).

One problem with data transformations is that translating results on the transformed scale back to the original scale is not always straightforward. Estimating quantities such as means, variances, and confidence limits in the transformed scale and then transforming them back to the original scale usually leads to biased and inconsistent estimates (Gilbert, 1987, p. 149; Fisher and van Belle, 1993, p. 466). For example, exponentiating the confidence limits for a mean based on log-transformed data does not yield a confidence interval for the mean on the original scale. Instead, this yields a confidence interval for the median (see Chapter 5). It should be noted, however, that quantiles (percentiles) and rank-based procedures are invariant to monotonic transformations (Helsel and Hirsch, 1992, p. 12).

You can use ENVIRONMENTALSTATS for S-PLUS to determine an "optimal" Box-Cox transformation, based on one of three possible criteria:

- Probability Plot Correlation Coefficient (PPCC)
- Shapiro-Wilk Goodness-of-Fit Test Statistic (W)
- Log-Likelihood Function

You can also compute the value of the selected criterion for a range of values of the transform power λ.

We will explain the idea of a correlation coefficient in Chapter 9. For now, all you need to know is that a correlation coefficient ranges between −1 and 1, and that if the data come from a normal distribution then the probability plot correlation coefficient (PPCC) should be "close" to 1. We will also explain the Shapiro-Wilk goodness-of-fit test statistic (denoted W) in Chapter 7. For now, all you need to know is that W is closely related to the PPCC, it is constrained to lie between 0 and 1, and that if the data come from a normal distribution then W should be "close" to 1. Finally, the idea of a likelihood function is discussed in Chapter 5. All you need to know here is that larger values of the likelihood function should indicate a better fit to a normal distribution.

Figure 3.24 displays a plot of the probability plot correlation coefficient vs. various values of the transform power λ for the Reference area TcCB data. For this data set, the PPCC reaches its maximum at about $\lambda=0$, which corresponds to a log transformation. Besides plotting the objective function vs. λ, you can also generate Q-Q plots and Tukey mean-difference Q-Q plots for each of the values of λ.

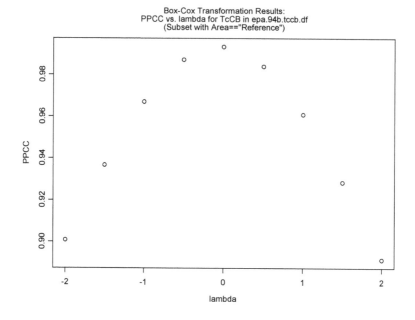

Figure 3.24   Probability plot correlation coefficient vs. Box-Cox transform power (λ) for the Reference area TcCB data

Menu

To create the plot of the PPCC vs. the transform power λ for the Reference area TcCB data shown in Figure 3.24 using the ENVIRONMENTALSTATS for S-PLUS pull-down menu, follow these steps.

1. In the Object Explorer, highlight the **Data** folder in the left-hand column. In the right-hand (Object) column, highlight the shortcut **epa.94b.tccb.df**.
2. On the S-PLUS menu bar, make the following menu choices: **EnvironmentalStats>EDA>Box-Cox Transformations**. This will bring up the Box-Cox Transformations dialog box.
3. For Data to Use, make sure the **Pre-Defined Data** button is selected.
4. The Data Set box should display **epa.94b.tccb.df**.
5. In the Variable box, choose **TcCB**.
6. In the Subset Rows with box, type **Area=="Reference"**.
7. Click **OK** or **Apply**.

You will see a print-out of the results showing the value of the PPCC for each of the nine values of the transform power λ. You will also see a graph-sheet with 19 pages. The first page is the same as Figure 3.24. The next nine pages display the normal Q-Q plots for each of the values of λ, and the last nine pages display the corresponding Tukey mean-difference Q-Q plots.

Command

To create the plot of the PPCC vs. the transform power λ for the Reference area TcCB data shown in Figure 3.24 using the ENVIRONMENTALSTATS for S-PLUS Command or Script Window, type these commands.

```
attach(epa.94b.tccb.df)
boxcox.list <- boxcox(TcCB[Area=="Reference"])
plot(boxcox.list)
detach()
```

At this point, you will see a menu selection. You can select 2 to plot the PPCC vs. λ (Figure 3.24). Selecting 3 will produce the nine Q-Q plots for each of the values of λ, and selecting 4 will produce the nine Tukey mean-difference Q-Q plots. You can create all of these plots by selecting 1. Select 0 to exit the menu.

Alternatively, to create all of the plots, you can type

```
plot(boxcox.list, plot.type="All")
```

To print the results, type

```
boxcox.list
```

To detach the data frame `epa.94b.tccb.df` from your search list, type

```
detach()
```

GRAPHS FOR TWO OR MORE VARIABLES

This main section discusses graphs you can use when you are interested in looking at the relationship between two or more variables. Table 3.11 summarizes the plots we will discuss in this section. Two kinds of plots that we will postpone discussion of until Chapter 11 are time series plots and autocorrelation plots. Three kinds of plots for multivariate data that are not discussed in this section include trilinear diagrams (Helsel and Hirsch, 1992, pp. 56–57), profile plots (Chambers et al., 1983, pp. 157–163; Helsel and Hirsch, 1992, pp. 51–52), and star plots (Chambers et al., 1983, pp. 157–161; Helsel and Hirsch, 1992, p. 53).

Plot Type	Data Characteristics	Plot Characteristics
Scatterplot	Numeric data collected in (x,y) pairs (two variables)	Points on two-dimensional graph indicating the values of each pair
Three-dimensional Scatter Plot or Cloud Plot	Numeric data collected in (x,y,z) triplets (three variables)	Points on three-dimensional graph indicating the value of each triplet
Scatterplot with Text	"	Scatterplot that uses the values of the z variable for the plotting symbols
Bubble Plot	"	Scatterplot with size of symbol proportional to z variable
Contour Plot	"	Interpolated smooth lines on two-dimensional graph indicating changes in z variable
Image Plot or Level Plot	"	Interpolated, grided color or gray-scale picture on two-dimensional graph indicating changes in z variable
Filled Contour Plot	"	Combination contour plot and level plot
Surface Plot or Wireframe Plot or Perspective Plot, and Filled versions	"	Multiple interpolated smooth lines on a three-dimensional graph indicating changes in z variable; filled versions have color or gray-scale between the lines
Scatterplot Matrix	Numeric data collected on three or more variables	All possible pairwise scatterplots
Multi-Panel Conditioning Plot or Trellis Plot	Data collected on two or more variables	Any of the plots for one variable or two variables, conditioned on one or more other variables

Table 3.11   A description of plots for two or more variables

Scatterplots and Loess Smooths

A *scatterplot* is used to look at the relationship between two variables, say x and y. The variables somehow occur "naturally" in pairs (e.g., ozone and temperature readings taken at the same time from the same place). We take a sample of n (x, y) pairs from the population of "all possible" (x, y) pairs, then plot the n pairs of points on Cartesian coordinates. Often, we are interested in determining whether there is some sort of deterministic relationship between the two variables (e.g., linear or curvilinear) that may be hidden by "noise" in the data.

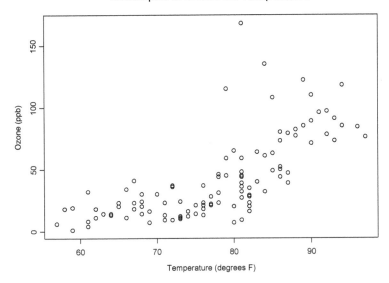

Figure 3.25   Scatterplot of ozone vs. temperature for data in `environmental`

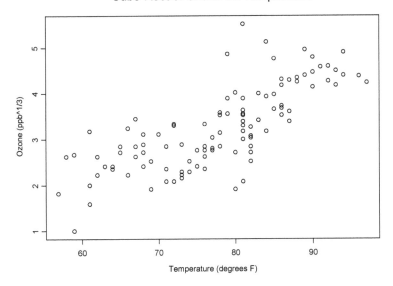

Figure 3.26   Scatterplot of cure-root of ozone vs. temperature for data in `air`

S-PLUS contains two built-in data frames called `air` and `environ-mental` that contain observations on ozone (ppb for `environmental`, ppb$^{1/3}$ for `air`), solar radiation (langleys), temperature (degrees Fahrenheit), and wind speed (mph) for 111 days between May and September 1973. Figure 3.25 shows the scatterplot of ozone vs. temperature. This scatterplot show that ozone is positively correlated with temperature: as temperature increases, so does the ozone concentration. It also shows that values of ozone become more variable as temperature increases. Of course, here we are ignoring information in the other variables (wind speed and radiation). Figure 3.26 shows the cube-root of ozone vs. temperature using the data from the `air` data frame. It looks like this transformation of the ozone data tends to linearize the relationship with temperature and also keeps the variability of the ozone values about the same across the range of temperatures.

To assess whether the relationship between cube-root ozone and temperature is truly linear, we could fit a straight line using simple linear regression and assess the goodness of the fit by doing the usual things like looking at the residuals and testing whether the slope is significantly different from 0 (see Chapter 9 for an explanation of linear regression models). A very handy method of looking for a "signal" in the "noise" without having to impose a specific model on the data is to use the smoother called *loess* invented by William Cleveland, and whose name comes from the phrase "locally weighted regression" (Chambers et al., 1983, pp. 94–104; Cleveland, 1993, Chapter 3; Cleveland, 1994, pp. 168–180; Helsel and Hirsch, 1992, pp. 45–46). A loess curve is constructed as follows.

1. For each x-value, a window (local neighborhood or bin) is constructed about that value.
2. A straight line (or optionally a quadratic line) is fit only to the observations within the window using weighted regression so that observations close to the x-value receive more weight than observations farther away from the x-value.
3. Once the line is fit, the y-value is "predicted" based on the current x-value. Often, a robust regression procedure is used in which during an iterative fitting process observations with large residuals (observed value minus fitted value) are given less weight than the other observations, and the fit is computed again.

Figure 3.27 shows that in fact the relationship between the cube-root of ozone and temperature appears not to be linear, but slightly curvilinear. Perhaps it could even be modeled as two separate straight lines, with one line describing the relationship when the temperature is less than 75 degrees and a second line describing the relationship when the temperature is over 75 degrees.

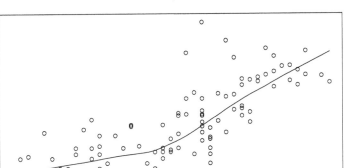

Figure 3.27   Cube-root of ozone vs. temperature with a loess smooth added

Menu

To produce the scatterplot of ozone vs. temperature shown in Figure 3.25 using the S-PLUS pull-down menus or toolbars, follow these steps.

1. Open the Object Explorer, and click on the **Find S-Plus Objects** button (the binoculars icon).
2. In the Pattern box, type **environmental**, then click **OK**.
3. Highlight the shortcut **environmental** in the Object column of the Object Explorer.
4. On the S-PLUS menu bar, make the following menu choices: **Graph>2D Plot**. This will bring up the Insert Graph dialog box. Under the Axes Type column, **Linear** should be highlighted. Under the Plot Type column, select **Scatter Plot** and click **OK**. (Alternatively, left-click on the **2D Plots** button, then left-click on the **Scatter** button.)
5. The Line/Scatter Plot dialog box should appear. The Data Set box should display **environmental**.
6. In the x Columns box, choose **temperature**.
7. In the y Columns box, choose **ozone**.
8. Click **OK**.

To produce the scatterplot of the cube-root of ozone vs. temperature shown in Figure 3.26, repeat the above steps, but in Steps 2, 3, and 5, replace environmental with **air**. To produce the scatterplot and added loess smooth shown in Figure 3.27, follow Steps 4 to 8 above (using **air** instead of environmental), except in Step 4, select **Smoothing – Loess Plot** instead of Scatter Plot. (Alternatively, left-click on the **2D Plots** button, then left-click on the **Loess** button.)

Command

To produce the scatterplot shown in Figure 3.25 using the S-PLUS Command or Script Window, type this command.

```
plot(environmental$temperature, environmental$ozone,
    xlab = "Temperature (degrees F)", ylab = "Ozone",
    main = "Scatterplot of Ozone vs. Temperature")
```

To produce the scatterplot shown in Figure 3.26, type this command.

```
plot(air$temperature, air$ozone,
    xlab = "Temperature (degrees F)",
    ylab = "Ozone (ppb^1/3)",
    main = "Cube-Root of Ozone vs. Temperature")
```

To produce the scatterplot and added loess smooth shown in Figure 3.27, type this command.

```
scatter.smooth(air$temperature, air$ozone,
    xlab = "Temperature (degrees F)",
    ylab = "Ozone (ppb^1/3)",
    main = paste("Cube-Root of Ozone vs. Temperature",
      "with Loess Smooth"))
```

You can also use the function xyplot to produce these plots (see the S-PLUS documentation).

Three-Dimensional Plots

Three-dimensional scatterplots (also called cloud plots), scatterplots with text, bubble plots, contour plots, image plots (also called level plots), and surface plots (also called wireframe plots or perspective plots) are all used to display the relationship between three variables. Contour plots, image plots, and surface plots are best suited for data in which two of the variables are observed on a regular grid in the x-y plane, because z-values at (x,y) pairs not on the grid must be interpolated in some fashion. Even if the (x,y) pairs do not fall on a regular grid, however, it is fairly easy in S-PLUS to produce smoothed z-values, but you must be careful, as we will see.

A *three-dimensional scatterplot* or *cloud plot* plots (x, y, z) triplets on a rendering of a three-dimensional graph. Figure 3.28 displays a cloud plot of ozone vs. wind speed and temperature for the data in the `environmental` data frame. From this perspective, it looks like perhaps the relationship between ozone and temperature changes as wind speed changes, and that for higher values of wind speed ozone does necessarily increase with temperature. Since we are rendering a three-dimensional picture in two dimensions, the plot will look different from different perspectives. In the Windows version of S-PLUS, you can easily rotate a cloud plot by simply clicking on the middle of the plot and then dragging one of the three filled circles that appears.

3D Scatterplot for Ozone Data

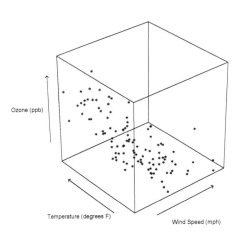

Figure 3.28   Three-dimensional scatterplot (cloud plot) of ozone vs. wind speed and temperature for the data in the `environmental` data frame

A *scatterplot with text* is a scatterplot in which the plotting symbol is the value of a third variable. It is a simple, crude tool to convey information about how the z variable changes with the x and y variables. Figure 3.29 shows a scatterplot with text for temperature vs. wind speed for which the plotting symbol is the value of ozone. Here we see that ozone ranges from 1 ppb to 168 ppb, and in general ozone is largest for low wind speeds and high temperatures, although there are a few points where ozone is also high at moderate and high wind speeds. Also note that there are no observations for simultaneously low values of wind speed and low values of temperature, nor are there any for simultaneously high values of wind speed and high values of temperature.

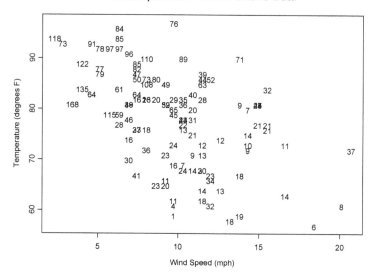

Figure 3.29   Scatterplot of temperature vs. wind speed, with plotting symbols equal to the value of ozone, using the data in environmental

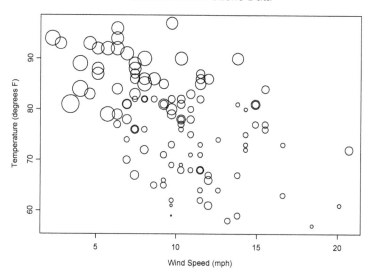

Figure 3.30   Bubble plot of temperature vs. wind speed, with size of plotting symbols proportional to the value of ozone, using the data in environmental

A **bubble plot** is simply a scatterplot where the size (area) of the plotting symbol is proportional to the value of some third variable. Figure 3.30 shows a bubble plot for temperature vs. wind speed, where the area of the plotting symbol is proportional to the value of ozone. Here we see the same features as in Figure 3.29, but our eyes do not have to compare numeric values of ozone.

Just as we fit a smooth curve to ozone as a function of temperature in Figure 3.27, we can fit a smooth surface to ozone as a function of temperature and wind speed. There are several different kinds of plots we can use to display this smooth fitted surface, including a contour plot, a level plot, and a surface plot. A **contour plot** uses lines denoting constant height to show how the z variable changes over the x-y plane. Figure 3.31 displays a contour plot showing how ozone changes with wind speed and temperature. An **image plot** or **level plot** uses colored or grayscale grids to show how the z variable changes over the x-y plane. Figure 3.32 shows a **filled contour plot**, which is a level plot combined with a contour plot. A **surface plot** (also called **wireframe plot** or **perspective plot**) uses multiple interpolated smooth lines on a three-dimensional graph to indicate changes in the z variable, and a filled version has color or grayscale between the lines. Figure 3.33 shows a **filled surface plot** for the ozone data. All of these plots were created by feeding the original (wind speed, temperature, ozone) triplets into a triangular interpolation algorithm to produce estimated values of ozone for points on an evenly spaced grid in the x-y plane.

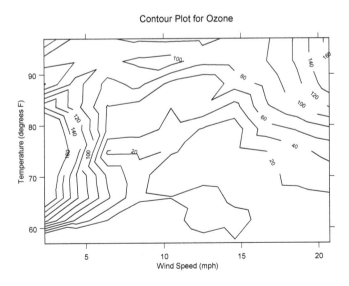

Contour Plot for Ozone

Figure 3.31   Contour plot of ozone (ppb) as a function of wind speed and temperature for the data in `environmental`

Filled Contour Plot of Ozone

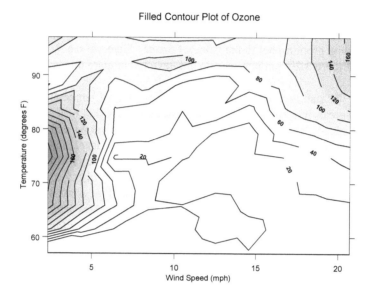

Figure 3.32   Filled contour plot of ozone (ppb) as a function of wind speed and temperature for the data in `environmental`

Filled Surface Plot for Ozone

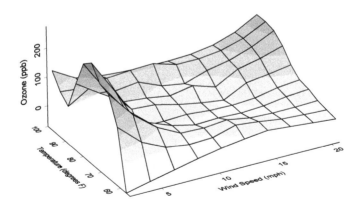

Figure 3.33   Filled surface plot of ozone (ppb) as a function of wind speed and temperature for the data in `environmental`

The plots in Figures 3.31 to 3.33 show estimated values of ozone over the entire combined ranges of wind speed and temperature. Note for example that it looks like ozone increases quickly as both wind speed and temperature increase, and also as wind speed decreases towards 0 and the temperature is around 75 degrees. This illustrates the *danger* of using contour, level, and surface plots when your data do not span a square area. Look at Figure 3.29 and recall that there are no observed (x,y) pairs of wind speed and temperature where both of these variables are large, nor are there any for very low values of wind speed when the temperature is around 75 degrees. We therefore have no idea how ozone behaves in these two areas of the plots, yet the plots might make us think there are very real trends here. (In S-PLUS you can display just the part of the plot that contains the (x,y) observations.)

Another important point to note about these three-dimensional plots is that we are looking at a *fitted surface*, not the original data. In Figure 3.27 we can see both the original data relating ozone to temperature, and the loess fit. In Figures 3.31 to 3.33, we only see the fitted surface, so we do not have any kind of feel for the observed variability in ozone.

Menu

To produce the three-dimensional scatterplot of ozone vs. wind speed and temperature shown in Figure 3.28 using the S-PLUS pull-down menus or toolbars, follow these steps.

1. In the Object Explorer, highlight the shortcut **environmental** in the Object column.
2. On the S-PLUS menu bar, make the following menu choices: **Graph>3D Plot**. This will bring up the Insert Graph dialog box. Under the Axes Type column, **3D** should be highlighted. Under the Plot Type column, select **3D Scatter Plot** and click **OK**. (Alternatively, left-click on the **3D Plots** button, then left-click on the **3D Scatter** button.)
3. The 3D Line/Scatter Plot dialog box should appear. The Data Set box should display **environmental**.
4. In the x Columns box, choose **wind**.
5. In the y Columns box, choose **temperature**.
6. In the z Columns box, choose **ozone**.
7. Click **OK**.

To produce the scatterplot with text shown in Figure 3.29, follow these steps.

1. On the S-PLUS menu bar, make the following menu choices: **Graph>2D Plot**. This will bring up the Insert Graph dialog box. Under the Axes Type column, **Linear** should be highlighted. Under the Plot Type column, select **Scatter Plot** and click **OK**. (Alterna-

tively, left-click on the **2D Plots** button, then left-click on the **Scatter** button.)

2. The Line/Scatter Plot dialog box should appear. The Data Set box should display **environmental**.
3. In the x Columns box, choose **wind**.
4. In the y Columns box, choose **temperature**.
5. In the z Columns box, choose **ozone**.
6. Click on the **Symbol** tab.
7. In the Height box, type **0.25**.
8. Click on the **Use Text As Symbol** box to select this option.
9. In the Text to Use box, select **z Column**.
10. Click **OK**.

To produce the bubble plot shown in Figure 3.30, follow these steps.

1. On the S-PLUS menu bar, make the following menu choices: **Graph>2D Plot**. This will bring up the Insert Graph dialog box. Under the Axes Type column, **Linear** should be highlighted. Under the Plot Type column, select **Bubble Plot** and click **OK**. (Alternatively, left-click on the **2D Plots** button, then left-click on the **Bubble** button.)
2. The Line/Scatter Plot dialog box should appear. The Data Set box should display **environmental**.
3. In the x Columns box, choose **wind**.
4. In the y Columns box, choose **temperature**.
5. In the z Columns box, choose **ozone**.
6. Click **OK**.

To produce the contour plot shown in Figure 3.31, follow these steps.

1. On the S-PLUS menu bar, make the following menu choices: **Graph>2D Plot**. This will bring up the Insert Graph dialog box. Under the Axes Type column, **Linear** should be highlighted. Under the Plot Type column, select **Contour Plot** and click **OK**. (Alternatively, left-click on the **2D Plots** button, then left-click on the **Contour** button.)
2. The **Contour Plot** dialog box should appear. The Data Set box should display **environmental**.
3. In the x Columns box, choose **wind**.
4. In the y Columns box, choose **temperature**.
5. In the z Columns box, choose **ozone**.
6. Click on the **Labels** tab.
7. In the Precision box, type **0**.
8. Click on the **Contour/Fills** tab.

9. Under Contour Specs, in the Minimum Level box, type **20**, in the Maximum Level box type **180**, and in the # of Levels box type **9**.

10. Click **OK**.

To produce the filled contour plot shown in Figure 3.32, follow these steps.

1. On the S-PLUS menu bar, make the following menu choices: **Graph>2D Plot**. This will bring up the Insert Graph dialog box. Under the Axes Type column, **Linear** should be highlighted. Under the Plot Type column, select **Contour - Filled** and click **OK**. (Alternatively, left-click on the **2D Plots** button, then left-click on the **Filled Contour** button.)

2. The Contour Plot dialog box should appear. The Data Set box should display **environmental**.

3. In the x Columns box, choose **wind**.

4. In the y Columns box, choose **temperature**.

5. In the z Columns box, choose **ozone**.

6. Click on the **Labels** tab.

7. Under Styles, click on the **Bold** box to select this option.

8. In the Precision box, type **0**.

9. Click on the **Contour/Fills** tab.

10. Under Contour Specs, in the Minimum Level box, type **20**, in the Maximum Level box type **280**, and in the # of Levels box type **14**.

11. In the Fill Type box, select **3 Color Range**.

12. In the Start Color box, choose **Bright White**.

13. In the Middle Color 2 box, choose **Lt Cyan**.

14. In the End Color box, choose **Lt Blue**.

15. Click **OK**.

To produce the filled surface plot shown in Figure 3.33, follow these steps.

1. On the S-PLUS menu bar, make the following menu choices: **Graph>3D Plot**. This will bring up the Insert Graph dialog box. Under the Axes Type column, **3D** should be highlighted. Under the Plot Type column, select **Surface – Filled, Data Grid** and click **OK**. (Alternatively, left-click on the **3D Plots** button, then left-click on the **Filled Data Grid Surface** button.)

2. The **Surface/3D Bar Plot** dialog box should appear. The Data Set box should display **environmental**.

3. In the x Columns box, choose **wind**.

4. In the y Columns box, choose **temperature**.

5. In the z Columns box, choose **ozone**.

6. Click on the **Fills** tab.

7. In the Fill Type box, select **3 Color Range**.

8. In the Start Color box, choose **Bright White**.

9. In the Middle Color 2 box, choose **Lt Cyan**.
10. In the End Color box, choose **Lt Blue**.
11. Click **OK**.

You may modify the steps listed above that produce the contour plots and surface plot so that the plots only show estimated values of ozone for the area of the *x-y* plane that contains observations of the (wind, temperature) pairs. To do this, add the following step before you click on **OK** or **Apply**: Click on the **Gridding/Hist** tab, and uncheck the **Extrapolate** box.

Command

To produce the three-dimensional plots shown in Figures 3.27 to 3.32 using the S-PLUS Command or Script Window, first attach the `environmental` data frame to the search list with the following command.

```
attach(environmental)
```

To produce the three-dimensional scatterplot of ozone vs. wind speed and temperature shown in Figure 3.28 using the S-PLUS Command or Script Window, type this command.

```
cloud(ozone ~ wind * temperature,
    data = environmental, xlab = "Wind Speed (mph)",
    ylab = "Temperature (degrees F)",
    zlab = "Ozone (ppb)", main = list(cex = 1.25,
    "3D Scatterplot for Ozone Data"))
```

To produce the scatterplot with text shown in Figure 3.29, type these commands.

```
plot(wind, temperature, type = "n",
    xlab = "Wind Speed (mph)",
    ylab = "Temperature (degrees F)",
    main = "Scatterplot with Text for Ozone Data")
text(wind, temperature, labels = ozone)
```

To produce the bubble plot shown in Figure 3.30, type these commands.

```
plot(wind, temperature, type = "n",
    xlab = "Wind Speed (mph)",
    ylab = "Temperature (degrees F)",
    main = "Bubble Plot for Ozone Data")
symbols(wind, temperature,
    circles = sqrt(ozone), add = T, inches = 0.25)
```

To produce the contour plots and surface plot shown in Figures 3.30 to 3.32 using the S-PLUS Command or Script Windows, we must first create a matrix with interpolated values of ozone (this is done automatically when we use the menu). To create the matrix type these commands. (Note: These commands are a bit complicated because there are some duplicate values of the pairs (wind, temperature), so we will take the median value of ozone for these duplicate pairs, then call the `interp` function.)

```
xy <- paste(wind, temperature, sep = ",")

i <- match(xy, xy)

z.smooth <- unlist(lapply(split(ozone, i), median))

ord <- !duplicated(xy)

ozone.smooth <- data.frame(wind = wind[ord],
    temperature = temperature[ord], ozone = z.smooth)

ozone.list <-  interp(ozone.smooth$wind,
    ozone.smooth$temperature, ozone.smooth$ozone,
    extrap = T, ncp = 2)
```

The object `ozone.list` is a list with three components called x, y, and z. The components x and y are numeric vectors with 40 evenly spaced points between the minimum and maximum values of wind and temperature, respectively. The component z is a 40 by 40 matrix containing the interpolated values of ozone. If you do not wish to plot estimated values of ozone that are extrapolated outside the area of the x-y plane that contains the observed (wind, temperature) pairs, then set `extrap=F` in the call to `interp`.

To produce the contour plot shown in Figure 3.31, type this command:

```
contour(ozone.list, xlab = "Wind Speed (mph)",
    ylab = "Temperature (degrees F)",
    main = "Contour Plot of Ozone")
```

or this command:

```
contourplot(c(z) ~ rep(x, 40) * rep(y, rep(40,40)),
    data = ozone.list, xlab = "Wind Speed (mph)",
    ylab = "Temperature (degrees F)",
    main = "Contour Plot of Ozone")
```

To produce the filled contour plot shown in Figure 3.32, type this command.

```
contourplot(c(z) ~ rep(x, 40) * rep(y, rep(40,40)),
    data = ozone.list, region = T,
    xlab = "Wind Speed (mph)",
    ylab = "Temperature (degrees F)",
    main = "Filled Contour Plot of Ozone")
```

To produce the surface plot shown in Figure 3.33 (but not filled), type these commands:

```
persp(ozone.list, xlab = "Wind Speed (mph)",
    ylab = "Temperature (degrees F)",
    zlab = "Ozone (ppb)")
title(main = "Surface Plot of Ozone")
```

or this command:

```
wireframe(c(z) ~ rep(x, 40) * rep(y, rep(40,40)),
    data = ozone.list, xlab = "Wind Speed (mph)",
    ylab = "Temperature (degrees F)",
    main = "Surface Plot of Ozone")
```

Scatterplot Matrix and Brushing

A *scatterplot matrix* displays all possible pairwise scatterplots for three or more variables, where the plots share some axes. Figure 3.34 shows the scatterplot matrix for all of the variables in the environmental data frame. These plots show that ozone is positively correlated with radiation and temperature, and negatively correlated with wind speed. Also, temperature and wind speed are negatively correlated. There is no apparent relationship between radiation and temperature.

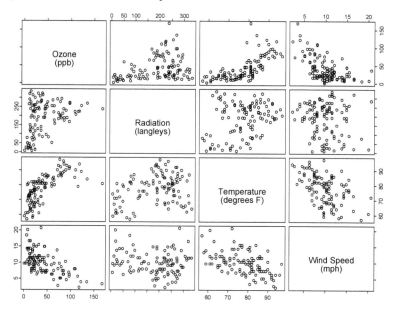

Figure 3.34   Scatterplot matrix for the variables in the environmental data frame

Cleveland (1994, p. 196) points out the inverted V-shape pattern of the plot of ozone vs. radiation and notes that for low values of radiation, high values of ozone never occur because the photochemical reactions that produce ozone require a certain amount of solar radiation. Also, for the highest values of radiation, high values of ozone do not occur either because wind speed tends to be moderate to high and temperature tends to be moderate to low. In fact, very high levels of radiation occur only when the air pollution (including ozone) is minimal.

In S-PLUS, you can also brush a scatterplot matrix interactively (Cleveland, 1994, pp. 206–209). That is, you can highlight points in one of the scatterplots and these same cases will show up highlighted in the other scatterplots.

Menu

To produce the scatterplot matrix shown in Figure 3.34 using the S-PLUS pull-down menus or toolbars, follow these steps.

1. In the Object Explorer, highlight the shortcut **environmental** in the Object column.
2. On the S-PLUS menu bar, make the following menu choices: **Graph>2D Plot**. This will bring up the Insert Graph dialog box. Under the Axes Type column, select **Matrix**. Under the Plot Type column, select **Scatter Plot Matrix** and click **OK**. (Alternatively, left-click on the **2D Plots** button, then left-click on the **Scatter Matrix** button.)
3. The Scatter Plot Matrix dialog box should appear. The Data Set box should display **environmental**.
4. Click **OK**.

You can brush points in the scatterplot matrix one of two ways. The first way is to create the scatterplot matrix following the above steps, and then do the following.

5. Click on the **Graph Tools** button, then click on the **Select Data** button. Move the mouse cursor to one of the scatterplots, then hold the left mouse button down and drag it to envelope the points you want to select within the growing rectangle. Once you have selected the points you want, lift up on the left mouse button, and the points will be highlighted in all of the scatterplots.

The other way to brush points in the scatterplot matrix is to do the following.

1. On the S-PLUS menu bar, make the following menu choices: **Graph>Brush and Spin**. This will bring up the Brush Properties dialog box.
2. The Data Set box should display **environmental**.

3. Click **OK**.
4. In the Brush window, left-click outside of any plots to adjust the size and shape of the brush.
5. Within any of the scatterplots, left-click and drag the brushing area over the points you want to highlight.
6. See the S-PLUS documentation for more information on how to use brush and spin.

Command

To produce the scatterplot matrix shown in Figure 3.34 using the S-PLUS Command or Script Windows, type this command:

```
pairs(environmental, labels = c("Ozone\n(ppb)",
   "Radiation\n(langleys)",
   "Temperature\n(degrees F)", "Wind Speed\n(mph)"))
```

or this command:

```
splom(~environmental)
```

To brush the scatterplot matrix, type this command.

```
brush(environmental)
```

Multi-Panel Conditioning Plot (Coplot) or Trellis Plot

Multi-panel conditioning plots, also called *coplots* or *Trellis plots*, let you view the behavior of one variable or the relationship between two or more variables conditioned on the value of one or more other variables (Cleveland, 1993, Chapters 4 and 5; Cleveland, 1994, pp. 198–205). Figure 3.10 is a histogram conditioning plot showing the histograms of the logarithm of TcCB conditioned on the value of the variable Area. Figure 3.35 shows a scatterplot conditioning plot of ozone vs. temperature conditioned on wind speed, with a loess smooth added to each plot. Here you can see that the highest values of ozone occur at the lowest values of wind speed, and that in general ozone increases with temperature, but the form of the relationship appears to vary with wind speed. For example, ozone rises quickly with temperature for low wind speeds, but rises more gradually with temperature at the highest wind speeds.

The coplot in Figure 3.35 divides wind speed into four categories that contain about the same number of observations. There is not that much difference in wind speed between the category 7.4 to 9.7 vs. 9.7 to 11.5, but the last category 11.5 to 20.7 contains much higher wind speeds. We can create a coplot with only three categories of wind speed that are equal in range instead of number of observations. Figure 3.36 shows this plot.

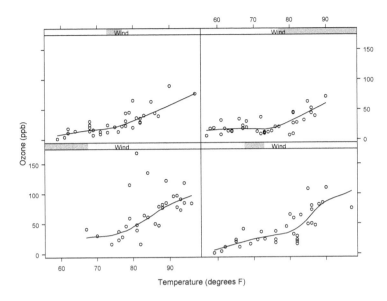

Figure 3.35   Coplot of ozone vs. temperature conditioned on wind speed, with a
loess smooth added, using the data in `environmental`

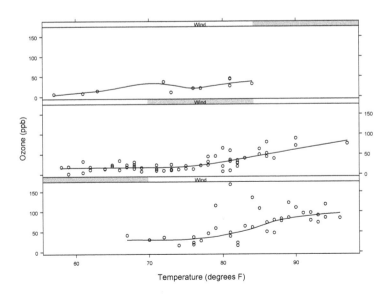

Figure 3.36   Coplot of ozone vs. temperature conditioned on wind speed, with three
categories of wind speed of equal range

Menu

To produce the scatterplot coplot shown in Figure 3.35 using the S-PLUS pull-down menus or toolbars, follow these steps.

1. In the Object Explorer, left-double-click the shortcut **environmental** in the Object column. This will bring up a data window.
2. On the S-PLUS menu bar, make the following menu choices: **Graph>2D Plot**. This will bring up the Insert Graph dialog box. Under the Axes Type column, **Linear** should be highlighted. Under the Plot Type column, select **Smoothing – Loess Plot** and click **OK**. (Alternatively, left-click on the **2D Plots** button, then left-click on the **Loess** button.)
3. The Line/Scatter Plot dialog box should appear. The Data Set box should display **environmental**.
4. In the x Columns box, select **temperature**.
5. In the y Columns box, select **ozone**. Click **OK**. This will produce a single scatter plot of ozone vs. temperature with a loess smooth.
6. **Click** on the data window to bring it forward. **Left click** on the top of the wind column to highlight that column, then **left-click** in a cell of the column, then **drag** the column to the top of the graphsheet and **drop** it.

To produce the scatterplot coplot shown in Figure 3.36, follow these steps.

1. Double-click outside the graph area in the plot you just produced. This will bring up the Graph2D Properties dialog box..
2. Click on the **Multipanel** tab.
3. Under Continuous Conditioning, in the # of Panels box, type **3**, and in the Interval Type box select **Equal Ranges**. Click **OK**.

Command

To produce the scatterplot coplot shown in Figure 3.35 using the S-PLUS Command or Script Windows, type these commands.

```
wind <- environmental$wind
Wind <- equal.count(wind, number = 4, overlap = 0)
xyplot(ozone ~ temperature | Wind,
   data = environmental,
   xlab = "Temperature (degrees F)",
   ylab = "Ozone (ppb)", panel = function(x,y) {
   panel.xyplot(x,y); panel.loess(x,y)})
```

To produce the scatterplot coplot shown in Figure 3.36, type these commands.

```
cut.points <- seq(min(wind), max(wind), length = 4)
cut.mat <- matrix(c(cut.points[-4], cut.points[-1]),
   nrow = 3, ncol = 2)
Wind <- shingle(wind, intervals = cut.mat)
xyplot(ozone ~ temperature | Wind,
   data = environmental,
   xlab = "Temperature (degrees F)",
   ylab = "Ozone (ppb)", panel = function(x,y) {
   panel.xyplot(x,y); panel.loess(x,y)},
   layout = c(1, 3, 1))
```

SUMMARY

- Summary or descriptive statistics can be classified by what they measure (location, spread, skew, kurtosis, etc.) and also how they behave when unusually extreme observations are present (sensitive vs. robust).
- Because environmental data usually involve measures of chemical concentrations, and concentrations cannot fall below 0, environmental data often tend to be positively skewed.
- Graphical displays are usually far superior to summary statistics for conveying information in a data set.
- For univariate data in which there is a label for each value, use dot plots, bar charts, and pie charts. Dot plots are usually superior to pie charts.
- For conveying the distribution of univariate data, use strip plots, histograms, density plots, boxplots, and quantile plots (also called empirical cumulative distribution function plots).
- To compare two data sets or to compare a data set to a theoretical probability distribution, use quantile-quantile (Q-Q) plots (also called probability plots), and Tukey mean-difference Q-Q plots.
- To look at the relationship between two numeric variables, use a scatterplot. A fitted line or curve such as a loess smooth is often helpful to see the signal among the "noise."
- To look at the relationship between three variables, use scatterplots with text, bubble plots, three-dimensional scatter plots, contour plots, level plots (also called image plots), and surface plots (also called wireframe or perspective plots).
- Be aware that contour, level, and surface plots usually show only fitted or smooth z values, not the actual observations, and also that the fitted z values may have been extrapolated outside the area of the x-y plane that contain the actual observations.

- To look at the relationship between three or more variables, use matrix scatterplots, and multi-panel conditioning plots (also called coplots or Trellis plots). Brushing matrix scatterplots may reveal important cases.

EXERCISES

3.1.   Table 3.12 below shows nickel concentrations (ppb) collected at four groundwater monitoring wells (USEPA, 1992c, p. 7). In ENVIRONMENTALSTATS for S-PLUS these data are stored in the data frame `epa.92c.nickel1.df`.

      a.   Plot the observations by month, then plot them by well. Do you see any major differences between months or wells?

      b.   Compute summary statistics for these data (combine wells and months). How do the mean and median compare?

      c.   Create a histogram with a density plot overlaid for these data.

      d.   Create a normal Q-Q plot and a Tukey mean-difference Q-Q plot for these data.

      e.   Do theses data appear to come from a normal population?

      f.   Try using another kind of Q-Q plot to fit these data.

Month	Well 1	Well 2	Well 3	Well 4
1	58.8	19	39	3.1
2	1.0	81.5	151	942
3	262	331	27	85.6
4	56	14	21.4	10
5	8.7	64.4	578	637

Table 3.12   Nickel data (ppb) from groundwater monitoring wells (USEPA, 1992c, p. 7)

3.2.   Table 3.13 below shows arsenic concentrations (ppb) collected quarterly at two groundwater monitoring wells. These data are modified from benzene data given in USEPA (1992c, p. 56). In ENVIRONMENTALSTATS for S-PLUS these data are stored in the data frame `epa.92c.arsenic3.df`.

      a.   For each well, plot the observations by year. Do you see any major differences between years?

      b.   Compute summary statistics for each well (combine years).

      c.   Compare the observed distribution of arsenic at each well. Use whatever types of plots you wish.

d. Does the compliance well appear to show any evidence of contamination? Why or why not?
e. Does it appear that the background well data may be modeled as coming from a normal distribution?
f. Note that sampling ceased at the background well after year 3. What problems might this cause in a statistical analysis, especially if the compliance well had showed evidence of contamination?

Well	Year	Observed Arsenic (ppb)			
Background	1	12.6	30.8	52.0	28.1
	2	33.3	44.0	3.0	12.8
	3	58.1	12.6	17.6	25.3
Compliance	4	48.0	30.3	42.5	15.0
	5	47.6	3.8	2.6	51.9

Table 3.13   Arsenic data from groundwater monitoring wells

3.3. Using the Reference area TcCB data show in Table 3.2 on page 55, create the four different kinds of histograms: number (counts), percent of total, fraction of total, and density. Explain what the bars mean for each one.

3.4. The built-in data set halibut in S-PLUS contains annual catch per unit effort (CPUE) and exploitable biomass of Pacific halibut for the years 1935 through 1989.

a. Compute summary statistics for each variable.
b. For each variable, create a histogram, boxplot, strip plot, quantile plot, normal Q-Q plot, and Tukey mean-difference Q-Q plot.
c. Do both variables appear to come from a normal population?

3.5. Repeat the steps in the previous exercise for the S-PLUS data set rain.nyc1, which contains annual total New York City precipitation, in inches, from 1869 to 1957.

3.6. Using the data in rain.nyc1, look at a set of Box-Cox transformations by varying the transform power λ between –2 and 2 in increments of 0.5. That is, follow the steps shown in the section Box-Cox Data Transformations and Q-Q Plots. Based on the plot of PPCC vs. λ and the Q-Q plots, which transformation appears to be best?

3.7. Table 3.14 below shows copper concentrations (ppb) collected at three groundwater monitoring wells: two background wells and one compliance well (USEPA, 1992c, p. 47). These data are stored in `epa.92c.copper1.df` in ENVIRONMENTALSTATS for S-PLUS.

    a. Compute summary statistics by well type (background vs. compliance).

    b. Use plots to compare the compliance well with the data for both background wells combined.

    c. Does there appear to be any evidence of contamination?

Month	Well 1 (Background)	Well 2 (Background)	Well 3 (Compliance)
1	4.2	5.2	9.4
2	5.8	6.4	10.9
3	11.3	11.2	14.5
4	7.0	11.5	16.1
5	7.3	10.1	21.5
6	8.2	9.7	17.6

Table 3.14   Copper data (ppb) from groundwater monitoring wells (USEPA, 1992c, p. 47)

3.8. Table 3.15 below shows arsenic concentrations (ppm) collected at six groundwater monitoring wells (USEPA, 1992c, p. 6). In ENVIRONMENTALSTATS for S-PLUS these data are stored in the data frame `epa.92c.arsenic1.df`.

Month	Well 1	Well 2	Well 3	Well 4	Well 5	Well 6
1	22.90	2.0	2.0	7.84	24.9	0.34
2	3.09	1.25	109.4	9.3	1.3	4.78
3	35.7	7.8	4.5	25.9	0.75	2.85
4	4.18	52	2.5	2.0	27	1.2

Table 3.15   Arsenic data (ppm) from groundwater monitoring wells (USEPA, 1992c, p. 21)

    a. Compute summary statistics for each well.

    b. Use strip plots to compare the distribution of observations at each well.

    c. Use boxplots to compare the distribution at each well. Do you think it is a good idea to use boxplots with only four observations per well? Why or why not?

3.9. Using the median daily maxima ozone concentration for June-August, 1974 and latitudes and longitudes stored in the data frame `ozone.NE` you created on page 70, create a contour plot, level plot, and surface plot of ozone concentrations. Can you figure out how to add a map of this part of the U.S. to the contour plot? (Hint: Use the function `usa`.)

3.10. The built-in data set `quakes.bay` in S-PLUS contains the location, time and magnitude of earthquakes in the San Francisco Bay Area from 1962 to 1981.

   a. Create a scatterplot using just longitude (on the *x*-axis) and latitude (on the *y*-axis). What kind of pattern do you see? Can you think of an explanation for it?

   b. Create a scatterplot with text for these data, putting the longitude on the *x*-axis, the latitude on the *y*-axis, and plotting the magnitudes of the earthquakes. Is this kind of plot helpful for these data? Why or why not?

   c. Create a bubble plot for these data. Is this kind of plot helpful for these data? Why or why not?

   d. Create a contour plot and surface plot, both with and without extrapolation. What happens when you extrapolate outside the *x-y* plane where the quakes were observed?

3.11. The built-in data sets `evap.y` and `evap.x` in S-PLUS contain observations on soil evaporation and related variables, including soil temperature, air temperature, relative humidity, and total wind (see the S-PLUS help file for evap).

   a. Create a data frame to combine these data using the following command.

```
evap <- data.frame(evap=evap.y, evap.x)
```

   b. Plot evaporation vs. average relative humidity, and add a loess smooth.

   c. Use scatterplot matrices and brushing to look at the relationship between evaporation and some of the other variables (e.g., average soil temperature, average air temperature, average relative humidity, and total wind). What patterns do you see?

   d. Use multi-panel conditioning plots to look at these data.

4 PROBABILITY DISTRIBUTIONS

Approximations to the Real World

As we stated in Chapter 3, a population is defined as the entire collection of measurements about which we want to make a statement, such as all possible measurements of dissolved oxygen in a specific section of a stream within a certain time period. Probability distributions are idealized mathematical models that are used to model the variability inherent in a population (e.g., all measures of dissolved oxygen will not exactly match each other). Certain probability distributions come up again and again in environmental statistics. This chapter discusses the concepts of a random variable and a probability distribution, talks about important probability distributions in environmental statistics, and shows you how to use various functions in S-PLUS and ENVIRONMENTALSTATS for S-PLUS to plot and generate quantities associated with probability distributions.

WHAT IS A RANDOM VARIABLE?

When you perform an environmental study, you take a sample from a population and record the measurements associated with the sample. For example, the population may be all possible measurements of dissolved oxygen in a certain section of a stream within a given time period, and you may take 50 physical samples of water and measure the dissolved oxygen in each sample. The actual numbers, the actual concentrations of dissolved oxygen, are realizations of a concept called a random variable.

A *random variable* is "the value of the next observation in an experiment" (Watts, 1991, as quoted by Berthoux and Brown, 1994, p. 7). It is the measurement associated with the next physical sample from the population. Outcomes of a random variable usually differ in value from observation to observation. This is because there are usually several sources of variability that contribute to the final value (see Figure 1.1).

Example 4.1: Flipping a Fair Coin 10 Times

One of the simplest examples of a random variable involves flipping a coin and observing whether the coin lands "heads" or "tails." In this case, the population is the set of outcomes from flipping the coin an infinite number of times, and the random variable is the outcome from flipping the coin.

Figure 4.1 shows the results of flipping a fair coin 10 times. Here, the term "fair" means that a head is just as likely to occur as a tail. The histogram (really a bar chart) in Figure 4.1 shows the outcome from such an experiment, where 0 represents a tail and 1 represents a head. We can see that there were six tails and four heads in our experiment.

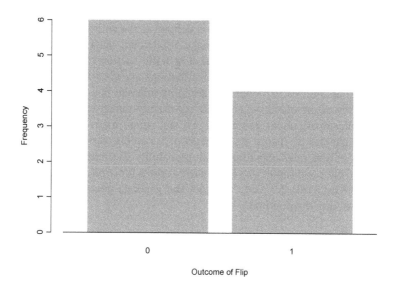

Figure 4.1   Results of flipping a fair coin 10 times

DISCRETE VS. CONTINUOUS RANDOM VARIABLE

The outcome in the coin-tossing experiment is an example of a ***discrete*** random variable; it can take on only two possible values (either 0 or 1). In general, a discrete random variable can take on only a finite or countably infinite number of values.

On the other hand, a ***continuous*** random variable can take on an infinite number of values. For example, the concentration of 1,2,3,4-tetrachlorobenze (TcCB) in a physical sample of soil may theoretically take on an infinite number of values. In reality, however, the values of a "continuous" random variable are limited by the precision of the instrument used to measure the random variable and the level of mathematical rounding or truncation used to report the results.

Example 4.2: TcCB Measurements Modeled as Continuous

The TcCB data we talked about in Chapter 3 are associated with an experiment in which soil samples were taken from two different areas and the amount of TcCB (in ppb) was measured in each physical soil sample. In this case, the population is the set of measurements from all possible soil samples, and the random variable is the concentration of TcCB that *will be* observed in a soil sample. The 47 observed measures of TcCB in the Reference area are summarized in the histogram below.

Histogram of Reference Area TcCB Data

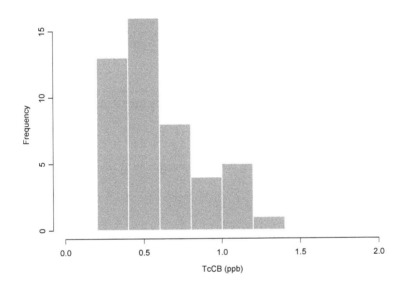

Figure 4.2   Histogram of the Reference area TcCB data

WHAT IS A PROBABILITY DISTRIBUTION?

In our coin-tossing experiment (Figure 4.1), 40% of our tosses (4/10) resulted in a head. We can represent this with the **relative frequency histogram** or **density histogram** shown in Figure 4.3. For each class (bar) of a density histogram, the proportion of observations falling in that class is equal to the area of the bar; that is, it is the width of the bar times the height of the bar. In Figure 4.3, the width of each bar happens to be 1, so Figure 4.3 is exactly the same as Figure 4.1 except we have divided the numbers on the y-axis by the total number of times we flipped the coin.

Now let us repeat the coin-tossing experiment, except we will flip the coin 100 times. The result of this experiment is shown in Figure 4.4. In this

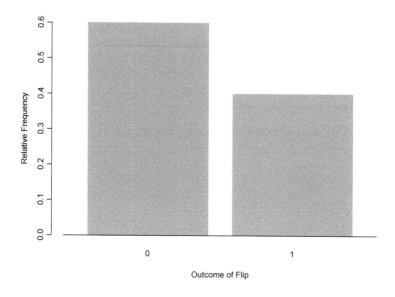

Figure 4.3   Relative frequency (density) histogram for 10 coin flips

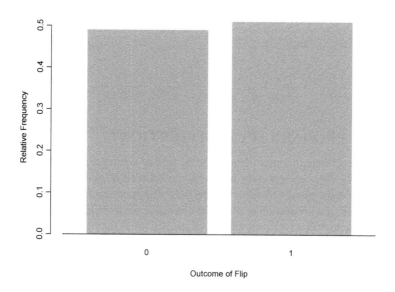

Figure 4.4   Relative frequency (density) histogram for 100 coin flips

case a little over 50% of our tosses resulted in a head. Because we are toss-
ing a fair coin, we would expect that as we toss the coin more and more
times, the percentage of heads should "get close" to 50%. This is called the
"Law of Large Numbers."

For a discrete random variable that takes on integer values, a ***probability
distribution*** can be thought of as what a density (relative frequency) histogram of outcomes would look like if you could take a very large or infinite
number of samples. Figure 4.5 shows the probability distribution for the
coin-tossing experiment.

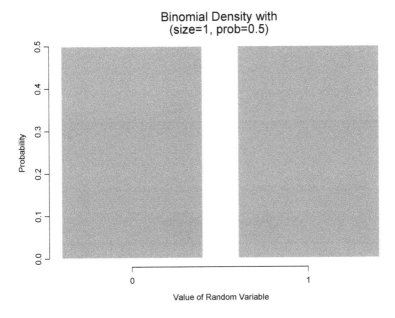

Figure 4.5   Probability distribution for the coin-tossing experiment

Figure 4.6 shows the relative frequency (density) histogram for the Reference area TcCB data. As we stated before, for each class (bar) of this histogram, the proportion of observations falling in that class is equal to the
area of the bar; that is, it is the width of the bar times the height of the bar. If
we could take many, many more samples and create relative frequency histograms with narrower and narrower classes, we might end up with a picture
that looks like Figure 4.7, which shows the probability density function of a
lognormal random variable with a mean of 0.6 and a coefficient of variation
of 0.5. For a continuous random variable, a ***probability distribution*** can be
thought of as what a density (relative frequency) histogram of outcomes
would look like if you could keep taking more and more samples and making the histogram bars narrower and narrower.

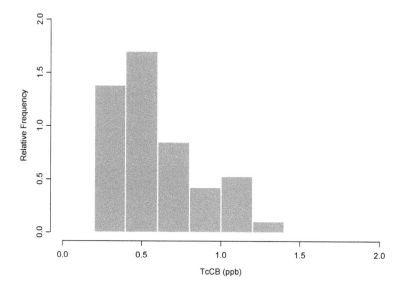

Figure 4.6   Relative frequency (density) histogram of Reference area TcCB data

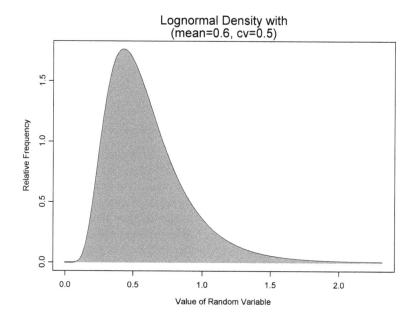

Figure 4.7   A lognormal probability distribution

PROBABILITY DENSITY FUNCTION (PDF)

A *probability density function (pdf)* is a mathematical formula that describes the relative frequency of a random variable. Sometimes the picture of this formula (e.g., Figures 4.5 and 4.7) is called the pdf.

If a random variable is discrete, its probability density function is sometimes called a *probability mass function*, since it shows the "mass" of probability at each possible value of the random variable. The probability density (mass) function for the coin-tossing experiment shown in Figure 4.5 is given by:

$$f(x) = \Pr(X = x) = \begin{cases} 0.5, x = 0 \\ 0.5, x = 1 \end{cases} \qquad (4.1)$$

In this equation, the letter f stands for the probability density (mass) function, the letters $\Pr$ stand for the phrase "the probability that," the uppercase X stands for the random variable (the outcome of the coin toss), and the lower case x stands for the actual value that the random variable takes on after one toss of the coin (0 = tail, 1 = head).

The probability density function for the lognormal distribution shown in Figure 4.7 is given by:

$$f(x) = \frac{1}{x\sigma\sqrt{2\pi}} \exp\left\{-\frac{1}{2\sigma^2}\left[\log(x) - \mu\right]^2\right\}, \quad x > 0 \quad (4.2)$$

where

$$\mu = \log\left(\frac{\theta}{\sqrt{\tau^2 + 1}}\right)$$

$$\sigma = \left[\log(\tau^2 + 1)\right]^{1/2}$$

$$\theta = 0.6$$

$$\tau = 0.5$$

(We will talk more about what all these Greek letters mean later in this chapter when we talk about the lognormal distribution. Also, in the next chapter we will talk about how we came up with a mean of 0.6 and a coefficient of variation of 0.5).

For a relative frequency (density) histogram, the area of the bar is the probability of falling in that interval. Similarly, for a continuous random variable, the probability that the random variable falls into some interval, say between 0.75 and 1, is simply the area under the pdf between these two interval endpoints. Mathematically, this is written as:

$$\Pr\left(0.75 \leq X \leq 1\right) = \int_{0.75}^{1} f\left(x\right) dx \qquad (4.3)$$

For the lognormal pdf shown in Figure 4.7, the area under the curve between 0.75 and 1 is about 0.145, so there is a 14.5% chance that the random variable will fall into this interval.

Plotting Probability Density Functions

Figure 4.8 display examples of all of the available probability distributions in S-PLUS and ENVIRONMENTALSTATS for S-PLUS. These probability distributions can be used as models for populations. Almost all of these distributions can be derived from some kind of theoretical mathematical model (e.g., the binomial distribution for binary outcomes, the Poisson distribution for "rare" events, the Weibull distribution for extreme values, the normal distribution for sums of several random variables, etc.). Later in this chapter we will discuss in detail probability distributions that are commonly used in environmental statistics.

Menu

To produce the binomial pdf shown in Figure 4.5 using the ENVIRONMENTALSTATS for S-PLUS pull-down menu, follow these steps.

1. On the S-PLUS menu bar, make the following menu choices: **EnvironmentalStats>Probability Distributions and Random Numbers>Plot Distribution**. This will bring up the Plot Distribution Function dialog box.
2. In the Distribution box, choose **Binomial**. In the size box, type **1**. In the prob box, type **0.5**.
3. Click **OK** or **Apply**.

To produce the lognormal pdf shown in Figure 4.7, follow these steps.

1. On the S-PLUS menu bar, make the following menu choices: **EnvironmentalStats>Probability Distributions and Random Numbers>Plot Distribution**. This will bring up the Plot Distribution Function dialog box.
2. In the Distribution box, choose **Lognormal (Alternative)**. In the mean box, type **0.6**. In the cv box, type **0.5**. Click **OK** or **Apply**.

Command

To produce the binomial pdf shown in Figure 4.5 using the ENVIRONMENTALSTATS for S-PLUS Command or Script Window, type this command.

```
pdfplot("binom", list(size=1, prob=0.5))
```

To produce the lognormal pdf shown in Figure 4.7, type this command.

```
pdfplot("lnorm.alt", list(mean=0.6, cv=0.5))
```

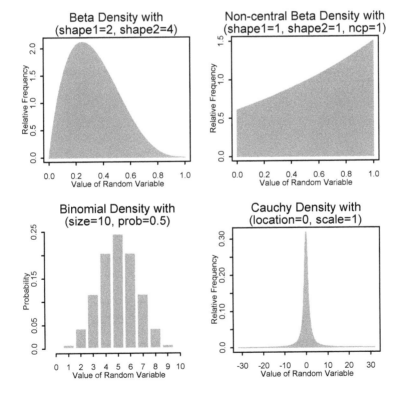

Figure 4.8   Probability distributions in S-PLUS and ENVIRONMENTALSTATS for S-PLUS

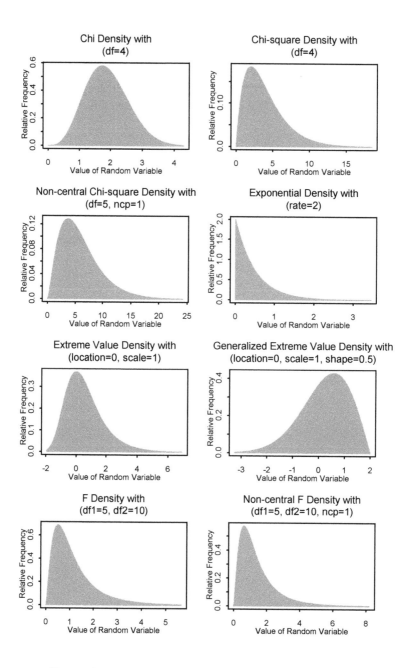

Figure 4.8 (continued)   Probability distributions in S-Plus and
ENVIRONMENTALSTATS for S-Plus

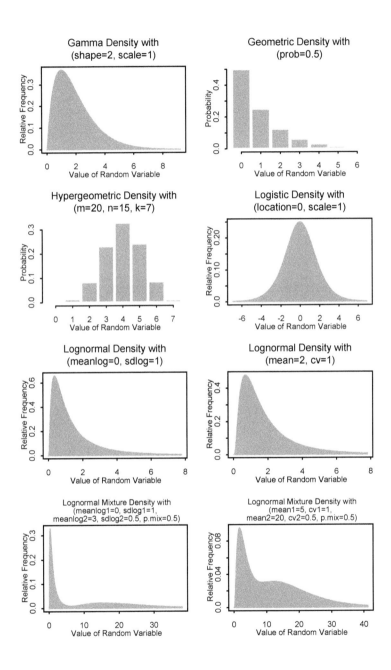

Figure 4.8 (continued)   Probability distributions in S-PLUS and
ENVIRONMENTALSTATS for S-PLUS

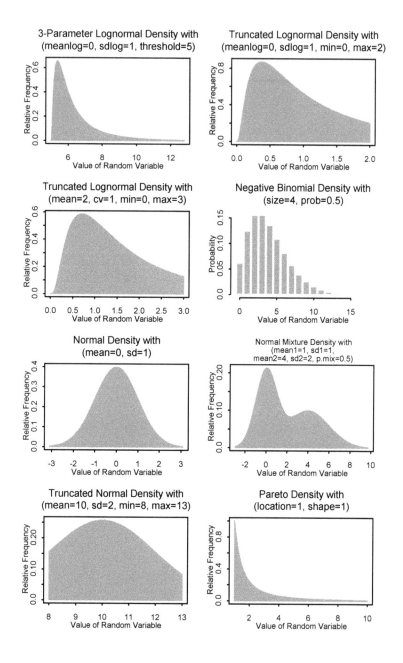

Figure 4.8 (continued)   Probability distributions in S-Plus and
ENVIRONMENTALSTATS for S-Plus

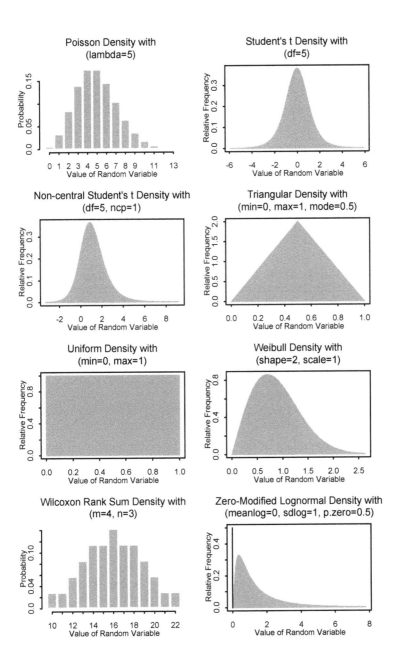

Figure 4.8 (continued) Probability distributions in S-PLUS and
ENVIRONMENTALSTATS for S-PLUS

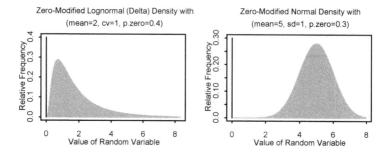

Figure 4.8 (continued)   Probability distributions in S-PLUS and
ENVIRONMENTALSTATS for S-PLUS

Computing Values of the Probability Density Function

You can use S-PLUS and ENVIRONMENTALSTATS for S-PLUS to compute the value of the pdf for any of the built-in probability distributions. As we saw in Equation (4.1), the value of the pdf for the binomial distribution shown in Figure 4.5 is 0.5 for $x = 0$ (a tail) and 0.5 for $x = 1$ (a head). From Equation (4.2), you can show that for the lognormal distribution shown in Figure 4.7, the values of the pdf evaluated at 0.5, 0.75, and 1 are about 1.67, 0.88, and 0.35, respectively.

Menu

To compute the values of the pdf of the binomial distribution shown in Figure 4.5 using the ENVIRONMENTALSTATS for S-PLUS pull-down menu, follow these steps.

1. On the S-PLUS menu bar, make the following menu choices: **EnvironmentalStats>Probability Distributions and Random Numbers>Density, CDF, Quantiles**. This will bring up the Densities, Cumulative Probabilities, or Quantiles dialog box.
2. For the Data to Use buttons, choose **Expression**. In the Expression box, type **0:1**. In the Distribution box, choose **Binomial**. In the size box type **1**. In the prob box type **0.5**. Under the Probability or Quantile group, make sure the **Density** box is checked.
3. Click **OK** or **Apply**.

To compute the values of the pdf of the lognormal distribution shown in Figure 4.7 for the values 0.5, 1, and 1.5, follow these steps.

1. On the S-PLUS menu bar, make the following menu choices: **EnvironmentalStats>Probability Distributions and Random**

Numbers>Density, CDF, Quantiles. This will bring up the Densities, Cumulative Probabilities, or Quantiles dialog box.

2. For the Data to Use buttons, choose **Expression**. In the Expression box, type **c(0.5, 0.75, 1)**. In the Distribution box, choose **Lognormal (Alternative)**. In the mean box type **0.6**. In the cv box type **0.5**. Under the Probability or Quantile group, make sure the Density box is checked.

3. Click **OK** or **Apply**.

Command

To compute the values of the pdf of the binomial distribution shown in Figure 4.5 using the S-PLUS Command or Script Window, type this command.

```
dbinom(0:1, size=1, prob=0.5)
```

To compute the values of the pdf of the lognormal distribution shown in Figure 4.7 for the values 0.5, 0.75, and 1, type this command using ENVIRONMENTALSTATS for S-PLUS.

```
dlnorm.alt(c(0.5, 0.75, 1), mean=0.6, cv=0.5)
```

CUMULATIVE DISTRIBUTION FUNCTION (CDF)

The *cumulative distribution function (cdf)* of a random variable X, sometimes called simply the distribution function, is the function F such that

$$F(x) = \Pr(X \le x) \tag{4.4}$$

for all values of x. That is, $F(x)$ is the probability that the random variable X is less than or equal to some number x. The cdf can also be defined or computed in terms of the probability density function (pdf) f as

$$F(x) = \Pr(X \le x) = \int_{-\infty}^{x} f(t)\, dt \tag{4.5}$$

for a continuous distribution, and for a discrete distribution it is

$$F(x) = \Pr(X \le x) = \sum_{x_i \le x} f(x_i) \tag{4.6}$$

Figure 4.9 illustrates the relationship between the probability density function and the cumulative distribution function for the lognormal distribution shown in Figure 4.7.

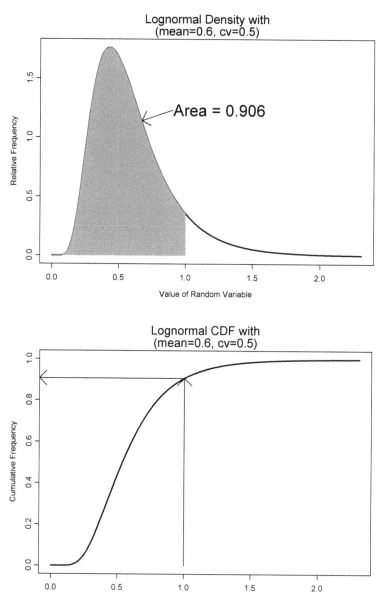

Figure 4.9   Relationship between the pdf and the cdf for a lognormal distribution

You can use the cdf to compute the probability that a random variable will fall into some specified interval. For example, the probability that a random variable X falls into the interval $[0.75, 1]$ is given by:

$$\Pr\left(0.75 \leq X \leq 1\right) = \int_{0.75}^{1} f\left(x\right) dx$$

$$= \Pr\left(X \leq 1\right) - \Pr\left(X \leq 0.75\right) +$$
$$\Pr\left(X = 0.75\right) \qquad\qquad \textbf{(4.7)}$$

$$= F\left(1\right) - F\left(0.75\right) +$$
$$\Pr\left(X = 0.75\right)$$

For a continuous random variable, the probability that X is exactly equal to 0.75 is 0 (because the area under the pdf between 0.75 and 0.75 is 0), but for a discrete random variable there may be a positive probability of X taking on the value 0.75.

Plotting Cumulative Distribution Functions

Figure 4.10 displays the cumulative distribution function for the binomial random variable whose pdf was shown in Figure 4.5. Figure 4.11 displays the cdf for the lognormal random variable whose pdf was shown in Figure 4.7.

We can see from Figure 4.10 that the cdf of a binomial random variable is a step function (which is also true of any discrete random variable). The cdf is 0 until it hits $x = 0$, at which point it jumps to 0.5 and stays there until it hits $x = 1$, at which point it stays at 1 for all values of x at 1 and greater. On the other hand, the cdf for the lognormal distribution shown in Figure 4.11 is a smooth curve that is 0 below $x = 0$, and rises towards 1 as x increases.

Menu

To produce the binomial cdf shown in Figure 4.10 using the ENVIRONMENTALSTATS for S-PLUS pull-down menu, follow these steps.

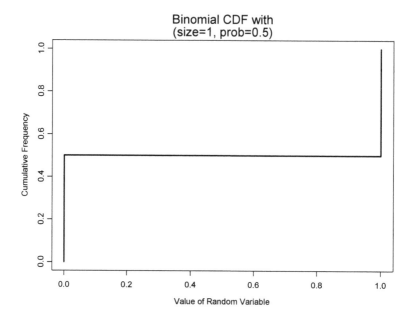

Figure 4.10   Cumulative distribution function for the coin-tossing experiment

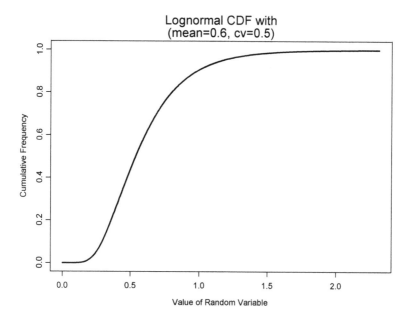

Figure 4.11   Cumulative distribution function for a lognormal probability distribution

1. On the S-PLUS menu bar, make the following menu choices:
    **EnvironmentalStats>Probability Distributions and Random
    Numbers>Plot Distribution**. This will bring up the Plot Distribu-
    tion Function dialog box.
2. In the Distribution box, choose **Binomial**. In the size box, type **1**. In
    the prob box, type **0.5**.
3. Under Plotting Information, for the Plot button, choose **CDF**.
4. Click **OK** or **Apply**.

To produce the lognormal cdf shown in Figure 4.11, follow these steps.

1. On the S-PLUS menu bar, make the following menu choices:
    **EnvironmentalStats>Probability Distributions and Random
    Numbers>Plot Distribution**. This will bring up the Plot Distribu-
    tion Function dialog box.
2. In the Distribution box, choose **Lognormal (Alternative)**. In the
    mean box, type **0.6**. In the cv box, type **0.5**.
3. Under Plotting Information, for the Plot button, choose **CDF**.
4. Click **OK** or **Apply**.

Command

To produce the binomial cdf shown in Figure 4.10 using the
ENVIRONMENTALSTATS for S-PLUS Command or Script Window, type this
command.

```
cdfplot("binom", list(size=1, prob=0.5))
```

To produce the lognormal cdf shown in Figure 4.11, type this command.

```
cdfplot("lnorm.alt", list(mean=0.6, cv=0.5))
```

Computing Values of the Cumulative Distribution Function

You can use S-PLUS and ENVIRONMENTALSTATS for S-PLUS to compute
the value of the cdf for any of the built-in probability distributions. As we
saw in Figure 4.10, the value of the cdf for the binomial distribution shown
in that figure is 0.5 for x between 0 and 1, and 1 for x greater than or equal
to 1. From Equations (4.2) and (4.5), you can show that for the lognormal
distribution shown in Figure 4.7, the values of the cdf evaluated at 0.5, 0.75,
and 1 are about 0.44, 0.76, and 0.91, respectively.

Menu

To compute the values of the cdf of the binomial distribution shown in
Figure 4.10 using the ENVIRONMENTALSTATS for S-PLUS pull-down menu,
follow these steps.

1. On the S-Plus menu bar, make the following menu choices: **EnvironmentalStats>Probability Distributions and Random Numbers>Density, CDF, Quantiles**. This will bring up the Densities, Cumulative Probabilities, or Quantiles dialog box.
2. For the Data to Use buttons, choose **Expression**. In the Expression box, type **0:1**. In the Distribution box, choose **Binomial**. In the size box type **1**. In the prob box type **0.5**. Under the Probability or Quantile group, make sure the **Cumulative Probability** box is checked.
3. Click **OK** or **Apply**.

To compute the values of the cdf of the lognormal distribution shown in Figure 4.11 for the values 0.5, 0.75, and 1, follow these steps.

1. On the S-Plus menu bar, make the following menu choices: **EnvironmentalStats>Probability Distributions and Random Numbers>Density, CDF, Quantiles**. This will bring up the Densities, Cumulative Probabilities, or Quantiles dialog box.
2. For the Data to Use buttons, choose **Expression**. In the Expression box, type **c(0.5, 0.75, 1)**. In the Distribution box, choose **Lognormal (Alternative)**. In the mean box type **0.6**. In the cv box type **0.5**. Under the Probability or Quantile group, make sure the **Cumulative Probability** box is checked.
3. Click **OK** or **Apply**.

Command

To compute the values of the cdf of the binomial distribution shown in Figure 4.10 using the S-Plus Command or Script Window, type this command.

```
pbinom(0:1, size=1, prob=0.5)
```

To compute the values of the cdf of the lognormal distribution shown in Figure 4.11 for the values 0.5, 1, and 1.5, type this command using ENVIRONMENTALSTATS for S-Plus.

```
plnorm.alt(c(0.5, 0.75, 1), mean=0.6, cv=0.5)
```

QUANTILES AND PERCENTILES

Loosely speaking, the pth *quantile* of a population is the (a) number such that a fraction p of the population is less than or equal to this number. The pth quantile is the same as the $100p^{th}$ *percentile*; for example, the 0.5 quantile is the same as the 50th percentile. In Figure 4.9, you can see that for a lognormal distribution with a mean of 0.6 and a coefficient of variation of 0.5, the number 1 is the 0.906 quantile, or the 90.6 percentile.

Here is a more technical definition of a quantile. If X is a random variable with some specified distribution, the pth *quantile* of the distribution of X, denoted x_p, is a (the) number that satisfies:

$$\Pr\left(X < x_p\right) \leq p \leq \Pr\left(X \leq x_p\right) \qquad (4.8)$$

where p is a number between 0 and 1 (inclusive).

If there is more than one number that satisfies the above condition, the pth quantile of X is often taken to be the average of the smallest and largest numbers that satisfy the condition. The S-PLUS functions for computing quantiles, however, return the smallest number that satisfies the above condition (see **Computing Quantiles and Percentiles** below).

If X is a continuous random variable, the pth quantile of X is simply defined as the value such that the cdf of that value is equal to p:

$$\Pr\left(X \leq x_p\right) = F\left(x_p\right) = p \qquad (4.9)$$

The $100p^{th}$ *percentile* is another name for the pth quantile. That is, the $100p^{th}$ percentile is the (a) number such that $100p\%$ of the distribution lies below this number.

Quantiles Are Invariant under Monotonic Transformations

An important property of quantiles is that *quantiles are invariant under monotonic transformations*. That is, let X denote a random variable and set $Y = g(X)$, where g is some monotonic function. Then the relationship between the pth quantile for X and the pth quantile for Y is:

$$y_p = g\left(x_p\right)$$

$$\qquad (4.10)$$

$$x_p = g^{-1}\left(y_p\right)$$

For example, if X has a lognormal distribution, then $Y = \log(X)$ has a normal distribution (see the section Lognormal Distribution below). If y_p is the pth quantile of the distribution of Y, then the pth quantile of the distribution of

X is $x_p = \exp(y_p)$. So if the median of Y is 10, then the median of X is $\exp(10)$.

Computing Quantiles and Percentiles

A plot of the cumulative distribution function makes it easy to visually pick out important quantiles, such as the median (50th percentile) or the 95th percentile. Looking at the cdf of the binomial distribution shown in Figure 4.10, it is easy to see that any number less than 0 is a 0th percentile, any number greater than or equal to 0 and less than 1 is a 50th percentile, and any number greater than or equal to 1 is a 100th percentile. (By convention, most textbooks and practitioners would say that 0.5 is the 50th percentile.) Similarly, looking at the cdf of the lognormal distribution shown in Figure 4.11, the median (50th percentile) is about 0.5 and the 95th percentile is about 1.1 (the actual values are 0.537 and 1.167, respectively).

Menu

To compute the 0th , 25th, 50th, 75th, and 100th percentiles of the binomial distribution shown in Figure 4.10 using the ENVIRONMENTALSTATS for S-PLUS pull-down menu, follow these steps.

1. On the S-PLUS menu bar, make the following menu choices: **EnvironmentalStats>Probability Distributions and Random Numbers>Density, CDF, Quantiles**. This will bring up the Densities, Cumulative Probabilities, or Quantiles dialog box.
2. For the Data to Use buttons, choose **Expression**. In the Expression box, type **c(0, 0.25, 0.5, 0.75, 1)**. In the Distribution box, choose **Binomial**. In the size box type **1**. In the prob box type **0.5**. Under the Probability or Quantile group, make sure the **Quantile** box is checked.
3. Click **OK** or **Apply**.

To compute the 50th and 95th percentiles of the lognormal distribution shown in Figure 4.11, follow these steps.

1. On the S-PLUS menu bar, make the following menu choices: **EnvironmentalStats>Probability Distributions and Random Numbers>Density, CDF, Quantiles**. This will bring up the Densities, Cumulative Probabilities, or Quantiles dialog box.
2. For the Data to Use buttons, choose **Expression**. In the Expression box, type **c(0.5, 0.95)**. In the Distribution box, choose **Lognormal (Alternative)**. In the mean box type **0.6**. In the cv box type **0.5**. Under the Probability or Quantile group, make sure the **Quantile** box is checked.
3. Click **OK** or **Apply**.

Command

To compute the 0th, 25th, 50th, 75th, and 100th percentiles of the binomial distribution shown in Figure 4.10 using the S-PLUS Command or Script Window, type this command.

```
qbinom(c(0, 0.25, 0.5, 0.75, 1), size=1, prob=0.5)
```

To compute the 50th and 95th percentiles of the lognormal distribution shown in Figure 4.11, type this command using ENVIRONMENTALSTATS for S-PLUS.

```
qlnorm.alt(c(0.5, 0.95), mean=0.6, cv=0.5)
```

GENERATING RANDOM NUMBERS FROM PROBABILITY DISTRIBUTIONS

With the advance of modern computers, experiments and simulations that just a decade ago would have required an enormous amount of time to complete using large-scale computers can now be easily carried out on personal computers. Simulation is fast becoming an important tool in environmental statistics and all fields of statistics in general (see Chapter 13).

For all of the distributions shown in Figure 4.8, you can generate random numbers (actually, pseudo-random numbers) from these distributions using S-PLUS and ENVIRONMENTALSTATS for S-PLUS. The density histograms in Figure 4.3 and Figure 4.4 are based on random numbers that were generated with S-PLUS. We will talk more about how random numbers are generated in S-PLUS in Chapter 13. For now, all you need to know is that when you set the random number seed to a specific value, you will always get the same sequence of "random" numbers.

Menu

To generate the random numbers from a binomial distribution that were used to produce Figure 4.3 (the outcomes of 10 coin flips) using the ENVIRONMENTALSTATS for S-PLUS pull-down menu, follow these steps.

1. On the S-PLUS menu bar, make the following menu choices: **EnvironmentalStats>Probability Distributions and Random Numbers>Random Numbers>Univariate**. This will bring up the Univariate Random Number Generation dialog box.
2. In the Sample Size box, type **10**. In the Set Seed with box, type **482**. In the Distribution box, choose **Binomial**. In the size box type **1**. In the prob box type **0.5**.
3. Click **OK** or **Apply**.

To generate the random numbers that were used to produce Figure 4.4 (the outcomes of 100 coin flips), follow the same steps as above, except in the Sample Size box type **100**, and in the Set Seed with box type **256**.

To generate 100 random numbers from the lognormal distribution shown in Figure 4.7, follow these steps.

1.   On the S-PLUS menu bar, make the following menu choices: **EnvironmentalStats>Probability Distributions and Random Numbers>Random Numbers>Univariate**. This will bring up the Univariate Random Number Generation dialog box.
2.   In the Sample Size box, type **100**. In the Distribution box, choose **Lognormal (Alternative)**. In the mean box type **0.6**. In the cv box type **0.5**.
3.   Click **OK** or **Apply**.

Command

To compute the random numbers from a binomial distribution that were used to produce Figure 4.3 (the outcomes of 10 coin flips) using the S-PLUS Command or Script Window, type these commands.

```
set.seed(482)
rbinom(10, size=1, prob=0.5)
```

To compute the random numbers that were used to produce Figure 4.4 (the outcomes of 100 coin flips), type these commands.

```
set.seed(256)
rbinom(100, size=1, prob=0.5)
```

To generate 100 random numbers from the lognormal distribution shown in Figure 4.7, type these commands using ENVIRONMENTALSTATS for S-PLUS.

```
rlnorm.alt(100, mean=0.6, cv=0.5)
```

CHARACTERISTICS OF PROBABILITY DISTRIBUTIONS

Every probability distribution has certain characteristics. Statisticians have defined various quantities to describe the characteristics of a probability distribution, such as the mean, median, mode, variance, standard deviation, coefficient of variation, skew, and kurtosis. The following sub-sections present detailed definitions of these quantities. Note that in Chapter 3 when we discussed these quantities, we were actually talking about *sample* statistics

(statistics based on a random sample). Here, we are talking about quantities associated with a *population*.

Mean, Median, and Mode

The mean, median, and mode are all measures of the central tendency of a probability distribution (although the mode is used very little in practice). When someone talks about the *average* of a population, he or she is usually referring to the mean, but sometimes to the median or mode.

The *median* of a probability distribution is simply the 50th percentile of that distribution. That is, it is the (a) number such that half of the area under the pdf is to the left of that number and half of the area is to the right. Note that for discrete distributions, percentiles are not necessarily unique so neither is the median necessarily unique. As we saw earlier, a median of the binomial distribution shown in Figure 4.5 and Figure 4.10 is 0, and the median of the lognormal distribution shown in Figure 4.7 and Figure 4.11 is 0.54.

The *mode* of a probability distribution is simply the (a) quantile associated with the largest value(s) of the pdf. A probability distribution may have more than one mode, such as the normal mixture distribution shown in Figure 4.8 on page 150. The mode of the lognormal distribution shown in Figure 4.7 is about 0.5. The binomial distribution shown in Figure 4.5 does not have a unique mode.

With respect to a probability distribution, the terms *mean*, *average*, and *expected value* all denote the same quantity. If you think of a picture of a probability density function and imagine it is made out of a thin sheet of metal, the mean is the point along the bottom of the distribution (the x-axis) at which the distribution balances.

The mean of a probability distribution is often denoted by the Greek letter μ (mu). Mathematically, the mean of a continuous random variable is defined as

$$\mu = E\left(X\right) = \int_{-\infty}^{\infty} x\, f\left(x\right) dx \qquad (4.11)$$

and the mean of a discrete random variable is defined as

$$\mu = E\left(X\right) = \sum_{x} x\, f\left(x\right) \qquad (4.12)$$

Note that the mean may be equal to a value that the random variable never assumes. For example, for the binomial distribution shown in Figure 4.5, the mean of the distribution is 0.5, even though the random variable can only take on the values 0 (tails) or 1 (heads).

Figure 4.12 illustrates the mean, median, and mode for the lognormal distribution shown in Figure 4.7. Note that in this case the mean is larger than the median.

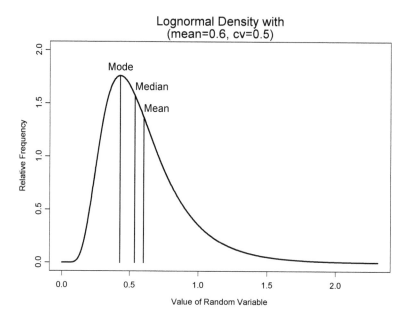

Figure 4.12   Mean, median, and mode for a lognormal distribution

Variance and Standard Deviation

The **variance** of a probability distribution is the average squared distance between an observed value of X and the mean μ. It is a measure of the spread of the distribution about the mean. Because the variance is an average *squared* distance, it is in squared units (e.g., ppb^2). The **standard deviation** is the square root of the variance and its units are the same as the original units of the random variable. The standard deviation is usually denoted by the Greek letter σ (sigma), and the variance is thus usually denoted by σ^2. Note that both the variance and standard deviation are always positive quantities.

Mathematically, the variance of a continuous random variable is defined as:

$$\sigma^2 = Var\left(X\right) = E\left[\left(X - \mu\right)^2\right]$$

(4.13)

$$= \int_{-\infty}^{\infty} \left(x - \mu\right)^2 f\left(x\right) dx$$

and the variance of a discrete random variable is defined as:

$$\sigma^2 = Var\left(X\right) = E\left[\left(X - \mu\right)^2\right]$$

(4.14)

$$= \sum_{x} \left(x - \mu\right)^2 f\left(x\right)$$

Figure 4.15 illustrates two normal distributions with different standard deviations (and hence different variances).

Coefficient of Variation

The *coefficient of variation* of a distribution (sometimes denoted CV) is defined as the ratio of the standard deviation to the mean:

$$CV = \sigma/\mu$$ (4.15)

The coefficient of variation is a measure of the spread of the distribution relative to the size of the mean. Note that since μ and σ have identical units (e.g., ppb, degrees, mg/L, etc.), the coefficient of variation is a unit-less measure. It is usually used to characterize positive, right skewed distributions such as the lognormal distribution. Figure 4.20 illustrates two lognormal distributions with different means and coefficients of variation.

Skew and Kurtosis

In the past, the skew and kurtosis of a distribution were often used to compare the distribution to a normal (Gaussian) distribution. These quantities and their estimates are not so widely reported anymore, although they are still used to fit distributions in the system of Pearson curves (Johnson et al., 1994, pp. 15–25).

Skew

The rth central moment of a random variable X is defined as:

$$\mu_r = E\left[(x - \mu)^r\right] \tag{4.16}$$

The first central moment of any distribution is 0. The second central moment is the same thing as the variance.

The *coefficient of skewness* of a distribution is defined to be the third central moment scaled by the cube of the standard deviation:

$$Skew = \mu_3 / \sigma^3 \tag{4.17}$$

The skew measures the distribution of values about the mean. If small values of the distribution are bunched close to the mean and large values of the distribution extend far above the mean (as in Figures 4.7, 4.9, and 4.12), the skew will be positive and the distribution is said to be *positively skewed* or *right skewed*. Conversely, if small values of the distribution extend well below the mean and large values of the distribution tend to bunch close to the mean, the skew will be negative and the distribution is said to be *negatively skewed* or *left skewed*.

Kurtosis

The *coefficient of kurtosis* of a distribution is the fourth central moment scaled by the square of the variance:

$$Kurtosis = \mu_4 / \sigma^4 \tag{4.18}$$

The kurtosis is a measure of the proportion of values in the tails (extreme ends) of the distribution. The kurtosis is independent of the mean and variance of the distribution.

For a normal distribution, the coefficient of kurtosis is 3. Distributions with kurtosis less than 3 are called *platykurtic* — they have shorter tails than a normal distribution. Distributions with kurtosis greater than 3 are called *leptokurtic* — they have heavier tails than a normal distribution. Sometimes, the term "kurtosis" is used to mean the "coefficient of excess kurtosis," which is the coefficient of kurtosis minus 3.

IMPORTANT DISTRIBUTIONS IN ENVIRONMENTAL STATISTICS

Random variables are random because they are affected by physical, chemical, and/or biological processes that we cannot measure or control with absolute precision. Certain probability distributions are used extensively in environmental statistics because they model the outcomes of physical, chemical, and/or biological processes fairly well. The rest of this section discusses some of these distributions.

Normal Distribution

The *normal* or *Gaussian distribution* is by far the most frequently used distribution in probability and statistics. It is sometimes a good model for environmental data, but in a lot of cases the lognormal distribution or some other positively skewed distribution provides a better model. Johnson et al. (1994, Chapter 13) discuss the normal distribution in detail.

The probability density function of the normal distribution is given by:

$$f\left(x\right) = \frac{1}{\sigma\sqrt{2\pi}} \exp\left[-\frac{1}{2}\left(\frac{x-\mu}{\sigma}\right)^2\right],$$

(4.19)

$$-\infty < x < \infty$$

where μ denotes the mean of the distribution and σ denotes the standard deviation. This distribution is often denoted by $N(\mu, \sigma)$.

A *standard normal distribution* is a normal distribution with a mean of 0 and a standard deviation of 1, denoted $N(0, 1)$. Figure 4.13 shows the pdf of the standard normal distribution, and Figure 4.14 shows the corresponding cdf. Note that a normal distribution is *symmetric* about its mean.

The Mean and Standard Deviation of a Normal Distribution

A normal probability distribution (probability density function) is a bell-shaped curve, regardless of the values of the mean μ or the standard deviation σ. The value of μ determines where the center of the distribution lies along the x-axis, and the value of σ determines the spread of the distribution. Figure 4.15 compares a $N(0, 1)$ pdf with a $N(3, 2)$ pdf.

The Sum or Average of Normal Random Variables is Normal

A very special and important property of the normal distribution is that the sum or average of several normal random variables is itself a normal ran-

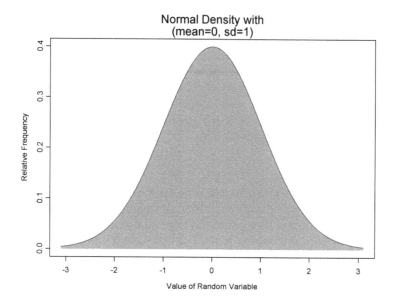

Figure 4.13   Probability density function of the standard normal distribution

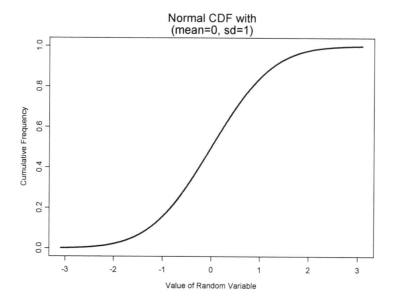

Figure 4.14   Cumulative distribution function of the standard normal distribution

Two Normal Distributions

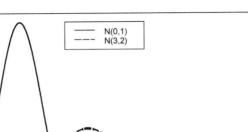

Figure 4.15   Two normal probability density functions

dom variable. Technically speaking, let X_1, X_2, ... , X_n denote n independent normal random variables, where X_i follows a normal distribution with mean μ_i and standard deviation σ_i. Then the sum of the random variables

$$Y = \sum_{i=1}^{n} X_i \qquad (4.20)$$

has a normal distribution with mean given by:

$$E\left(Y\right) = E\left(\sum_{i=1}^{n} X_i\right) = \sum_{i=1}^{n} E\left(X_i\right) = \sum_{i=1}^{n} \mu_i \qquad (4.21)$$

and a variance given by:

$$Var\left(Y\right) = Var\left(\sum_{i=1}^{n} X_i\right) = \sum_{i=1}^{n} Var\left(X_i\right) = \sum_{i=1}^{n} \sigma_i^2 \quad \textbf{(4.22)}$$

That is, the mean of the sum is simply the sum of the means, and the variance of the sum is the sum of the variances.

The average of the random variables

$$\overline{X} = \frac{1}{n} \sum_{i=1}^{n} X_i \quad \textbf{(4.23)}$$

also has a normal distribution with mean and variance given by

$$E\left(\overline{X}\right) = \frac{1}{n} \sum_{i=1}^{n} \mu_i \quad \textbf{(4.24)}$$

$$Var\left(\overline{X}\right) = \frac{1}{n^2} \sum_{i=1}^{n} \sigma_i^2 \quad \textbf{(4.25)}$$

We can use the above information to derive the **distribution of the sample mean** when we are sampling from a population that follows a normal distribution with mean μ and standard deviation σ. In this case, all of the means of the n random variables are the same value μ, and all of the standard deviations are the same value σ, so the sample mean follows a normal distribution with the following mean, variance, and standard deviation:

$$E\left(\overline{X}\right) = \mu \quad \textbf{(4.26)}$$

$$Var\left(\overline{X}\right) = \frac{\sigma^2}{n} \quad \textbf{(4.27)}$$

$$Sd\left(\overline{X}\right) = \frac{\sigma}{\sqrt{n}} \qquad\qquad (4.28)$$

The formulas in the above equations for the mean and variance of the sum of random variables, and the mean, variance, and standard deviation of the sample mean actually hold no matter what the underlying distribution of the population.

The Central Limit Theorem

The normal distribution often provides a good model for data because it can be shown that this is the distribution that results when you add up or average lots of independent random variables, *no matter what the underlying distribution* of any of these random variables (Ott, 1995, Chapter 7). The technical name for this phenomenon is the ***Central Limit Theorem***.

Figure 4.16 illustrates the Central Limit Theorem. Here, the random variable X is assumed to come from an exponential distribution with a mean of 5 (which implies a rate parameter of 1/5). The pdf of this distribution is shown in the top graph of Figure 4.16. Note that this distribution is extremely skewed to the right and does not look like a normal distribution at all. In the first experiment, we generate 10 observations from the exponential distribution and then compute the sample mean of these 10 observations. We then repeat this experiment 999 more times. The histogram of these 1,000 sample means is shown in the middle graph of Figure 4.16. You can see that this histogram is certainly not as skewed to the right as the original exponential distribution, but it does not really look bell-shaped either. The bottom graph in Figure 4.16 shows the results of a new experiment in which, for each of the 1,000 trials, we generate 100 observations from the exponential distribution and then average these observations. You can see that here the histogram of the 1,000 sample means does look like a bell-shaped curve.

How well you can approximate the distribution of a sum or average of random variables with a normal distribution depends on the sample size (how many observations you have to sum or average) and how different the underlying distributions of the random variables are from a normal distribution. For example, if all the random variables come from a normal distribution, then we have seen that their sum or average automatically follows a normal distribution. On the other hand, with our simulation experiment using the exponential distribution, we see that when we average only 10 outcomes instead of 100 outcomes, the distribution of the sample mean does not really look bell-shaped.

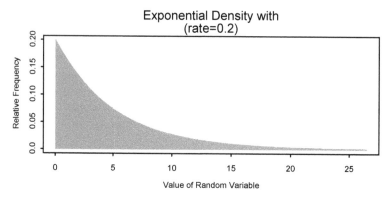

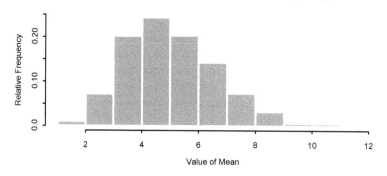

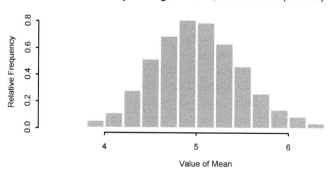

Figure 4.16   Illustration of the Central Limit Theorem

Probabilities and Deviations from the Mean

Figure 4.17 shows the relationship between the distance from the mean (in units of the standard deviation σ) and the area under the curve for any normal distribution. You can see that there is about a 68% chance of falling within one standard deviation of the mean, about a 95% chance of falling within two standard deviations of the mean, and almost a 100% chance of falling within three standard deviations of the mean. This is an important property, because even though Equation (4.19) shows that the pdf is positive for all values of x between $-\infty$ and $+\infty$ (hence there is always a possibility of seeing very negative or very positive values), the probability of observing values outside three standard deviations of the mean is miniscule. Thus, even though environmental data almost always take on only positive values, you can sometimes use the normal distribution to model them.

Transforming Back to a Standard Normal Distribution: The z-transformation

Any normal random variable can be transformed back to a standard normal random variable by subtracting its mean and dividing by its standard deviation. That is, if X is a random variable with a $N(\mu, \sigma)$ distribution, then

$$Z = \frac{X - \mu}{\sigma} \qquad (4.29)$$

is a random variable with a $N(0, 1)$ distribution. This is sometimes called the *z-transformation*. One important use of the z-transformation is to transform the sample mean to produce a standard normal random variable:

$$Z = \frac{\overline{X} - \mu}{\sigma/\sqrt{n}} \qquad (4.30)$$

When we sample from a population that follows a normal distribution, the random variable Z in Equation (4.30) exactly follows a standard normal distribution. When the population does not follow a normal distribution, the Central Limit Theorem tells us that the random variable Z in Equation (4.30) still follows a standard normal distribution approximately. We will return to these ideas when we discuss confidence intervals and hypothesis tests in Chapters 5 and 7.

Note that for an observed value x of a normal random variable X, the z-transformation tells us how far away x is from the mean in units of the stand-

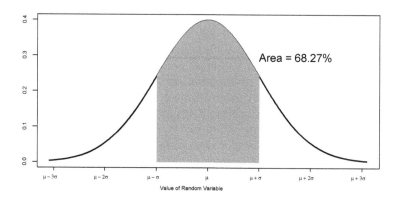

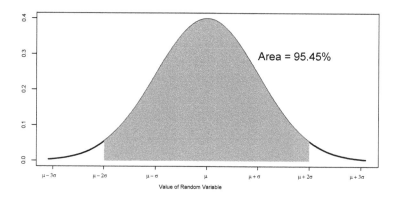

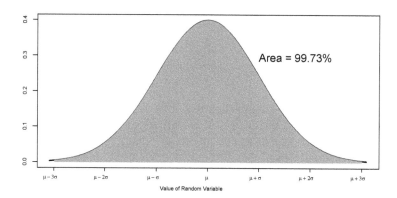

Figure 4.17   Relationship between distance from mean and area under the curve for a normal distribution

ard deviation (and also whether it is larger than or smaller than the mean). For example, there is only about a 5% chance that the z value will fall outside the interval [−2, 2], and it is very unlikely that the z value will fall outside the interval [−3, 3].

Lognormal Distribution

The two-parameter *lognormal distribution* is the distribution of a random variable whose logarithm is normally distributed. That is, if $Y = \log(X)$ (where $\log$ denotes the natural logarithm) and Y has a normal distribution with mean μ and standard deviation σ, then X is said to have a lognormal distribution. The distribution of X is often denoted as $\Lambda(\mu, \sigma)$ where μ and σ denote the mean of the transformed random variable Y. (Note: Λ denotes the uppercase Greek letter lambda.) Figure 4.18 shows the pdf of a lognormal distribution with parameters $\mu = 0$ and $\sigma = 1$; that is, $X = \exp(Z)$, where Z is a standard normal random variable. Figure 4.19 shows the corresponding cdf.

The probability density function of a lognormal distribution was shown in Equation (4.2). A slightly different way to write the pdf of a lognormal distribution is shown below in Equation (4.31). Compare this equation with Equation (4.19) showing the pdf of a normal distribution.

$$f(x) = \frac{1}{x\sigma\sqrt{2\pi}} \exp\left\{-\frac{1}{2}\left[\frac{\log(x) - \mu}{\sigma}\right]^2\right\},$$

(4.31)

$$x > 0$$

The two major characteristics of the lognormal distribution (probability density function) are that it is bounded below at 0 and it is skewed to the right. These characteristics are typical of environmental data. Several authors have illustrated various mechanisms to explain the occurrence of the lognormal distribution in nature (e.g., Ott, 1990, Chapters 8 and 9). The basic idea is that since the Central Limit Theorem states that the result of summing up random variables will produce a normal distribution, multiplying random variables will produce a lognormal distribution (recall that $\log(ab) = \log(a) + \log(b)$).

Unfortunately, just because the values of a data set are bounded below at 0 and appear to be skewed to the right, this does not automatically imply that the lognormal distribution is a good model. Several other probability distributions are bounded below at 0 and skewed to the right as well (for example

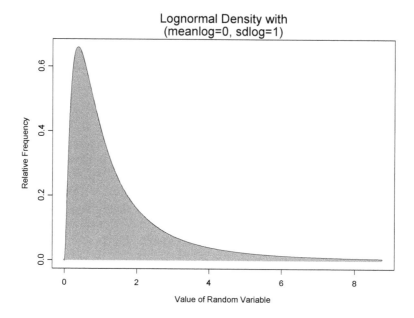

Figure 4.18   Probability density function of a lognormal distribution

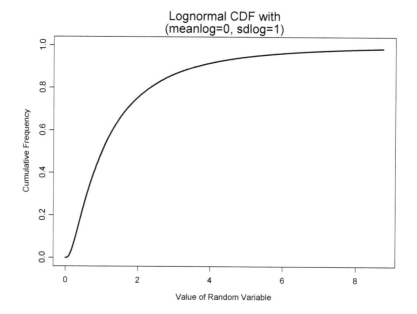

Figure 4.19   Cumulative distribution function of a lognormal distribution

gamma, generalized extreme value, Weibull, mixture of lognormals, etc.). These distributions can look very similar around the median of the distribution but be very different looking in the extreme tail of the distribution (e.g., 90th and 95th percentiles). Because distinguishing the shape of the tail of a distribution usually requires an extensive number of observations, it is often difficult to determine the best fitting model to a skewed set of environmental data. We will talk more about fitting data to distributions in Chapters 5 and 7.

Alternative Parameterization: Mean and CV

Sometimes, a lognormal distribution is characterized by its mean and coefficient of variation rather than the mean and standard deviation of the log-transformed random variable. Let $Y = \log(X)$ be distributed as a N(μ, σ) random variable, so that X is distributed as a $\Lambda(\mu, \sigma)$ random variable. Let θ (theta) denote the mean of X and let τ (tau) denote the coefficient of variation of X. The relationship between μ, σ, θ, and τ is given by:

$$\theta = E(X) = \exp\left(\mu + \frac{\sigma^2}{2}\right) \qquad (4.32)$$

$$\tau = CV(X) = \sqrt{\exp(\sigma^2) - 1} \qquad (4.33)$$

$$\mu = E(Y) = \log\left(\frac{\theta}{\sqrt{\tau^2 + 1}}\right) \qquad (4.34)$$

$$\sigma = \sqrt{Var(Y)} = \sqrt{\log(\tau^2 + 1)} \qquad (4.35)$$

There are two very important points to note about these relationships. The first is that even though $X = \exp(Y)$, it is *not* true that $\theta = \exp(\mu)$, that is, the mean of a lognormal distribution is not equal to e raised to the mean of the log-transformed random variable. In fact, the quantity $\exp(\mu)$

is the *median* of the lognormal distribution (see Equation (4.10)), so the median is always smaller than the mean (see Figure 4.12). The second point is that τ, the coefficient of variation of X, only depends on σ, the standard deviation of Y.

The formula for the skew of a lognormal distribution can be written as:

$$Skew = 3\,CV + CV^3 \qquad \textbf{(4.36)}$$

(Stedinger et al., 1993). This equation shows that large values of the CV correspond to very skewed distributions. As τ gets small, the distribution becomes less skewed and starts to resemble a normal distribution. Figure 4.20 shows two different lognormal distributions characterized by the mean θ and CV τ.

Two Lognormal Distributions

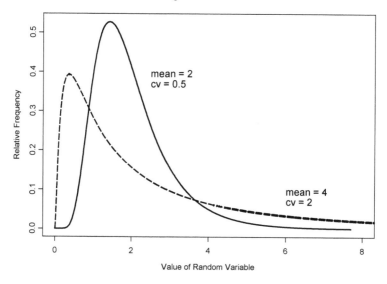

Figure 4.20   Probability density functions for two lognormal distributions

Three-Parameter Lognormal Distribution

The two-parameter lognormal distribution is bounded below at 0. The *three-parameter lognormal distribution* includes a threshold parameter γ (gamma) that determines the lower boundary of the random variable. That is, if $Y = \log(X-\gamma)$ has a normal distribution with mean μ and standard deviation σ, then X is said to have a three-parameter lognormal distribution.

The threshold parameter γ affects only the location of the three-parameter lognormal distribution; it has no effect on the variance or the shape of the distribution. Note that when $\gamma = 0$, the three-parameter lognormal distribution reduces to the two-parameter lognormal distribution. The three-parameter lognormal distribution is sometimes used in hydrology to model rainfall, stream flow, pollutant loading, etc. (Stedinger et al., 1993).

Binomial Distribution

After the normal distribution, the **binomial distribution** is one of the most frequently used distributions in probability and statistics. It is used to model the number of occurrences of a specific event in n independent trials. The outcome for each trial is binary: yes/no, success/failure, 1/0, etc. The binomial random variable X represents the number of "successes" out of the n trials. In environmental monitoring, sometimes the binomial distribution is used to model the proportion of observations of a pollutant that exceed some ambient or cleanup standard, or to compare the proportion of detected values at background and compliance units (USEPA, 1989a, Chapters 7 and 8; USEPA, 1989b, Chapter 8; USEPA, 1992b, p. 5-29; Ott, 1995, Chapter 4).

The probability density (mass) function of a binomial random variable X is given by:

$$f\left(x\right) = \binom{n}{x} p^x \left(1 - p\right)^{n-x}, \quad x = 0, 1, 2, \ldots, n \quad \textbf{(4.37)}$$

where n denotes the number of trials and p denotes the probability of "success" for each trial. It is common notation to say that X has a B(n, p) distribution.

The first quantity on the right-hand side of Equation (4.37) is called the binomial coefficient. It represents the number of different ways you can arrange the x "successes" to occur in the n trials. The formula for the binomial coefficient is:

$$\binom{n}{x} = \frac{n!}{x!\left(n - x\right)!} \quad \textbf{(4.38)}$$

The quantity $n!$ is called "n factorial" and is the product of all of the integers between 1 and n. That is,

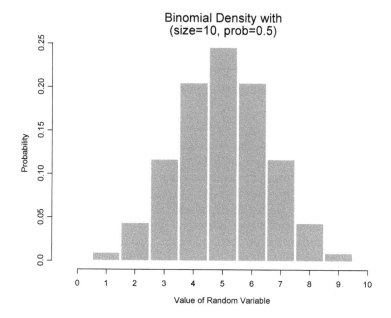

Figure 4.21   Probability density function of a B(10, 0.5) random variable

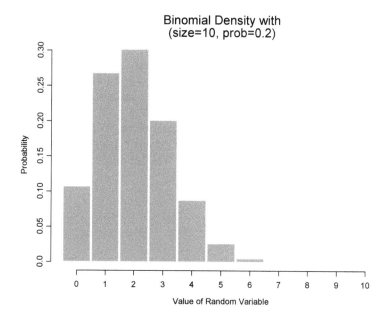

Figure 4.22   Probability density function of a B(10, 0.2) random variable

$$n! = n(n-1)(n-2)\cdots 2 \; 1 \qquad \textbf{(4.39)}$$

Figure 4.5 shows the pdf of a B(1, 0.5) random variable and Figure 4.10 shows the associated cdf. Figure 4.21 and Figure 4.22 show the pdf's of a B(10, 0.5) and B(10, 0.2) random variable, respectively.

The Mean and Variance of the Binomial Distribution

The mean and variance of a binomial random variable are:

$$E(X) \; = \; np$$

$$\qquad \textbf{(4.40)}$$

$$Var(X) \; = \; np(1-p)$$

The average number of successes in n trials is simply the probability of a success for one trial multiplied by the number of trials. The variance depends on the probability of success. Figure 4.23 shows the function $f(p) = p(1-p)$ as a function of p. The variance of a binomial random variable is greatest when the probability of success is ½, and the variance decreases to 0 as the probability of success decreases to 0 or increases to 1.

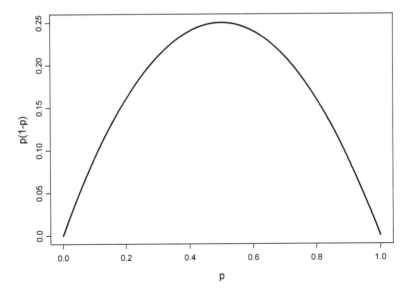

Figure 4.23   The variance of a B(1, p) random variable as a function of p

Hypergeometric Distribution

One way the binomial distribution arises is when you sample a *finite* population *with replacement*, and for each sampling unit you select you note whether it does or does not exhibit some characteristic. For example, suppose you pick a card at random from a standard deck of 52 playing cards, note whether the card is a face card or not, then return the card to the deck. If you repeat this experiment n times and let X denote the total number of face cards (jack, queen, or king) you picked, then X has a binomial distribution with parameters n and $p = 12/52$. The **hypergeometric distribution** arises in exactly this same situation, except that sampling is done *without* replacement. The hypergeometric distribution is not used as often in environmental studies as the binomial distribution, but we did use it in the case study in Chapter 2 that dealt with verifying the percentage of employees in a building who were familiar with the four emergency preparedness (EP) facts.

The probability density (mass) function of a hypergeometric random variable X is given by:

$$f(x) = \frac{\binom{m}{x}\binom{n}{k-x}}{\binom{m+n}{k}} , \quad x = 0, 1, \ldots, k \qquad (4.41)$$

where m denotes the number of "successes" in the population, x denotes the number of "successes" in the sample, n denotes the number of "failures" in the population, and k denotes the total number of sampling units selected for the sample. Note that the total population size is $m+n$. This distribution is sometimes denoted by H(m, n, k).

For our playing card example, if we choose k cards from the deck at random without replacement and let X denote the number of face cards in the sample, then X has a hypergeometric distribution with parameters $m = 12$, $n = 40$, and k. Figure 4.24 shows the pdf of this distribution for $k = 5$. When the sample size k is relatively small (e.g., $k/(m+n) < 0.1$), the effect of sampling without replacement is slight and the hypergeometric distribution can be approximated by the binomial distribution with parameters $n = k$ and $p = m/(m+n)$ (Evans et al., 1993, p. 86; Johnson et al., 1992, p. 257). Figure 4.25 shows the binomial approximation for the playing card example with $n = 5$ and $p = 12/52$.

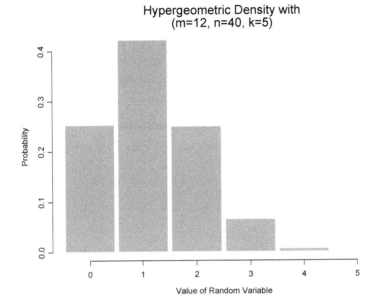

Figure 4.24   Probability density function of a H(12, 40, 5) random variable

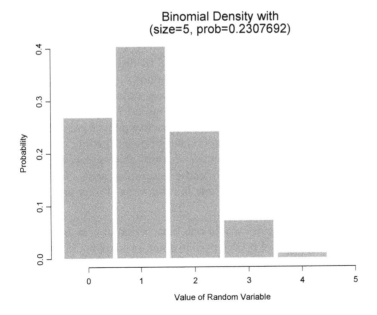

Figure 4.25   Probability density function of a B(5, 12/52) random variable

For the case study in Chapter 2, we created the tables of required samples sizes for each building (Tables 2.9 to 2.12) by using the hypergeometric distribution where m denotes the number of employees in the building who are not familiar with the EP facts (i.e., the number of DUDs), and n denotes the number of employees who are familiar with the EP facts. We will talk more about using the hypergeometric distribution to determine the required sample sizes in Chapter 8.

The Mean and Variance of the Hypergeometric Distribution

The mean and variance of the hypergeometric distribution are given by:

$$E(X) = k\left(\frac{m}{m+n}\right)$$

$$(4.42)$$

$$Var(X) = k\left(\frac{m}{m+n}\right)\left[1-\left(\frac{m}{m+n}\right)\right]\left(\frac{m+n-k}{m+n-1}\right)$$

Note that by setting $n = k$ and $p = m/(m+n)$ these formulas are exactly the same as for a binomial distribution (Equation (4.40)), except that the formula for the variance has a finite population correction factor.

Poisson Distribution

The Poisson distribution is named after one of its discoverers, who in 1837 derived this distribution as the limiting distribution of a B(n, p) distribution, where n tends to infinity, p tends to 0, and np stays constant (Johnson et al., 1992, Chapter 4). This distribution is useful for modeling the number of occurrences of a "rare" event (e.g., radioactive particle emissions, number of cars arriving at an intersection, number of people getting in line, etc.) within a given time period, or the number of "rare" objects (e.g., plants, animals, microorganisms, etc.) within a specific geographic area. The requirements for a Poisson distribution include the following (Fisher and van Belle, 1993, p. 212):

1. The number of occurrences within one time period or area is independent of the number of occurrences in a non-overlapping time period or area.
2. The expected number of occurrences within one time period or area is constant over all time periods or areas.

3. The expected number of occurrences within one time period or area decreases to 0 as the length of the time period or size of the area decreases to 0.

In environmental monitoring, the Poisson distribution has been used to model the number of times a pollution standard is violated over a given time period (Ott, 1995, Chapter 5). Gibbons (1987b) used the Poisson distribution to model the number of detected compounds per scan of the 32 volatile organic priority pollutants (VOC), and also to model the distribution of chemical concentration (in ppb).

The probability density (mass) function of a Poisson random variable is given by:

$$f(x) = \frac{\lambda^x e^{-\lambda}}{x!} \quad , \quad x = 0, 1, 2, \ldots \quad \quad (4.43)$$

where λ (lambda) denotes the average number of events that occur within the given time period or area. This distribution is often denoted by Poisson(λ) or P(λ). Figures 4.26 and 4.27 show the pdf's of a Poisson(1) and Poisson(3) random variable, respectively.

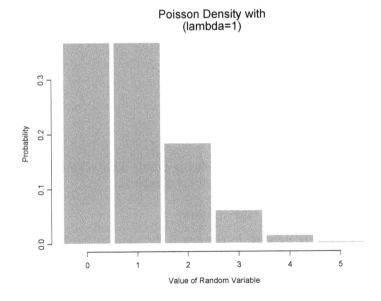

Figure 4.26   Probability density function of a Poisson(1) random variable

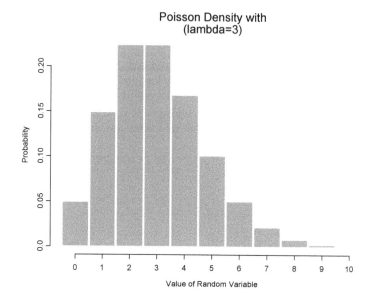

Figure 4.27   Probability density function of a Poisson(3) random variable

The Mean and Variance of the Poisson Distribution

The mean and variance of a Poisson distribution are given by:

$$E(X) = \lambda$$

$$\text{(4.44)}$$

$$Var(X) = \lambda$$

Note that the variance depends upon (in fact is exactly equal to) the mean. Thus, the larger the mean, the more variable the distribution.

Gamma (Pearson Type III) Distribution

The *three-parameter gamma distribution* is the same as the *Pearson Type III distribution* (Johnson et al., 1994, p. 337; Stedinger et al., 1993). Like the three-parameter lognormal distribution, the gamma distribution is sometimes used in hydrology to model rainfall, stream flow, pollutant loading, etc. (Stedinger et al., 1993).

The probability density function of the three-parameter gamma distribution is given by:

$$f\left(x\right) = \frac{\left(x - \gamma\right)^{\alpha-1} \exp\left[-\dfrac{\left(x - \gamma\right)}{\beta}\right]}{\beta^{\alpha}\ \Gamma\left(\alpha\right)},$$

(4.45)

$$\alpha > 0, \quad \beta > 0, \quad x > \gamma$$

(Johnson et al., 1994, p. 337) where $\Gamma\left(\ \right)$ denotes the gamma function and is defined as:

$$\Gamma\left(x\right) = \begin{cases} \displaystyle\int_{0}^{\infty} t^{x-1}e^{-t}dt\ , & x > 0 \\[2em] \dfrac{\Gamma\left(x + 1\right)}{x} & , \quad x < 0,\ x \neq -1, -2, \ldots \end{cases}$$

(4.46)

(Johnson et al., 1992, p. 5). The parameter α determines the shape of the pdf, the parameter β determines the scale of the pdf, and the parameter γ is the threshold parameter.

Two Gamma Distributions

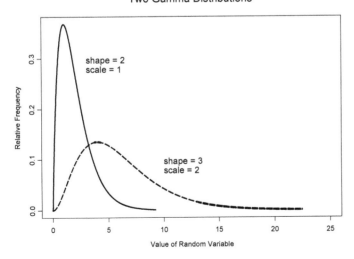

Figure 4.28   Two different gamma distributions

The *two-parameter gamma distribution* is a special case of the three-parameter gamma distribution with the location parameter set to 0 (i.e., $\gamma = 0$). The two-parameter gamma distribution is often simply called the gamma distribution. Figure 4.28 shows two different gamma distributions.

Mean, Variance, Skew, and Coefficient of Variation

The mean, variance, skew, and CV of a two-parameter gamma distribution are:

$$E\left(X\right) \quad = \alpha\beta$$

$$Var\left(X\right) \quad = \alpha\beta^2$$

$$Skew\left(X\right) \quad = 2/\sqrt{\alpha} \qquad (4.47)$$

$$CV\left(X\right) \quad = 1/\sqrt{\alpha}$$

Note that the shape parameter α completely determines both the skew and the CV.

Extreme Value Distribution

The *extreme value distribution* is used to model extreme values (maxima or minima) of some process or series of measurements, such as daily maximum temperatures, daily maximum concentrations of air pollutants (e.g., Kuchenhoff and Thamerus, 1996), or annual maximum of daily precipitation (e.g., Stedinger et al., 1993). The extreme value distribution gets its name from the fact that this distribution is the limiting distribution (as n approaches infinity) of the greatest (smallest) value among n independent random variables each having the same continuous distribution (Johnson et al., 1995, Chapter 22).

There are three families of extreme value distributions. The one described here is the Type I, also called the Gumbel extreme value distribution or simply *Gumbel distribution*. This distribution is sometimes denoted $EV(\eta,\theta)$.

The probability density function of the two-parameter extreme value distribution is given by:

$$f\left(x\right) = \frac{1}{\theta}\,e^{-(x-\eta)/\theta}\,\exp\left[-e^{-(x-\eta)/\theta}\right],$$

$$-\infty < x < \infty \qquad \textbf{(4.48)}$$
$$-\infty < \eta < \infty$$
$$\theta > 0$$

where η (eta) denotes the location parameter and θ (theta) denotes the scale parameter. This distribution is the largest extreme value distribution; it is used to model maxima. (A different pdf is used for the smallest extreme value distribution.) Figure 4.29 shows the pdf of an EV(10,1) distribution and Figure 4.30 shows the corresponding cdf.

The Mode, Mean, and Variance of an Extreme Value Distribution

The mode, mean, and variance of the two-parameter extreme value distribution are:

$$Mode\left(X\right) = \eta$$

$$E\left(X\right) = \eta + \varepsilon\theta \qquad \textbf{(4.49)}$$

$$Var\left(X\right) = \frac{\theta^2\pi^2}{6}$$

where ε denotes a number called Euler's constant and is approximately equal to 0.5772157. The skew (about 1.14) and kurtosis (5.4) are constant.

Generalized Extreme Value Distribution

A *generalized extreme value (GEV) distribution* includes a shape parameter κ (kappa), as well as the location and scale parameters η and θ (Johnson et al., 1995, Chapter 22). The three families of extreme value distributions are all special kinds of GEV distributions. When the shape parameter $\kappa = 0$, the GEVD reduces to the Type I extreme value (Gumbel) distribution. When $\kappa > 0$, the GEVD is the same as the Type II extreme value distribution, and when $\kappa < 0$ it is the same as the Type III extreme value distribution. See the help file on GEVD in ENVIRONMENTALSTATS for S-PLUS for more information on this distribution.

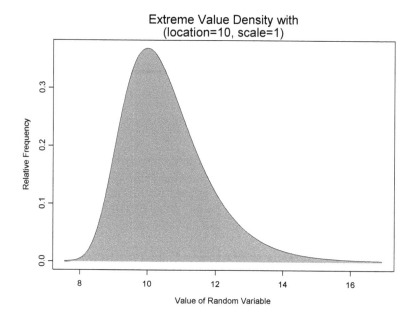

Figure 4.29   Probability density function of an extreme value distribution

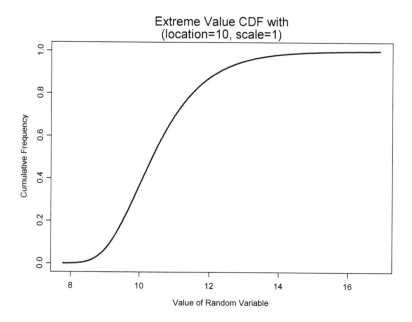

Figure 4.30   Cumulative distribution function of an extreme value distribution

Mixture Distributions

A *mixture distribution* results when some portion of the population comes from one probability distribution and the rest of the population comes from another distribution. Mixture distributions are often used to model data sets that contain a few or several gross "outliers"; that is, most of the observations appear to come from one distribution, but some appear to come from a distribution with a shifted mean and possibly shifted standard deviation.

In environmental statistics, mixture distributions are useful for modeling data from a remediated site that contains residual contamination, such as the Cleanup area TcCB data shown in Figure 3.2 of Chapter 3. Gilliom and Helsel (1986) used a lognormal mixture distribution (as well as several others) to assess the performance of various estimators of population parameters.

The general form of the pdf for the mixture of two distributions is:

$$f\left(x\right) = \left(1 - p\right) f_1\left(x\right) + p\, f_2\left(x\right) \qquad \textbf{(4.50)}$$

where $f_1(\ )$ denotes the pdf of the first distribution, $f_2(\ )$ denotes the pdf of the second distribution, and p is a number between 0 and 1 that denotes the mixing proportion.

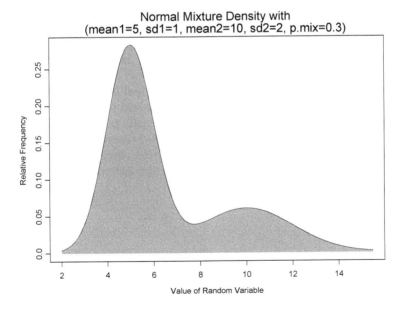

Figure 4.31   Probability density function of a mixture of two normal distributions

With ENVIRONMENTALSTATS for S-PLUS, you can compute the pdf, cdf, quantiles, and random numbers from the mixture of two normal distributions or the mixture of two lognormal distributions. Figure 4.31 shows the pdf of a normal mixture distribution where the first distribution is a N(5, 1), the second distribution is a N(10, 2), and the mixing proportion is 30% (i.e., 70% of the first distribution and 30% of the second distribution).

Mean and Variance

The mean and variance of the mixture of two distributions are given by:

$$E\left(X\right) = \left(1 - p\right)\mu_1 + p\,\mu_2$$

$$(4.51)$$

$$Var\left(X\right) = \left(1 - p\right)^2 \sigma_1^2 + p^2\,\sigma_2^2$$

where μ_1 and μ_2 denote the means of the first and second distribution, σ_1 and σ_2 denote the standard deviations of the first and second distribution, and p denotes the mixing proportion.

Zero-Modified Distributions

A *zero-modified distribution* is a distribution that has been modified to put extra probability mass at 0. That is, it is a special kind of mixture distribution in which a portion of the population comes from the original distribution, but the rest of the population is all zeros (Johnson et al., 1992, p. 312). In environmental statistics, the *zero-modified lognormal distribution* (also called the *delta distribution*) is sometimes used as a model for data sets with nondetects, where the nondetects are assumed to actually be 0 values (e.g., Owen and DeRouen, 1980). USEPA (1992c, pp. 28–34) contains an example where the data appear to come from a zero-modified normal distribution.

The general form of the pdf for a zero-modified distribution is:

$$h\left(x\right) = \begin{cases} p & , \; x = 0 \\ \left(1 - p\right)f\left(x\right) & , \; x \neq 0 \end{cases}$$

$$(4.52)$$

where $h()$ denotes the pdf of the zero-modified distribution, $f()$ denotes the pdf of the original distribution, and p denotes the proportion of the population with the value of 0.

With ENVIRONMENTALSTATS for S-PLUS, you can compute the pdf, cdf, quantiles, and random numbers from the zero-modified normal and zero-modified lognormal distributions. Figure 4.32 shows the pdf of a zero-modified lognormal distribution (delta distribution) with a mean of 3 and a cv of 0.5 for the lognormal part of the distribution, and a probability mass of 20% at 0.

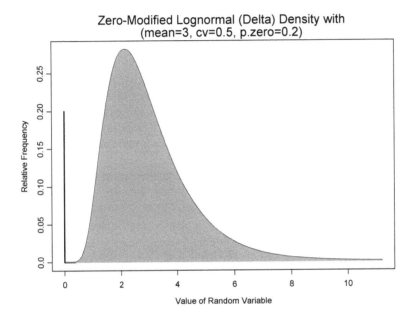

Figure 4.32   Probability density function of a zero-modified lognormal distribution

Mean and Variance

Aitchison (1955) shows that the mean and variance of a zero-modified distribution are given by:

$$E\left(X\right) \quad = \left(1 - p\right)\mu$$

$$(4.53)$$

$$Var\left(X\right) = \left(1 - p\right)\sigma^2 + p\left(1 - p\right)\mu^2$$

where μ and σ denote the mean and standard deviation of the original distribution.

MULTIVARIATE PROBABILITY DISTRIBUTIONS

We will not discuss multivariate probability distributions and multivariate statistics in this book, except in the context of linear regression, time series analysis, and spatial statistics. A good introduction to multivariate statistical analysis is Johnson and Wichern (1998). We do, however, need to discuss the concepts of covariance and correlation, but we will postpone this discussion until Chapter 9 on Linear Models.

SUMMARY

- A *random variable* is the measurement associated with the next physical sample from the population. Outcomes of a random variable usually differ in value from observation to observation.
- A *discrete* random variable can take on only a finite or countably infinite number of values. A *continuous* random variable can take on an infinite number of values.
- A *probability density function (pdf)* is a mathematical formula that describes the relative frequency of a random variable. Sometimes the picture of this formula (e.g., Figure 4.8) is called the pdf. If a random variable is discrete, its probability density function is sometimes called a probability mass function.
- The *cumulative distribution function (cdf)* is a mathematical formula that describes the probability that the random variable is less than or equal to some number. Sometimes the picture of this formula (e.g., Figure 4.9) is called the cdf.
- The pth *quantile*, also called the $100p^{th}$ *percentile*, is the (a) number such that 100p% of the distribution lies below this number.
- You can use S-PLUS and ENVIRONMENTALSTATS for S-PLUS to compute values of the pdf, cdf, quantiles, and random numbers from several probability distributions (see Figure 4.8). You can also plot these distributions.
- Every probability distribution has certain characteristics. Statisticians have defined various quantities to describe the characteristics of a probability distribution, such as the mean, median, mode, variance, standard deviation, coefficient of variation, skew, and kurtosis.
- Important probability distributions that are used in environmental statistics include the normal, lognormal, binomial, Poisson, gamma, and extreme value distributions, as well as mixture and zero-modified distributions.

EXERCISES

4.1. Describe some experiment or observational study. Define the population, and how you will take physical samples. What is the random variable? What kind of shape and characteristics do you think the probability distribution of this random variable will have? Why?

4.2. For a gamma random variable with parameters shape = 3 and scale = 2, do the following:

    a. Generate 10 random numbers from this distribution and create a density histogram.

    b. Generate 20 random numbers from this distribution and create a density histogram.

    c. Generate 100 random numbers from this distribution and create a density histogram.

    d. Generate 1000 random numbers from this distribution and create a density histogram.

    e. Add a plot of the probability distribution function to the last plot that you created.

4.3. Compute the value of the probability density function for the following normal distributions:

    a. $N(0, 1)$ at $-3, -2, -1, 0, 1, 2$, and 3.

    b. $N(20, 6)$ at $17, 18, 19, 20, 21, 22$, and 23.

    c. Based on the results in parts **a** and **b** above, can you come up with a formula involving the pdf $f()$ that expresses how a normal distribution is symmetric about its mean?

4.4. Plot the cumulative distribution function of a standard normal distribution. Looking at the plot, estimate the values of the quantiles for the following percentages: 1%, 2.5%, 5%, 10%, 50%, 90%, 95%, 97.5%, and 99%.

4.5. Repeat Exercise 4.4, but compute the exact values of the quantiles (to the 2nd decimal place).

4.6. Repeat Exercises 4.4 and 4.5 for a $N(20, 6)$ distribution.

4.7. Let X denote a $N(\mu, \sigma)$ random variable, and let x_p denote the pth quantile from this distribution. Also, let z_p denotes the pth quantile of a standard normal distribution. Show that the results in Exercises 4.5 and 4.6 above satisfy the following formula.

$$x_p = \mu + \sigma z_p \qquad\qquad (4.54)$$

4.8. Compute the value of the cdf at 23 for a N(20, 6) random variable.

4.9. Repeat Exercise 4.8, but use only a function that assumes a standard normal (i.e., N(0, 1)) random variable. (Hint: Use the z-transformation shown in Equation (4.29)).

4.10. A hazardous waste facility monitors the groundwater adjacent to the facility to ensure no chemicals are seeping into the groundwater. Several wells have been drilled, and once a month groundwater is pumped from each well and tested for certain chemicals. The facility has a permit specifying an Alternate Concentration Limit (ACL) for aldicarb of 30 ppb. This limit cannot be exceeded more than 5% of the time.

  a. If the natural distribution of aldicarb at a particular well can be modeled as a normal distribution with a mean of 20 and a standard deviation of 5, how often will this well exceed the ACL (i.e., what is the probability that the aldicarb level at that well will be greater than 30 ppb on any given sampling occasion)?

  b. What is the z-value of the ACL?

4.11. Repeat Exercise 4.10, assuming the aldicarb concentrations follow a N(28, 2) distribution.

4.12. Repeat part **a** of Exercise 4.10 assuming the aldicarb concentrations follow a lognormal distribution with a mean of 20 and a CV of 1.

4.13. Suppose the concentration of arsenic (in ppb) in groundwater sampled quarterly from a particular well can be modeled with a normal distribution with a mean of 5 and a standard deviation of 1.

  a. What is the distribution of the average of four observations taken in one year?

  b. What is the probability that a single observation will be greater than 7 ppb?

  c. What is the probability that the average of four observations will be greater than 7 ppb?

  d. Plot the distribution of a single observation and the distribution of the sample mean (based on four observations) on the same graph.

4.14. The sum of several binomial random variables that have the same parameter p (the probability of "success") is also a binomial random variable. For example, suppose you flip a coin 10 times and record the number of heads. Let X_1 denote the number of heads. This is a B(10, p) random variable (where $p = 0.5$ if the coin is fair). Now suppose you repeat this experiment, and let X_2 denote the number of heads in this second experiment. Then X_2 is also a B(10, p) random variable. If you combine the results of these two experiments, this is exactly the same as if you had flipped the coin 20 times and recorded the number of heads. That is, the random variable $Y = X_1 + X_2$ has a B(20, p) distribution.

   a. Suppose the coin is not fair, and that the probability of a head is only 10%. Plot the distribution of a B(10, 0.1) random variable.

   b. Now plot the distribution of a B(20, 0.1) random variable.

   c. Now plot the distribution of a B(500, 0.1) random variable. (Use a right-tail cutoff of 0.0001.)

   d. How does the shape of the probability density function change as the number of flips changes? Why can you use the Central Limit Theorem to explain the change in the shape of the distribution?

4.15. *Normal Approximation to the Binomial.* Suppose that the random variable Y has a B(n, p) distribution.

   a. Find the probability that Y is greater than 2 when $n = 10$ and $p = 0.1$.

   b. Find the probability that Y is greater than 60 when $n = 200$ and $p = 0.2$.

   c. Let $X_1, X_2, \ldots, X_n$ denote n independent B(1, p) random variables. Write Y as a function of these n random variables.

   d. What are the mean and variance of each of the B(1, p) random variables?

   e. Using Equations (4.21) and (4.22), determine the mean and variance of Y.

   f. Using Equation (4.40), determine the mean and variance of Y.

   g. The Central Limit Theorem says that the distribution of Y can be modeled as approximately a normal distribution. What are the mean and variance of this distribution?

   h. Repeat parts **a** and **b** assuming Y is approximately normally distributed.

     i.   For parts **a** and **b**, create a plot of the binomial pdf, then overlay the approximating normal pdf.

     j.   *Note:* A rule of thumb for how large n should be in order to use the normal approximation to the binomial is: $np(1-p) \geq 10$ (Fisher and van Belle, 1993, p. 183). You can also use a continuity correction to achieve slightly better results (see Fisher and van Belle, 1993, p. 184).

4.16. Suppose you can model the concentration of TcCB in soil at a background site as coming from a lognormal distribution with some mean θ and some CV τ. Suppose further that a regulation states that if any concentrations of TcCB at a remediated site are larger than the 95th percentile at the background site, then the remediated site will be declared to be still contaminated. Suppose the probability distribution of TcCB in the remediated site is exactly the same as the distribution of TcCB in the background site.

     a.   If you take $n = 1$ soil sample from the remediated site, what is the probability that the TcCB concentration will exceed the 95th percentile of the background site?

     b.   If you take $n = 10$ soil samples, what is the probability that at least one of the 10 TcCB concentrations will exceed the 95th percentile of the background site? (Hint: Use the binomial distribution.)

     c.   Repeat part **b** for $n = 100$ soil samples.

     d.   Do the results in parts **a**, **b**, and **c** depend on the assumption of a lognormal distribution?

     e.   If you owned the land, had to follow this regulation, and wanted to show the remediated site is "clean," would you want to take lots of soil samples or very few soil samples?

4.17. Suppose you can model the distribution of a chemical in surface water as a lognormal distribution, where the mean of the log-transformed data is 1 and the standard deviation of the log-transformed data is 2. What is the median of this distribution? What is the mean of this distribution? What is the CV of this distribution?

4.18. Generate 10 random numbers from a N(10, 2) distribution. Compute the sample mean, median, standard deviation, skew, and kurtosis and compare them to the true values.

4.19. Repeat Exercise 4.18 using 20, 50, 100, and 1000 random numbers.

4.20. Suppose a regulation is promulgated specifying that an air quality standard cannot be exceeded more than once per year *on average*. Further, suppose that for a particular emission from a factory, the probability that the emission concentration will exceed the standard on any given day is 1/365.

    a. Consider a sample of one year of emissions data. Assume emission concentrations are independent from day to day. Using the binomial distribution to model these data, what is the expected number of exceedances?

    b. Given your answer in part **a**, does the distribution of emissions satisfy the regulation?

    c. What is the probability that the standard will be exceeded at least twice in any given year?

4.21. Redo Exercise 4.20, but assume a Poisson model in which the unit of time is one year and the mean is given by $\lambda = np = 365*(1/365) = 1$ (i.e., one exceedance per year).

4.22. Using the Poisson model of Exercise 4.21, what is the probability that the standard will be exceeded at least four times in 3 years?

4.23. In the field of hydrology, the $100p^{th}$ percentile x_p is often called the $100(1-p)$ *percent exceedance event* because it will be exceeded with probability $1-p$ (Stedinger et al., 1993). For example, the 99th percentile is the 1% exceedance event. Often, hydrologists specify the *return period* (sometimes called the *recurrence interval*) instead of the exceedance probability. For example, the 99th percentile of annual maximum flood-flow is the same as the 100-year flood, because each year there is a 1 in 100 chance that it will be exceeded, and the expected number of times it will be exceeded in 100 years is once. In general, x_p is the T-year flood for $T = 1/(1-p)$ (Stedinger et al., 1993).

    a. Derive the formula for p, given the value of T.

    b. Assuming annual stream flow can be modeled as a normal distribution with a mean of 4,500 (cfs) and a standard deviation of 1,000 (cfs), find the values of the 25-year, 50-year, and 100-year flood.

5   ESTIMATING DISTRIBUTION PARAMETERS AND QUANTILES

Letting the Data Shape the Model

Chapters 1 and 2 discussed the ideas of a population and a sample. Chapter 4 described probability distributions, which are used to model populations. Based on using the graphical tools discussed in Chapter 3 to look at your data, and based on your knowledge of the mechanism producing the data, you can model the data from your sampling program as having come from a particular kind of probability distribution. Once you decide on what probability distribution to use (if any), you usually need to estimate the parameters associated with that distribution. For example, you may need to compare the mean or 95th percentile of the concentration of a chemical in soil, groundwater, surface water, or air with some fixed standard. This chapter discusses methods of estimating distribution parameters and quantiles for various probability distributions (as well as constructing confidence intervals for these quantities), and how to use ENVIRONMENTALSTATS for S-PLUS to implement these methods.

METHODS FOR ESTIMATING DISTRIBUTION PARAMETERS

The three most frequently used methods for estimating the parameters of probability distributions are: the method of moments estimator (MME), the maximum likelihood estimator (MLE), and the minimum variance unbiased estimator (MVUE). More recently, *L*-moment estimators have been introduced and used in the hydrological literature. Sometimes these methods produce slightly different formulas for the estimators, but often they produce the same formula. This section briefly explains the ideas behind MMEs, MLEs, and MVUEs. The theory of *L*-moments is discussed in detail in the ENVIRONMENTALSTATS for S-PLUS help file Glossary: *L*-Moments.

Estimators Are Random Variables

Throughout this section, we will assume that X denotes a random variable and x denotes the observed value of this random variable based on a sample. Thus, x_1, x_2, ..., x_n denote n observations from a particular probability distribution. All estimators use the observed values from the sample to estimate the population parameter(s) (e.g., the sample mean estimates the population mean). Formally, if we let $\hat{\theta}$ denote an estimator of some population parameter θ, then

$$\hat{\theta} = h\left(x_1, x_2, \ldots, x_n\right) \tag{5.1}$$

where $h()$ denotes some function. For the sample mean, this function $h()$ is given by:

$$h\left(x_1, x_2, \ldots, x_n\right) = \frac{1}{n} \sum_{i=1}^{n} x_i \tag{5.2}$$

Because an estimator is a function of random variables, it itself is a random variable and has an associated probability distribution. This is something to keep in mind while we discuss the various methods of estimating parameters. We will talk more about this later in the section Comparing Different Estimators.

Method of Moments Estimators

A *method of moments estimator (MME)* is based on the simple idea of equating the sample moments based on the data with the moments of the probability distribution.

The Moments of a Distribution

The rth *moment* of the distribution of a random variable X is often denoted by:

$$\mu_r' = E\left(X^r\right) \tag{5.3}$$

(e.g., Evans et al., 1993, p. 13). The first moment of X is called the *mean* of X and is usually denoted simply by μ. That is:

$$\mu = \mu_1' = E\left(X\right) \tag{5.4}$$

The formula for the rth *sample moment* is:

$$\hat{\mu}'_r = \frac{1}{n} \sum_{i=1}^{n} x_i^r \qquad (5.5)$$

Thus, the method of moments estimator of the mean of a distribution is simply the sample mean:

$$\hat{\mu}_{mme} = \bar{x} \qquad (5.6)$$

where

$$\bar{x} = \frac{1}{n} \sum_{i=1}^{n} x_i \qquad (5.7)$$

The rth *central moment* of a random variable X is defined as:

$$\mu_r = E\left[(X - \mu)^r\right] \qquad (5.8)$$

(Evans et al., 1993, p. 13). The second central moment of X is called the *variance* of X and is usually denoted by σ^2. That is:

$$\sigma^2 = \mu_2 = E\left[(X - \mu)^2\right] \qquad (5.9)$$

The square root of the second central moment is called the **standard deviation** of X and is usually denoted by σ.

The formula for the rth **sample central moment** is:

$$\hat{\mu}_r = \frac{1}{n} \sum_{i=1}^{n} (x_i - \bar{x})^r \qquad (5.10)$$

Thus, the method of moments estimator of the variance of a distribution is:

$$\hat{\sigma}^2_{mme} = s^2_m = \frac{1}{n} \sum_{i=1}^{n} (x_i - \bar{x})^2 \qquad (5.11)$$

and the method of moments estimator of the standard deviation of a distribution is:

$$\hat{\sigma}_{mme} = s_m = \sqrt{\frac{1}{n} \sum_{i=1}^{n} (x_i - \bar{x})^2} \qquad (5.12)$$

The quantity in Equation (5.11) is sometimes called the sample variance, but more often the term "sample variance" refers to the unbiased estimator of variance shown in Equation (5.49) below. Similarly, the quantity in Equation (5.12) is sometimes called the sample standard deviation, but more often the term "sample standard deviation" refers to the square root of the unbiased estimator of variance.

Example 5.1: Method of Moment Estimators for a Normal Distribution

A normal distribution is characterized by its mean (μ) and variance (σ^2), or by its mean and standard deviation (σ). The method of moments estimators for the mean, variance, and standard deviation are given by:

$$\hat{\mu}_{mme} = \bar{x} \qquad (5.13)$$

$$\hat{\sigma}^2_{mme} = s^2_m \qquad (5.14)$$

$$\hat{\sigma}_{mme} = s_m \qquad (5.15)$$

Example 5.2: Method of Moments Estimators for a Lognormal Distribution

A lognormal distribution can be characterized by its mean (θ) and standard deviation (η), by its mean (θ) and coefficient of variation (τ), or by the mean of the log-transformed random variable (μ) and the standard deviation of the log-transformed random variable (σ). (See Equations (4.32) to (4.35) in Chapter 4 for the relationship between these different parameters.) The method of moments estimators for the mean, variance, and standard deviation are the same as for a normal distribution:

$$\hat{\theta}_{mme} = \bar{x} \tag{5.16}$$

$$\hat{\eta}^2_{mme} = s^2_m \tag{5.17}$$

$$\hat{\eta}_{mme} = s_m \tag{5.18}$$

The method of moments estimators for the mean, variance, and standard deviation of the log-transformed random variable are exactly the same, except they are based on the log-transformed observations:

$$\hat{\mu}_{mme} = \bar{y} \tag{5.19}$$

$$\hat{\sigma}^2_{mme} = s^2_{y,m} \tag{5.20}$$

$$\hat{\sigma}_{mme} = s_{y,m} \tag{5.21}$$

where

$$y_i = \log\left(x_i\right), \quad i = 1, 2, \ldots, n \tag{5.22}$$

$$\bar{y} = \frac{1}{n} \sum_{i=1}^{n} y_i \qquad (5.23)$$

$$s_{y,m}^2 = \frac{1}{n} \sum_{i=1}^{n} (y_i - \bar{y})^2 \qquad (5.24)$$

$$s_{y,m} = \sqrt{\frac{1}{n} \sum_{i=1}^{n} (y_i - \bar{y})^2} \qquad (5.25)$$

Example 5.3: Method of Moment Estimators for a Binomial Distribution

A binomial distribution is characterized by two parameters: n, the number of trials, and p, the probability of success for each trial. The single random variable X denotes the number of successes in n trials. The parameter n is set in advance, so the only parameter to estimate is p. As we saw in Chapter 4, if X has a $B(n, p)$ distribution, then the mean of X is

$$E(X) = \mu = np \qquad (5.26)$$

which can be rewritten as:

$$p = \frac{\mu}{n} \qquad (5.27)$$

For a single observation x of the random variable X, the sample mean is the same as x, so the method of moments estimator of p is:

$$\hat{p}_{mme} = \frac{x}{n} \qquad (5.28)$$

That is, the method of moments estimator of the probability of "success" is simply the number of successes divided by the number of trials.

Another way to look at this estimator is as follows. The random variable X can be thought of as the sum of n independent $B(1, p)$ random variables. That is

$$X = \sum_{i=1}^{n} X_i \qquad (5.29)$$

where X_i denotes the outcome for the ith trial (1 = success, 0 = failure). Using this notation, the method of moments estimator of p can be written as:

$$\hat{P}_{mme} = \frac{X}{n} = \frac{\sum_{i=1}^{n} x_i}{n} = \overline{x} \qquad (5.30)$$

Maximum Likelihood Estimators

A *maximum likelihood estimator (MLE)* is based on the simple idea of using the value of the parameter that yields the largest probability of seeing that particular outcome. For example, if you flip a (possibly weighted) coin 100 times and record 80 heads and 20 tails, does it seem more likely that the probability of a head is 50% or 80%? If in fact the probability of a head is 0.5, the likelihood (probability) of seeing exactly 80 heads out of 100 flips is about 4×10^{-10} (i.e., 4 in 10 billion). On the other hand, if the probability of a head is 0.8, the likelihood of seeing exactly 80 heads out of 100 flips is about 0.1.

Figure 5.1 shows the probability densities of a B(100, 0.5) and a B(100, 0.8) distribution. You can see from this figure that getting something close to 80 heads out of 100 flips is much more likely for the B(100, 0.8) distribution than for a B(100, 0.5) distribution.

The Likelihood Function

To compute probabilities using probability density functions, we assume the distribution parameter(s) are known and the observations have yet to be taken. The method of maximum likelihood estimation reverses these assumptions: the observations have been taken and are known, and the distribution parameter(s) are unknown. In this case, the probability density func-tion turns into the *likelihood function*.

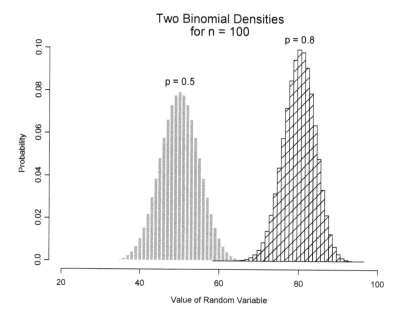

Figure 5.1   Probability density of a B(100, 0.5) and B(100, 0.8) random variable

For example, consider the binomial distribution. As we saw in Chapter 4, the probability density function for a B(n, p) random variable is:

$$f\left(x \mid p\right) = \binom{n}{x} p^x \left(1 - p\right)^{n-x} \tag{5.31}$$

where x takes on integer values between 0 and n, and p is a fixed number between 0 and 1. In the above equation, the left-hand side can be read as "the probability that $X = x$, given the value of p." Now assume we have flipped a coin n times and recorded x heads. The likelihood function is:

$$f\left(p \mid x\right) = \binom{n}{x} p^x \left(1 - p\right)^{n-x} \tag{5.32}$$

where p is some number between 0 and 1, and x is a fixed integer. For example, if we flipped a coin 100 times and recorded 80 heads, the likelihood function becomes:

$$f\left(p \mid x = 80\right) = \binom{100}{80} p^{80} \left(1 - p\right)^{20} \qquad \text{(5.33)}$$

We can use calculus to maximize the likelihood function relative to p. For our example with 80 heads out of 100 flips, the maximum likelihood estimate (MLE) of p is 80%. In general, for a binomial distribution, the MLE of p is given by:

$$\hat{p}_{mle} = \frac{x}{n} \qquad \text{(5.34)}$$

which is exactly the same as the method of moments estimator.

The above discussion uses only one observation from a binomial distribution. In general, when we have n independent observations from some distribution with pdf $f()$, the likelihood function is simply the product of the individual likelihood functions. (Recall that the probability of two independent events is the product of the probabilities.) So a general formula for the likelihood function is:

$$f\left(\theta \mid x_1, x_2, \ldots, x_n\right) = \prod_{i=1}^{n} f\left(\theta \mid x_i\right) \qquad \text{(5.35)}$$

where θ denotes the vector of parameters associated with the distribution (e.g., mean and standard deviation). Often it is easier to maximize the logarithm of the likelihood function (called the *log-likelihood*), rather than the likelihood function itself. A general formula for the log-likelihood is:

$$\log\left[f\left(\theta \mid x_1, x_2, \ldots, x_n\right) \right] = l\left(\theta \mid x_1, x_2, \ldots, x_n\right)$$

$$\qquad \text{(5.36)}$$

$$= \log\left[\prod_{i=1}^{n} f\left(\theta \mid x_i\right) \right] = \sum_{i=1}^{n} \log\left[f\left(\theta \mid x_i\right) \right]$$

Example 5.4: Maximum Likelihood Estimators for a Normal Distribution

Suppose we have n observations from a $N(\mu, \sigma)$ distribution and we want to estimate the parameters of this distribution. The log-likelihood is:

$$\log \left[f \left(\mu, \sigma \mid x_1, x_2, \ldots, x_n \right) \right] = \log \left[\prod_{i=1}^{n} f \left(\mu, \sigma \mid x_i \right) \right]$$

$$(5.37)$$

$$= -\frac{n}{2} \log \left(2\pi\sigma^2 \right) - \frac{1}{2\sigma^2} \sum_{i=1}^{n} \left(x_i - \mu \right)^2$$

We can use calculus to maximize this function first relative to μ, then relative to σ^2. The MLEs of the mean and variance are:

$$\hat{\mu}_{mle} = \overline{x} \qquad\qquad (5.38)$$

$$\hat{\sigma}^2_{mle} = s^2_m \qquad\qquad (5.39)$$

The MLE of any monotonic function of a distribution parameter is simply the value of the function evaluated at the MLE. Since the square root function is a monotonic function, the MLE of the standard deviation is:

$$\hat{\sigma}_{mle} = \sqrt{\hat{\sigma}^2_{mle}} = s_m \qquad\qquad (5.40)$$

Thus, for a normal distribution, the maximum likelihood estimators are exactly the same as the method of moment estimators.

Example 5.5: Maximum Likelihood Estimators for a Lognormal Distribution

For a lognormal distribution, we can compute the MLEs of the mean and variance for the log-transformed observations, then use the relationship between these parameters and the mean and CV for the untransformed observations (see Equations (4.32) to (4.35) in Chapter 4) to compute the MLEs on the original scale. Recall the relationship between the mean (μ) and standard deviation (σ) on the log-scale and the mean (θ), coefficient of variation (τ), and standard deviation (η) on the original scale:

$$\theta = \exp\left(\mu + \frac{\sigma^2}{2}\right) \tag{5.41}$$

$$\tau = \sqrt{\exp\left(\sigma^2\right) - 1} \tag{5.42}$$

$$\eta = \theta\,\tau \tag{5.43}$$

The maximum likelihood estimators of the mean (θ), coefficient of variation (τ), and standard deviation (η) on the original scale are given by:

$$\hat{\theta}_{mle} = \exp\left(\hat{\mu}_{mle} + \frac{\hat{\sigma}^2_{mle}}{2}\right)$$

$$\tag{5.44}$$

$$= \exp\left(\bar{y} + \frac{s^2_{y,m}}{2}\right)$$

$$\hat{\tau}_{mle} = \sqrt{\exp\left(\hat{\sigma}^2_{mle}\right) - 1}$$

$$\tag{5.45}$$

$$= \sqrt{\exp\left(s^2_{y,m}\right) - 1}$$

$$\hat{\eta}_{mle} = \hat{\theta}_{mle}\,\hat{\tau}_{mle} \tag{5.46}$$

where $\bar{y}$ and $s^2_{y,m}$ are defined in Equations (5.22) to (5.24). Unlike the normal distribution, for the lognormal distribution the MLEs of the mean and standard deviation are not the same as the method of moments estimators shown in Equations (5.16) and (5.18).

Minimum Variance Unbiased Estimators

The **bias** or **systematic error** of an estimator is defined to be the difference between the average value of the estimator (remember, estimators are themselves random variables) and the parameter it is estimating. Technically, if $\hat{\theta}$ is an estimator of a population parameter θ, then

$$Bias\left(\hat{\theta}\right) = E\left(\hat{\theta}\right) - \theta \qquad \textbf{(5.47)}$$

If the average value of an estimator equals the population parameter, then the bias is equal to 0 and the estimator is said to be **unbiased**. If the average value of the estimator is less than the population parameter, the estimator is said to be negatively biased. If the average value of the estimator is greater than the population parameter, the estimator is said to be positively biased. An estimator with negative or positive bias is called **biased**.

Unbiased Estimators of the Mean and Variance

As we saw in Chapter 4, the sample mean is an unbiased estimator of the population mean. That is

$$E\left(\bar{x}\right) = E\left(X\right) = \mu \qquad \textbf{(5.48)}$$

An unbiased estimator of the population variance is given by:

$$s^2 = \frac{1}{n-1} \sum_{i=1}^{n} \left(x_i - \bar{x}\right)^2 \qquad \textbf{(5.49)}$$

The quantity s^2 is often called the **sample variance**. The square root of this quantity is often called the **sample standard deviation** and is denoted s. Note that s is *not* an unbiased estimator of the population standard deviation σ.

$$s = \sqrt{\frac{1}{n-1} \sum_{i=1}^{n} (x_i - \bar{x})^2}$$ (5.50)

The unbiased estimator of the variance is always larger than the method of moments estimator of variance shown in Equation (5.11):

$$s^2 = \frac{n}{n-1} s_m^2$$ (5.51)

Definition of Minimum Variance Unbiased Estimator

For a particular probability distribution, there may be several possible estimators of a parameter that are unbiased. A **minimum variance unbiased estimator (MVUE)** has the smallest variance in the class of all unbiased estimators. The technical details of deriving MVUEs are discussed in basic text books on mathematical statistics.

Example 5.6: Minimum Variance Unbiased Estimators for a Normal Distribution

The minimum variance unbiased estimators for the mean and variance of a normal distribution are given by the sample mean and variance defined in Equations (5.7) and (5.49). That is,

$$\hat{\mu}_{mvue} = \bar{x}$$ (5.52)

$$\hat{\sigma}^2_{mvue} = s^2$$ (5.53)

Example 5.7: Minimum Variance Unbiased Estimators for a Lognormal Distribution

The minimum variance unbiased estimators of the mean and variance of a lognormal distribution (original scale) were derived by Finney (1941) and are discussed in Gilbert (1987, pp. 164–167) and Cohn et al. (1989). These estimators are computed as:

$$\hat{\theta}_{mvue} = e^{\bar{y}} g_{n-1} \left(\frac{s_y^2}{2} \right) \qquad (5.54)$$

$$\hat{\eta}_{mvue}^2 = e^{2\bar{y}} \left\{ g_{n-1} \left(2s_y^2 \right) - g_{n-1} \left[\frac{(n-2) s_y^2}{n-1} \right] \right\} \qquad (5.55)$$

where $\bar{y}$ is defined in Equations (5.22) and (5.23), and

$$s_y^2 = \frac{1}{n-1} \sum_{i=1}^{n} \left(y_i - \bar{y} \right)^2 \qquad (5.56)$$

$$g_m(z) = \sum_{i=0}^{\infty} \frac{m^i (m + 2i)}{m (m + 2)(m + 2i)} \left(\frac{m}{m+1} \right)^i \left(\frac{z^i}{i!} \right) \qquad (5.57)$$

Note that s_y^2 is simply the MVUE of σ^2 (the variance of the log-transformed random variable).

Example 5.8: Minimum Variance Unbiased Estimator for a Binomial Proportion

The MVUE of the parameter p of a binomial distribution is exactly the same as the method of moments and the maximum likelihood estimator:

$$\hat{p}_{mvue} = \frac{x}{n} \qquad (5.58)$$

(See Equations (5.28) and (5.34).)

USING ENVIRONMENTALSTATS FOR S-PLUS TO ESTIMATE DISTRIBUTION PARAMETERS

With ENVIRONMENTALSTATS for S-PLUS, you can estimate distribution parameters based on a sample for almost all of the probability distributions discussed in Chapter 4. Here are three examples.

Example 5.9: Estimating Parameters of a Normal Distribution

Recall that in Chapter 3 we saw that the Reference area TcCB data appeared to come from a lognormal distribution. Using the data frame `new.epa.94b.tccb.df` we created in that chapter, we can estimate the mean and standard deviation of the log-transformed data. These estimates are approximately –0.62 and 0.47 log(ppb). Figure 5.2 shows a density histogram of the log-transformed Reference area TcCB data, along with the fitted normal distribution based on these estimates.

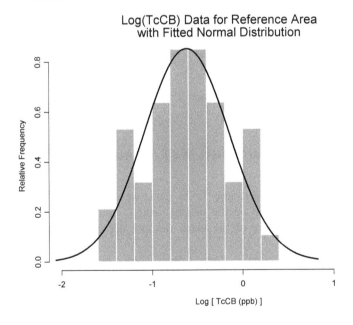

Figure 5.2   Histogram of log-transformed Reference area TcCB data with fitted normal distribution

Menu

To estimate the mean and standard deviation of the log-transformed Reference area TcCB concentrations using the ENVIRONMENTALSTATS for S-PLUS pull-down menu, follow these steps.

1. In the Object Explorer, make sure **new.epa.94b.tccb.df** is high-lighted.
2. On the S-PLUS menu bar, make the following menu choices: **EnvironmentalStats>Estimation>Parameters**. This will bring up the Estimate Distribution Parameters dialog box.
3. For Data to Use, make sure the **Pre-Defined Data** button is selected.
4. In the Data Set box, make sure **new.epa.94b.tccb.df** is selected.
5. In the Variable box, select **log.TcCB**.
6. In the Subset Rows with box, type **Area=="Reference"**.
7. In the Distribution/Estimation section, make sure **Normal** is selected in the distribution box, and **mvue** is selected in the Estimation Method box.
8. Under the Confidence Interval section, **uncheck** the Confidence Interval box. Click **OK** or **Apply**.

Command

To estimate the mean and standard deviation of the log-transformed Reference area TcCB concentrations using the ENVIRONMENTALSTATS for S-PLUS Command or Script Window, type these commands.

```
attach(new.epa.94b.tccb.df)
enorm(log.TcCB[Area=="Reference"])
detach()
```

Example 5.10: Estimating Parameters of a Lognormal Distribution

Again looking at the Reference area TcCB data, we can estimate the mean and coefficient of variation on the original scale. Table 5.1 below compares the different estimators of the mean (θ) and the coefficient of variation (τ).

Rounding to the nearest hundredth, there is no difference between any of the estimates of the mean, and all of the estimates of the coefficient of variation are the same except for the MME. The method denoted QMLE stands for "Quasi-Maximum Likelihood Estimator" and is exactly the same as the MLE, except that the MLE of σ^2 shown in Equation (5.24) and used in Equations (5.44) and (5.45) is replaced with the more commonly used MVUE of σ^2 shown in Equation (5.56) (see Cohn et al., 1989). Cohn et al. (1989) and Parkin et al. (1988) have shown that the QMLE and the MLE of the mean can be severely biased for typical environmental data, and suggest always using the MVUE. Figure 5.3 shows a density histogram of the Reference area TcCB data along with the fitted lognormal distribution based on the MVUE estimates.

Parameter	Estimator	Value
Mean (θ)	MME	0.60
"	MLE	0.60
"	QMLE	0.60
"	MVUE	0.60
CV (τ)*	MME	0.47
"	MLE	0.49
"	QMLE	0.49
"	MVUE	0.49

Table 5.1   Comparison of different estimators for the mean and coefficient of variation, assuming a lognormal distribution for the Reference area TcCB
data. Estimates are rounded to the nearest hundredth.

* Estimated CV is the ratio of the estimated standard deviation to the
estimated mean, with these estimates computed based on the specified
method.

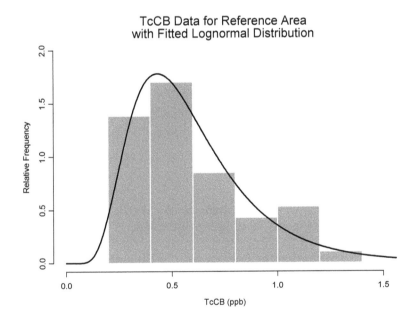

Figure 5.3   Histogram of Reference area TcCB data with fitted lognormal distribution based on MVUEs

Menu

To estimate the mean and coefficient of variation of the Reference area TcCB concentrations using the ENVIRONMENTALSTATS for S-PLUS pull-down menu, follow these steps.

1. In the Object Explorer, make sure **epa.94b.tccb.df** is highlighted.
2. On the S-PLUS menu bar, make the following menu choices: **EnvironmentalStats>Estimation>Parameters**. This will bring up the Estimate Distribution Parameters dialog box.
3. For Data to Use, make sure the **Pre-Defined Data** button is selected.
4. In the Data Set box, make sure **epa.94b.tccb.df** is selected.
5. In the Variable box, select **TcCB**.
6. In the Subset Rows with box, type **Area=="Reference"**.
7. In the Distribution/Estimation section, make sure **Lognormal (Alternative)** is selected in the distribution box, and **mvue** is selected in the Estimation Method box.
8. Under the Confidence Interval section, **uncheck** the Confidence Interval box. Click **OK** or **Apply**.

To use the other methods of estimation, in Step 7 above, select something other than **mvue** in the Estimation Method box.

Command

To estimate the mean and coefficient of variation of the Reference area TcCB concentrations using the ENVIRONMENTALSTATS for S-PLUS Command or Script Window, type these commands.

```
attach(epa.94b.tccb.df)
elnorm.alt(TcCB[Area=="Reference"])
detach()
```

By default, the function `elnorm.alt` computes the MVUEs. You can use the argument `method` to specify other types of estimators.

Example 5.11: Estimating the Parameter of a Binomial Distribution

The guidance document USEPA (1992c, p. 36) contains observations on benzene concentrations (ppb) in groundwater from six background wells sampled monthly for 6 months. The data are shown in Table 5.2 and are stored in the data frame `epa.92c.benzene1.df` in ENVIRONMENTALSTATS for S-PLUS. Nondetect values are reported as "<2." Of the 36 values, 33 are nondetects. Here, we will use these data to estimate the probability of observing a nondetect value at any of the six wells, which is about 92%.

Month	Well 1	Well 2	Well 3	Well 4	Well 5	Well 6
1	<2	<2	<2	<2	<2	<2
2	<2	<2	<2	15.0	<2	<2
3	<2	<2	<2	<2	<2	<2
4	<2	12.0	<2	<2	<2	<2
5	<2	<2	<2	<2	<2	10.0
6	<2	<2	<2	<2	<2	<2

Table 5.2   Benzene data (ppb) from groundwater monitoring wells (USEPA, 1992c, p. 36)

Menu

To estimate the probability of observing a nondetect value using the ENVIRONMENTALSTATS for S-PLUS pull-down menu, follow these steps.

1. In the Object Explorer, find the data frame **epa.92c.benzene1.df** and then make sure it is highlighted.
2. On the S-PLUS menu bar, make the following menu choices: **EnvironmentalStats>Estimation>Parameters**. This will bring up the Estimate Distribution Parameters dialog box.
3. For Data to Use, make sure the **Pre-Defined Data** button is selected.
4. In the Data Set box, make sure **epa.92c.benzene1.df** is selected.
5. In the Variable box, select **Censored**.
6. In the Distribution/Estimation section, select **Binomial** in the distribution box.
7. Under the Confidence Interval section, **uncheck** the Confidence Interval box. Click **OK** or **Apply**.

Command

To estimate the probability of observing a nondetect value using the ENVIRONMENTALSTATS for S-PLUS Command or Script Window, type this command.

```
ebinom(epa.92c.benzene1.df$Censored)
```

COMPARING DIFFERENT ESTIMATORS

As we stated earlier in this chapter, all estimators are functions of random variables (the observations in the sample), and are therefore themselves random variables. They will rarely be exactly equal to the parameter they are trying to estimate. Instead, they will "bounce around" the parameter they are estimating. Figure 5.4 displays the distribution of the sample mean for a N(10, 2) distribution when the sample sizes are $n = 4$ and $n = 30$.

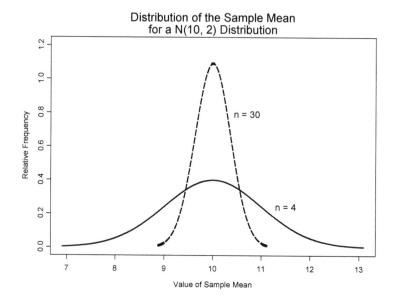

Figure 5.4   Distribution of the sample mean for a N(10, 2) distribution based on sample sizes of n = 4 and n = 30

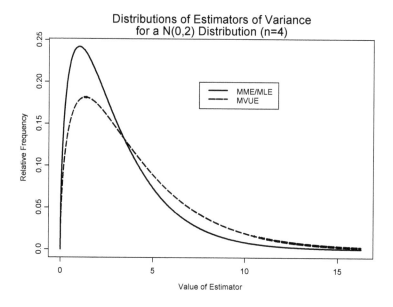

Figure 5.5   Distribution of the method of moments/maximum likelihood estimator of variance for a N(0, 2) distribution, based on a sample size of 4

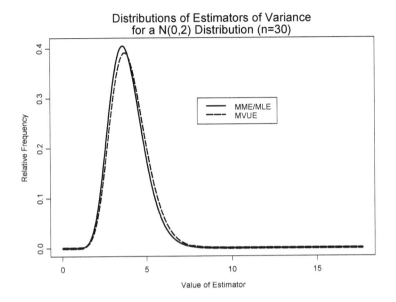

Figure 5.6   Distribution of the minimum variance unbiased estimator of variance for
a N(0, 2) distribution, based on a sample size of 30

Figure 5.5 displays the distributions of the two different estimators of variance for a N(0, 2) distribution (the MME/MLE and the MVUE), based on a sample size of $n = 4$. Figure 5.6 shows the same thing for a sample size of $n = 30$. (In these two figures, the true value of the parameter we are trying to estimate is the variance, which is $2^2 = 4$.) Here we see that for the small sample size there is a bit of difference in the distributions of the two estimators. By the time the sample size reaches 30, however, there is very little difference in the distributions. As the sample size increases, the two distributions converge (see Equation (5.51)).

Mean Square Error (MSE)

Given that there are several ways to estimate the parameters of a probability distribution, how should you decide which method to use? One criterion that is commonly used to compare estimators is the ***mean square error (MSE)*** (Fisher and van Belle, 1993, p. 126). If $\hat{\theta}$ is an estimator of some parameter θ, its mean square error is defined as:

$$MSE\left(\hat{\theta}\right) = E\left[\left(\hat{\theta} - \theta\right)^2\right] = Var\left(\hat{\theta}\right) + \left[Bias\left(\hat{\theta}\right)\right]^2 \quad \textbf{(5.59)}$$

That is, the mean square error is the average of the square of the distance between the estimator and the parameter it is estimating. Note that for an unbiased estimator, the MSE is the same as its variance.

Example 5.12: A Study from the Environmental Literature

Gilliom and Helsel (1986) studied the performance of various estimators of the mean, standard deviation, median, and interquartile range for a number of parent distributions where the data sets were subjected to Type I left single censoring (we will talk more about censoring in Chapter 10). Based on an analysis of hundreds of data sets for trace constituents, the four parent distributions they considered were lognormal, contaminated lognormal (mixture of two lognormals), gamma, and delta (zero-modified lognormal). They used the root mean square error (RMSE) to compare the performance of the estimators.

Example 5.13: MSE of the Sample Mean vs. the Sample Median for a Normal Distribution

Suppose we have a random sample of n observations from a N(μ, σ) distribution. We can estimate the population mean μ with either the sample mean or the sample median (remember that for a normal distribution, the population mean and median are the same). Which estimator is better?

We saw in Chapter 4 and this chapter that the expected value of the sample mean is the same as the population mean (no matter what the underlying distribution), so the bias of the sample mean is 0. Thus, the mean squared error of the sample mean is the same as the variance of the sample mean and is given by:

$$MSE\left(\overline{x}\right) \ = \ Var\left(\overline{x}\right) \ = \ \frac{\sigma^2}{n} \tag{5.60}$$

(See Chapter 4 for the derivation of the variance of the sample mean, no matter what the underlying distribution.)

For a normal distribution, the sample median is also an unbiased estimator of the population mean, so its MSE is also equal to its variance, which is approximately equal to:

$$MSE\left(Median\right) \ = \ Var\left(Median\right) \ \approx \ \frac{\pi}{2}\frac{\sigma^2}{n} \tag{5.61}$$

(Johnson et al., 1994, p. 123). Therefore, the ratio of the MSE of the sample mean to the MSE of the sample median is approximately:

$$\frac{MSE\left(\bar{x}\right)}{MSE\left(Median\right)} \approx \frac{2}{\pi} \tag{5.62}$$

Thus, for a normal distribution the sample median is only about 63.7% as efficient as the sample mean for estimating the population mean.

Example 5.14: MSEs of the Sample Variances for a Normal Distribution

Suppose we have a random sample of n observations from a $N(\mu, \sigma)$ distribution. We can estimate the population variance σ^2 with either the minimum variance unbiased estimator s^2 shown in Equation (5.49), or the method of moments/maximum likelihood estimator s_m^2 shown in Equation (5.11). Which estimator is better?

The MVUE s^2 is unbiased, so its MSE is the same as its variance, which is given by:

$$MSE\left(s^2\right) = Var\left(s^2\right) = \frac{2\sigma^4}{n-1} \tag{5.63}$$

(Johnson et al., 1994, p. 128). The MME/MLE s_m^2 is not unbiased. Its expected value is given by:

$$E\left(s_m^2\right) = \frac{\left(n-1\right)}{n}\sigma^2 \tag{5.64}$$

hence its bias is given by:

$$Bias\left(s_m^2\right) = \frac{\left(n-1\right)}{n}\sigma^2 - \sigma^2 = -\frac{\sigma^2}{n} \tag{5.65}$$

The variance of this estimator is given by:

$$Var\left(s_m^2\right) = \frac{(n-1)}{n^2} 2\sigma^4 \qquad (5.66)$$

Therefore the mean square error of s_m^2 is:

$$MSE\left(s_m^2\right) = \frac{(n-1)}{n^2} 2\sigma^4 + \left(-\frac{\sigma^2}{n}\right)^2 \qquad (5.67)$$

$$= \frac{(2n-1)}{n^2} \sigma^4$$

The ratio of the MSE of s^2 compared to the MSE of s_m^2 is:

$$\frac{MSE\left(s^2\right)}{MSE\left(s_m^2\right)} = \left(\frac{2\sigma^4}{n-1}\right)\Big/ \left[\left(\frac{2n-1}{n^2}\right)\sigma^4\right] \qquad (5.68)$$

$$= \frac{2n^2}{2n^2 - 3n + 1}$$

Sample Size	Ratio of MSEs
2	2.67
3	1.80
4	1.52
5	1.39
10	1.17
25	1.06
50	1.03
100	1.02

Table 5.3   Ratio of mean square error for the MVUE to the mean square error for the MME/MLE for estimating the population variance of a normal distribution

Table 5.3 shows the values of this ratio for various values of the sample size n. Note that this ratio is always larger than 1, but converges to 1 as the sample size gets larger and larger. Thus, in terms of the mean square error, the estimator s_m^2 is slightly more efficient than s^2, even though the latter is usually the estimator that is used in practice.

Comparing Estimators Based on Confidence Intervals

In practice, a decision maker needs an estimate of a parameter *and* a measure of how much "wiggle" is in that estimate. You can use confidence intervals to quantify the degree of "wiggle" in a point estimate (see below). Given that confidence intervals are really what are needed and used in practice, it makes more sense to compare estimators by how well confidence intervals based on these estimators perform. One possible measure of performance is the mean or median width of the confidence interval.

There is not necessarily a one-to-one correspondence between how well an estimator does based on its MSE vs. how well it does based on measuring the performance of its confidence interval. This is because the confidence interval associated with the estimator depends on not only how closely the estimator "bounces" around the true parameter value, but also on how well the estimator of the variance of the estimator performs.

ACCURACY, BIAS, MEAN SQUARE ERROR, PRECISION, RANDOM ERROR, SYSTEMATIC ERROR, AND VARIABILITY

The terms accuracy, bias, mean square error, precision, random error, systematic error, and variability are used quite often in scientific papers and reports. It is very important to understand what the author really means when he or she uses these terms, and it is very important for the author to use these terms correctly. In this section we will review the meaning of the terms we know already, and talk about the new terms we have not yet introduced.

Bias and Systematic Error

The *bias* or *systematic error* of an estimator is a measure of how close, on average, that estimator comes to the true population parameter (Equation (5.47)). If the average value of the estimator equals the population parameter, the bias or systematic error is 0 and the estimator is called *unbiased*. If the average value of the estimator is less than the population parameter, the bias is negative and the estimator is said to be *negatively biased*. If the average value of the estimator is greater than the population parameter, the bias is positive and the estimator is said to be *positively biased*. An estimator with negative or positive bias is called *biased* and possesses systematic error.

Precision, Random Error, and Variability

The *precision* or *random error* or *variability* of an estimator is measured by its *variance*; it is a measure of how much the estimator "bounces around" its average value. An estimator is *precise* if it has a small variability, and is *imprecise* if it has a large amount of variability.

Mean Square Error and Root Mean Square Error

The *mean square error (MSE)* of an estimator is the average or mean of the squared distance between the estimator and parameter it is trying to estimate. It is a measure of how much the estimator "bounces around" the parameter it is supposed to be estimating. It can be shown that the mean square error of an estimator is equal to its variability (random error) plus the square of its bias (systematic error). The *root mean square error (RMSE)* is simply the square root of the mean square error and is preferred by some authors because it retains the original units.

Accuracy

There are two conflicting definitions for the term *accuracy* (Clark and Whitfield, 1994). Some definitions say accuracy measures the same thing as bias or systematic error: an estimator has high accuracy if it is unbiased or has a very small bias. An estimator has low accuracy if it has a very large bias (e.g., Sokal and Rolfe, 1981, p. 13; Fisher and van Belle, 1993, p. 125).

Other definitions say that accuracy is a measure of both bias and precision (systematic and random error): an estimator has high accuracy only if it has both small or no bias *and* high precision (e.g., Gilbert, 1987, pp. 11–12; Berthoux and Brown, 1994, pp. 11–12). Note that this definition is equivalent to the measure mean square error given above. That is, an estimator has high accuracy if it has a small mean square error.

Because of the ambiguity of the term "accuracy," it is safer to use the terms bias and precision and their associated meanings instead, as recommended by the American Society for Testing Materials (ASTM) and Clark and Whitfield (1994).

Example 5.15: Estimators with Varying Bias and Precision

Figure 5.7 illustrates the four possible combinations of extremes in bias (systematic error) and precision (random error or variability). In this simulation experiment, a random sample of 100 observations was taken from a N(10, 1) distribution, and four estimators of the population mean were computed. The simulation was run 500 times. The four estimators are:

$$\hat{\mu}_1 = \bar{x} = \frac{1}{100} \sum_{i=1}^{100} x_i$$

$$\hat{\mu}_2 = \frac{x_{(1)} + x_{(100)}}{2}$$

$$\hat{\mu}_3 = \frac{1}{50} \sum_{i=51}^{100} x_{(i)}$$

$$\hat{\mu}_4 = \frac{x_{(51)} + x_{(100)}}{2}$$

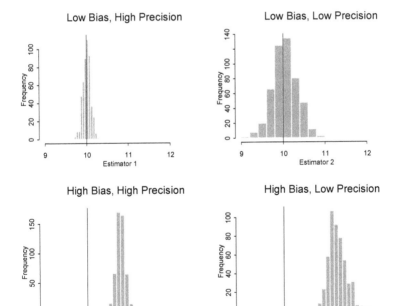

Figure 5.7   Result of simulation experiment showing the distribution of four different estimators of the mean of a N(10, 1) distribution

where $x_{(i)}$ denotes the ith ordered value in the sample. The first estimator is the sample mean, which is an unbiased estimator of the true mean. The second estimator is the average of the smallest and largest value, which is also an unbiased estimator of the true mean, but is more variable than the sample mean. The third estimator is the average of the 50 largest values. This is a biased estimator of the true mean. The fourth estimator is the average of the 51st largest value and the largest value. This is also a biased estimator of the true mean, and is more variable than the third estimator.

PARAMETRIC CONFIDENCE INTERVALS FOR DISTRIBUTION PARAMETERS

Decision makers need statistics to help them make intelligent decisions. Examples of decision makers in the environmental field include:

- A Federal, state, or local regulator in charge of soil cleanup, groundwater monitoring, or air quality.
- The manager of a plant trying to increase the efficiency of the process and decrease the amount of air pollution the plant emits.
- A consultant trying to decide whether a site owned by his or her client needs to be cleaned up.
- A member of the public who is attending a meeting concerning the risk of a proposed incinerator to be built in the neighborhood.

No matter what the circumstances, numbers should always be presented with some indication of their bias and precision (how much "wiggle" they have in them).

One way to help decision makers is to present an estimate of the population parameter (e.g., mean) along with a confidence interval for the parameter. A *confidence interval* for some population parameter θ is an interval on the real line constructed from a sample of n observations so that it *will* contain the parameter θ with some specified probability $(1-\alpha)100\%$, where α is some fraction between 0 and 1 (usually α is less than 0.5). The quantity $(1-\alpha)100\%$ is called the *confidence coefficient* or *confidence level* associated with the confidence interval.

Most confidence intervals used in practice are valid under the following assumptions:

1. The sample is some type of random sample.
2. The observations are independent of one another.

Violations of the second assumption are discussed in Chapters 11 and 12, which deal with methods for analyzing time series and spatial data.

A common third assumption used in constructing confidence intervals is:

3. The observations come from a specified probability distribution
    (e.g., normal, lognormal, binomial, Poisson, etc.).

Confidence intervals based on this last assumption are called ***parametric
confidence intervals***; they assume you know what kind of distribution describes the population. Confidence intervals that do not require this last assumption are called ***nonparametric confidence intervals***. We will talk more
about nonparametric confidence intervals in the section Nonparametric Confidence Intervals Based on Bootstrapping.

A confidence interval is a random interval; that is, the lower and/or upper bounds are random variables that are computed based on sample statistics. Prior to taking one specific sample, the probability that the confidence
interval *will* contain θ is $(1-\alpha)100\%$. Once a specific sample is taken and a
particular confidence interval is computed, the particular confidence interval
associated with that sample either contains θ or it does not. You usually do
not know whether the confidence interval actually contains θ or not because
you usually do not know what the value of θ is, which is why you are constructing a confidence interval for θ in the first place!

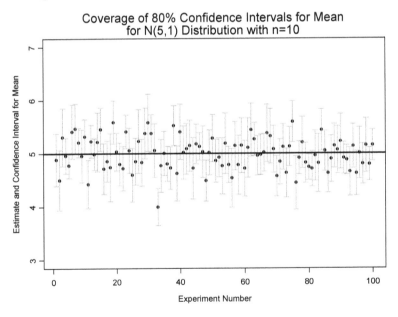

Figure 5.8   Results of simulation experiment showing the estimates and confidence
intervals for the population mean. The true mean is $\mu = 5$.

If an experiment is repeated N times, and for each experiment a sample is
taken and a $(1-\alpha)100\%$ confidence interval for θ is computed, then the number of confidence intervals that actually contain θ is a binomial random vari-

able with parameters $n = N$ and $p = 1-\alpha$, that is, it follows a B(N, $1-\alpha$) distribution. Figure 5.8 shows the results of such a simulated experiment in which a random sample of $n = 10$ observations was taken from a N(5, 1) distribution and an 80% confidence interval for the mean was constructed based on these 10 observations. The experiment was repeated 100 times. In this case, the actual number of confidence intervals that contained the true mean of $\mu = 5$ was 84.

Choosing the Confidence Level

So why bother with anything less than a 100% confidence interval? That way, you know that your interval will always contain the population parameter. As the saying goes, "there is no such thing as a free lunch." A 100% confidence interval must span the whole possible range of the population parameter (e.g., 0 to ∞ for the mean of a lognormal distribution, and 0 to 1 for the probability of success associated with a binomial distribution).

In general, the width of a confidence interval depends on the sample size, the amount of variability inherent in the population, and the confidence level that you choose (see Chapter 8). Increasing the sample size or decreasing the confidence level will make the confidence interval smaller. If you can improve your sampling scheme and handling protocol, you may be able to decrease the variability you see in the data and decrease the size of the confidence interval this way as well. There is no set rule for what confidence level you should choose; this is a subjective decision, but often 90% or 95% is used.

Confidence Interval for the Mean of a Normal Distribution

We saw in Chapter 4 that if we take a sample of size n from a normal distribution with mean μ and standard deviation σ, then the sample mean also follows a normal distribution with the following mean and standard deviation:

$$\mu_{\bar{x}} = E\left(\bar{x}\right) = \mu \tag{5.69}$$

$$\sigma_{\bar{x}} = Sd\left(\bar{x}\right) = \frac{\sigma}{\sqrt{n}} \tag{5.70}$$

Therefore, based on the z-transformation, the quantity

$$z = \frac{\overline{x} - \mu_{\overline{x}}}{\sigma_{\overline{x}}} = \frac{\overline{x} - \mu}{\sigma/\sqrt{n}} \qquad (5.71)$$

follows a standard normal distribution.

Letting z_p denote the pth quantile of the standard normal distribution and assuming $p < 0.5$, we can write the following probability statement.

$$\Pr\left[z_p \leq \frac{\overline{x} - \mu}{\sigma/\sqrt{n}} \leq z_{1-p}\right] = (1 - p) - p$$

$$(5.72)$$

$$= 1 - 2p$$

Setting $p = \alpha/2$, and recalling that the normal distribution is symmetric about its mean so that for a standard normal distribution $z_p = -z_{1-p}$, we can rewrite the above equation as:

$$\Pr\left[-z_{1-\alpha/2} \leq \frac{\overline{x} - \mu}{\sigma/\sqrt{n}} \leq z_{1-\alpha/2}\right] = 1 - \alpha \qquad (5.73)$$

Furthermore, using algebra, we can rewrite the above equation as:

$$\Pr\left[\overline{x} - z_{1-\alpha/2}\frac{\sigma}{\sqrt{n}} \leq \mu \leq \overline{x} + z_{1-\alpha/2}\frac{\sigma}{\sqrt{n}}\right] =$$

$$(5.74)$$

$$1 - \alpha$$

Therefore, when we know the value of the population standard deviation σ, a $(1-\alpha)100\%$ confidence interval for the mean μ is:

$$\left[\overline{x} - z_{1-\alpha/2}\frac{\sigma}{\sqrt{n}} , \overline{x} + z_{1-\alpha/2}\frac{\sigma}{\sqrt{n}}\right] \qquad (5.75)$$

Equation (5.75) is of little use in practice because we usually do not know the value of the population standard deviation σ. We can, however, use the sample standard deviation s of Equation (5.50) to estimate σ and modify the z-statistic in Equation (5.71) as follows:

$$t = \frac{\overline{X} - \mu_{\overline{X}}}{\hat{\sigma}_{\overline{X}}} = \frac{\overline{X} - \mu}{s/\sqrt{n}} \tag{5.76}$$

This statistic is called the t-statistic and follows a Student's t-distribution with $n-1$ degrees of freedom (Johnson et al., 1995, p. 362; Fisher and van Belle, 1993, p. 143; Zar, 1999, p. 92). Figure 5.9 shows the pdf of the t-distribution for various values of the degrees of freedom parameter ν. Like the standard normal distribution, the t-distribution is symmetric about 0. Also, as the degrees of freedom get larger and larger, the t-distribution approaches the standard normal distribution.

Probability Density of Student's t-Distribution

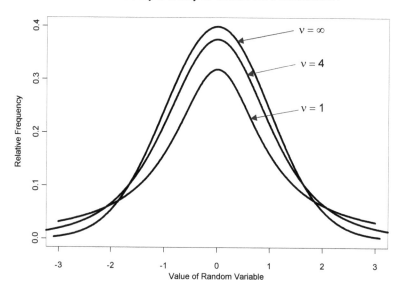

Figure 5.9   The probability density function of Student's t-distribution for 1, 4, and ∞ degrees of freedom

Letting $t_{\nu,p}$ denote the pth quantile of the t-distribution with ν degrees of freedom, we can modify Equation (5.74) to use s and t instead of σ and z to get:

$$\Pr\left[\bar{x} \; - \; t_{n-1,1-\alpha/2} \; \frac{s}{\sqrt{n}} \; \le \; \mu\right.$$

(5.77)

$$\left.\le \; \bar{x} \; + \; t_{n-1,1-\alpha/2} \; \frac{s}{\sqrt{n}}\right] \; = \; 1 \; - \; \alpha$$

so that a $(1-\alpha)100\%$ confidence interval for the mean μ is:

$$\left[\bar{x} \; - \; t_{n-1,1-\alpha/2} \; \frac{s}{\sqrt{n}} \; , \; \bar{x} \; + \; t_{n-1,1-\alpha/2} \; \frac{s}{\sqrt{n}}\right] \quad \text{(5.78)}$$

Equation (5.78) shows the formula for what is called a *two-sided confidence interval*. We can also create one-sided confidence intervals as well. The formula for the *one-sided upper confidence interval* is:

$$\left[-\infty \; , \; \bar{x} \; + \; t_{n-1,1-\alpha} \; \frac{s}{\sqrt{n}}\right] \quad \text{(5.79)}$$

and the formula for the *one-sided lower confidence interval* is:

$$\left[\bar{x} \; - \; t_{n-1,1-\alpha} \; \frac{s}{\sqrt{n}} \; , \; \infty\right] \quad \text{(5.80)}$$

Example 5.16: Confidence Interval for the Mean of Log-Transformed TcCB Concentrations

As an example of computing a confidence interval for the mean of a normal distribution, we will look at the Reference area TcCB data again, which appeared to come from a lognormal distribution. Using the data frame new.epa.94b.tccb.df we created in Chapter 3, we can estimate the mean and standard deviation of the log-transformed data, and also compute a confidence interval for the mean of the log-transformed data. As we saw in Example 5.9, the estimated mean and standard deviation are approximately −0.62 and 0.47 log(ppb). A 95% confidence interval for the mean is [−0.76, −0.48]. Figure 5.10 shows a density histogram of the log-transformed Reference area TcCB data, along with the fitted normal distribution *and* the confidence interval for the mean.

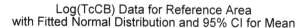

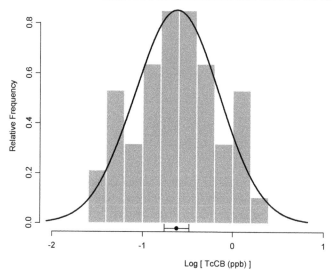

Figure 5.10   Histogram and fitted normal distribution for the log-transformed Reference area TcCB data, along with 95% confidence interval for the mean

Menu

To estimate the mean and standard deviation of the log-transformed Reference area TcCB concentrations and compute a 95% confidence interval for the mean using the ENVIRONMENTALSTATS for S-PLUS pull-down menu, follow these steps.

1. In the Object Explorer, make sure **new.epa.94b.tccb.df** is highlighted.
2. On the S-PLUS menu bar, make the following menu choices: **EnvironmentalStats>Estimation>Parameters**. This will bring up the Estimate Distribution Parameters dialog box.
3. For Data to Use, make sure the **Pre-Defined Data** button is selected.
4. In the Data Set box, make sure **new.epa.94b.tccb.df** is selected.
5. In the Variable box, select **log.TcCB**.
6. In the Subset Rows with box, type **Area=="Reference"**.
7. In the Distribution/Estimation section, make sure **Normal** is selected in the Distribution box, and **mvue** is selected in the Estimation Method box.

8. Under the Confidence Interval section, **check** the Confidence Interval box. In the CI Type box, select **two-sided**. In the CI Method box, select **exact**. In the Confidence Level (%) box, type **95**. In the Parameter box, select **mean**. Click **OK** or **Apply**.

Command

To estimate the mean and standard deviation of the log-transformed Reference area TcCB concentrations and create a 95% confidence interval for the mean using the ENVIRONMENTALSTATS for S-PLUS Command or Script Window, type these commands.

```
attach(new.epa.94b.tccb.df)
enorm(log.TcCB[Area=="Reference"], ci=T,
   conf.level=0.95)
detach()
```

Confidence Interval for the Variance of a Normal Distribution

Usually we are interested in estimating the population variance or standard deviation in order to quantify the amount of uncertainty we have in our estimate of the population mean. Sometimes, however, the estimate of variance and a confidence interval for the variance is of interest in itself. For example, in Chapter 8 we will talk about computing required sample sizes for various kinds of intervals and hypothesis tests, and these formulas involve the population variance. It is usually a good idea to use some kind of upper bound for the population variance when computing required sample sizes. One possible upper bound is the upper confidence limit for the variance.

For n independent observations from a normal distribution, it can be shown that the quantity

$$\chi^2 = \frac{(n-1)\,s^2}{\sigma^2} \tag{5.81}$$

follows a chi-square distribution with $n - 1$ degrees of freedom, where s^2 denotes the sample variance shown in Equation (5.49) (Fisher and van Belle, 1993, p. 117; Sheskin, 1997, p. 73; Zar, 1999, p. 111). Letting $\chi^2_{v,p}$ denote the pth quantile of a chi-square distribution with v degrees of freedom, and assuming $p < 0.5$, we can write the following probability statement:

$$1 - \alpha = Pr\left[\chi^2_{n-1,\alpha/2} \le \frac{(n-1)\,s^2}{\sigma^2} \le \chi^2_{n-1,1-\alpha/2}\right]$$

(5.82)

$$= Pr\left[\frac{(n-1)\,s^2}{\chi^2_{n-1,1-\alpha/2}} \le \sigma^2 \le \frac{(n-1)\,s^2}{\chi^2_{n-1,\alpha/2}}\right]$$

Hence, a two-sided $(1-\alpha)100\%$ confidence interval for the population variance σ^2 is given by:

$$\left[\frac{(n-1)\,s^2}{\chi^2_{n-1,1-\alpha/2}}\,,\,\frac{(n-1)\,s^2}{\chi^2_{n-1,\alpha/2}}\right]$$

(5.83)

Similarly, a one-sided upper $(1-\alpha)100\%$ confidence interval for the population variance is given by:

$$\left[-\infty\,,\,\frac{(n-1)\,s^2}{\chi^2_{n-1,\alpha}}\right]$$

(5.84)

and a one-sided lower $(1-\alpha)100\%$ confidence interval for the population variance is given by:

$$\left[\frac{(n-1)\,s^2}{\chi^2_{n-1,1-\alpha}}\,,\,\infty\right]$$

(5.85)

Because of the central limit theorem, the formulas for the confidence interval for the mean (Equations (5.78) to (5.80)) work well (i.e., the confidence level is maintained) even when the underlying population is not normally distributed, as long as the sample size is reasonably "large." The formulas for the confidence interval for the population variance, however, are extremely sensitive to departures from normality. For example, if the underlying population is skewed, the confidence interval will be too wide.

Example 5.17: Confidence Interval for the Variance of Log-Transformed TcCB Concentrations

We saw in Example 5.16 that the estimated mean and standard deviation for the log-transformed Reference area TcCB data are about −0.62 and 0.47 log(ppb). The estimated variance is 0.22, and a 95% upper confidence bound for the variance is 0.32.

Menu

To estimate the mean and standard deviation of the log-transformed Reference area TcCB concentrations and compute a one-sided upper 95% confidence interval for the variance using the ENVIRONMENTALSTATS for S-PLUS pull-down menu, follow the exact same steps as in Example 5.16, except in Step 8, in the CI Type box select **upper** and in the Parameter box select **Variance**.

Command

To estimate the mean and standard deviation of the log-transformed Reference area TcCB concentrations and create a one-sided upper 95% confidence interval for the variance using the ENVIRONMENTALSTATS for S-PLUS Command or Script Window, type these commands.

```
attach(new.epa.94b.tccb.df)
enorm(log.TcCB[Area=="Reference"], ci=T,
   ci.type="upper", conf.level=0.95,
   ci.param="variance")
detach()
```

General Formula for a Parametric Confidence Interval

The formula for the confidence interval for the mean of a normal distribution shown in Equations (5.78) to (5.80) are exact because the distribution of the sample mean for a normal distribution is itself normal, and because of the relationship between the normal distribution and Student's t-distribution. Formulas for exact confidence intervals are also available for other distributions such as the binomial, exponential, and Poisson distributions.

Often, however, a confidence interval for a parameter is based on a "normal approximation." That is, the distribution of the estimator is assumed to be approximately normal even though the estimator is a function of observations from some distribution other than a normal distribution. (Under certain conditions, which usually hold in practice, the distribution of any maximum likelihood estimator is asymptotically normal.)

In this case, the general formula for an approximate $(1-\alpha)100\%$ confidence interval for the population parameter θ is:

$$\left[\; \hat{\theta} - z_{1-\alpha/2} \; \hat{\sigma}_{\hat{\theta}} \; , \; \hat{\theta} + z_{1-\alpha/2} \; \hat{\sigma}_{\hat{\theta}} \; \right] \qquad (5.86)$$

where $\hat{\theta}$ denotes the estimator of θ, and $\hat{\sigma}_{\hat{\theta}}$ denotes an estimator of the standard deviation of $\hat{\theta}$. The quantity $\hat{\sigma}_{\hat{\theta}}$ is often called the **standard error** of the estimator $\hat{\theta}$.

A slightly more conservative formula for the confidence interval replaces the quantile from the normal distribution with the quantile from the Student's t-distribution:

$$\left[\; \hat{\theta} - t_{v,1-\alpha/2} \; \hat{\sigma}_{\hat{\theta}} \; , \; \hat{\theta} + t_{v,1-\alpha/2} \; \hat{\sigma}_{\hat{\theta}} \; \right] \qquad (5.87)$$

Here v denotes the degrees of freedom associated with the estimate of θ. This confidence interval will always be wider than the one shown in Equation (5.86). The difference in width diminishes as the sample size increases.

Confidence Interval for the Mean of a Lognormal Distribution

As we saw in Chapter 4, the lognormal distribution can be characterized by the mean (μ) and standard deviation (σ) of the log-transformed random variable, or by the mean (θ) and coefficient of variation (τ) of the original random variable. If you are interested in characterizing the log-transformed distribution, then simply take the logarithms of the original data and treat them like they come from a normal distribution: use the sample mean to estimate μ, and use the formula in Equation (5.78) to construct a two-sided confidence interval for μ.

Often, however, we want an estimate of the mean on the original scale, because this is what the decision maker wants to see or because this is what we need to do further analysis. For example, in modeling exposure to a toxic chemical from contact with soil at a contaminated site, it is often assumed that an individual will crisscross the site in a random pattern over time. In this way, the individual is exposed to the spatially averaged mean concentration (vs. say the median concentration), so the mean on the original scale is the parameter of interest (USEPA, 1992d; 1996c, p. 85).

Some textbooks (e.g., Zar, 1999, p. 275) suggest computing the confidence interval for the log-transformed data, then exponentiating the confidence limits to produce a confidence interval for the mean on the original scale. This method actually produces a confidence interval for the *median* of the distribution, not the mean (see the section Estimates and Confidence Intervals for Distribution Quantiles (Percentiles), below).

Several researchers have studied the problem of estimating and producing a confidence interval for the mean of a lognormal distribution, including Finney (1941), Bradu and Mundlak (1970), Land (1971), Land (1972), Land (1975), Cohn et al. (1989), Parkin et al. (1990), and Singh et al. (1997). We will review some of these methods here, and then look at some examples to discuss their advantages and disadvantages.

Land's Exact Method

Land (1971, 1975) derived a method for computing one-sided (lower or upper) uniformly most accurate unbiased confidence intervals for the mean θ. A two-sided confidence interval can be constructed by combining an optimal lower confidence limit with an optimal upper confidence limit. This procedure for two-sided confidence intervals is only asymptotically optimal, but for most purposes should be acceptable (Land, 1975, p. 387).

A one-sided upper $(1-\alpha)100\%$ confidence interval for the mean is given by:

$$\left\{ 0, \ \exp\left[\left(\bar{y} + \frac{s_y^2}{2} \right) + s_y \frac{H_{1-\alpha}}{\sqrt{n-1}} \right] \right\} \qquad (5.88)$$

a one-sided lower $(1-\alpha)100\%$ confidence interval for the mean is given by:

$$\left\{ \exp\left[\left(\bar{y} + \frac{s_y^2}{2} \right) + s_y \frac{H_\alpha}{\sqrt{n-1}} \right], \ \infty \right\} \qquad (5.89)$$

and a two-sided $(1-\alpha)100\%$ confidence interval for the mean is given by:

$$\left\{ \exp\left[\left(\bar{y} + \frac{s_y^2}{2} \right) + s_y \frac{H_{\alpha/2}}{\sqrt{n-1}} \right], \right.$$

$$(5.90)$$

$$\left. \exp\left[\left(\bar{y} + \frac{s_y^2}{2} \right) + s_y \frac{H_{1-\alpha/2}}{\sqrt{n-1}} \right] \right\}$$

where $\bar{y}$ and s_y^2 are defined in Equations (5.23) and (5.56), and the quantity H_p involves some complicated mathematics. Tabled values of this "H-statistic" are given in Land (1975) and Gilbert (1987, pp. 264–265). Land (1975) denotes this value with a C, but Gilbert (1987, pp. 169–171, 264–265) uses H, and some guidance documents refer to this quantity as the H-statistic (e.g., USEPA, 1992d), so H is used here for consistency with the environmental literature.

Other Methods

Other methods for computing a confidence interval for the mean of a lognormal distribution include those suggested by Parkin et al. (1990), Cox (Land, 1972; El-Shaarawi, 1989), and Gilbert (1987, p. 167). These methods are described in detail in the ENVIRONMENTALSTATS for S-PLUS help file for elnorm.alt.

Parkin et al. (1990) studied the performance of various methods for constructing a confidence interval for the mean of a lognormal distribution via Monte Carlo simulation. They compared approximate methods to Land's optimal method. They used four parent lognormal distributions to generate observations; all had mean 10, but differed in coefficient of variation: 50, 100, 200, and 500%. They also generated sample sizes from 6 to 100 in increments of 2. For each combination of parent distribution and sample size, they generated 25,000 Monte Carlo trials. Parkin et al. (1990) found that for small sample sizes ($n \leq 20$), none of the approximate methods worked very well. For $n > 20$, their method (Parkin et al., 1990) provided coverage reasonably close to the assumed coverage. Cox's method (Land, 1972) worked well for $n > 60$, and performed slightly better than Parkin et al.'s method for highly skewed populations.

Example 5.18: Confidence Interval for the Mean of Reference Area TcCB Concentrations

We saw in Example 5.10 that the estimated mean TcCB concentration in the Reference area is 0.60 ppb, based on the observed $n = 47$ concentrations. Table 5.4 displays two-sided 95% confidence intervals for the mean concentration based on various methods. These confidence intervals are almost identical to each other.

Menu

To estimate the mean and coefficient of variation of the Reference area TcCB concentrations and compute a 95% confidence interval for the mean using the ENVIRONMENTALSTATS for S-PLUS pull-down menu, follow these steps.

Confidence Interval	Method
[0.52, 0.70]	Land
[0.50, 0.74]	Parkin et al.
[0.52, 0.69]	Cox
[0.51, 0.68]	Normal Approximation

Table 5.4   Two-sided 95% confidence intervals for the mean TcCB concentration in the Reference area

1. In the Object Explorer, make sure **epa.94b.tccb.df** is highlighted.
2. On the S-PLUS menu bar, make the following menu choices: **EnvironmentalStats>Estimation>Parameters**. This will bring up the Estimate Distribution Parameters dialog box.
3. For Data to Use, make sure the **Pre-Defined Data** button is selected.
4. In the Data Set box, make sure **epa.94b.tccb.df** is selected.
5. In the Variable box, select **TcCB**.
6. In the Subset Rows with box, type **Area=="Reference"**.
7. In the Distribution/Estimation section, make sure **Lognormal (Alternative)** is selected in the distribution box, and **mvue** is selected in the Estimation Method box.
8. Under the Confidence Interval section, **check** the Confidence Interval box. In the CI Type box, select **two-sided**. In the CI Method box, select **land**. In the Confidence Level (%) box, type **95**. Click **OK** or **Apply**.

To use a different method to construct the confidence interval, in Step 8 select something other than **land** in the CI Method box.

Command

To estimate the mean and coefficient of variation of the Reference area TcCB concentrations and create a 95% confidence interval for the mean using the ENVIRONMENTALSTATS for S-PLUS Command or Script Window, type these commands.

```
attach(epa.94b.tccb.df)
elnorm.alt(TcCB[Area=="Reference"], ci=T,
   conf.level=0.95)
detach()
```

By default, `elnorm.alt` uses Land's method. You can change the method for constructing the confidence interval by using the `ci.method` argument.

Example 5.19: Upper Confidence Limit for the Mean of Chromium Concentrations

The guidance document USEPA (1992d) contains an example of 15 observations of chromium concentrations (mg/kg) which are assumed to come from a lognormal distribution. The example shows how to compute the 95% upper confidence limit for the mean of the distribution (θ) using Land's (1972) method.   The data are shown in Table 5.5 below.   In ENVIRONMENTALSTATS for S-PLUS, these data are stored in the vector epa.92d.chromium.vec.

Chromium Concentrations (mg/kg)							
10	13	20	36	41	59	67	110
110	136	140	160	200	230	1300	

Table 5.5   Chromium concentrations assumed to come from a lognormal distribution (USEPA, 1992d)

Table 5.6 displays one-sided upper 95% confidence intervals for the mean concentration of chromium based on various methods. In this case, there is quite a bit of difference among these methods, with the upper confidence limit based on Land's method about twice the size of the limits based on Parkin et al.'s (1990) method and the normal approximation.

Confidence Interval	Method
[0, 497]	Land
[0, 230]	Parkin et al.
[0, 366]	Cox
[0, 261]	Normal Approximation

Table 5.6   One-sided upper 95% confidence limits for the mean concentration of chromium, based on data from USEPA (1992d)

Menu

To estimate the mean and coefficient of variation of the chromium concentrations and compute a one-sided upper 95% confidence interval for the mean using the ENVIRONMENTALSTATS for S-PLUS pull-down menu, follow these steps.

1. In the Object Explorer, find the vector **epa.92d.chromium.vec**.
2. On the S-PLUS menu bar, make the following menu choices: **EnvironmentalStats>Estimation>Parameters**. This will bring up the Estimate Distribution Parameters dialog box.
3. For Data to Use, make sure the **Pre-Defined Data** button is selected.
4. In the Data Set box, make sure **epa.92d.chromium.vec** is selected.

5. In the Variable box, **epa.92d.chromium.vec** should be automatically selected.
6. In the Distribution/Estimation section, make sure **Lognormal (Alternative)** is selected in the distribution box, and **mvue** is selected in the Estimation Method box.
7. Under the Confidence Interval section, **check** the Confidence Interval box. In the CI Type box, select **upper**. In the CI Method box, select **land**. In the Confidence Level (%) box, type **95**. Click **OK** or **Apply**.

To use a different method to construct the confidence interval, in Step 7 select something other than **land** in the CI Method box.

Command

To estimate the mean and coefficient of variation of the chromium concentrations and create a one-sided upper 95% confidence interval for the mean using the ENVIRONMENTALSTATS for S-PLUS Command or Script Window, type this command.

```
elnorm.alt(epa.92d.chromium.vec, ci=T,
    ci.type="upper")
```

By default, `elnorm.alt` uses Land's. You can change the method for constructing the confidence interval by using the `ci.method` argument.

Example 5.20: Upper Confidence Limit for the Mean Based on a Small Sample Size

We saw in Example 5.18 that the confidence limits for the mean TcCB concentration are fairly close, but in Example 5.19 the confidence limits for the mean chromium concentration do not agree so well. The TcCB data contain $n = 47$ observations, while the chromium data contain $n = 15$ observations. Here, we will generate 10 observations from a lognormal distribution with a mean of 10 and a coefficient of variation of 2, then compare the one-sided upper 95% confidence limits based on the various methods. Table 5.7 displays the 10 simulated observations, and Table 5.8 displays the upper confidence limits.

10 Random Numbers From a Lognormal Distribution				
0.2764807	0.7303524	2.1939532	5.3785007	6.8565868
7.2216128	8.2545299	9.1507064	9.7874597	95.8133610

Table 5.7   10 random numbers from a lognormal distribution with a mean of 10 and coefficient of variation of 2

Confidence Interval	Method
[0, 182]	Land
[0, 96]	Parkin et al.
[0, 68]	Cox
[0, 28]	Normal Approximation

Table 5.8 One-sided upper 95% confidence limits for the mean based on 10 observations from a lognormal distribution with a mean of 10 and a coefficient of variation of 2

This example illustrates a disturbing characteristic of Land's method for computing confidence intervals: the upper confidence limit is about twice as large as the largest observed value! In fact, for small sample sizes, the upper confidence limit based on Land's method can be one or more orders of magnitude larger than the largest observed value. Furthermore, although Land's method is exact and has optimal properties, it is extremely sensitive to the assumption that the data come from a lognormal distribution. Deviations from this distribution (e.g., a mixture distribution that appears similar to a lognormal distribution) can also result in extremely large values for the upper confidence limit.

Singh et al. (1997) refer to a simulation study they performed in which they found that for sample sizes of 30 or less, Land's method produces upper confidence limits that are "unacceptably high," especially for populations with a CV greater than 1. They recommend using nonparametric methods, such as the bootstrap (see the section Nonparametric Confidence Intervals Based on Bootstrapping, below).

Menu

To reproduce this example using the ENVIRONMENTALSTATS for S-PLUS pull-down menu, first we have to create the data.

1. On the S-PLUS menu bar, make the following menu choices: **EnvironmentalStats>Probability Distributions and Random Numbers>Random Numbers>Univariate**. This will bring up the Univariate Random Number Generation dialog box.
2. In the Sample Size box, type **10**.
3. In the Set Seed With box, type **23**.
4. In the Distribution box, select **Lognormal (Alternative)**. In the mean box, type **10**. In the cv box, type **2**.
5. In the Save As box, type **lognormal.df**. Click **OK** or **Apply**.

Now we can create the confidence interval based on Land's method.

1. In the Object Explorer, select the **Data** folder in the left-hand column and highlight **lognormal.df** in the right-hand column.

2. On the S-PLUS menu bar, make the following menu choices:
    EnvironmentalStats>Estimation>Parameters. This will bring up
    the Estimate Distribution Parameters dialog box.
3. For Data to Use, make sure the **Pre-Defined Data** button is selected.
4. In the Data Set box, make sure **lognormal.df** is selected.
5. In the Variable box, select **Sample**.
6. In the Distribution/Estimation section, make sure **Lognormal (Alternative)** is selected in the distribution box, and **mvue** is selected in
    the Estimation Method box.
7. Under the Confidence Interval section, **check** the Confidence Interval box. In the CI Type box, select **upper**. In the CI Method box,
    select **land**. In the Confidence Level (%) box, type **95**. Click **OK** or
    Apply.

To use a different method to construct the confidence interval, in Step 7 select something other than **land** in the CI Method box.

Command

To reproduce this example using the ENVIRONMENTALSTATS for S-PLUS
Command or Script Window, type these commands.

```
set.seed(23)
lognormal.data <- rlnorm.alt(10, mean=10, cv=2)
elnorm.alt(lognormal.data, ci=T, ci.type="upper")
```

This produces the confidence interval based on Land's method. You can
change the method for constructing the confidence interval by using the
`ci.method` argument.

One-Sided Confidence Interval for the Mean of a Skewed Distribution

Chen (1995) developed a modified t-statistic for performing a one-sided
test of hypothesis on the mean of a skewed distribution (see Chapter 7).
Based on the relationship between hypothesis tests and confidence intervals
(see Chapter 7), Chen's method can be used to construct a one-sided confidence interval for the mean of a skewed distribution. With this method, you
can construct a one-sided lower confidence interval for the mean of a positively skewed distribution, or a one-sided upper confidence interval for the
mean of a negatively skewed distribution. Because environmental data are
usually positively skewed, and because we are usually interested in a two-sided or one-sided upper confidence limit on the mean, this method for constructing confidence intervals will usually not be used in practice. The associated test of hypothesis, however, was developed to solve a specific problem in environmental statistics (see Chapter 7).

Example 5.21: Lower Confidence Limit for the Mean of Chromium Concentrations

Table 5.9 displays one-sided lower 95% confidence intervals for the mean concentration of chromium (see Example 5.19), based on several methods, including Chen's general method for positively skewed distributions. In this case, the lower confidence limit based on Chen's method agrees closely with that based on Land's method.

Confidence Interval	Method
[95, ∞]	Land
[110, ∞]	Parkin et al.
[82, ∞]	Cox
[58, ∞]	Normal Approximation
[97, ∞]	Chen

Table 5.9   One-sided lower 95% confidence limits for the mean concentration of chromium, based on data from USEPA (1992d)

Menu

To produce the confidence limits shown in Table 5.9 for the first four methods using the ENVIRONMENTALSTATS for S-PLUS pull-down menu, follow the steps shown under the Menu section in Example 5.19, except in Step 9 select **lower** instead of upper in the CI Type box. To produce the confidence interval based on Chen's method, follow these steps.

1. In the Object Explorer, make sure **epa.92d.chromium.vec** is highlighted.
2. On the S-PLUS menu bar, make the following menu choices: **EnvironmentalStats>Hypothesis Tests>Compare Samples>One Sample>t-Test for Skewed Data**. This will bring up the One-Sample t-Test for Skewed Data dialog box.
3. For Data to Use, make sure the **Pre-Defined Data** button is selected.
4. In the Data Set box, make sure **epa.92d.chromium.vec** is selected.
5. In the Variable box, **epa.92d.chromium.vec** should be automatically selected.
6. In the Alternative Hypothesis box, make sure **greater** is selected.
7. In the Confidence Level (%) box, type **95**. In the CI Method box, select **z**. Click **OK** or **Apply**.

Command

To produce the confidence limits shown in Table 5.9 for the first four methods using the ENVIRONMENTALSTATS for S-PLUS Command or Script Window, type this command.

```
elnorm.alt(epa.92d.chromium.vec, ci=T,
   ci.type="lower")
```

By default, `elnorm.alt` uses Land's method to construct the confidence interval. You can change the method for constructing the confidence interval by using the `ci.method` argument.

To produce the confidence interval based on Chen's method, type this command.

```
chen.t.test(epa.92d.chromium.vec)$interval$limits
```

Confidence Interval for the Probability of Success for a Binomial Distribution

We saw earlier that when the random variable X follows a $B(n, p)$ distribution, the estimated probability of "success" is simply the observed number of successes divided by the total number of trials:

$$\hat{p} = \frac{x}{n} \tag{5.91}$$

There are two ways to construct a confidence interval for the probability of success: an exact method and a method based on a normal approximation.

Exact Confidence Interval

The exact $(1-\alpha)100\%$ confidence interval for p is given by:

$$\left[LCL_{\alpha/2} , \quad UCL_{\alpha/2} \right] \tag{5.92}$$

where the confidence limits are computed such that:

$$Pr\left(X \geq x \mid p = LCL_{\alpha/2}\right) = \frac{\alpha}{2} \tag{5.93}$$

$$Pr\left(X \leq x \mid p = UCL_{\alpha/2}\right) = \frac{\alpha}{2} \tag{5.94}$$

That is, the lower confidence limit is the smallest value of p such that the probability of observing at least as many successes as were actually observed is no less than (in fact equal to) $\alpha/2$. Similarly, the upper confidence limit is the largest value of p such that the probability of observing no more than the number of successes actually observed is no greater than (in fact equal to) $\alpha/2$.

It can be shown (Zar, 1999, pp. 527–529, Fisher and Yates, 1963) that:

$$LCL_{\alpha/2} = \frac{x}{x + (n - x + 1)\, F_{v_1, v_2, 1-\alpha/2}} \qquad (5.95)$$

$$UCL_{\alpha/2} = \frac{(x + 1)\, F_{v_2 + 2, v_1 - 2, 1 - \alpha/2}}{n - x + (x + 1)\, F_{v_2 + 2, v_1 - 2, 1 - \alpha/2}} \qquad (5.96)$$

where

$$v_1 = 2(n - x + 1) \qquad (5.97)$$

$$v_2 = 2x \qquad (5.98)$$

and $F_{v_1, v_2, r}$ denotes the rth quantile of the F-distribution with v_1 and v_2 degrees of freedom.

To construct a one-sided lower confidence interval, $\alpha/2$ is replaced with α in Equation (5.95), and the upper confidence limit is set to 1. To construct a one-sided upper confidence interval, $\alpha/2$ is replaced with α in Equation (5.96), and the lower confidence limit is set to 0.

Note that an exact lower confidence bound can be computed even when all of the observations are "successes," and an exact upper confidence bound can be computed even when all of the observations are "failures." In fact, when all the observations are failures (so $x = 0$), we cannot use Equations (5.95) and (5.96), but it is easy to show in this case that the upper $(1-\alpha)100\%$ confidence limit is given by:

$$UCL_\alpha = 1 - \alpha^{1/n} \qquad (5.99)$$

Similarly, when all the observations are successes (so $x = n$), the lower $(1-\alpha)100\%$ confidence limit is given by:

$$LCL_\alpha = \alpha^{1/n} \qquad (5.100)$$

Confidence Interval Based on Normal Approximation

Following Equation (5.86), an approximate $(1-\alpha)100\%$ confidence interval for p, based on the normal approximation to the distribution of the estimate of p, is given by:

$$\left[\hat{p} - z_{1-\alpha/2} \, \hat{\sigma}_{\hat{p}} \, , \; \hat{p} + z_{1-\alpha/2} \, \hat{\sigma}_{\hat{p}} \right] \qquad (5.101)$$

where

$$\hat{\sigma}_{\hat{p}} = \sqrt{\frac{\hat{p}(1 - \hat{p})}{n}} \qquad (5.102)$$

Most authors (e.g., Fleiss, 1981, p. 15) recommend using a correction for continuity to bring the normal curve probabilities in closer agreement with the binomial probabilities.

$$\left[\hat{p} - z_{1-\alpha/2} \, \hat{\sigma}_{\hat{p}} - \frac{1}{2n} \, , \; \hat{p} + z_{1-\alpha/2} \, \hat{\sigma}_{\hat{p}} + \frac{1}{2n} \right] \qquad (5.103)$$

The normal approximation does not work well for small sample sizes, especially when np or $n(1-p)$ is less than 5, or when p is small ($p < 0.2$) or large ($p > 0.8$) (Zar, 1984, p. 379). As Zar (1999, p. 529) points out, with the availability of the exact method, there is really no need to use the normal approximation.

Example 5.22: Confidence Interval for the Probability of a Nondetect

Recall Example 5.11 in which we computed the probability of observing a "nondetect" for benzene concentrations (ppb) in groundwater based on observations from six background wells sampled monthly for 6 months. The data are shown in Table 5.2, and in ENVIRONMENTALSTATS for S-PLUS they are stored in the data frame `epa.92c.benzene1.df`. Nondetect values are reported as "<2." Of the 36 values, 33 are nondetects. The estimated probability of observing a nondetect is therefore 33/36, or about 92%. Table 5.10 displays 95% confidence intervals for the probability of observing a nondetect value at any of the six wells, based on the exact method and the normal approximation method. You can see that both methods based on the normal approximation yield an impossible upper limit of 1. If in fact p were equal to 1, then we would not have been able to observe any detects!

Confidence Interval	Method
[0.78, 0.98]	Exact
[0.83, 1]	Normal Approximation (No Continuity Correction)
[0.81, 1]	Normal Approximation (Continuity Correction)

Table 5.10   Two-sided 95% confidence limits for the probability of observing a nondetect for benzene, based on data from USEPA (1992c)

Menu

To produce the confidence intervals shown in Table 5.10 using the ENVIRONMENTALSTATS for S-PLUS pull-down menu, follow these steps.

1. In the Object Explorer, make sure **epa.92c.benzene1.df** is highlighted.
2. On the S-PLUS menu bar, make the following menu choices: **EnvironmentalStats>Estimation>Parameters**. This will bring up the Estimate Distribution Parameters dialog box.
3. For Data to Use, make sure the **Pre-Defined Data** button is selected.
4. In the Data Set box, make sure **epa.92c.benzene1.df** is selected.
5. In the Variable box, select **Censored**.
6. In the Distribution box, select **Binomial**.
7. Make sure the Confidence Interval box is **checked**. In the CI Type box, select **two-sided**. In the CI Method box, select **exact**. In the Confidence Level (%) box, type **95**. Click **OK** or **Apply**.

These steps will produce the exact confidence interval. To produce the confidence interval based on the normal approximation with continuity correction, in Step 7, in the CI Method box, select **approx**.

Command

To produce the confidence limits shown in Table 5.10 using the ENVIRONMENTALSTATS for S-PLUS Command or Script Window, type these commands.

```
ebinom(epa.92c.benzene1.df$Censored, ci=T)

ebinom(epa.92c.benzene1.df$Censored, ci=T,
   ci.method="approx", correct=F)

ebinom(epa.92c.benzene1.df$Censored, ci=T,
   ci.method="approx", correct=T)
```

Confidence Interval for the Mean of a Poisson Distribution

As we discussed in Chapter 4, the Poisson distribution has been used to model the number of times a pollution standard is violated over a given time period (Ott, 1995, Chapter 5), as well as to model the distribution of chemical concentration for a chemical that shows up as mostly nondetects with a few detected values (Gibbons, 1987b).

Given n independent observations from a Poisson distribution with parameter λ (the average number of events that occur within the given time period or area), the population value of λ is estimated as:

$$\hat{\lambda} = \overline{x} = \frac{1}{n} \sum_{i=1}^{n} x_i \qquad (5.104)$$

This estimator is the maximum likelihood, method of moments, and minimum variance unbiased estimator of λ. There are three ways to compute a confidence interval for the mean λ. Each of these is discussed below.

Exact Confidence Interval

An important property of the Poisson distribution is that the sum of independent Poisson random variables is also a Poisson random variable. In particular, the random variable Y, where Y is defined as

$$Y = \sum_{i=1}^{n} X_i \qquad (5.105)$$

has a Poisson distribution with mean $\lambda_Y = n\lambda$. (Note that Y is simply the total number of events observed.) This property is used to construct an exact confidence interval for λ.

Similar to the formula for an exact confidence interval for the probability of success for a binomial distribution (Equation (5.92)), the exact $(1-\alpha)100\%$ confidence interval for λ is given by:

$$\left[\; LCL_{\alpha/2} \; , \; UCL_{\alpha/2} \; \right] \tag{5.106}$$

where the confidence limits are computed such that:

$$\Pr\left(Y \geq y \; \mid \; \lambda = LCL_{\alpha/2} \right) = \frac{\alpha}{2} \tag{5.107}$$

$$\Pr\left(Y \leq y \; \mid \; \lambda = UCL_{\alpha/2} \right) = \frac{\alpha}{2} \tag{5.108}$$

That is, the lower confidence limit is the smallest value of λ such that the probability of observing at least as many total events as were actually observed is no less than (in fact equal to) $\alpha/2$. Similarly, the upper confidence limit is the largest value of λ such that the probability of observing no more than the number of total events actually observed is no greater than (in fact equal to) $\alpha/2$.

To construct a one-sided lower confidence interval, $\alpha/2$ is replaced with α in Equation (5.107), and the upper confidence limit is set to ∞. To construct a one-sided upper confidence interval, $\alpha/2$ is replaced with α in Equation (5.108), and the lower confidence limit is set to 0.

Note that an exact upper confidence bound can be computed even when all of the n observations are 0. In this case, the one-sided upper $(1-\alpha)100\%$ confidence limit for λ is:

$$UCL_\alpha = \frac{-\log(\alpha)}{n} \tag{5.109}$$

Confidence Interval Based on Pearson-Hartley Approximation

For a two-sided $(1-\alpha)100\%$ confidence interval for λ, the Pearson and Hartley approximation (Zar, 1999, p. 574; Pearson and Hartley, 1970, p. 81) is given by:

$$\left[\frac{\chi^2_{2y,\alpha/2}}{2n} \, , \, \frac{\chi^2_{2y+2,1-\alpha/2}}{2n} \right] \qquad (5.110)$$

where $\chi^2_{\nu,p}$ denotes the pth quantile of a chi-square distribution with ν degrees of freedom. One-sided confidence intervals are computed in a similar fashion.

Confidence Interval Based on Normal Approximation

Following Equation (5.86), an approximate $(1-\alpha)100\%$ confidence interval for λ, based on the normal approximation to the distribution of the estimate of λ, is given by:

$$\left[\hat{\lambda} - z_{\alpha/2}\, \hat{\sigma}_{\hat{\lambda}} \, , \, \hat{\lambda} + z_{\alpha/2}\, \hat{\sigma}_{\hat{\lambda}} \right] \qquad (5.111)$$

where

$$\hat{\sigma}_{\hat{\lambda}} = \sqrt{\frac{\hat{\lambda}}{n}} \qquad (5.112)$$

This approximation is inferior to the Pearson-Hartley approximation. As for the case of the binomial distribution, since the exact method is available, there is really no need to use either of the approximations.

Example 5.23: Confidence Interval for the Average Benzene Concentration

Let us look again at the benzene in groundwater data that we looked at in Examples 5.11 and 5.22. The data are shown in Table 5.2, and in ENVIRONMENTALSTATS for S-PLUS they are stored in the data frame epa.92c.benzene1.df. Nondetect values are reported as "<2." Of the 36 values, 33 are nondetects. Following Gibbons (1987, p. 574), USEPA (1992c, p. 36) assumes that these data can be modeled as having come from a Poisson process. The three detected values represent a count of the number of molecules of benzene observed out of a much larger number of water molecules. The 33 nondetect values are also assumed to represent counts, where the count must be equal to 0 or 1, since they are recorded as "<2" ppb.

Following USEPA (1992c), for this example we will set all of the values of the nondetects to 1. The estimated value of λ is 1.94 ppb. Table 5.11 displays one-sided upper 95% confidence intervals for the average concentration of benzene, based on the exact method and the two approximate methods. Here the Pearson-Harley method yields the same answer as the exact method, and the normal approximation method gives a very similar upper limit.

Confidence Interval	Method
[0, 2.37]	Exact
[0, 2.37]	Pearson-Hartley Approximation
[0, 2.33]	Normal Approximation

Table 5.11   One-sided upper 95% confidence limits for the average benzene concentration (ppb), based on data from USEPA (1992c, p. 36)

As a sensitivity analysis, we can determine the exact upper confidence limit when we set the nondetects to 0 or 2 as well. For the first case, the upper confidence limit is 1.35 ppb, and for the second case it is 3.37 ppb.

It is by no means clear, nor commonly accepted, that modeling concentration data with a Poisson distribution is valid. Furthermore, the method of dealing with nondetects is purely subjective. A Poisson Q-Q plot of the benzene data with the nondetects set equal to 1 is shown in Figure 5.11. In this plot, the numbers on the plot represent the number of observations plotted at that particular (x, y) coordinate. Based on looking at many "typical" Poisson Q-Q plots, it is quite apparent that these data do not appear to fit a Poisson model too well.

Menu

To produce the confidence intervals shown in Table 5.11 using the ENVIRONMENTALSTATS for S-PLUS pull-down menu, it will be helpful to first create a modified version of epa.92c.benzen.df that contains a version of the Benzene variable with all of the nondetects set to 1.

1. In the Object Explorer, make sure **epa.92c.benzene1.df** is highlighted.
2. On the S-PLUS menu bar, make the following menu choices: **Data>Transform**. This will bring up the Transform dialog box.

Poisson Q-Q Plot for Benzene

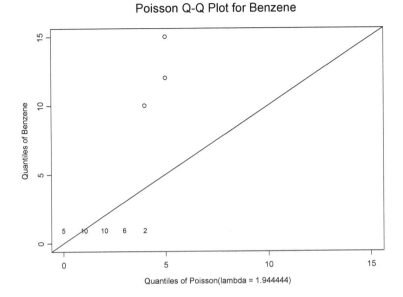

Figure 5.11   Poisson Q-Q plot for the benzene data, with nondetects set equal to 1

3. In the Data Frame box, **epa.92c.benzene1.df** should be selected.
4. In the New Column Name box, type **new.Benzene**.
5. Under the Add to Expression section, in the Variable box, choose **Censored**. In the Function box, choose **ifelse**. Click on the **Add** button.
6. Modify the Expression box so that the expression is as follows: **ifelse(Censored, 1, Benzene)**. Click **OK**. (At this point, you will get a warning message telling you that you have created a new copy of the data frame that masks the original copy.)
7. In the left column of the Object Explorer, click on **Data**.
8. In the Object column of the Object Explorer, right-click on the short-cut **epa.92c.benzene1.df** and choose **Properties**.
9. In the Name box, rename this data frame to **new.epa.92c.benzene1.df**, then click **OK**.

Now we can produce the confidence intervals by following these steps.

1. In the Object Explorer, make sure **new.epa.92c.benzene1.df** is highlighted.
2. On the S-PLUS menu bar, make the following menu choices: **EnvironmentalStats>Estimation>Parameters**. This will bring up the Estimate Distribution Parameters dialog box.
3. For Data to Use, make sure the **Pre-Defined Data** button is selected.

4. In the Data Set box, make sure **new.epa.92c.benzene1.df** is selected.
5. In the Variable box, select **new.Benzene**.
6. In the Distribution box, select **Poisson**.
7. Make sure the Confidence Interval box is **checked**. In the CI Type box, select **upper**. In the CI Method box, select **exact**. In the Confidence Level (%) box, type **95**. Click **OK** or **Apply**.

These steps will produce the exact confidence interval. To produce the confidence intervals based on the Pearson-Hartley and normal approximations, in Step 7, in the CI Method box, select **pearson.hartley.approx** or **normal.approx**.

To produce several Poisson Q-Q plots from a Poisson distribution with parameter $\lambda = 2$, based on a sample size of $n = 36$, follow these steps.

1. On the S-PLUS menu bar, make the following menu choices: **EnvironmentalStats>EDA>Q-Q Plot>Q-Q Plot Gestalt**. This will bring up the Q-Q Plot Gestalt dialog box.
2. **Check** the Estimate Parameters box.
3. In the Distribution box, select **Poisson**.
4. In the lambda box, type **2**.
5. In the Sample Size box, type **36**.
6. In the # Pages box, type **5**.
7. Click on the **Plotting** tab, and select **Number** by the Duplicate Points buttons. Click **OK**.

You can look at even more "typical" plots by increasing the value of the number you type in the # Pages box in Step 6.

Command

To produce the confidence limits shown in Table 5.11 using the ENVIRONMENTALSTATS for S-PLUS Command or Script Window, type these commands.

```
attach(epa.92c.benzene1.df)

Benzene[Censored] <- 1

epois(Benzene, ci=T, ci.type="upper")

epois(Benzene, ci=T, ci.type="upper",
   ci.method="pearson.hartley.approx")

epois(Benzene, ci=T, ci.type="upper",
   ci.method="normal.approx")

rm(Benzene)

detach()
```

To produce several Poisson Q-Q plots from a Poisson distribution with parameter $\lambda = 2$, based on a sample size of $n = 36$, type this command.

```
qqplot.gestalt(dist="pois", param.list=list(lambda=2),
    sample.size=36, num.pages=5, estimate.params=T,
    qq.line.type="0-1", add.line=T,
    duplicate.points.method="number")
```

You can look at even more "typical" plots by increasing the value of the num.pages argument.

NONPARAMETRIC CONFIDENCE INTERVALS BASED ON BOOTSTRAPPING

So far, we have talked about constructing parametric confidence intervals, which require the following three assumptions:

1. The sample is some type of random sample.
2. The observations are independent of one another.
3. The observations come from a specified probability distribution (e.g., normal, lognormal, binomial, Poisson, etc.).

Parametric confidence intervals assume you know what kind of distribution describes the population (the third assumption above). *Nonparametric confidence intervals* do not require this last assumption. If we are interested in constructing a confidence interval for the median or another percentile, there is an easy way to do this based on the ranks of the observations, rather than the actual data itself (see the section Nonparametric Estimates for Quantiles later in this chapter). In this section, we will show you a more general method to construct a nonparametric confidence interval for any kind of population parameter.

To construct parametric confidence intervals like the one shown in Equation (5.78) for the mean of a normal distribution or the one shown in Equation (5.92) for the probability of success for a binomial distribution, we have to make an assumption about the form of the population distribution. Once we assume a particular type of population distribution, we can usually figure out the distribution of the estimator of the population parameter (e.g., for a $N(\mu, \sigma)$ distribution, the distribution of the sample mean is normal with mean μ and standard deviation $\sigma/\sqrt{n}$), and based on the distribution of the estimator we can construct a confidence interval.

To construct a nonparametric confidence interval, we do not want to make any assumptions about the population distribution, but in that case, what can we do? Bradly Efron (1979a,b) had the following idea. He reasoned that since you can use the empirical cumulative distribution function (also called the empirical cdf, see Chapter 3) to estimate the population cdf, you can also use the empirical cdf to estimate the distribution of the estimator by randomly sampling from the empirical cdf several times, and each time constructing an estimate based on this random sample. Once you have this estimated distribution of the estimator, you can construct a confidence interval for the population parameter.

Efron called his idea the ***bootstrap*** (Efron, 1979a,b; Efron and Tibshirani, 1993), after the saying of "pulling yourself up by your own bootstraps." One definition of bootstrap is "to cause to succeed without the help of others," and here we are not relying on the help of assuming a particular parametric distribution. The basic steps of the bootstrap are:

1. Estimate the parameter based on the data.
2. Sample the data with replacement B times, and each time estimate the parameter based on this bootstrap sample.
3. Use the estimated parameter created in Step 1 and the bootstrap distribution of the estimator created in Step 2 to create a confidence interval for the parameter.

(Efron, 1979a,b; Efron and Tibshirani, 1993; Shao and Tu, 1995; Manly, 1997; Davison and Hinkley, 1997). Note that in Step 2 above, we are using the empirical probabilities estimator shown in Equation (3.20) for the plotting positions for the empirical cdf (see Chapter 3).

Methods of Constructing Bootstrap Confidence Intervals

There are several ways to construct a bootstrap confidence interval (Efron and Tibshirani, 1993; Manly, 1997; Davison and Hinkley, 1997). Here we will discuss just four:

1. Standard Normal Method
2. Empirical Percentile Method
3. Bias-Corrected Percentile Method
4. Bias-Corrected and Adjusted Percentile Method

Of these four, S-PLUS computes confidence intervals based on the second and fourth method.

In these discussions, we will assume the parameter we are estimating is θ, we will let $\hat{\theta}$ denote the estimate of θ based on the original sample, we will let $\hat{\theta}_b$ denote the estimate of θ based on the bth bootstrap sample, and we will let $\hat{\theta}^*$ denote a random variable that has the bootstrap distribution of $\hat{\theta}$ based on the given data.

Standard Normal Method

This method is based on the assumption that the distribution of the estimator $\hat{\theta}$ is approximately normal with mean θ, and therefore we can use Equation (5.86) to construct an approximate confidence interval for θ (Efron and Tibshirani, 1993, p. 168; Manly, 1997, p. 35). That is, the two-sided $(1-\alpha)100\%$ for θ is given by:

$$\left[\hat{\theta} - z_{1-\alpha/2} \, \hat{\sigma}_{\hat{\theta}} \, , \, \hat{\theta} + z_{1-\alpha/2} \, \hat{\sigma}_{\hat{\theta}} \right] \qquad (5.113)$$

In this case, all we need is $\hat{\sigma}_{\hat{\theta}}$, the estimated standard deviation of $\hat{\theta}$. To get this estimate, we generate B bootstrap samples, compute B estimates of θ based on each of the B bootstrap samples, then compute the sample standard deviation of $\hat{\theta}$ based on these B estimates of θ. More technically, for each bootstrap sample we compute $\hat{\theta}_b$, the estimate of θ based on the bth bootstrap sample. The estimated standard deviation of $\hat{\theta}$ is then computed as:

$$\hat{\sigma}_{\hat{\theta}} = \sqrt{\frac{\sum\limits_{b=1}^{B} \left(\hat{\theta}_b - \overline{\hat{\theta}} \right)^2}{B-1}} \qquad (5.114)$$

where

$$\overline{\hat{\theta}} = \frac{1}{B} \sum_{b=1}^{B} \hat{\theta}_b \qquad (5.115)$$

Efron and Tibshirani (1993, p. 52) and Manly (1997, p. 87) suggest that usually no more than $B = 200$ bootstraps are needed to get a good estimate of the standard deviation of the estimator.

Empirical Percentile Method

What if we do not want to assume the distribution of $\hat{\theta}$ is approximately normal? For example, typical environmental data is skewed to the right, and the distribution of the sample mean may be skewed even for moderately large sample sizes if the underlying population distribution is extremely skewed. The empirical percentile method is based on the assumption that the distribution of the estimator $\hat{\theta}$ is not necessarily approximately normal with mean θ, but rather that the distribution of some ***monotonic transformation*** of $\hat{\theta}$, say $f\left(\hat{\theta}\right)$, is approximately normal with mean $f\left(\theta\right)$ and standard deviation 1 (Efron and Tibshirani, 1993, p. 170; Manly, 1997, p. 39). In this case, a two-sided $(1-\alpha)100\%$ confidence interval for $f\left(\theta\right)$ is:

$$\left[\; f\left(\hat{\theta}\right) - z_{1-\alpha/2}\;,\;\; f\left(\hat{\theta}\right) + z_{1-\alpha/2}\;\right] \tag{5.116}$$

But we want a confidence interval for θ, and we do not know the form of f.

If we knew the form of f, we could create the bootstrap distribution of $f\left(\hat{\theta}\right)$. Since the true distribution of $f\left(\hat{\theta}\right)$ is normal with mean $f\left(\theta\right)$ and standard deviation 1, we would expect the bootstrap distribution of $f\left(\hat{\theta}\right)$ to be approximately normal with mean $f\left(\hat{\theta}\right)$ and standard deviation 1. Recalling Problem 4.7 and Equation (4.54) in Chapter 4, we can see that the confidence interval in Equation (5.116) consists of the $\alpha/2$ and $1-\alpha/2$ quantiles of the bootstrap distribution of $f\left(\hat{\theta}\right)$. Furthermore, since f is monotonic and percentiles are invariant under monotonic transformations (see Chapter 4), an approximate two-sided $(1-\alpha)100\%$ confidence interval for θ can be based on the empirical percentiles of the bootstrap distribution of $\hat{\theta}$:

$$\left[\; \hat{\theta}^{*}_{\alpha/2}\;,\;\; \hat{\theta}^{*}_{1-\alpha/2}\;\right] \tag{5.117}$$

where $\hat{\theta}^{*}_{p}$ denotes the pth quantile of the bootstrap distribution of $\hat{\theta}$. For example, a two-sided 95% confidence interval for θ is simply the interval between the observed 2.5 percentile and 97.5 percentile of the bootstrap distribution of $\hat{\theta}$. Note that we do not need to know the form of the transformation function f in order to use this method.

Efron and Tibshirani (1993, pp. 188, 275) recommend using at least $B = 1,000$ bootstraps for this method of constructing 90% confidence intervals. Manly (1997, p. 88) suggests using at least $B = 2,000$ for a 95% confidence interval. These recommendations also apply to the "bias-corrected percentile method" and the "bias-corrected and adjusted percentile method," both discussed below.

Bias-Corrected Percentile Method

The empirical percentile method does not work well in practice if $f\left(\hat{\theta}\right)$ is a biased estimator of $f\left(\theta\right)$. In this case, the median value of the boot-strap distribution of $\hat{\theta}$ is not "close" to $\hat{\theta}$. The bias-corrected percentile method is a modification of the empirical percentile method to overcome this problem. It makes the more general assumption that the distribution of $f\left(\hat{\theta}\right)$ is approximately normal with mean $f\left(\theta\right) - z_0$ and standard deviation 1, where the bias z_0 is a constant (Efron, 1981; Manly, 1997, p. 46). In this case, a two-sided $(1-\alpha)100\%$ confidence interval for $f\left(\theta\right)$ is:

$$\left[\; f\left(\hat{\theta}\right) + z_0 - z_{1-\alpha/2}\; ,\;\; f\left(\hat{\theta}\right) + z_0 + z_{1-\alpha/2}\; \right] \quad \text{(5.118)}$$

But we want a confidence interval for θ, and now we not only do not know the form of f, but we do not know the value of z_0 either.

If we knew the form of f, we could create the bootstrap distribution of $f\left(\hat{\theta}\right)$. Since the true distribution of $f\left(\hat{\theta}\right)$ is normal with mean $f\left(\theta\right) - z_0$ and standard deviation 1, we would expect the bootstrap distri-bution of $f\left(\hat{\theta}\right)$ to be approximately normal with mean $f\left(\hat{\theta}\right) - z_0$ and standard deviation 1. Recalling Problem 4.7 and Equation (4.54) in Chapter 4, we can see that the confidence interval in Equation (5.118) consists of the $p_L{}^{th}$ and $p_U{}^{th}$ quantiles of the bootstrap distribution of $f\left(\hat{\theta}\right)$, where

$$p_L = \Phi\left(2z_0 + z_{\alpha/2}\right) \quad\quad\quad \text{(5.119)}$$

$$p_U = \Phi\left(2z_0 + z_{1-\alpha/2}\right) \quad\quad\quad \text{(5.120)}$$

Again, since f is monotonic and percentiles are invariant under monotonic transformations, an approximate two-sided $(1-\alpha)100\%$ confidence interval for θ can be based on the empirical percentiles of the bootstrap distribution of $\hat{\theta}$:

$$\left[\hat{\theta}^*_{p_L} , \hat{\theta}^*_{p_U} \right] \tag{5.121}$$

That is, the lower confidence bound is the p_L^{th} quantile of the bootstrap distribution of $\hat{\theta}$, and the upper confidence bound is the p_U^{th} quantile of the bootstrap distribution of $\hat{\theta}$, where p_L and p_U are defined in Equations (5.119) and (5.120), respectively.

The confidence interval in Equation (5.121) requires knowing the value of the bias z_0. In practice, we must estimate this quantity. Since the quantity $f(\hat{\theta}) - f(\theta) + z_0$ has a standard normal distribution, the true value of z_0 is simply:

$$z_0 = \Phi^{-1}(p) \tag{5.122}$$

where

$$p = \Pr(\hat{\theta} \le \theta) \tag{5.123}$$

(Efron and Tibshirani, 1993, p. 327; Manly, 1997, p. 46). That is, z_0 is the pth quantile of a standard normal distribution, where p is defined in Equation (5.123). This is because of the following:

$$p = \Pr(\hat{\theta} \le \theta) = \Pr\left[f(\hat{\theta}) \le f(\theta) \right]$$

$$= \Pr\left[f(\hat{\theta}) - f(\theta) + z_0 \le z_0 \right] \tag{5.124}$$

$$= \Pr(Z \le z_0) = \Phi(z_0)$$

where Z denotes a standard normal random variable, and Φ denotes the cdf of a standard normal distribution. Note that if θ is the median of the distribution of $\hat{\theta}$, so that $f(\theta)$ is the median of the distribution of $f(\hat{\theta})$, then z_0 = 0 and we are back to the case of the empirical percentile method.

We can estimate the probability p in Equation (5.123) by computing the proportion of times the bootstrap estimate of θ is less than or equal to $\hat{\theta}$. Thus, we can estimate z_0 with

$$\hat{z}_0 = \Phi^{-1}\left(\hat{p}\right) \tag{5.125}$$

where

$$\hat{p} = \frac{1}{B}\sum_{b=1}^{B} I\left(\hat{\theta}_b \le \hat{\theta}\right) \tag{5.126}$$

$$I\left(x\right) = \begin{cases} 0 \text{ if } x \text{ is false} \\ 1 \text{ if } x \text{ is true} \end{cases} \tag{5.127}$$

(I is often called the indicator function.)

Once we have an estimate of z_0, the two-sided $(1-\alpha)100\%$ confidence interval for θ is computed as:

$$\left[\ \hat{\theta}^\star_{\hat{p}_L}\ ,\ \hat{\theta}^\star_{\hat{p}_U}\ \right] \tag{5.128}$$

where

$$\hat{p}_L = \Phi\left(2\hat{z}_0 + z_{\alpha/2}\right) \tag{5.129}$$

$$\hat{p}_U = \Phi\left(2\hat{z}_0 + z_{1-\alpha/2}\right) \tag{5.130}$$

Bias-Corrected and Adjusted Percentile Method

This method is also called the bias-corrected and accelerated (BCa) method (Efron and Tibshirani, 1993, p. 185), or the accelerated bias-corrected percentile method (Manly, 1997, p. 49). Here we call it the bias-corrected and adjusted percentile method to be consistent with S-PLUS documentation.

The bias-corrected and adjusted percentile method is a modification of the bias-corrected method. It makes the more general assumption that the distribution of the monotonic transformation of $\hat{\theta}$, $f\left(\hat{\theta}\right)$, is approximately normal with mean $f\left(\theta\right) - z_0 \left[\, 1 + a\, f\left(\theta\right) \,\right]$ and standard deviation $1 + af\left(\theta\right)$, where z_0 and a are constants (Efron and Tibshirani, 1993, pp. 184, 326; Manly, 1997, p. 49). The addition of the acceleration constant a allows the standard deviation of $f\left(\hat{\theta}\right)$ to vary linearly with $f\left(\theta\right)$. (Note that when $a = 0$, we are back to the bias-corrected percentile method just discussed.)

For this case, a two-sided $(1-\alpha)100\%$ confidence interval for θ is given by:

$$\left[\ \hat{\theta}^*_{\hat{p}_L}\ ,\ \ \hat{\theta}^*_{\hat{p}_U}\ \right] \tag{5.131}$$

where

$$\hat{p}_L = \Phi\left[\, \hat{z}_0 + \frac{\hat{z}_0 + z_{\alpha/2}}{1 - \hat{a}\left(\hat{z}_0 + z_{\alpha/2}\right)} \,\right] \tag{5.132}$$

$$\hat{p}_U = \Phi\left[\, \hat{z}_0 + \frac{\hat{z}_0 + z_{1-\alpha/2}}{1 - \hat{a}\left(\hat{z}_0 + z_{1-\alpha/2}\right)} \,\right] \tag{5.133}$$

and z_0 is estimated by Equation (5.125). One way to estimate the acceleration constant a is by:

$$\hat{a} = \frac{\sum\limits_{i=1}^{n} \left(\hat{\theta} - \hat{\theta}_{-i} \right)^3}{6 \left[\sum\limits_{i=1}^{n} \left(\hat{\theta} - \hat{\theta}_{-i} \right)^2 \right]^{3/2}} \qquad (5.134)$$

where $\hat{\theta}_{-i}$ denotes the estimate of θ based on leaving out the ith observation. See Efron and Tibshirani (1993, pp. 184–188, 326–328) and Manly (1997, pp. 49–55) for details.

Which Bootstrap Method Should I Use?

The four methods of constructing bootstrap confidence intervals that we have just discussed were presented in order of the one with the most restrictive assumptions to the one with the least restrictive assumptions. As stated earlier, of these four methods, S-PLUS computes confidence intervals based on the empirical percentile method and the bias-corrected and adjusted (BCa) percentile method. The BCa method requires more computations than the empirical percentile method, but it provides coverage closer to the assumed coverage over a broad range of parent distributions (Efron and Tibshirani, 1993, Chapter 14).

When Does the Bootstrap Work Well?

If we construct a 95% confidence interval for the mean based on the bootstrap, is this a "good" confidence interval? That is, if we repeated the experiment many, many times, and each time we created a confidence interval based on the bootstrap, would the confidence interval include the true mean about 95% of the time?

The answer to the above question depends on at least three things:

- The shape of the parent distribution.
- The sample size.
- The parameter you are estimating.

The bootstrap works well if the empirical cdf is a "good" estimate of the population cdf, which is determined by the first two factors. For a fairly symmetric parent distribution, the bootstrap should work well for moderate sample sizes (e.g., $n \geq 20$), and perhaps even smaller sample sizes, if you are constructing a confidence interval for a measure of central location such as the population mean or median. If the parent distribution is highly skewed, you will probably need a bigger sample size. Manly (1997, p. 59) found that the BCa bootstrap method (with $B = 1{,}000$ bootstraps) works fairly well for constructing a 95% confidence interval for the mean of an exponential distribution with true mean of 1 when the sample size is $n = 20$.

How well the bootstrap works also depends heavily on what kind of parameter you are estimating. Manly (1997, Chapter 3) gives an example where 95% bootstrap confidence intervals for the standard deviation of an exponential distribution provide way too little coverage (i.e., they are too narrow). The sample size in this example was again $n = 20$.

As Manly (1997, p. 58) states, bootstrap methods should not be assumed to work with small sample sizes. If you are going to create a confidence interval based on a bootstrap and a "small" sample size, it is a good idea to run a simulation to determine the coverage of the confidence interval for various assumed parent distributions and sample sizes.

Nonparametric Confidence Intervals for the Mean

Table 5.12 shows 95% BCa bootstrap confidence intervals for the mean of the Reference area TcCB concentrations of Example 5.18, the chromium concentrations of Example 5.19, and the lognormal distribution of Example 5.20. For the TcCB concentrations, looking at Table 5.4 we see that the bootstrap confidence interval is identical to the one based on Cox's method, and of course for this example all of the confidence intervals are almost all identical. For the chromium concentrations, looking at Table 5.6 we see that the upper bootstrap confidence limit of 418 lies between the limit based on Cox's method (366) and the one based on Land's method (497). For the random sample of 10 observations from a lognormal distribution with a mean of 10 and a CV of 2, looking at Table 5.8 we see that the upper bootstrap confidence limit of 41 is smaller than any of the other ones except the one based on the normal approximation.

Example	Data	Confidence Interval	Confidence Interval Type
5.17	Reference area TcCB	[0.52, 0.69]	Two-sided
5.18	Chromium	[0, 418]	Upper
5.19	Random Sample	[0, 41]	Upper

Table 5.12   95% BCa bootstrap confidence intervals for three example data sets

Figure 5.12 displays the bootstrap distribution of the estimated mean for the chromium data in the form of a histogram and an overlaid density plot. The solid vertical line shows the observed mean of chromium (175.5), and the dotted vertical line shows the mean of the bootstrap distribution (172.2). The distance between these two lines is an estimate of the bias of the original estimator for the parameter it is estimating. Of course, in this case we know the statistic (the sample mean) is an unbiased estimator of the parameter (the population mean).

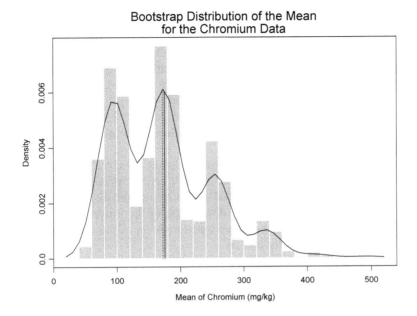

Figure 5.12   Bootstrap distribution of the sample mean for the chromium data

Menu

To produce the confidence intervals shown in Table 5.12 using the S-PLUS pull-down menu, follow these steps.

1. In the Object Explorer, make sure **epa.94b.tccb.df** is highlighted.
2. On the S-PLUS menu bar, make the following menu choices: **Statistics>Resample>Bootstrap**. This will bring up the Bootstrap Inference dialog box.
3. In the Data Set box, make sure **epa.94b.tccb.df** is selected.
4. In the Expression box, type **mean(TcCB[Area=="Reference"])**.
5. Click on the **Options** tab.
6. In the Number of Resamples box, type **2000**.
7. In the Random Number Seed box, type **47**. (Note: this is done just so that your results match the ones in this example. Leaving this box blank or explicitly supplying a different seed may yield very slightly different confidence limits.)
8. Click **OK** or **Apply**.

The above steps will produce the confidence interval for the mean of the Reference area TcCB concentrations.

To produce the confidence interval for the mean of the chromium data, repeat the above steps with the following exceptions: in Step 1, highlight the vector **epa.92d.chromium.vec** in the Object Explorer, in Step 3 **epa.92d.chromium.vec** should be selected in the Data Frame box, and in Step 4 type **mean(epa.92d.chromium.vec)** in the Expression box.

To produce the confidence interval for the mean of the lognormal data, repeat the above steps with the following exceptions: in Step 1, highlight the data frame **lognormal.df** in the Object Explorer, in Step 3 **lognormal.df** should be selected in the Data Frame box, and in Step 4 type **mean(Sample)** in the Expression box.

By default, the bootstrap distribution is also plotted. You can also create a normal Q-Q plot for the bootstrap distribution (see the options under the Plot tab), but remember the BCa method does not assume the distribution of the estimator (and hence the bootstrap distribution of the estimator) is approximately normal, but rather that some transformation of the distribution of the estimator is approximately normal.

Command

To produce the confidence limits shown in Table 5.12 using the S-PLUS Command or Script Window, type these commands.

```
TcCB.ref.boot.list <- bootstrap(data=epa.94b.tccb.df,
    statistic=mean(TcCB[Area=="Reference"]), B=2000,
    seed=47)

summary(TcCB.ref.boot.list)

chromium.boot.list <-
    bootstrap(data=epa.92d.chromium.vec,
    statistic=mean(epa.92d.chromium.vec), B=2000,
    seed=47)

summary(chromium.boot.list)

lognormal.boot.list <- bootstrap(data=lognormal.df,
    statistic=mean(Sample), B=2000, seed=47)

summary(lognormal.boot.list)
```

To produce plots of the bootstrap distribution like the one for the mean of chromium shown in Figure 5.12, type these commands.

```
plot(TcCB.ref.boot.list)

plot(chromium.boot.list)

plot(lognormal.boot.list)
```

A Simulation Study of the Coverage of the Upper Confidence Limit for the Mean Based on Positively Skewed Data

Table 5.13 shows the results of a small simulation study to assess how well the bootstrap BCa method works for constructing one-sided upper 95% confidence limits for the mean with small to moderate sample sizes when the underlying distribution is positively skewed. This study looked at just two types of parent distributions: a lognormal distribution and a lognormal mixture distribution. The first lognormal distribution has a mean of 10 and a CV of 1. The second lognormal distribution has a mean of 10 and a CV of 2. The lognormal mixture distribution is a mixture of a lognormal distribution with a mean of 5 and a CV of 1, and one with a mean of 30 and a CV of 0.5, with a mixing proportion is 30%. Thus the overall mean is $0.7 \times 5 + 0.3 \times 30 = 12.5$. This last parent distribution could be used to model chemical concentrations from an area that has been cleaned up but contains some residual contamination. Figure 5.13 displays the parent distributions.

Table 5.13 shows the actual coverage of the confidence intervals based on 1,000 simulations, thus the standard error on these estimates of coverage is about 1, since the estimated coverage is between about 80% to 93% (recall the formula for the standard error of an estimated proportion given in Equation (5.102)). For the first lognormal distribution ($CV = 1$), the coverage is almost adequate for $n = 30$, but the coverage is worse for the second lognormal distribution ($CV = 2$). For the lognormal mixture distribution, the coverage is not too much below the assumed coverage for n as small as 15.

Distribution	Sample size	% Coverage
Lognormal (10, 1)	10	84
"	15	85
"	20	87
"	25	89
"	30	90
Lognormal (10, 2)	10	77
"	15	80
"	20	83
"	25	82
"	30	85
Lognormal Mixture	10	90
"	15	92
"	20	92
"	25	92
"	30	93

Table 5.13   Actual percent coverage of one-sided upper 95% confidence intervals for the mean, based on 1,000 simulations using the BCa bootstrap method with $B = 2,000$ bootstraps

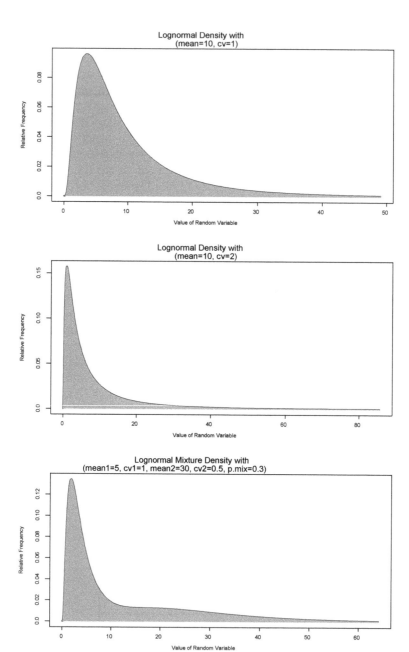

Figure 5.13   Probability density functions for the lognormal and lognormal mixture distributions used in the simulation study

These results illustrate that the bootstrap method does not work well for producing upper confidence limits for the mean of an extremely right skewed distribution until the sample size is fairly "large." This is because an extremely right skewed distribution has a small amount of probability concentrated on very high values, and the mean, unlike the median, is very sensitive to high values. These fairly high values will not show up too often in small sample sizes, so the empirical cdf will not be a good estimate of the population cdf for small sample sizes.

Nonparametric Confidence Intervals for the Difference between Two Means or Two Medians

The method of the bootstrap is not limited to constructing confidence intervals for a single population parameter. It is a quite general method that can be applied to more complicated problems such as comparing two means or medians, linear regression, and time series analysis (Efron and Tibshirani, 1993; Manly, 1997). Here we will illustrate how to apply the bootstrap to construct a confidence interval for the difference between two means and the difference between two medians.

The observed mean concentrations of TcCB in the Cleanup and Reference areas are 3.9 and 0.6 ppb, respectively, yielding a difference of 3.3 ppb. A two-sided 95% confidence interval for the difference between the two population mean concentrations, based on the BCa bootstrap method, is [0.75, 13]. Figure 5.14 shows the bootstrap distribution of the difference in the means.

The observed median concentrations of TcCB in the Cleanup and Reference areas are 0.43 and 0.54 ppb, respectively, yielding a difference of –0.11 ppb. A two-sided 95% confidence interval for the difference between the two population median concentrations, based on the BCa bootstrap method, is [–0.25, 0.6]. Figure 5.15 shows the bootstrap distribution of the difference in the medians.

The bootstrap distribution of the difference in the means and the resulting 95% confidence interval show that the difference in the true (population) means may be anywhere between about 1 and 13 ppb, reflecting the large skew in the Cleanup area TcCB data. On the other hand, the bootstrap distribution of the difference in the medians and the resulting 95% confidence interval show that the difference in the true (population) medians is probably relatively small. Recalling Figure 3.2 in Chapter 3, this is not surprising.

Menu

To produce the two-sided 95% confidence interval for the difference in the means using the S-PLUS pull-down menu, follow these steps.

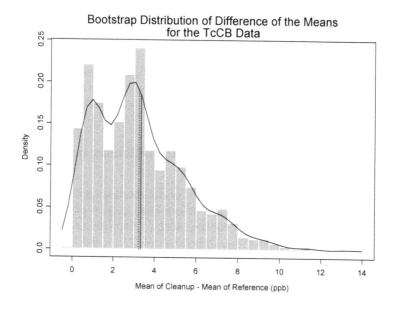

Figure 5.14   Bootstrap distribution of the difference in the mean TcCB concentration for the Cleanup and Reference areas

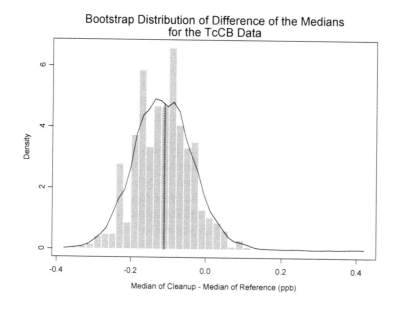

Figure 5.15   Bootstrap distribution of the difference in the median TcCB concentration for the Cleanup and Reference areas

1. In the Object Explorer, make sure **epa.94b.tccb.df** is highlighted.
2. On the S-PLUS menu bar, make the following menu choices: **Statistics>Resample>Bootstrap**. This will bring up the Bootstrap Inference dialog box.
3. In the Data Set box, make sure **epa.94b.tccb.df** is selected.
4. In the Expression box, type **mean(TcCB[Area=="Cleanup"]) - mean(TcCB[Area=="Reference"])**.
5. Click on the **Options** tab.
6. In the Number of Resamples box, type **2000**.
7. In the Grouping Variable box, select **Area**.
8. In the Random Number Seed box, type **47**. (Note: this is done just so that your results match the ones in this example. Leaving this box blank or explicitly supplying a different seed may yield very slightly different confidence limits.)
9. Click **OK** or **Apply**.

To produce the confidence interval for the difference of the median concentrations, repeat the above steps, except in Step 4, in the Expression box, replace "mean" with "median": **median(TcCB[Area=="Cleanup"]) - median(TcCB[Area=="Reference"])**.

Command

To produce the two-sided 95% confidence interval for the difference in the means using the S-PLUS Command or Script Window, type these commands.

```
TcCB.diff.means.boot.list <-
   bootstrap(data=epa.94b.tccb.df, statistic=
   mean(TcCB[Area=="Cleanup"])-
   mean(TcCB[Area=="Reference"]), group=Area, B=2000,
   seed=47)

summary(TcCB.diff.means.boot.list)

plot(TcCB.diff.means.boot.list,
   xlab="Mean of Cleanup - Mean of Reference (ppb)",
   main=paste("Bootstrap Distribution of ",
   "Difference of the Means\n", "for the TcCB Data",
   sep=""))
```

To produce the two-sided confidence interval for the difference in the medians, type these commands.

```
TcCB.diff.medians.boot.list <-
   bootstrap(data=epa.94b.tccb.df,
   statistic=median(TcCB[Area=="Cleanup"]) -
```

```
median(TcCB[Area=="Reference"]), group=Area,
B=2000, seed=47)
```

```
summary(TcCB.diff.medians.boot.list)
```

```
plot(TcCB.diff.medians.boot.list)
```

ESTIMATES AND CONFIDENCE INTERVALS FOR DISTRIBUTION QUANTILES (PERCENTILES)

Quantiles or percentiles are sometimes used in environmental standards and regulations (e.g., Berthouex and Brown, 1994, p. 65). For example, in order to determine compliance, you may be required to estimate an extreme percentile (e.g., the 95th percentile) for the "background level" distribution, and then compare observations at compliance wells or remediated areas to this upper percentile (or an upper confidence limit for this percentile). In the context of soil cleanup, USEPA (1994b, pp. 4.8–4.9) has called this the "Hot-Measurement Comparison." (There are some major problems with this technique that are discussed in Chapters 6 and 7.)

As another example, when monitoring groundwater around a RCRA landfill, the site may be in "compliance monitoring" with a fixed compliance limit for a particular chemical. The fixed compliance limit may be a maximum concentration limit (MCL) or an alternate concentration limit (ACL). Most MCLs in fact represent average levels, but sometimes an MCL or ACL may represent a limit that should be exceeded only a small fraction of the time, for example, the 99th percentile of the distribution. In this case you need to compare the 99th percentile of the distribution of the chemical's concentration in the groundwater with the MCL. If the estimated 99th percentile (or a lower confidence limit for this percentile) is greater than the MCL, then you are out of compliance.

As Berthoux and Brown (1994, p. 65) point out, even though using a 99th percentile for decision making sounds like we are doing the conservative or safe thing, we must remember that estimators are random variables, and estimates of extreme percentiles can be very, very random indeed! Berthouex and Brown (1994, p. 65) note that for a normal (and therefore a lognormal) distribution, the 50th percentile (i.e., median) can be estimated with greater precision than any other percentile, and the precision decreases rapidly for larger or smaller percentiles. Estimates of extreme percentiles are very imprecise, even for data sets with large sample sizes. Considering these facts, it is very important to quantify the precision of any estimate of an extreme percentile with a confidence interval.

Confidence intervals for quantiles are based on the idea of tolerance intervals. It can be shown (e.g., Conover, 1980, pp. 119–121) that an upper confidence interval for the pth quantile with confidence level $100(1-\alpha)\%$ is equivalent to an upper β-content tolerance interval with coverage $100p\%$ and confidence level $100(1-\alpha)\%$. Also, a lower confidence interval for the pth quantile with confidence level $100(1-\alpha)\%$ is equivalent to a lower β-content tolerance interval with coverage $100(1-p)\%$ and confidence level $100(1-\alpha)\%$. We will talk more about tolerance intervals in the next chapter.

Formula for the pth Quantile

Recall the definition of the pth quantile from Chapter 4: the pth *quantile* of a population is the (a) number such that a fraction p of the population is less than or equal to this number. If X is a continuous random variable, the pth quantile of X, denoted x_p, is simply defined as the value such that the cdf of that value is equal to p:

$$\Pr\left(X \le x_p\right) = F\left(x_p\right) = p \qquad (5.135)$$

For a discrete random variable, where several values may satisfy the definition for a specified value of p, the pth quantile of X is often taken to be the average of the smallest and largest numbers that satisfy the condition. The S-PLUS functions for computing quantiles, however, return the smallest number that satisfies the condition. A general formula for the pth quantile is therefore:

$$x_p = F^{-1}\left(p\right) \qquad (5.136)$$

where $F^{-1}(p)$ denotes the inverse cdf evaluated at p. For a discrete distribution, we will define this function to return the smallest value of all possible values.

Parametric Estimates of Quantiles

To estimate quantiles parametrically, all you need to do is estimate the parameters of the assumed distribution, then compute the quantiles based on the cdf of this distribution and Equation (5.136). For example, Figure 4.9 in Chapter 4 shows that the value 1 is about the 0.9 quantile (90th percentile) of a lognormal distribution with mean 0.6 and CV 0.5. Of course, to use this method you have to be fairly sure that the distribution you are assuming is a good model for the data.

Estimates and Confidence Intervals for Quantiles of a Normal Distribution

In Equation (4.54) of Chapter 4 (see Exercise 4.7), we stated that the pth quantile of a normal distribution with mean μ and standard deviation σ is:

$$x_p = \mu + \sigma\, z_p \qquad\qquad (5.137)$$

where z_p denotes the pth quantile of a standard normal distribution. Thus, a natural estimate for the pth quantile simply replaces μ and σ with their estimated values:

$$\hat{x}_p = \bar{x} + s\, z_p \qquad\qquad (5.138)$$

where $\bar{x}$ and s are defined in Equations (5.7) and (5.50), respectively. The estimator in Equation (5.138) is sometimes called the "quasi maximum likelihood estimator" (qmle; Cohn et al., 1989) of x_p because if the maximum likelihood estimator of standard deviation s_m (Equation (5.40)) were used in place of s, then this estimator would be the mle of x_p.

Based on the theory of tolerance intervals for a normal distribution (Guttman, 1970, pp. 87–89; Owen, 1962, p. 117; Odeh and Owen, 1980; Gibbons, 1994, pp. 85–86), the one-sided upper $(1-\alpha)100\%$ confidence interval for the pth quantile x_p is given by:

$$\left[-\infty,\ \ \bar{x} + t_{n-1,\, z_p\sqrt{n},\, 1-\alpha}\ \ \frac{s}{\sqrt{n}} \right] \qquad\qquad (5.139)$$

where $t_{\nu,\delta,p}$ denotes the pth quantile of the non-central Student's t-distribution with ν degrees of freedom and non-centrality parameter δ. Similarly, the one-sided lower $(1-\alpha)100\%$ confidence interval for the pth quantile x_p is given by:

$$\left[\bar{x} + t_{n-1,\, z_p\sqrt{n},\, \alpha}\ \ \frac{s}{\sqrt{n}}\ ,\ \infty \right] \qquad\qquad (5.140)$$

and the two-sided $(1-\alpha)100\%$ confidence interval for the pth quantile x_p is given by:

$$\left[\bar{x} + t_{n-1, z_p \sqrt{n}, \frac{\alpha}{2}} \frac{s}{\sqrt{n}} \right. ,$$

(5.141)

$$\left. \bar{x} + t_{n-1, z_p \sqrt{n}, 1-\frac{\alpha}{2}} \frac{s}{\sqrt{n}} \right]$$

Example 5.24: Comparing Lower Confidence Limits of the 95th Percentile to an ACL for Aldicarb

The guidance document USEPA (1989b) contains an example on pages 6-11 to 6-13 of aldicarb concentrations (ppm) at three groundwater monitoring compliance wells (four monthly samples at each well). These data are shown in Table 5.14, and in ENVIRONMENTALSTATS for S-PLUS they are stored in the data frame `epa.89b.aldicarb2.df`. In this example, it is assumed that the permit establishes an ACL of 50 ppm that should not be exceeded more than 5% of the time. Thus, for each well, we need to determine whether the 95th percentile of aldicarb concentrations is greater than the ACL of 50 ppm. To do this, we need to compute a one-sided *lower* confidence limit for the 95th percentile. If this confidence limit is above the ACL, then we can be fairly sure that the true 95th percentile is above the ACL. (Note: the EPA guidance document incorrectly computes a one-sided *upper* confidence limit for the 95th percentile.)

Month	Well 1	Well 2	Well 3
Jan	19.9	23.7	25.6
Feb	29.6	21.9	23.3
Mar	18.7	26.9	22.3
Apr	24.2	26.1	26.9

Table 5.14   Aldicarb concentrations (ppm) at three compliance wells, from USEPA (1989b, p. 6-13)

Statistic	Well 1	Well 2	Well 3
Mean	23.1	24.7	24.5
SD	4.9	2.3	2.1
95th Percentile	31.2	28.4	28.0
LCL on 95th Percentile	25.3	25.7	25.5

Table 5.15   Statistics for aldicarb concentrations at the three compliance wells.

Table 5.15 displays the estimated mean, standard deviation, 95th percentile, and lower one-sided 99% confidence limit for the 95th percentile for each of the three compliance wells. All of the lower confidence limits for the 95th percentile are well below the ACL of 50 ppm, so none of the wells is out of compliance.

To give you an idea of the amount of variability in these estimates of the 95th percentile, we can also create two-sided confidence intervals. The two-sided 99% confidence intervals for the 95th percentiles at each well (rounded to whole numbers) are: [25, 80], [25, 51], and [25, 49]. Of course, a major problem here is the very small sample size ($n = 4$) at each well.

Menu

To produce the one-sided lower 99% confidence limits for the 95th percentiles for each of the compliance wells using the ENVIRONMENTALSTATS for S-PLUS pull-down menu, follow these steps.

1. In the Object Explorer, find the data frame **epa.89b.aldicarb2.df**.
2. On the S-PLUS menu bar, make the following menu choices: **EnvironmentalStats>Estimation>Quantiles**. This will bring up the Estimate Distribution Quantiles dialog box.
3. For Data to Use, make sure the **Pre-Defined Data** button is selected.
4. In the Data Set box, make sure **epa.89b.aldicarb2.df** is selected.
5. In the Variable box, select **Aldicarb**.
6. In the Subset Rows with box, type **Well=="1"**.
7. In the Quantile(s) box, type **0.95**.
8. Under the Distribution/Estimation group, for Type, make sure the **Parametric** button is selected. In the Distribution box, select **Normal**. For Estimation Method, select **qmle**.
9. Under the Confidence Interval group, make sure the Confidence Interval box is **checked**. For CI Type, select **lower**. For CI Method, select **exact**. For Confidence Level (%), type **99**. Click **OK** or **Apply**.

This will produce the estimated 95th percentile and one-sided lower 99% confidence limit for this percentile for the first well. To produce these quantities for the other two wells, in Step 6 above, type **Well=="2"** or **Well=="3"**.

Command

To produce the one-sided lower 99% confidence limits for the 95th percentiles for each of the compliance wells using the ENVIRONMENTALSTATS for S-PLUS Command or Script Window, type these commands.

```
attach(epa.89b.aldicarb2.df)
lapply(split(Aldicarb, Well), eqnorm, p=0.95, ci=T,
  ci.type="lower", conf.level=0.99)
detach()
```

Estimates and Confidence Intervals for Quantiles of a Lognormal Distribution

As we stated in Chapter 4, quantiles are invariant under monotonic transformations (see Equation (4.10)). Therefore, the formulas to estimate percentiles and create confidence intervals for them for the normal distribution can be easily modified for the lognormal distribution.

Let X denote a lognormal random variable with mean θ and CV τ, and let μ and σ denote the mean and standard deviation of $Y = \log(X)$. Based on Equation (5.137) above, the pth quantile of the distribution of X is:

$$x_p = \exp\left(\mu + \sigma z_p\right) \qquad (5.142)$$

and the quasi-maximum likelihood estimator of this quantile is:

$$\hat{x}_p = \exp\left(\bar{y} + s_y z_p\right) \qquad (5.143)$$

where $\bar{y}$ is defined in Equations (5.22) and (5.23) and s_y is defined in Equation (5.56). Furthermore, the one-sided upper $(1-\alpha)100\%$ confidence interval for the pth quantile x_p is given by:

$$\left[-\infty, \quad \exp\left(\bar{x} + t_{n-1, z_p\sqrt{n}, 1-\alpha}\ \frac{s}{\sqrt{n}}\right)\right] \qquad (5.144)$$

the one-sided lower $(1-\alpha)100\%$ confidence interval for the pth quantile x_p is given by:

$$\left[\exp\left(\bar{x} + t_{n-1, z_p\sqrt{n}, \alpha}\ \frac{s}{\sqrt{n}}\right), \quad \infty\right] \qquad (5.145)$$

and the two-sided $(1-\alpha)100\%$ confidence interval for the pth quantile x_p is given by:

$$
\left[\; \exp\left(\bar{x} + t_{n-1,\, z_p\sqrt{n},\, \frac{\alpha}{2}}\; \frac{s}{\sqrt{n}}\right),\; \right.
$$

$$
\left. \exp\left(\bar{x} + t_{n-1,\, z_p\sqrt{n},\, 1-\frac{\alpha}{2}}\; \frac{s}{\sqrt{n}}\right)\; \right]
$$

(5.146)

Example 5.25: Comparing Lower Confidence Limits of the 95th Percentile to an ACL for Chrysene

The guidance document USEPA (1992c, pp. 52–54) contains an example of chrysene concentrations (ppb) at five groundwater monitoring compliance wells (four monthly samples at each well). These data are shown in Table 5.16, and in ENVIRONMENTALSTATS for S-PLUS they are stored in the data frame epa.92c.chrysene.df. They are assumed to come from a lognormal distribution.

In this example, it is assumed that the permit establishes an ACL of 80 ppb that should not be exceeded more than 5% of the time. Thus, for each well, we need to determine whether the 95th percentile of chrysene concentrations is greater than the ACL of 80 ppb. To do this, we need to compute a one-sided *lower* confidence limit for the 95th percentile. If this confidence limit is above the ACL, then we can be fairly sure that the true 95th percentile is above the ACL. (Note: The EPA guidance document incorrectly computes a one-sided *upper* confidence limit for the 95th percentile.)

Month	Well 1	Well 2	Well 3	Well 4	Well 5
1	19.7	10.2	68.0	26.8	47.0
2	39.2	7.2	48.9	17.7	30.5
3	7.8	16.1	30.1	31.9	15.0
4	12.8	5.7	38.1	22.2	23.4

Table 5.16   Chrysene concentrations (ppb) at five compliance wells, from USEPA (1992c, p. 52)

Statistic	Well 1	Well 2	Well 3	Well 4	Well 5
Mean	19.9	9.8	46.3	24.7	29.0
CV	0.7	0.5	0.4	0.2	0.5
95th Percentile	51.4	19.0	78.5	36.5	58.5
LCL on 95th Percentile	22.6	11.1	51.6	26.9	32.9

Table 5.17   Statistics for chrysene concentrations at the five compliance wells

Table 5.17 displays the estimated mean, coefficient of variation, 95th percentile, and lower one-sided 99% confidence limit for the 95th percentile for each of the five compliance wells. All of the lower confidence limits for the 95th percentile are well below the ACL of 80 ppb, so none of the wells is out of compliance.

To give you an idea of the amount of variability in these estimates of the 95th percentile, we can also create two-sided confidence intervals. The two-sided 99% confidence intervals for the 95th percentiles at each well (rounded to whole numbers) are: [25, 178], [11, 63], [52, 235], [27, 95], and [34, 185]. As in the previous example, a major problem here is the very small sample size ($n = 4$) at each well.

Menu

To produce the one-sided lower 99% confidence limits for the 95th percentiles for each of the compliance wells using the ENVIRONMENTALSTATS for S-PLUS pull-down menu, follow these steps.

1. In the Object Explorer find the data frame **epa.92c.chrysene.df**.
2. On the S-PLUS menu bar, make the following menu choices: **EnvironmentalStats>Estimation>Quantiles**. This will bring up the Estimate Distribution Quantiles dialog box.
3. For Data to Use, make sure the **Pre-Defined Data** button is selected.
4. In the Data Set box, make sure **epa.92c.chrysene.df** is selected.
5. In the Variable box, select **Chrysene**.
6. In the Subset Rows with box, type **Well=="1"**.
7. In the Quantile(s) box, type **0.95**.
8. Under the Distribution/Estimation group, for Type, make sure the **Parametric** button is selected. In the Distribution box, select **Lognormal**. For Estimation Method, select **qmle**.
9. Under the Confidence Interval group, make sure the Confidence Interval box is **checked**. For CI Type, select **lower**. For CI Method, select **exact**. For Confidence Level (%), type **99**. Click **OK** or **Apply**.

This will produce the estimated 95th percentile and one-sided lower 99% confidence limit for this percentile for the first well. To produce these quantities for the other four wells, in Step 6 above, type **Well=="2"**, **Well=="3"**, **Well=="4"**, or **Well=="5"**.

Command

To produce the one-sided lower 99% confidence limits for the 95th percentiles for each of the compliance wells using the ENVIRONMENTALSTATS for S-PLUS Command or Script Window, type these commands.

```
attach(epa.92c.chrysene.df)
lapply(split(Chrysene, Well), eqlnorm, p=0.95, ci=T,
   ci.type="lower", conf.level=0.99)
detach()
```

Estimates and Confidence Intervals for Quantiles of a Poisson Distribution

As we stated earlier, to estimate quantiles parametrically, all you need to do is estimate the parameters of the assumed distribution, then compute the quantiles based on the cdf of this distribution and Equation (5.136). For a Poisson distribution with mean λ, if we let $x_{p|\lambda}$ denote the pth quantile of this distribution, then the estimated p^{th} quantile is:

$$\hat{x}_{p|\lambda} = x_{p|\lambda=\hat{\lambda}} \qquad (5.147)$$

where $\hat{\lambda}$ is given in Equation (5.104). Because $\hat{\lambda}$ is the maximum likelihood estimator, the estimator in Equation (5.147) is the maximum likelihood estimator of the pth quantile.

Based on the theory of tolerance intervals for a Poisson distribution (Zacks, 1970), the one-sided upper $(1-\alpha)100\%$ confidence interval for the pth quantile is given by:

$$\left[0, \ x_{p|\lambda=UCL} \right] \qquad (5.148)$$

where UCL denotes the upper $(1-\alpha)100\%$ confidence limit for λ (see the section Confidence Interval for the Mean of a Poisson Distribution above). Similarly, a one-sided lower $(1-\alpha)100\%$ confidence interval for the pth quantile is given by:

$$\left[x_{p|\lambda=LCL}, \ \infty \right] \qquad (5.149)$$

where LCL denotes the lower $(1-\alpha)100\%$ confidence limit for λ, and a two-sided $(1-\alpha)100\%$ confidence interval for p^{th} quantile is given by:

$$\left[x_{p|\lambda=LCL}, \ x_{p|\lambda=UCL} \right] \qquad (5.150)$$

where *LCL* and *UCL* denote the two-sided lower and upper $(1-\alpha)100\%$ confidence limits for λ.

Example 5.26: Computing an Upper Confidence Limit on the 95th Percentile of Benzene Concentrations

Let us look again at the benzene concentrations in groundwater data that we looked at in Examples 5.11, 5.22, and 5.23. The data are shown in Table 5.2, and in ENVIRONMENTALSTATS for S-PLUS they are stored in the data frame epa.92c.benzene1.df. Nondetect values are reported as "<2." Of the 36 values, 33 are nondetects. Following Gibbons (1987, p. 574), USEPA (1992c, p. 36) assumes that these data can be modeled as having come from a Poisson process. The three detected values represent a count of the number of molecules of benzene observed out of a much larger number of water molecules. The 33 nondetect values are also assumed to represent counts, where the count must be equal to 0 or 1, since they are recorded as "<2" ppb. Following USEPA (1992c), for this example we will set all of the values of the nondetects to 1.

USEPA (1992c, pp. 38–40) describes how to compute an upper confidence limit for the 95th percentile of the benzene concentrations at the six background wells, with the idea that this limit will be used as a threshold value for concentrations observed in downgradient monitoring wells (i.e., if the concentration at a downgradient well exceeds this limit, this indicates there may be contamination in the groundwater). This is the "Hot-Measurement Comparison," and, as already stated, there are some major problems with this technique that are discussed in Chapters 6 and 7.

For this data set, the estimated mean is 1.94, the estimated 95th percentile is 4, and the one-sided upper 95% confidence limit for the 95th percentile is 5. As we stated earlier in Example 5.23, these data do not appear to follow a Poisson distribution too well.

Menu

To produce the one-sided upper 95% confidence limit for the 95th percentile of benzene concentrations using the ENVIRONMENTALSTATS for S-PLUS pull-down menu, follow these steps.

1. In the Object Explorer, highlight **new.epa.92c.benzene1.df**.
2. On the S-PLUS menu bar, make the following menu choices: **EnvironmentalStats>Estimation>Quantiles**. This will bring up the Estimate Distribution Quantiles dialog box.
3. For Data to Use, make sure the **Pre-Defined Data** button is selected.
4. In the Data Set box, make sure **new.epa.92c.benzene1.df** is selected.
5. In the Variable box, select **new.Benzene**.
6. In the Quantile(s) box, type **0.95**.

7. Under the Distribution/Estimation group, for Type, make sure the **Parametric** button is selected. In the Distribution box, select **Poisson**. For Estimation Method, select **mle**.
8. Under the Confidence Interval group, make sure the Confidence Interval box is **checked**. For CI Type, select **upper**. For CI Method, select **exact**. For Confidence Level (%), type **95**. Click **OK** or **Apply**.

Command

To produce the one-sided upper 95% confidence limit for the 95th percentile of the benzene concentrations using the ENVIRONMENTALSTATS for S-PLUS Command or Script Window, type these commands.

```
attach(epa.92c.benzene1.df)

Benzene[Censored] <- 1

eqpois(Benzene, p=0.95, ci=T, ci.type="upper")

rm(Benzene)

detach()
```

Nonparametric Estimates for Quantiles

To estimate quantiles nonparametrically, all you need to do is estimate the cdf nonparametrically using the empirical cdf, then use linear interpolation (if necessary). Graphically, this just means connecting the points in the quantile plot by straight lines, then performing the same inverse procedure as illustrated in Figure 4.9. For example, if you have $n = 10$ observations, you might consider the smallest observation an estimate of the 10th percentile and the largest observation an estimate of the 100th percentile. This implies that you are using the plotting position formula shown in Equation (3.21) in Chapter 3. Note, however, that because of the plotting positions you are using, you cannot estimate the 5th percentile (unless you want to assume the smallest value is an estimate of any percentile between the 0th and 10th percentile).

In S-PLUS, quantiles are estimated nonparametrically by first creating the empirical cdf with the following plotting positions:

$$\hat{p}_i = \frac{i-1}{n-1} \qquad (5.151)$$

This is the plotting position formula of Equation (3.22) in Chapter 3, with $a = 1$. Once the empirical cdf is computed, linear interpolation is used to estimate the pth quantile:

$$\hat{x}_p = x_{(i+1)} - \left[x_{(i+1)} - x_{(i)} \right] \left(\frac{\hat{p}_{i+1} - p}{\hat{p}_{i+1} - \hat{p}_i} \right) \qquad (5.152)$$

where $x_{(i)}$ denotes the ith largest value in the sample. For this method, the smallest observation estimates the 0th percentile and the largest observation estimates the 100th percentile.

One problem with estimating quantiles nonparametrically vs. parametrically is that you need many more observations to estimate extreme quantiles with good precision. In fact, even though S-PLUS will give us estimates for the 5th and 95th percentiles, it does not make sense intuitively that we should be able to estimate anything less than the 10th percentile or anything more than the 90th percentile with any kind of precision if $n = 10$. This characteristic becomes clear when we create nonparametric confidence intervals for quantiles (see the next section). On the other hand, an advantage to estimating quantiles nonparametrically is that it is often easy to deal with censored values since all you have to do is rank them.

Nonparametric Confidence Intervals for Quantiles

It can be shown (e.g., Conover, 1980, pp. 114–116) that for the ith order statistic:

$$\Pr \left[X_{(i)} > x_p \right] = F_{n,p} \left(i - 1 \right) \qquad (5.153)$$

for $i = 1, 2, \dots, n$, where $F_{n,p}(y)$ denotes the cumulative distribution function of a binomial random variable based on n trials and probability of "success" p evaluated at y. This fact is used to construct exact nonparametric confidence intervals for quantiles.

A two-sided nonparametric confidence interval for the pth quantile is constructed as:

$$\left[x_{(r)}, \ x_{(s)} \right] \qquad (5.154)$$

where $1 \le r \le n-1$, $2 \le s \le n$, and $r < s$. This confidence interval has an associated confidence level given by:

$$F_{n,p}(s-1) - F_{n,p}(r-1) \qquad (5.155)$$

(actually, for a discrete distribution, it is at least this large, and may be larger). This is because:

$$\Pr\left[X_{(r)} \le x_p \le X_{(s)}\right]$$

$$= \Pr\left[x_p \le X_{(s)}\right] - \Pr\left[x_p < X_{(r)}\right] \qquad (5.156)$$

$$\ge \Pr\left[x_p < X_{(s)}\right] - \Pr\left[x_p < X_{(r)}\right]$$

Looking at Equation (5.153), you can see that the last line of Equation (5.156) is the same as Equation (5.155).

Similarly, a one-sided lower nonparametric confidence interval for the pth quantile is given by:

$$\left[\, X_{(r)}\, ,\quad \infty \,\right] \qquad (5.157)$$

and has an associated confidence level of:

$$1 - F_{n,p}(r-1) \qquad (5.158)$$

and a one-sided upper nonparametric confidence interval for the pth quantile is given by:

$$\left[\, -\infty\, ,\quad X_{(s)} \,\right] \qquad (5.159)$$

and has an associated confidence level of:

$$F_{n,p}\,(s-1) \qquad\qquad\qquad \textbf{(5.160)}$$

Because nonparametric confidence intervals for percentiles are based on the ranks of the data, the confidence level is discrete in nature. That is, for a particular percentile, you can only get certain confidence levels, and these confidence levels are limited by the sample size. Table 5.18 illustrates the confidence levels associated with a one-sided upper confidence interval for the 95th percentile, based on various sample sizes. For this table, the upper confidence limit is the maximum value (Equation (5.159) with $s = n$). You can see that a confidence level greater than 95% cannot be achieved until the sample size is larger than $n = 50$.

Sample Size (n)	Confidence Level (%)
5	23
10	40
15	54
20	64
25	72
50	92
75	98
100	99

Table 5.18   Confidence levels for one-sided upper nonparametric confidence intervals for the 95th percentile, based on using the maximum value as the upper confidence limit

Example 5.27: Constructing a Nonparametric Upper Confidence Limit for the 95th Percentile of Copper Concentrations

The guidance document USEPA (1992c, pp. 55–56) contains an example of copper concentrations (ppb) at five groundwater monitoring wells: three background wells and two compliance wells (eight monthly samples at each well, except the first four missing at the two compliance wells). These data are shown in Table 5.19, and in ENVIRONMENTALSTATS for S-PLUS they are stored in the data frame epa.92c.copper2.df. Note that 15 out of the 24 observations at the background wells are nondetects recorded as "<5."

USEPA (1992c, pp. 55–56) describes how to compute a nonparametric upper confidence limit for an extreme percentile of the copper concentrations at the three background wells, with the idea that this limit will be used as a threshold value for concentrations observed in the compliance wells (i.e., if the concentration at a compliance well exceeds this limit, this indicates there may be contamination in the groundwater). This is the "Hot-Measurement

Comparison," and, as already stated, there are some major problems with this technique that are discussed in Chapters 6 and 7.

	Background Wells			Compliance Wells	
Month	Well 1	Well 2	Well 3	Well 4	Well 5
1	<5	9.2	<5		
2	<5	<5	5.4		
3	7.5	<5	6.7		
4	<5	6.1	<5		
5	<5	8.0	<5	6.2	<5
6	<5	5.9	<5	<5	<5
7	6.4	<5	<5	7.8	5.6
8	6.0	<5	<5	10.4	<5

Table 5.19   Copper concentrations (ppb) at five groundwater monitoring wells, from USEPA (1992c, p. 55)

For this data set, 15/24 = 62.5% of the values at the background wells are nondetects, so the estimated median is some value less than 5. The estimated 95th percentile is 7.925, and using the largest value of 9.2 ppb as the upper confidence limit for the 95th percentile yields a one-sided upper confidence interval with a confidence level of only 71%! If we are willing to use 9.2 as a threshold value, then Well 4 indicates possible contamination, whereas Well 5 does not.

Menu

To nonparametrically estimate the 95th percentile of copper concentrations at the background wells, and produce the one-sided upper confidence limit for the 95th percentile using the ENVIRONMENTALSTATS for S-PLUS pull-down menu, follow these steps.

1. In the Object Explorer, find the data frame **epa.92c.copper2.df**.
2. On the S-PLUS menu bar, make the following menu choices: **EnvironmentalStats>Estimation>Quantiles**. This will bring up the Estimate Distribution Quantiles dialog box.
3. For Data to Use, make sure the **Pre-Defined Data** button is selected.
4. In the Data Set box, make sure **epa.92c.copper2.df** is selected.
5. In the Variable box, select **Copper**.
6. In the Subset Rows with box, type **Well.type=="Background"**.
7. In the Quantile(s) box, type **0.95**.
8. Under the Distribution/Estimation group, for Type, select the **Nonparametric** button.
9. Under the Confidence Interval group, make sure the Confidence Interval box is **checked**. For CI Type, select **upper**. For CI Method,

select **exact**. For Confidence Level (%), type **95**. For Lower Bound, type **0**. Click **OK** or **Apply**.

Command

To nonparametrically estimate the 95th percentile of the copper concentrations at the background wells, and produce the one-sided upper confidence limit for the 95th percentile using the ENVIRONMENTALSTATS for S-PLUS Command or Script Window, type these commands.

```
attach(epa.92c.copper2.df)

eqnpar(Copper[Well.type=="Background"], p=0.95, ci=T,
    lb=0, ci.type="upper", approx.conf.level=0.95)

detach()
```

Nonparametric Confidence Intervals for Quantiles Based on Bootstrapping

Earlier in this chapter, we discussed the general method of bootstrapping to create nonparametric confidence intervals for a population parameter (e.g., mean, difference between two means, etc.). Why not use the bootstrap to create nonparametric confidence intervals for percentiles? Well, we could, but it turns out that using the bootstrap to create confidence intervals for percentiles gives us essentially the same result as the rank-based method we just discussed (Efron, 1979a; Efron, 1982, Chapter 10).

A CAUTIONARY NOTE ABOUT CONFIDENCE INTERVALS

A confidence interval is a form of statistical inference, where we infer characteristics about the population based on a sample from the population. All the methods that we have discussed for estimating distribution parameters and creating confidence intervals for them assume we have a representative sample from the population, and that the observations are independent of one another. It is often the case, however, that our sample represents a sample from a smaller population than the one for which we wish to make an inference. For example, we may wish to make an inference about the 95th percentile of copper concentrations in groundwater at specific background monitoring wells. This distribution may very well have temporal components, so that there is a specific pattern of variability between month and years. If we take a sample that covers only part of a year, then we have not sampled from a population of copper concentrations that includes monthly and yearly variations.

Look at Figure 1.1 in Chapter 1 again to remind yourself of all of the factors that contribute to variability in environmental data. An extensive sampling program that takes into account *all* of these possible factors is rare and usually not feasible. Therefore, a general rule of thumb is that most confidence intervals are too narrow.

SUMMARY

- Whether you are conducting a preliminary, descriptive study of the environment or monitoring the environment for contamination under a specific regulation, you usually need to characterize the distribution of whatever you are looking at (e.g., a chemical in the environment), which involves ***estimating distribution parameters*** such as the mean, median, standard deviation, 95th percentile, etc.

- Estimators are functions of the observations from the sample, so ***estimators are random variables*** and have associated probability distributions.

- The three main methods of estimating distribution parameters are the method of moments estimator (***MME***), maximum likelihood estimator (***MLE***), and minimum variance unbiased estimator (***MVUE***).

- The ***bias*** or ***systematic error*** of an estimator is defined to be the difference between the average value of the estimator and the parameter it is estimating. If the expected value of the estimator equals the parameter it is estimating, it is called an ***unbiased estimator***.

- The ***precision*** or ***random error*** or ***variability*** of an estimator is measured by its ***variance***; it is a measure of how much the estimator "bounces around" its average value. An estimator is ***precise*** if it has a small variability, and it is ***imprecise*** if it has a large amount of variability.

- The ***mean square error (MSE)*** of an estimator is the average or mean of the squared distance between the estimator and the parameter it is trying to estimate. It is a measure of how much the estimator "bounces around" the parameter it is supposed to be estimating. The mean square error of an estimator is equal to its variability (random error) plus the square of its bias (systematic error).

- The term ***accuracy*** is ambiguous. Some authors use it to refer to the bias of an estimator, and some authors use it to refer to the mean square error of an estimator. The term accuracy should be avoided and instead the terms bias, precision, and mean square error should be used.

- There are several ways to compare the performance of different estimators of the same quantity. One common method is to compare estimators based on their mean square errors. Another possible way to compare estimators is by how well confidence intervals based on each estimator perform.

- A *confidence interval* for some population parameter θ is an interval on the real line constructed from a sample of n observations so that it *will* contain the parameter θ with some specified probability $(1-\alpha)100\%$, where α is some fraction between 0 and 1. The quantity $(1-\alpha)100\%$ is called the *confidence coefficient* or *confidence level* associated with the confidence interval.

- *Parametric confidence intervals* assume you know what kind of distribution describes the population. Confidence intervals that do not require this assumption are called *nonparametric confidence intervals*.

- *A confidence interval is a random interval*; that is, the lower and/or upper bounds are random variables that are computed based on sample statistics. Prior to taking one specific sample, the probability that the confidence interval *will* contain θ is $(1-\alpha)100\%$. Once a specific sample is taken and a particular confidence interval is computed, the particular confidence interval associated with that sample either contains θ or it does not.

- The width of a confidence interval depends on the sample size, the amount of variability inherent in the population, and the confidence level that you choose.

- You can use *bootstrapping* to construct nonparametric confidence intervals for a population parameter. How well bootstrapping works depends on how well the empirical cdf estimates the population cdf.

- You can compute quantiles by using the inverse of the cumulative distribution function, so you can estimate quantiles by using the inverse of the estimated cdf. Parametric estimates of quantiles assume a particular probability distribution, while nonparametric estimates of quantiles use the empirical cdf.

- You can create confidence intervals for quantiles. Confidence intervals for quantiles are based on the idea of tolerance intervals, which are discussed in the next chapter.

- You rarely can take a random sample from the full population to which you really want to extrapolate, so confidence intervals are usually too narrow because you underestimate the amount of variability in the population.

- You can use S-PLUS and ENVIRONMENTALSTATS for S-PLUS to estimate distribution parameters and quantiles and create confidence intervals.

EXERCISES

5.1.   Consider the nickel concentrations in groundwater shown in Table 3.12 of Exercise 3.1 in Chapter 3. In ENVIRONMENTALSTATS for S-PLUS these data are stored in the data frame `epa.92c.nickel1.df`. Assume these data come from a lognormal distribution (combine all months and wells).

   a.   Estimate the parameters of this distribution.
   b.   Create a histogram of the data, then overlay the fitted distribution. You can either use the log-transformed data and overlay the fitted normal distribution, or use the original data and overlay the fitted lognormal distribution.
   c.   Create a two-sided 95% confidence interval for the mean using all of the available methods (Land's, Parkin et al.'s, Cox's, normal approximation) and compare the results.
   d.   Use the bootstrap method to create a nonparametric two-sided 95% confidence interval for the mean, and compare it to the other confidence intervals you created in part **a**.
   e.   Compute a 99% upper confidence limit for the 95th percentile.
   f.   Compute a nonparametric upper confidence limit for the 95th percentile by using the maximum value as the upper limit. What is the confidence level associated with this confidence interval?

5.2.   Consider the arsenic concentrations in groundwater shown in Table 3.13 of Exercise 3.2 in Chapter 3. In ENVIRONMENTALSTATS for S-PLUS these data are stored in the data frame `epa.98.arsenic3.df`. Assume the data at the background well come from a normal distribution (combine all years). Repeat parts **a-f** of Exercise 5.1. (Note: for part **c** of this exercise there is only one method of constructing the parametric confidence interval.)

5.3.   Consider the Reference area TcCB data show in Table 3.2 of Chapter 3 (in ENVIRONMENTALSTATS for S-PLUS these data are stored in the data frame `epa.94b.tccb.df`). Assume these data come from a lognormal distribution.

   a.   Estimate the parameters of this distribution.
   b.   Create a histogram of the data, then overlay the fitted distribution. You can either use the log-transformed data and overlay the fitted normal distribution, or use the original data and overlay the fitted lognormal distribution.

    c.   Create an upper 99% confidence limit for the 95th percen-
tile   Do any of the observations from the Cleanup area
fall above this limit?

5.4.   The built-in data set `halibut` in S-PLUS contains annual catch
per unit effort (CPUE) and exploitable biomass of Pacific halibut
for the years 1935 through 1989. Assume the CPUE data come
from a normal distribution, but do not make any assumptions
about the distribution of biomass.

    a.   Estimate the parameters of the assumed normal distribu-
tion for CPUE. Create a histogram of the data, then overlay the fitted distribution.

    b.   For the CPUE data, compute a two-sided 95% confidence
interval for the mean.

    c.   For the biomass data, compute a nonparametric two-sided
95% confidence interval for the mean using the bootstrap
method.

    d.   For the biomass data, try to create a two-sided 95% confi-
dence interval for the median. Is the confidence level for
this confidence interval 95%?

    e.   Compare the two intervals you created in parts **c** and **d**.

5.5.   Consider the results of the simulation experiment shown in Table
5.13 for the lognormal distribution with mean 10 and CV 1, for
the sample size $n = 10$ (first row of the table). The estimated proportion of times the bootstrap confidence interval will actually
contain the mean is 84%. This result is based on 1,000 simulations.

    a.   What is the estimated standard error of the estimated pro-
portion?

    b.   Plot the value of the estimated standard error vs. the num-
ber of simulations, assuming 10, 20, 50, 100, 250, 500,
and 1,000 simulations were run. (Assume the same estimated proportion of 84% for each case.)

    c.   Based on the 1,000 simulations, construct a two-sided
95% confidence interval for the true proportion of times
the bootstrap confidence interval will contain the mean.

6 PREDICTION INTERVALS, TOLERANCE INTERVALS, AND CONTROL CHARTS

Is That Observation "Out of Bounds"?

The groundwater monitoring example discussed in Chapter 1 involves comparing chemical concentrations at the downgradient wells with "background" values based on upgradient wells (for detection monitoring) or comparing them with Ground Water Protection Standards (for assessment monitoring). The soil cleanup example discussed in Chapter 1 involves comparing chemical concentrations in soil with a "soil screening level." Any activity that requires comparing new values to "background" or "standard" values creates a decision problem: If the new values greatly exceed the background or standard value, is this evidence of a true difference (i.e., is there contamination)? Or are the true underlying concentrations the same as background or the standard value and this is just a "chance" event? Table 6.1 illustrates this decision problem (see Tables 2.1 and 2.2 in Chapter 2).

	Reality	
Your Decision	No Contamination	Contamination
Contamination	Mistake: Type I Error (Probability = α)	Correct Decision (Probability = $1-\beta$)
No Contamination	Correct Decision	Mistake: Type II Error (Probability = β)

Table 6.1   The decision problem of comparing new values to a "background" or "standard" value

The new values you are obtaining are random quantities and therefore have an associated probability distribution (recall Figure 1.2 in Chapter 1 for a list of factors that influence the new values). Because the new values are random quantities, any hard and fast rule you make, any dividing line you choose between deciding "Contamination" vs. "No Contamination," will result in instances when you make the correct decision and in instances when you make a mistake. Statisticians call the two kinds of mistakes you can make a *Type I error* and a *Type II error*. Of course, in the real world, once you make a decision, you take an action (e.g., clean up the site or do noth-

295

ing), and you hope to find out eventually whether the decision you made was the correct decision.

For a specific decision rule, the probability of making a Type I error is usually denoted with the Greek letter α (alpha), and the probability of making a Type II error is usually denoted with the Greek letter β (beta). With this nomenclature, the probability $1-\beta$ denotes the probability of correctly deciding there is contamination when in fact it is present. This probability is called the *power* of the decision rule.

One way to come up with an objective decision rule to decide whether there is contamination or not is to base it on some kind of statistical interval or test. In this chapter, we will discuss three statistical tools you can use to create background intervals: prediction intervals, tolerance intervals, and control charts. Here, the decision rule about whether contamination has occurred is based on whether the new observations fall inside or outside the background interval. This decision rule is a specific kind of hypothesis test. In this chapter, we will concentrate only on making sure we have some kind of control over the probability of a Type I error (the probability of declaring contamination when in fact none is present). We will discuss hypothesis tests in more detail in Chapter 7, and in Chapter 8 we will talk about sample size calculations and controlling the power of a test.

In the context of groundwater monitoring, one aspect of this kind of testing that cannot be over-emphasized is adequately accounting for spatial variability. Because of the ubiquitous presence of spatial variability in chemical concentrations in groundwater, it is usually best to generate background data specific to each monitoring well (i.e., use intra-well comparisons). That is, the ideal situation is to construct the background intervals based on pre-landfill data for each well, and compare future observations at a well to the interval for that particular well. Often, however, observations at downgradient wells are not available prior to the construction and operation of the landfill. In this case, upgradient well data can be combined to create a background interval, and observations at each downgradient well can be compared to this interval. If spatial variability is present and a major source of variation, however, you have to somehow account for it (e.g., Davis, 1994; Davis and McNichols, 1999).

PREDICTION INTERVALS

A *prediction interval* for some population is an interval on the real line constructed so that it will contain k future observations or averages from that population with some specified probability $(1-\alpha)100\%$, where α is some fraction between 0 and 1 (usually α is less than 0.5), and k is some pre-specified positive integer. (Note: We will see in a bit how this α is really the same as the α we used above to denote the probability of a Type I error.) Just as for confidence intervals (see Chapter 5), the quantity $(1-\alpha)100\%$ is called the **confidence coefficient** or **confidence level** associated with the prediction interval.

The basic idea of a prediction interval is to assume a particular probability distribution (e.g., normal, lognormal, etc.) for some process generating the data (e.g., quarterly observations of chemical concentrations in groundwater), compute sample statistics from a baseline sample, and then use these sample statistics to construct a prediction interval, assuming the distribution of the data does not change in the future. For example, if X denotes a random variable from some population, and we know what the population looks like (e.g., N(10, 2)) so we can compute the quantiles of the population, then a $(1-\alpha)100\%$ two-sided prediction interval for the next $k = 1$ observation of X is given by:

$$\left[\; x_{\alpha/2} \; , \; x_{1-\alpha/2} \; \right]$$

(6.1)

where x_p denotes the pth quantile of the distribution of X. Similarly, a $(1-\alpha)100\%$ one-sided upper prediction interval for the next observation is given by:

$$\left[\; -\infty \; , \; x_{1-\alpha} \; \right]$$

(6.2)

and a $(1-\alpha)100\%$ one-sided lower prediction interval for the next observation is given by:

$$\left[\; x_{\alpha} \; , \; \infty \; \right]$$

(6.3)

For general values of k, the two-sided, one-sided upper, and one-sided lower prediction intervals are given, respectively, by:

$$\left[\; x_{p_L} \; , \; x_{p_U} \; \right]$$

(6.4)

$$\left[\; -\infty \; , \; x_{(1-\alpha)^{1/k}} \; \right]$$

(6.5)

$$\left[x_{1-(1-\alpha)^{1/k}} \, , \quad \infty \right] \tag{6.6}$$

where

$$p_L \;=\; \frac{1 - (1 - \alpha)^{1/k}}{2} \tag{6.7}$$

$$p_U \;=\; \frac{1 + (1 - \alpha)^{1/k}}{2} \tag{6.8}$$

These equations can be derived using the binomial distribution. That is, the probability that all of the next k future observations will fall into the prediction interval is simply p^k, where p denotes the probability of one future observation falling into the prediction interval. Setting $p^k = 1-\alpha$ yields the prediction intervals in Equations (6.4) to (6.6). Figure 6.1 shows the probability density function of a N(10, 2) distribution along with one-sided upper 95% prediction limits for various values of k. (Note that the upper prediction limit increases with k, the number of future observations the prediction limit should contain.)

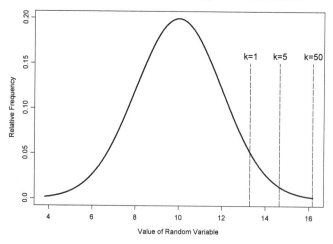

Figure 6.1   One-sided upper 95% prediction limits for a N(10, 2) distribution

Usually the true distribution of X is unknown, so the values of the prediction limits have to be estimated based on estimating the parameters of the distribution of X. For the usual case when the exact distribution of X is unknown, a prediction interval is thus a random interval; that is, the lower and upper bounds are random variables computed based on sample statistics in the baseline sample. Prior to taking one specific baseline sample, the probability that the prediction interval *will* contain the next k observations is $(1-\alpha)100\%$. Once a specific baseline sample is taken and the prediction interval based on that sample is computed, the probability that that prediction interval will contain the next k observations is not necessarily $(1-\alpha)100\%$, but it should be close to this value.

Suppose an experiment is performed N times, and suppose that for each experiment:

1. A sample is taken and a $(1-\alpha)100\%$ prediction interval for $k = 1$ future observation is computed.
2. One future observation is generated and compared to the prediction interval.

Then the number of times a prediction interval generated in Step 1 above will contain a future observation generated in step 2 above is a binomial random variable with parameters $n = N$ and $p = 1-\alpha$, that is, it follows a B(N, $1-\alpha$) distribution. Figure 6.2 shows the results of such a simulated experiment in which a random sample of $n = 10$ observations was taken from a N(5, 1) distribution and an 80% prediction interval for $k = 1$ future observation was constructed based on these 10 observations. Then one future observation was generated. The experiment was repeated 100 times. In this case, the actual number of times the prediction interval contained the future observation was 78.

It is important to note that if only one baseline sample is taken and only one prediction interval for $k = 1$ future observation is computed, then the number of future observations out of a total of N future observations that will be contained in that one prediction interval is a binomial random variable with parameters $n = N$ and $p = 1-\alpha^*$, where α^* depends on the true population parameters and the computed bounds of the single prediction interval. The rest of this section explains how to construct prediction intervals for the normal, lognormal, and Poisson distributions, and also how to construct nonparametric prediction intervals. As in Chapter 5, throughout this chapter we will assume that X denotes a random variable and x denotes the observed value of this random variable based on a sample. Thus, $x_1, x_2, \ldots, x_n$ denote n independent observations from a particular probability distribution.

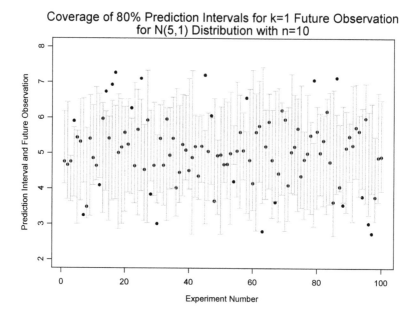

Figure 6.2   Results of simulation experiment showing the 80% prediction interval and one future observation for 100 simulations

Prediction Intervals for a Normal Distribution

For a normal distribution with mean μ and standard deviation σ, the form of a two-sided $(1-\alpha)100\%$ prediction interval for the next k future observations is:

$$\left[\ \bar{x} \ - \ Ks \ , \ \ \bar{x} \ + \ Ks \ \right] \tag{6.9}$$

Similarly, the form of a one-sided upper prediction interval is:

$$\left[\ - \ \infty \ , \ \ \bar{x} \ + \ Ks \ \right] \tag{6.10}$$

and the form of a one-sided lower prediction interval is:

$$\left[\ \bar{x} \ - \ Ks \ , \ \ \infty \ \right] \tag{6.11}$$

In these prediction intervals, $\bar{x}$ and s denote the sample mean and standard deviation, and K is a constant that depends on the sample size (n), the confidence level (($1-\alpha$)100%), the number of future observations (k), and the kind of prediction interval (two-sided vs. one-sided). Do not confuse the constant K (uppercase K) with the number of future observations (lowercase k). The symbol K is used here to be consistent with the notation used for tolerance intervals (see below).

We will discuss how K is computed in a bit. Note that the form of prediction intervals for a normal distribution is similar to the form of confidence intervals for the population mean (see Equations (5.78) to (5.80) in Chapter 5).

We can actually compute a prediction interval for k future *averages*, where m denotes the sample size associated with each of the k future averages. In this case, the constant K also depends on the value of m. When $m = 1$, each average is really just a single observation, so in the rest of this section we will sometimes use the term "average" instead of "observation or average."

Derivation of K for One Future Observation (k = 1)

As we noted earlier in this chapter, if we know the true distribution of X, that is, if we know the value of the mean μ and the standard deviation σ, then we can construct a two-sided ($1-\alpha$)100% prediction interval for the next $k = 1$ observation using Equation (6.1). Recall from Chapter 4 (Equation (4.54) in Exercise 4.7) that the pth quantile of a N(μ, σ) distribution can be written as:

$$x_p = \mu + z_p \sigma \qquad (6.12)$$

where z_p denotes the pth quantile of a standard normal distribution. Thus, when we know the values of μ and σ the prediction interval becomes:

$$\left[x_{\alpha/2} , \ x_{1-\alpha/2} \right]$$

$$= \left[\mu + z_{\alpha/2} \, \sigma , \ \mu + z_{1-\alpha/2} \, \sigma \right] \qquad (6.13)$$

$$= \left[\mu - z_{1-\alpha/2} \, \sigma , \ \mu + z_{1-\alpha/2} \, \sigma \right]$$

More generally, a two-sided $(1-\alpha)100\%$ prediction interval for the next $k = 1$ average based on a sample of size m is given by:

$$\left[\mu - z_{1-\alpha/2} \frac{\sigma}{\sqrt{m}}, \ \mu + z_{1-\alpha/2} \frac{\sigma}{\sqrt{m}} \right] \tag{6.14}$$

Because the values of μ and σ are almost always unknown, they must be estimated and the prediction interval must be constructed based on the estimated values of μ and σ. For a two-sided $(1-\alpha)100\%$ prediction interval for the next $k = 1$ average, the constant K is computed as:

$$K = t_{n-1,1-\alpha/2} \sqrt{\frac{1}{m} + \frac{1}{n}} \tag{6.15}$$

where $t_{v,p}$ denotes the pth quantile of the Student's t-distribution with v degrees of freedom. Similarly, for a one-sided lower or upper prediction interval, the constant K is computed as:

$$K = t_{n-1,1-\alpha} \sqrt{\frac{1}{m} + \frac{1}{n}} \tag{6.16}$$

These formulas for K are derived as follows. Let $\bar{y}$ denote the future average based on m observations. Then the quantity $\bar{y} - \bar{x}$ has a normal distribution with mean and variance given by:

$$E\left(\bar{y} - \bar{x}\right) = 0 \tag{6.17}$$

$$Var\left(\bar{y} - \bar{x}\right) = Var\left(\bar{y}\right) + Var\left(\bar{x}\right) - 2\,Cov\left(\bar{y}, \bar{x}\right) \tag{6.18}$$

$$= \frac{\sigma^2}{m} + \frac{\sigma^2}{n} - 0 = \sigma^2\left(\frac{1}{m} + \frac{1}{n}\right)$$

so the quantity

$$t = \frac{\overline{y} - \overline{x}}{s\sqrt{\dfrac{1}{m} + \dfrac{1}{n}}} \tag{6.19}$$

has a Student's t-distribution with $n-1$ degrees of freedom. Hence, to derive the formula for K for a one-sided upper prediction limit (Equations (6.10) and (6.16)), we note that

$$1 - \alpha = Pr\left[\frac{\overline{y} - \overline{x}}{s\sqrt{\dfrac{1}{m} + \dfrac{1}{n}}} \leq t_{n-1,1-\alpha}\right]$$

$$= Pr\left[\overline{y} - \overline{x} \leq t_{n-1,1-\alpha}\, s\sqrt{\dfrac{1}{m} + \dfrac{1}{n}}\right] \tag{6.20}$$

$$= Pr\left[\overline{y} \leq \overline{x} + t_{n-1,1-\alpha}\, s\sqrt{\dfrac{1}{m} + \dfrac{1}{n}}\right]$$

$$= Pr\left[\overline{y} \leq \overline{x} + t_{n-1,1-\alpha}\, \sqrt{\dfrac{1}{m} + \dfrac{1}{n}}\, s\right]$$

Similar reasoning yields the value of K for a one-sided lower prediction limit or a two-sided prediction limit.

Derivation of K for Several Future Observations (k>1)

When $k > 1$, there are at least two possible ways to compute K: an exact method due to Dunnett (1955), and an approximate (conservative) method based on the Bonferroni inequality (Miller, 1981a, pp. 8, 67–70; Gibbons, 1994, pp. 11–15). Each of these methods is explained below.

Exact Value of K

Dunnett (1955) derived the value of K in the context of the multiple comparisons problem of comparing several treatment means to one control mean. The value of K is computed as:

$$K = d \sqrt{\frac{1}{m} + \frac{1}{n}} \qquad\text{(6.21)}$$

where d is a constant that depends on the sample size (n), the confidence level (($1-\alpha$)100%), the number of future observations or averages (k), the sample size associated with the k future averages (m), and the kind of prediction interval (two-sided vs. one-sided).

For a one-sided upper or lower prediction interval, the value of d is the number that satisfies the following equation (Gupta and Sobel, 1957; Hahn, 1970a):

$$1 - \alpha =$$

$$\int_0^\infty F\left(d\,s, k, \rho\right) \; h\left(s\sqrt{n-1}, n-1\right) \; \sqrt{n-1} \; ds \qquad\text{(6.22)}$$

where

$$F\left(x, k, \rho\right) = \int_{-\infty}^\infty \left[\Phi\left(\frac{x + y\sqrt{\rho}}{\sqrt{1-\rho}}\right) \right]^k \; \phi\left(y\right) dy \qquad\text{(6.23)}$$

$$\rho = \frac{1}{\dfrac{n}{m} + 1} \qquad\text{(6.24)}$$

$$h\left(x, \nu\right) = \frac{x^{\nu-1} \exp\left(\dfrac{-x^2}{2}\right)}{2^{(\nu/2)-1} \; \Gamma\left(\dfrac{\nu}{2}\right)} \qquad\text{(6.25)}$$

and $\Phi()$ and $\phi()$ denote the cumulative distribution function and probability density function, respectively, of the standard normal distribution. Note that the function $h()$ defined in Equation (6.25) denotes the probability density function of a chi random variable with n degrees of freedom.

For a two-sided prediction interval, the value of d is the number that satisfies Equation (6.22), except that the function $F()$ in Equation (6.23) becomes:

$$F\left(x, k, \rho\right) =$$

$$\int_{-\infty}^{\infty} \left[\Phi\left(\frac{x + y\sqrt{\rho}}{\sqrt{1 - \rho}}\right) - \Phi\left(\frac{-x + y\sqrt{\rho}}{\sqrt{1 - \rho}}\right)\right]^{k} \phi\left(y\right) dy \qquad (6.26)$$

Approximate Value of K Based on the Bonferroni Inequality

When k is larger than one, a conservative way to construct a $(1-\alpha*)100\%$ prediction interval for the next k averages is to set $\alpha = \alpha*/k$ in Equation (6.15) or (6.16) for K (Chew, 1968). This value of K is based on the **Bonferroni inequality** (Miller, 1981a, p. 8; see the section The Multiple Comparisons Problem in Chapter 7), and will be conservative in that the computed prediction intervals will be wider than the exact predictions intervals. Hahn (1969, 1970a) compared the exact values of K with those based on the Bonferroni inequality for the case of $m = 1$ and found the approximation to be quite satisfactory except when n is small, k is large, and α is large. For example, Gibbons (1987a) notes that for a 99% prediction interval (i.e., $\alpha = 0.01$) for the next k observations, if $n>4$, the bias of K based on the Bonferroni Inequality is never greater than 1% no matter what the value of k.

Relationship between K, Sample Size, Number of Future Observations, and Confidence Level

Figure 6.3 displays a graph of K as a function of background sample size (n) and number of future observations (k) for a one-sided 99% prediction interval based on the exact value of K. Figure 6.4 displays the same thing, except in this figure the number of future observations (k) is set to 2 and the confidence level is allowed to vary. Figure 6.5 compares the exact and Bonferroni approximation methods of computing K for $k = 2$ future observations for a one-sided 99% prediction interval. These figures show that K decreases asymptotically with increasing sample size (n), and K increases with increasing number of future observations (k) and/or confidence level (($1-\alpha)100\%$).

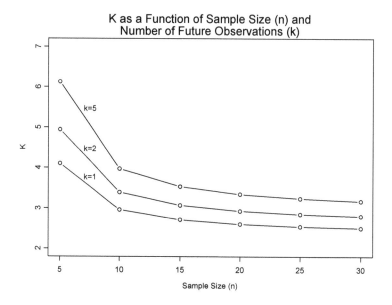

Figure 6.3: K as a function of sample size (n) and number of future observations (k) for a one-sided 99% prediction interval

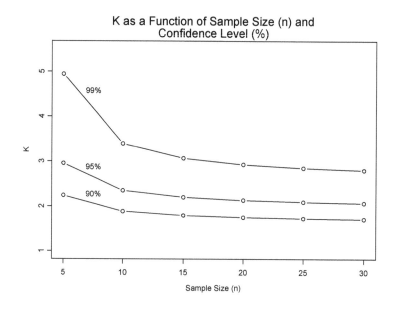

Figure 6.4: K as a function of sample size (n) and confidence level$((1-\alpha)100\%)$ for $k = 2$ future observations for a one-sided prediction interval

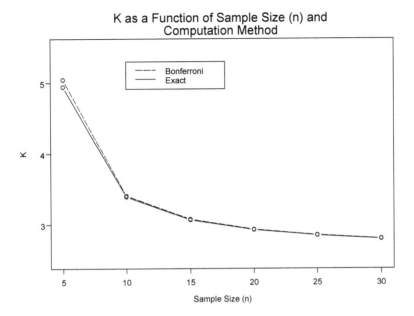

Figure 6.5: K as a function of sample size (n) and computation method for a one-sided 99% prediction interval

Example 6.1: Normal Prediction Interval for k = 4 Future Observations of Arsenic

Table 6.2 below shows arsenic concentrations (ppb) collected quarterly at two groundwater monitoring wells. These data are modified from benzene data given in USEPA (1992c, p. 56). In ENVIRONMENTALSTATS for S-PLUS these data are stored in the data frame arsenic3.df. In this example, we will use the data from the background well to construct a prediction interval for the next $k = 4$ observations.

Well	Year	Observed Arsenic (ppb)			
Background	1	12.6	30.8	52.0	28.1
	2	33.3	44.0	3.0	12.8
	3	58.1	12.6	17.6	25.3
Compliance	4	48.0	30.3	42.5	15.0
	5	47.6	3.8	2.6	51.9

Table 6.2 Arsenic data from groundwater monitoring wells

Combining all of the observations from the background well and assuming the data at the background well come from a normal distribution, the exact one-sided upper 95% prediction limit for the next $k = 4$ future observations is 72.9 ppb, and the one based on the Bonferroni method is 73.7. The four observed values of arsenic at the compliance well in year 4 are all below both of these prediction limits, as are all four values for year 5, so there is no evidence of contamination in either year.

Of course, this example ignores all sorts of possible design problems, including the fact that there is only one background well (so there is no way to estimate spatial variability), and the possibility of temporal correlation since there are four observations taken within a single month (that is, variability within a month may be smaller than variability between months). Also, sampling ceased at the background well in years 4 and 5 so there is no way to determine whether a change may have occurred in background concentrations during these years.

Menu

To construct a prediction interval for the arsenic concentrations using the ENVIRONMENTALSTATS for S-Plus pull-down menu, follow these steps.

1. In the Object Explorer, find the data frame **arsenic3.df**.
2. On the S-Plus menu bar, make the following menu choices: **EnvironmentalStats>Estimation>Prediction Intervals**. This will bring up the Prediction Interval dialog box.
3. For Data to Use, make sure the **Pre-Defined Data** button is selected.
4. In the Data Set box, make sure **arsenic3.df** is selected.
5. In the Variable box, select **Arsenic**.
6. In the Subset Rows with box, type **Well.type=="Background"**.
7. Click on the **Interval** tab.
8. In the Distribution/Sample Size section, make sure the Type button is set to **Parametric**, that **Normal** is selected in the Distribution box, and type **1** in the n(mean) box.
9. Under the Prediction Interval section, **uncheck** the Simultaneous box, type **4** in the # Future Obs box, in the PI Type box select **upper**, and in the PI Method box select **exact**.
10. Click **OK** or **Apply**.

These steps produce an exact upper prediction limit. To produce the upper prediction limit based on the Bonferroni method, in Step 9 above select **Bonferroni** in the PI Method box.

Command

To construct the exact prediction interval for the arsenic concentrations using the ENVIRONMENTALSTATS for S-Plus Command or Script Window, type these commands.

```
attach(arsenic3.df)
```

```
pred.int.norm(Arsenic[Well.type=="Background"], k=4,
    method="exact", pi.type="upper")
```

To compute the prediction interval based on the Bonferroni method, type this command.

```
pred.int.norm(Arsenic[Well.type=="Background"], k=4,
    method="Bonferroni", pi.type="upper")
```

Prediction Intervals for a Lognormal Distribution

A prediction interval for a lognormal distribution is constructed by simply taking the natural logarithm of the observations and constructing a prediction interval based on the normal distribution, then exponentiating the prediction limits to produce a prediction interval on the original scale of the data (Hahn and Meeker, 1991, p. 73). In fact, you can use any monotonic transformation of the observations that you think induces normality (e.g., a Box-Cox power transformation), compute the prediction interval on the transformed scale, and then use the inverse transformation on the prediction limits to produce a prediction interval on the original scale.

To construct a prediction interval for a lognormal distribution using the ENVIRONMENTALSTATS for S-PLUS pull-down menu, follow steps similar to those shown in the previous section for a normal distribution, except choose **Lognormal** in the distribution box. To construct a prediction interval for a lognormal distribution using the ENVIRONMENTALSTATS for S-PLUS Command or Script Window, type commands similar to those shown in the previous section for a normal distribution, except use the function `pred.int.lnorm` instead of `pred.int.norm`.

Prediction Intervals for a Poisson Distribution

Just as you can construct a prediction interval for the next k future observations or averages for a normal or lognormal distribution, you can construct a prediction interval for the next k future observations or *sums* for a Poisson distribution. Here, it is assumed the sums are based on a sample size of m. When $m = 1$, this is the same as a single observation.

Because of the discrete nature of the Poisson distribution, even if the true mean of the distribution λ were known exactly, the actual confidence level associated with a prediction interval will usually not be exactly equal to some desired value of $(1-\alpha)100\%$. For example, for the Poisson distribution with a mean of $\lambda = 2$, the interval [0, 3] contains 85.7% of this distribution and the interval [0, 4] contains 94.7% of this distribution. Thus, no interval can contain exactly 90% of this distribution, so it is impossible to construct an exact 90% prediction interval for the next $k = 1$ observation for this particular Poisson distribution.

For a Poisson distribution with mean λ, the form of a two-sided $(1-\alpha)100\%$ prediction interval for the next k future observations or sums is:

$$\left[\ m\bar{x}\ -\ K\ ,\ \ m\bar{x}\ +\ K\ \right]$$

$$=\ \left[\ cS_X\ -\ K\ ,\ \ cS_X\ +\ K\ \right]$$

(6.27)

Similarly, the form of a one-sided upper prediction interval is:

$$\left[\ 0\ ,\ \ m\bar{x}\ +\ K\ \right]\ =\ \left[\ 0\ ,\ \ cS_X\ +\ K\ \right]$$ (6.28)

and the form of a one-sided lower prediction interval is:

$$\left[\ m\bar{x}\ -\ K\ ,\ \ \infty\ \right]\ =\ \left[\ cS_X\ -\ K\ ,\ \ \infty\ \right]$$ (6.29)

In these prediction intervals, $\bar{x}$ denotes the sample mean, S_x denotes the sum of the observed values, and c denotes the ratio of the sample size for the future sums (m) to the sample size for the background data (n):

$$\bar{x}\ =\ \frac{1}{n}\sum_{i=1}^{n} x_i$$ (6.30)

$$S_X\ =\ \sum_{i=1}^{n} x_i$$ (6.31)

$$c\ =\ \frac{m}{n}$$ (6.32)

and K is a constant that depends on the sample size (n), the confidence level ($(1-\alpha)100\%$), the number of future observations or sums (k), the sample size associated with the future sums (m), and the kind of prediction interval (two-sided vs. one-sided). The derivation of K is explained below.

Derivation of K Based On Binomial Conditional Distribution

Let S_y denote the sum of m observations from a Poisson distribution with mean λ^*. That is,

$$S_y = \sum_{i=1}^{m} y_i \qquad (6.33)$$

where y_1, y_2, ..., y_m are m independent observations from a Poisson(λ^*) distribution. We want to construct a prediction interval for the next k values of S_y, assuming that the mean for the distribution of the future values is the same as the mean for the distribution of the background data (i.e., assuming $\lambda^* = \lambda$).

Nelson (1970) derives a prediction interval for the case $k = 1$ based on the conditional distribution of S_y given $S_x + S_y = w$, which has a binomial distribution ($B(n, p)$) with $n = w$ and

$$p = \frac{m\lambda^*}{m\lambda^* + n\lambda} \qquad (6.34)$$

(Johnson et al., 1992, p. 161). The prediction limits are computed as those most extreme values of S_y that still yield a non-significant test of the null hypothesis

$$H_0 : \lambda^* = \lambda \qquad (6.35)$$

which is equivalent to the null hypothesis

$$H_0 : p = \frac{m}{m + n} \qquad (6.36)$$

(see Chapter 7 for an explanation of hypothesis tests).

Using the relationship between the binomial and F-distribution (see the section Confidence Interval for the Probability of Success for a Binomial

Distribution in Chapter 5), Nelson (1982, p. 203) states that exact two-sided $(1-\alpha)100\%$ prediction limits $[LPL, \quad UPL]$ are the closest integer solutions to the following equations:

$$\frac{m}{LPL + 1} = \frac{n}{S_X} F_{(2\ LPL+2),\ 2S_X,\ 1-\alpha/2} \qquad (6.37)$$

$$\frac{UPL}{n} = \frac{S_X + 1}{n} F_{2S_X+2,\ 2\ UPL,\ 1-\alpha/2} \qquad (6.38)$$

where $F_{v_1, v_2, p}$ denotes the pth quantile of the F-distribution with v_1 and v_2 degrees of freedom. For a one-sided lower prediction limit, $\alpha/2$ is replaced with α in Equation (6.37) and UPL is set to ∞. For a one-sided upper prediction limit, $\alpha/2$ is replaced with α in Equation (6.38) and LPL is set to 0. Note that this method is not extended to the case $k > 1$.

Derivation of K Based on Normal Approximation to Binomial Conditional Distribution

Cox and Hinkley (1974, p. 245) derive an approximate prediction interval for the case $k = 1$, based on the same reasoning as Nelson (1970), but suggest using the normal approximation to the binomial distribution (without the continuity correction; see Zar, 1999, pp. 535–538 or Fisher and van Belle, 1993, pp. 183–184 for information on the continuity correction associated with the normal approximation to the binomial distribution). Under the null hypothesis shown in Equation (6.35), the quantity

$$z = \frac{S_y - \dfrac{c\left(S_x + S_y\right)}{1 + c}}{\sqrt{\dfrac{c\left(S_x + S_y\right)}{(1 + c)^2}}} \qquad (6.39)$$

is approximately distributed as a standard normal random variable. (Note that both USEPA, 1992c, p. 36, and McBean and Rovers, 1998, p. 114, incorrectly state the square of this quantity follows a standard normal distribution.)

For $k = 1$ future observation or sum, the two-sided prediction limits are computed by solving the equation

$$z^2 \leq z_{1-\alpha/2}^2 \qquad (6.40)$$

where z_p denotes the pth quantile of the standard normal distribution. In this case, Gibbons (1987b) notes that the quantity K in Equation (6.27) above is given by:

$$K = \frac{z_{1-\alpha/2}^2 \, c}{2} + z_{1-\alpha/2} \, c \sqrt{s_x \left(1 + \frac{1}{c}\right)} + \frac{z_{1-\alpha/2}^2}{4} \qquad (6.41)$$

For one-sided lower or upper prediction intervals, $\alpha/2$ is replaced with α in the above equation.

For more than one future observation or sum (i.e., $k > 1$), Gibbons (1987b) suggests using the Bonferroni inequality (see the section The Multiple Comparisons Problem in Chapter 7). That is, the value of K is computed exactly as for the case $k = 1$ described above, except for a two-sided prediction interval, $\alpha/2$ is replaced with $(\alpha/k)/2$, and for a one-sided prediction interval α is replaced with α/k.

Gibbons (1987b) also suggested using a Student's t approximation instead of a normal approximation. In this case, all occurrences of z_p are replaced with $t_{n-1,p}$ where $t_{v,p}$ denotes the pth quantile of the Student's t-distribution with v degrees of freedom.

Derivation of K Based on Normal Approximation

The normal approximation for Poisson prediction limits was given by Nelson (1970; 1982, p. 203) and is based on the fact that the mean and variance of a Poisson distribution are the same (see Equation (4.44) in Chapter 4), and for "large" values of n and m, both X and Y are approximately normally distributed. This method is discussed in the ENVIRONMENTALSTATS for S-PLUS help file for pred.int.pois.

Example 6.2: Poisson Prediction Interval for k = 1 Future Sum of Four Benzene Observations

Consider the benzene concentrations at six background groundwater monitoring wells shown in Table 5.2 in Chapter 5. In Example 5.23, we reproduced the example in USEPA (1992c, p. 36) of constructing a confidence interval for the mean concentration, assuming these data come from a Poisson distribution. In this example, the 33 values reported as "<2" were set to half the detection limit (i.e., 1). We saw that a Poisson Q-Q plot of these data indicates that the Poisson model is a not really a good fit.

Prediction Interval	Method
[0, 16.1]	Conditional Distribution
[0, 14.9]	Normal Approximation to Conditional Distribution
[0, 15.3]	Student's t Approximation to Conditional Distribution
[0, 14.9]	Normal Approximation

Table 6.3   One-sided upper 99% prediction limits for the sum of four benzene concentrations (ppb), based on data from USEPA (1992c, p. 36)

USEPA (1992c, pp. 36–38) uses these data to illustrate how to construct a one-sided upper 99% prediction interval for the next $k = 1$ sum of four observations (i.e., $m = 4$). This could therefore apply to the next four observations at a single downgradient well, or to the sum of the next single observations at four downgradient wells, but in either case you must assume the distribution of benzene follows the same Poisson distribution as for the combined six background wells. Table 6.3 displays the upper prediction limit based on the four methods discussed above. In this table, the prediction limits have not been rounded so that you can compare the outcome for the prediction limit based on Student's t approximation to the conditional distribution (third row) with the results in USEPA (1992c, p. 37). In practice, the prediction limits should be rounded because we are assuming the observations follow a Poisson distribution, which can only take on integer values.

Menu

To construct the Poisson prediction intervals shown in Table 6.3 using the ENVIRONMENTALSTATS for S-PLUS pull-down menu, follow these steps

1. In the Object Explorer, highlight **new.epa.92c.benzene1.df** (the data frame that you created in Example 5.23 in Chapter 5).
2. On the S-PLUS menu bar, make the following menu choices: **EnvironmentalStats>Estimation>Prediction Intervals**. This will bring up the Prediction Interval dialog box.
3. For Data to Use, make sure the **Pre-Defined Data** button is selected.
4. In the Data Set box, make sure **new.epa.92c.benzene1.df** is selected.
5. In the Variable box, select **new.Benzene**.
6. Click on the **Interval** tab.

7. In the Distribution/Sample Size section, make sure the Type button is set to **Parametric**. In the Distribution box select **Poisson**, and type **4** in the n(sum) box.
8. Under the Prediction Interval section, **uncheck** the Round Limits box, type **1** in the # Future Obs box, in the PI Type box select **upper**, in the PI Method box select **conditional**, and in the Confidence Level(%) box type **99**.
9. Click **OK** or **Apply**.

These steps produce the upper prediction limit based on the binomial conditional distribution. To produce the upper prediction limits based on the other three methods, in Step 8 above in the PI Method box, select **conditional.approx.normal**, **conditional.approx.t**, or **normal.approx**.

Command

To construct the Poisson prediction intervals shown in Table 6.3 using the ENVIRONMENTALSTATS for S-PLUS Command or Script Window, type these commands.

```
attach(epa.92c.benzene1.df)

Benzene[Censored] <- 1

pred.int.pois(Benzene, k=1, n.sum=4,
    method="conditional", pi.type="upper",
    conf.level=0.99, round.limits=F)

pred.int.pois(Benzene, k=1, n.sum=4,
    method="conditional.approx.normal",
    pi.type="upper", conf.level=0.99, round.limits=F)

pred.int.pois(Benzene, k=1, n.sum=4,
    method="conditional.approx.t", pi.type="upper",
    conf.level=0.99, round.limits=F)

pred.int.pois(Benzene, k=1, n.sum=4,
    method="normal.approx", pi.type="upper",
    conf.level=0.99, round.limits=F)

rm(Benzene)

detach()
```

Nonparametric Prediction Intervals

So far we have discussed how to construct prediction intervals assuming the background data come from a normal, lognormal, or Poisson distribution. You can also construct prediction intervals assuming other kinds of parametric distributions (e.g., Hahn and Meeker, 1991; Nelson, 1982) but we will not discuss these here. You can also construct prediction intervals without making any assumption about the distribution of the background data, except that the distribution is continuous. These kind of prediction intervals are called ***nonparametric prediction intervals***. Of course, nonparametric prediction intervals still require the assumption that the distribution of future observations is the same as the distribution of the observations used to create the prediction interval.

Derivation of Nonparametric Prediction Intervals

If we let $x_{(i)}$ denote the ith largest observation in our background sample of size n (i.e., the ith order statistic), then a two-sided nonparametric prediction interval is constructed as:

$$\left[\; x_{(u)} \; , \; x_{(v)} \; \right] \tag{6.42}$$

where u and v are positive integers between 1 and n, and $u < v$. That is, u denotes the rank of the lower prediction limit, and v denotes the rank of the upper prediction limit.

To make it easier to write some equations later on, we can also write the prediction interval in Equation (6.42) in a slightly different way as:

$$\left[\; x_{(u)} \; , \; x_{(n+1-w)} \; \right] \tag{6.43}$$

where

$$w \; = \; n \; + \; 1 \; - \; v \tag{6.44}$$

so that w is a positive integer between 1 and $n-1$, and $u < (n+1-w)$.

If we allow $u = 0$ and $w = 0$ and define lower and upper bounds as:

$$x_{(0)} \; = \; lb \tag{6.45}$$

$$x_{(n+1)} \; = \; ub \tag{6.46}$$

then Equation (6.43) above can also represent a one-sided lower or one-sided upper prediction interval as well. That is, a one-sided lower nonparametric prediction interval is constructed as:

$$\left[x_{(u)}, \ x_{(n+1)} \right] = \left[x_{(u)}, \ ub \right] \qquad (6.47)$$

and a one-sided upper nonparametric prediction interval is constructed as:

$$\left[x_{(0)}, \ x_{(n+1-w)} \right] = \left[lb, \ x_{(n+1-w)} \right] \qquad (6.48)$$

Usually, $lb = -\infty$ or $lb = 0$, and $ub = \infty$.

Danziger and Davis (1964) show that the probability that at least k out of the next m observations will fall in the interval defined in Equation (6.43) is given by:

$$1 - \alpha = \frac{\sum\limits_{i=k}^{m} \binom{m - i + u + w - 1}{m - i} \binom{i + n - u - w}{i}}{\binom{n + m}{m}} \qquad (6.49)$$

Setting $u = w = 1$ implies using the smallest and largest observed values as the prediction limits. In this case, it can be shown that the probability that at least k out of the next m observations will fall in the interval

$$\left[x_{(1)}, \ x_{(n)} \right] \qquad (6.50)$$

is given by:

$$1 - \alpha = \frac{\sum\limits_{i=k}^{m} (m - i + 1) \binom{n + i - 2}{i}}{\binom{n + m}{m}} \qquad (6.51)$$

Finally, setting $k = m$ in Equation (6.51), the probability that all of the next m observations will fall in the interval defined in Equation (6.50) is given by:

$$1 - \alpha = \frac{n\,(n - 1)}{(n + m)\,(n + m - 1)} \quad (6.52)$$

For one-sided prediction limits, the probability that all m future observations will fall below $x_{(n)}$ (upper prediction limit) is the same as the probability that all m future observations will fall above $x_{(1)}$ (lower prediction limit), which is given by:

$$1 - \alpha = \frac{n}{n + m} \quad (6.53)$$

Example 6.3: Nonparametric Prediction Interval for m = 2 Future Arsenic Concentrations

The guidance document USEPA (1992c, pp. 59–60) gives an example of constructing a nonparametric prediction interval for the next $k = m = 2$ monthly observations of arsenic concentrations (ppb) in groundwater at a downgradient well, based on observations from three background wells. The data are shown in Table 6.4 below and are stored in the data frame epa.92c.arsenic2.df in ENVIRONMENTALSTATS for S-PLUS.

Month	Well 1	Well 2	Well 3	Well 4
		Background Wells		Compliance Well
1	<5	7	<5	
2	<5	6.5	<5	
3	8	<5	10.5	
4	<5	6	<5	
5	9	12	<5	8
6	10	<5	9	14

Table 6.4   Arsenic data (ppb) from groundwater monitoring wells (USEPA, 1992c, p. 60)

The three background wells were sampled once per month for 6 months. The compliance well was only sampled in the 5th and 6th months. The EPA guidance document combines all of the observations from the three background wells ($n = 18$) and uses the maximum value 12 as an upper prediction limit for the next $k = m = 2$ observations at the compliance well. Based on Equation (6.53), this produces a 90% upper prediction interval. Since one of the values from the compliance well lies above the upper prediction limit, we might conclude there is evidence of contamination at the compliance well, but we should keep in mind that given the way we constructed our prediction interval, we would incorrectly declare contamination present when in fact it is not present about 10% of the time.

Menu

To construct the nonparametric prediction interval using the ENVIRONMENTALSTATS for S-PLUS pull-down menu, follow these steps.

1. In the Object Explorer find the data frame **epa.92c.arsenic2.df**.
2. On the S-PLUS menu bar, make the following menu choices: **EnvironmentalStats>Estimation>Prediction Intervals**. This will bring up the Prediction Interval dialog box.
3. For Data to Use, make sure the **Pre-Defined Data** button is selected.
4. In the Data Set box, make sure **epa.92c.arsenic2.df** is selected.
5. In the Variable box, select **Arsenic**.
6. In the Subset Rows with box, type **Well.type=="Background"**.
7. Click on the **Interval** tab.
8. In the Distribution/Sample Size section, set the Type button to **Nonparametric**.
9. Under the Prediction Interval section, in the # Future Obs box type **2.**, in the Min # Obs PI Should Contain box type **2**, in the PI Type box select **upper**, and in the Lower Bound box type **0**.
10. Click **OK** or **Apply**.

Command

To construct the nonparametric prediction interval using the ENVIRONMENTALSTATS for S-PLUS Command or Script Window, type these commands.

```
attach(epa.92c.arsenic2.df)

pred.int.npar(Arsenic[Well.type=="Background"], m=2,
    lb=0, pi.type="upper")

detach()
```

Parametric vs. Nonparametric Prediction Intervals

In the last example, we easily constructed an upper prediction limit for arsenic even though several values were reported as nondetects (<5). This is a major advantage of nonparametric prediction intervals over parametric prediction intervals. (It is in fact possible to construct parametric prediction intervals using data that contain nondetect values, but the mathematics and assumptions are complicated; see Chapter 10.) Also, with nonparametric prediction intervals, we do not have to worry about whether the distribution with which we have chosen to model the data is valid.

On the other hand, nonparametric prediction intervals often require large sample sizes to achieve an adequate confidence level. For example, if we want to construct an upper 95% prediction interval for the next $k = m = 2$ observations, we require a sample size of $n = 38$ (see Equation (6.53) and Chapter 8). There are no such requirements on sample size for parametric distributions, although the width of a parametric prediction interval certainly depends on the sample size.

Another important point to note is that usually nonparametric methods are "robust" against "outliers" (e.g., the median is unaffected by a few extremely large values), but this is not the case with nonparametric prediction limits based on the extreme values (i.e., based on the maximum and/or minimum values; Davis and McNichols, 1999). For example, if the background data contain a transcription error that makes the largest value 91 instead of 19, this would probably make the prediction interval unreasonably wide and cause us to miss contamination when it is present. To guard against this kind of problem, Davis and McNichols (1999) suggest deciding on which order statistic to use as the prediction limit (largest, next to largest, etc.) *after* looking at the background data but *before* looking at any future observations. Of course if the cause(s) of the outlier(s) can be identified and remedied, this is an even better approach.

SIMULTANEOUS PREDICTION INTERVALS

Analyzing data from a groundwater monitoring program involves several difficulties, including trying to control for natural spatial and temporal variability, and perhaps dealing with nondetect values. One of the main statistical problems that plague groundwater monitoring programs at hazardous and solid waste facilities is the requirement of testing several wells and several constituents at each well on each sampling occasion. The number of constituents monitored can range from around 5 to 60 or more, and some facilities may have as many as 150 monitoring wells (Davis and McNichols, 1999). This is an obvious multiple comparisons problem (see Chapter 7), and the naïve approach of using a prediction interval with a conventional confidence level (e.g., 95% or 99%) for each comparison of a compliance well with background for each chemical of concern leads to a very high probability of at least one declaration of contamination on each sampling occasion, when in fact no contamination has occurred at any of the wells at any time for any of the chemicals of concern. This problem was pointed out several years ago by Millard (1987a) and others.

Davis and McNichols (1987, 1994b, 1999) proposed simultaneous prediction intervals as a way of controlling the facility-wide false positive rate (FWFPR) while maintaining adequate power to detect contamination in the groundwater. A ***simultaneous prediction interval*** is a prediction interval that will contain a certain number of future observations with probability $(1-\alpha)100\%$ for each of r future sampling occasions, where r is some prespecified positive integer. The quantity r may refer to r distinct future sampling occasions in time at a single compliance well, or it may refer to sampling at r distinct compliance wells on one future sampling occasion. (Note: Current regulations prohibit evaluating false positive risks over more than one monitoring event, so for current practice r must refer to the number of compliance wells monitored, and "r future sampling occasions" refers to sampling each of the r wells once.) In either case, it is assumed that the distribution of concentrations is constant over all r future sampling occasions.

Although simultaneous prediction intervals help us control the Type I error rate (probability of declaring contamination when it is not present) over r monitoring wells, we need to control the Type I error rate over all monitoring wells *and* all constituents (chemicals and physical properties) we monitor. To do this, Davis and McNichols (1994b, 1999) suggest using the Bonferroni method and creating simultaneous prediction limits with confidence level $(1-\alpha/n_c)100\%$ for each of the n_c constituents being monitored. That is, for each constituent, the probability of mistakenly declaring contamination present at at least one of the compliance wells when in fact it is not present at any of the compliance wells (i.e., the significance level) is α/n_c.

There are several ways to define a rule for a simultaneous prediction limit. We will discuss the following three rules:

- **The *k*-of-*m* Rule**. For the k-of-m rule, at least k of the next m future observations will fall in the prediction interval with probability $(1-\alpha)100\%$ on each of the r future sampling occasions. If observations are being taken sequentially, for a particular sampling occasion (or monitoring well), up to m observations may be taken, but once k of the observations fall within the prediction interval, sampling can stop. If $m - (k-1)$ observations fall outside the prediction interval, then contamination is declared to be present. For example, suppose we have $r = 5$ monitoring wells and we want to use the 1-of-3 rule (i.e., $k = 1$ and $m = 3$). Then for the ith monitoring well ($i = 1, 2, 3, 4, 5$), if the first observation is in the interval, we can stop. If the first observation is outside the interval, we have to wait a specified time (e.g., a few weeks), and take a second observation. If the second observation is in the interval, we can stop. If the second observation is outside the interval, then we have to wait a specified time and take a third observation. If the third observation is in the interval, we can stop. If the third observation is outside the interval, then contamination is declared to be present. (Note that in the case $k = m$ and $r = 1$, a simultaneous prediction interval reduces to the simple prediction interval we have already discussed.)

- **California Rule**. This is the rule currently required in the state of California. For the California rule, with probability $(1-\alpha)100\%$, for each of the r future sampling occasions, either the first observation will fall in the prediction interval, or else all of the next $m - 1$ observations will fall in the prediction interval. That is, if the first observation falls in the prediction interval then sampling can stop. Otherwise, up to $m - 1$ more observations must be taken (with a sufficient waiting time between sampling occasions). If any of these subsequent $m - 1$ observations falls outside the interval, we declare contamination is present.

- **Modified California Rule**. For the Modified California rule, with probability $(1-\alpha)100\%$, for each of the r future sampling occasions, either the first observation will fall in the prediction interval, or else at least 2 out of the next 3 observations will fall in the prediction interval. That is, if the first observation falls in the prediction interval then sampling can stop. Otherwise, up to 3 more observations must be taken (with a sufficient waiting time between sampling occasions). If any two of these next three observations fall into the interval then sampling can stop. Otherwise, contamination is declared to be present.

Simultaneous Prediction Intervals for a Normal Distribution

The form of a simultaneous prediction interval based on background data from a normal distribution is the same as for a standard normal prediction interval (see Equations (6.9) to (6.11)). The derivation of the constant K for the three different rules is explained below.

Derivation of K for the k-of-m Rule

For the case when $r = 1$ future sampling occasion, both Hall and Prairie (1973) and Fertig and Mann (1977) discuss the derivation of K. Davis and McNichols (1987) extend the derivation to the case where r is a positive integer. They show that for a one-sided prediction interval, the probability p that at least k of the next m future observations will be contained in the interval given in Equation (6.10) or (6.11) for each of r future sampling occasions is given by:

$$p = \int_{0}^{1} T\left(\sqrt{nK};\ n - 1,\ \sqrt{n}\left[\Phi^{-1}(v) + \frac{\delta}{\sigma}\right]\right)$$

(6.54)

$$\times\ r\left[I\left(v;\ k,\ m + 1 - k\right)\right]^{r-1}\left[\frac{v^{k-1}\left(1 - v\right)^{m-k}}{B\left(k,\ m + 1 - k\right)}\right]dv$$

where $T(x;\ v,\ \Delta)$ denotes the cumulative distribution function of the non-central Student's t-distribution with v degrees of freedom and non-centrality parameter Δ evaluated at x, $\Phi()$ denotes the cdf of the standard normal distribution, $I(x;\ a,\ b)$ denotes the cdf of the beta distribution with shape parameters a and b evaluated at x, and $B(a,b)$ denotes the value of the beta function with parameters a and b:

$$B\left(a,\ b\right) = \frac{\Gamma(a)\,\Gamma(b)}{\Gamma(a + b)}$$

(6.55)

where $\Gamma()$ denotes the gamma function (see Equation (4.46) in Chapter 4). (See the ENVIRONMENTALSTATS for S-PLUS help files for **beta** and **Beta** for information on the beta function and beta distribution.)

For given values of the confidence level ($p = 1-\alpha$), sample size (n), minimum number of future observations to be contained in the interval per sampling occasion (k), maximum number of future observations per sampling occasion (m), number of future sampling occasions (r), and δ/σ, Equation (6.54) can be solved for K. The quantity δ (delta) denotes the difference between the mean of the population that was sampled to construct the prediction interval and the mean of the population that will be sampled to produce the future observations. The quantity σ (sigma) denotes the population standard deviation of both of these populations. Usually you assume $\delta = 0$ (hence $\delta/\sigma = 0$) unless you are interested in computing the "power" of the rule to detect a change in means between the populations (see Chapter 8).

Derivation of K for the California and Modified California Rule

For the California rule, the derivation of K is the same as for the k-of-m rule, except that Equation (6.54) becomes (Davis, 1998b):

$$p = \int_0^1 T\left(\sqrt{n}K; n-1, \sqrt{n}\left[\Phi^{-1}(v) + \frac{\delta}{\sigma}\right]\right)$$

$$\times \; r\left\{v\left[1 + v^{m-2}(1-v)\right]\right\}^{r-1} \qquad (6.56)$$

$$\times \left[1 + v^{m-2}(m-1-mv)\right]dv$$

For the Modified California rule, the derivation of K is the same as for the k-of-m rule, except that Equation (6.54) becomes (Davis, 1998b):

$$p = \int_0^1 T\left(\sqrt{n}K; n-1, \sqrt{n}\left[\Phi^{-1}(v) + \frac{\delta}{\sigma}\right]\right)$$

$$\times \; r\left\{v\left[1 + v\left(3 - v\left[5 - 2v\right]\right)\right]\right\}^{r-1} \qquad (6.57)$$

$$\times \left\{1 + v\left[6 - v\left(15 - 8v\right)\right]\right\}dv$$

Comparing the Three Rules

For the k-of-m rule, Davis and McNichols (1987) give tables with "optimal" choices of k (in terms of best power for a given overall confidence level) for selected values of m, r, and n. They found that the optimal ratios of k/m are generally small, in the range of 15 to 50%.

The California rule is mandated in that state for groundwater monitoring at waste disposal facilities when resampling verification is part of the statistical program (Barclay's Code of California Regulations, 1991). The California code mandates a "California" rule with $m \geq 3$. The motivation for this rule is a desire to have a majority of the observations in bounds. For example, for a k-of-m rule with $k = 1$ and $m = 3$, a monitoring well will pass if the first observation is out of bounds, the second resample is out of bounds, but the last resample is in bounds, so that 2 out of 3 observations are out of bounds. For the California rule with $m = 3$, either the first observation must be in bounds, or the next 2 observations must be in bounds in order for the monitoring location to pass. Davis (1998a) states that if the FWFPR is kept constant, then the California rule offers little increased power compared to the k-of-m rule, and can actually decrease the power of detecting contamination.

The Modified California Rule has been proposed as a compromise between a 1-of-m rule and the California rule. For a given FWFPR, the Modified California rule achieves better power than the California rule, and still requires at least as many observations in bounds as out of bounds, unlike a 1-of-m rule.

Different Notations between Different References

For the k-of-m rule described here, Davis and McNichols (1987) use the variable p instead of k to represent the minimum number of future observations the interval should contain on each of the r sampling occasions. Gibbons (1994, Chapter 1, Tables 1.5 to 1.13) presents extensive lists of the value of K for both the k-of-m rule and California rule, with values of n ranging from 4 to 100 and values of r ranging from 5 to 50. Gibbons's notation reverses the meaning of k and r compared to the notation used here. That is, in Gibbons's notation, k represents the number of future sampling occasions or monitoring wells, and r represents the minimum number of observations the interval should contain on each sampling occasion.

Example 6.4: Simultaneous Normal Prediction Intervals for Arsenic

Using the background well arsenic data in Table 6.2 of Example 6.1, Figure 6.6 displays the value of the 95% upper simultaneous prediction limit as a function of the number of future sampling occasions (r) for the 1-of-3 rule, the California rule with $m = 3$, and the Modified California rule. This figure shows that in this case the upper prediction limit is always largest for the California rule and smallest for the 1-of-3 rule. For example, if you have 10 compliance wells you need to compare to the background arsenic level, you would set the upper prediction limit to about 48 ppb for the 1-of-3 rule, about 53 ppb for the Modified California rule, and about 63 ppb for the California rule. The probability of declaring contamination at at least one of these 10 compliance wells when in fact no well is contaminated is about 5%. The naïve 95% upper prediction limit for one future observation at a single well is 59.4 ppb. If you used this limit for all 10 compliance wells (but did not allow any resampling at any of the wells), then the probability of declaring contamination at at least one of these wells when in fact no well is contaminated is around $1 - 0.95^{10} = 40\%$ or larger (see the section The Multiple Comparisons Problem in Chapter 7).

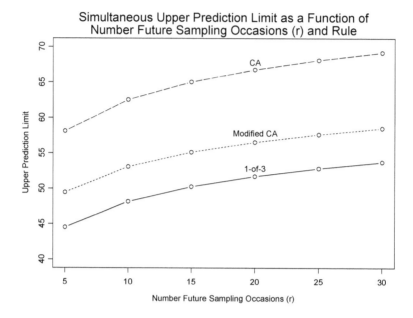

Simultaneous Upper Prediction Limit as a Function of Number Future Sampling Occasions (r) and Rule

Figure 6.6   95% upper prediction limit for arsenic as a function of the number of future sampling occasions (r) for the three different rules

Menu

To construct a simultaneous prediction interval for the arsenic concentrations using the ENVIRONMENTALSTATS for S-PLUS pull-down menu, follow these steps.

1. In the Object Explorer, find **arsenic3.df**.
2. On the S-PLUS menu bar, make the following menu choices: **EnvironmentalStats>Estimation>Prediction Intervals**. This will bring up the Prediction Interval dialog box.
3. For Data to Use, make sure the **Pre-Defined Data** button is selected.
4. In the Data Set box, make sure **arsenic3.df** is selected.
5. In the Variable box, select **Arsenic**.
6. In the Subset Rows with box, type **Well.type=="Background"**.
7. Click on the **Interval** tab.
8. In the Distribution/Sample Size section, make sure the Type button is set to **Parametric** and that **Normal** is selected in the Distribution box.
9. Under the Prediction Interval section, **check** the Simultaneous box, type **3** in the # Future Obs box, type **1** in the Min # Obs PI Should Contain box, type **10** in the # Future Sampling Occasions box, select

k.of.m in the Rule box, in the PI Type box select **upper**, and in the Conf Level (%) box type **95**.

10. Click **OK** or **Apply**.

These steps produce a simultaneous 95% upper prediction limit for the 1-of-3 rule for $r = 10$ future sampling occasions. To use different values of r, type a different number in the # Future Sampling Occasions box in Step 9 above. To use the California rule, select **California** in the Rule box in Step 9 above, and to use the Modified California rule, select **Modified California** in the Rule box in Step 9 above.

Command

To construct a simultaneous prediction interval for the arsenic concentrations using the ENVIRONMENTALSTATS for S-PLUS Command or Script Window, type these commands.

```
attach(arsenic3.df)
pred.int.norm.simultaneous(
    Arsenic[Well.type=="Background"], k=1, m=3, r=10,
    rule="k.of.m", pi.type="upper", conf.level=0.95)
```

These steps produce a simultaneous 95% upper prediction limit for the 1-of-3 rule for $r = 10$ future sampling occasions. To use different values of r, set the argument r to a different number.

To use the California and Modified California rules, type these commands.

```
pred.int.norm.simultaneous(
    Arsenic[Well.type=="Background"], m=3, r=10,
    rule="CA", pi.type="upper", conf.level=0.95)
pred.int.norm.simultaneous(
    Arsenic[Well.type=="Background"], r=10,
    rule="Modified.CA", pi.type="upper",
    conf.level=0.95)
detach()
```

What if Spatial or Temporal Variation is Present?

To compare background data with future observations, simultaneous prediction intervals require the assumption that the distribution of the future observations is the same as the background and that all observations are independent. On a temporal scale, for groundwater monitoring this means you must allow for at least a few weeks between observations, which essentially limits the value of m to at most about four, assuming you are taking quarterly samples (Davis and McNichols, 1999). On a spatial scale, Davis and

McNichols (1999) state that groundwater monitoring data almost always exhibit significant spatial variability in cases where the proportion of nondetect values is small enough that it is reasonable to consider parametric methods. In such cases, it is usually best to use intra-well prediction intervals at each well. That is, for each well construct the prediction interval based on background (pre-landfill) data from that well, and compare future observations at that well to the prediction interval for that particular well (Davis, 1998a). The individual α-level at each well is set to the FWFRP divided by the product of the number of wells and constituents, that is, $\alpha_{ind} = \alpha / (n_c \times r)$.

When you must combine background data from several background wells (instead of using intra-well comparisons) and substantial spatial variability is present, Davis (1994) and Davis and McNichols (1999) suggest performing a one-way analysis of variance on location and using the square root of the pooled variance estimator for the value of s in the prediction interval (see Equations (6.9) to (6.11)). In this case, the degrees of freedom associated with this pooled variance estimator is given by:

$$df_{nom} = n_w \left(n_t - 1 \right) \tag{6.58}$$

where n_w denotes the number of background wells and n_t denotes the number of observations at each background well. You would replace the quantity $n-1$ in Equations (6.54), (6.56), and (6.57) with the quantity df_{nom} defined in Equation (6.58). Of course, the pooled variance estimator assumes the variance is the same at each well. This is often not true in practice, but a variance-stabilizing transformation may help.

When temporal correlation is present, Davis (1994) and Davis and McNichols (1999) suggest performing a two-way mixed effects analysis of variance on location (fixed effect) and time (random effect), and deflating the degrees of freedom using the following formula:

$$df = df_{nom} \left[\frac{\left(n_w + F - 1 \right)^2}{n_w \left(n_w + F^2 - 1 \right)} \right] \tag{6.59}$$

where F denotes the F-statistic for the test of the significance of the time effect. Note that when $F = 1$, the degrees of freedom reduce to the expression in Equation (6.58).

Simultaneous Prediction Intervals for a Lognormal Distribution

Just as for a standard prediction interval for a lognormal distribution, a simultaneous prediction interval for a lognormal distribution is constructed by simply taking the natural logarithm of the observations and constructing a simultaneous prediction interval based on the normal distribution, then exponentiating the prediction limits to produce a simultaneous prediction interval on the original scale of the data. To construct a simultaneous prediction interval for a lognormal distribution using the ENVIRONMENTALSTATS for S-PLUS pull-down menu, follow steps similar to those shown in the previous section for a normal distribution, except choose **Lognormal** in the Distribution box. To construct a simultaneous prediction interval for a lognormal distribution using the ENVIRONMENTALSTATS for S-PLUS Command or Script Window, type commands similar to those shown in the previous section, except use the function `pred.int.lnorm.simultaneous`.

Simultaneous Nonparametric Prediction Intervals

Chou and Owen (1986) developed the theory for nonparametric simultaneous prediction limits for various rules, including the 1-of-m rule. Their theory, however, does not cover the California or Modified California rules, and uses an r-fold summation involving a minimum of 2^r terms. Davis and McNichols (1994b; 1999) extended the results of Chou and Owen (1986) to include the California and Modified California rule, and developed algorithms that involve summing far fewer terms.

Like a standard nonparametric prediction interval, a simultaneous nonparametric prediction interval is based on the order statistics from the sample (see Equations (6.42) to (6.48)). For a one-sided upper simultaneous nonparametric prediction interval, the upper prediction limit is usually the largest observation in the background data, but it could be the next largest or any other order statistic. Similarly, for a one-sided lower simultaneous nonparametric prediction interval, the lower prediction limit is usually the smallest observation.

Davis and McNichols (1999) give formulas for the probabilities associated with the one-sided upper simultaneous prediction interval shown in Equation (6.48). For the k-of-m rule, the probability that at least k of the next m future observations will be contained in the interval given in Equation (6.48) for each of r future sampling occasions is given by:

$$1 - \alpha = E\left[\sum_{i=0}^{m-k}\binom{k-1+i}{k-1}Y^k(1-Y)^i\right]^r$$

(6.60)

$$= \int_0^1\left[\sum_{i=0}^{m-k}\binom{k-1+i}{k-1}y^k(1-y)^i\right]^r f(y)\,dy$$

where Y denotes a random variable with a beta distribution with parameters v and $n+1-v$, and $f()$ denotes the pdf of this distribution. Note that v denotes the rank of the order statistic used as the upper prediction limit (e.g., v is usually equal to n).

For the California and Modified California rules, the probabilities are:

$$1 - \alpha = E\left[\sum_{i=0}^{r}\binom{r}{i}Y^{r-i+(m-1)i}(1-Y)^i\right]$$

(6.61)

$$= \int_0^1\left[\sum_{i=0}^{r}\binom{r}{i}y^{r-i+(m-1)i}(1-y)^i\right]f(y)\,dy$$

$$1 - \alpha = E\left\{Y^r\left[1 + Q + Q^2 - 2Q^3\right]^r\right\}$$

(6.62)

$$= \int_0^1 y^r\left[1 + q + q^2 - 2q^3\right]^r f(y)\,dy$$

where $Q = 1-Y$ and $q = 1-y$.

Davis and McNichols (1999) provide algorithms for computing the probabilities based on expanding polynomials and the formula for the expected value of a beta random variable. In the discussion section of Davis and McNichols (1999), however, Vangel points out that numerical integration is adequate, and this is how these probabilities are computed in ENVIRONMENTALSTATS for S-PLUS.

Comparing the Three Rules

As we stated earlier, although simultaneous prediction intervals help us control the Type I error rate (probability of declaring contamination when it is not present) over r monitoring wells, we need to control the Type I error rate over all monitoring wells *and* all constituents (chemicals and physical properties) we monitor. As in the case of parametric simultaneous prediction intervals, Davis and McNichols (1994b, 1999) suggest using the Bonferroni method and creating simultaneous prediction limits with confidence level $(1-\alpha/n_c)100\%$ for each of the n_c constituents being monitored. That is, for each constituent, the probability of mistakenly declaring contamination present at at least one of the compliance wells when in fact it is not present at any of the compliance wells (i.e., the significance level) is α/n_c.

Davis and McNichols (1999) note that the number of constituents monitored (n_c) can range from about 5 to 60, with more than 100 required annually in some states, which results in significance levels commonly around 0.001 and sometimes smaller than 0.0005. They note that for fixed values of background sample size (n), rank of the order statistic used as the upper prediction limit (v), and number of future sampling occasions (r), there always exists a 1-of-m plan that meets any pre-specified significance level α/n_c. With California plans, however, the smallest significance level is obtained when $m = 3$ and $v = n$. They show that for the California plan, very large background sample sizes (e.g., on the order of 100 or more) are needed to achieve acceptable significance levels. On the other hand, only moderate background sample sizes are required with the Modified California plan.

Davis and McNichols (1994b) give an example with $n = 40$ background observations, $r = 10$ monitoring wells, and $n_c = 5$ chemicals being monitored. For a FWFPR of 5%, the per-constituent Type I error rate should be $0.05/5 = 0.01$. A 1-of-3 plan yields a per-constituent Type I error rate of 0.008, whereas California plans with $m = 3$ or $m = 4$ yield per-constituent Type I error rates of 0.02 and 0.03, respectively. In order to use a California plan with $m = 3$, you would need a background sample size of at least $n = 60$.

Different Notations between Different References

As for the case of normal and lognormal simultaneous prediction intervals, for the k-of-m rule described here, Davis and McNichols (1994b, 1999) use the variable p instead of k to represent the minimum number of future observations the interval should contain on each of the r sampling occasions. Also, Davis and McNichols (1999) use the variable j instead of v to denote the rank of the order statistics used for the upper prediction limit.

Example 6.5: Nonparametric Simultaneous Prediction Intervals for Arsenic Concentrations

In Example 6.3 we showed how to create a one-sided upper nonparametric prediction interval for the next two observations of arsenic at a single downgradient well based on $n = 18$ observations at three upgradient wells (see Table 6.4). In that example, the significance level associated with this prediction interval is 10%.

Now suppose there are $r = 5$ downgradient wells, nine other constituents are being monitored besides arsenic (i.e., $n_c = 10$), and we want a FWFPR of 5% for each sampling occasion. Then the significance level for the simultaneous prediction interval for arsenic should be set to 0.05/10 = 0.005, which means the confidence level should be set to (1−0.005)100% = 99.5%. Using the maximum value of 12 ppb as the upper prediction limit (i.e., $v = n = 18$), Table 6.5 displays the associated confidence levels for various plans. The 1-of-3 and 1-of-4 plans satisfy the required confidence level, but neither of the California plans nor the Modified California plan do. Note that for k-of-m plans, the confidence level increases with increasing values of m, while the opposite is true for California plans. In order for a California plan with $m = 3$ to satisfy the required confidence level of 99.5%, a total of at least $n = 61$ background observations are needed. In order for the Modified California plan to satisfy the required confidence level, a total of at least $n = 24$ background observations are needed (see Chapter 8).

Rule	k	m	Confidence Level (%)
k-of-m	1	3	99.6
	1	4	99.9
CA		3	95.5
		4	93.9
Modified CA			99.1

Table 6.5   Confidence levels for various nonparametric simultaneous prediction intervals based on $n = 18$ background observations and $r = 5$ monitoring wells

Menu

To construct the simultaneous nonparametric prediction intervals and obtain their confidence levels (shown in Table 6.5) using the ENVIRONMENTALSTATS for S-PLUS pull-down menu, follow these steps

1.   In the Object Explorer, highlight the data frame **epa.92c.arsenic2.df**.

2. On the S-PLUS menu bar, make the following menu choices: **EnvironmentalStats>Estimation>Prediction Intervals**. This will bring up the Prediction Interval dialog box.
3. For Data to Use, make sure the **Pre-Defined Data** button is selected.
4. In the Data Set box, make sure **epa.92c.arsenic2.df** is selected.
5. In the Variable box, select **Arsenic**.
6. In the Subset Rows with box, type **Well.type=="Background"**.
7. Under the Results group, in the Conf Level Sig Digits box type **7**.
8. Click on the **Interval** tab.
9. In the Distribution/Sample Size section, set the Type button to **Non-parametric**.
10. Under the Prediction Interval section, **check** the Simultaneous box, in the # Future Obs box type **3**, in the Min # Obs PI Should Contain box type **1**, and in the # Future Sampling Occasions box type **5**. In the Rule box select **k of m**, in the PI Type box select **upper**, and in the Lower Bound box type **0**.
11. Click **OK** or **Apply**.

The above steps produce information for the 1-of-3 plan. To produce information for the 1-of-4 plan, Step 10 above becomes:

10. Under the Prediction Interval section, **check** the Simultaneous box, in the # Future Obs box type **4**, in the Min # Obs PI Should Contain box type **1**, and in the # Future Sampling Occasions box type **5**. In the Rule box select **k of m**, in the PI Type box select **upper**, and in the Lower Bound box type **0**.

To produce information for the California plan with $m = 3$, Step 10 above becomes:

10. Under the Prediction Interval section, **check** the Simultaneous box, in the # Future Obs box type **3**, and in the # Future Sampling Occasions box type **5**. In the Rule box select **CA**, in the PI Type box select **upper**, and in the Lower Bound box type **0**.

To produce information for the California plan with $m = 4$, Step 10 above becomes:

10. Under the Prediction Interval section, **check** the Simultaneous box, in the # Future Obs box type **4**, and in the # Future Sampling Occasions box type **5**. In the Rule box select **CA**, in the PI Type box select **upper**, and in the Lower Bound box type **0**.

Finally, to produce information for the Modified California plan, Step 10 above becomes:

10. Under the Prediction Interval section, **check** the Simultaneous box, and in the # Future Sampling Occasions box type **5**. In the Rule box select **Modified CA**, in the PI Type box select **upper**, and in the Lower Bound box type **0**.

Command

To construct the simultaneous nonparametric prediction intervals and obtain their confidence levels (shown in Table 6.5) using the ENVIRONMENTALSTATS for S-PLUS Command or Script Window, type these commands.

```
attach(epa.92c.arsenic2.df)

pred.list.1.of.3 <- pred.int.npar.simultaneous(
    Arsenic[Well.type=="Background"], k=1, m=3, r=5,
    rule="k.of.m", lb=0, pi.type="upper")

print(pred.list.1.of.3, conf.cov.sig.digits=7)

pred.list.1.of.4 <- pred.int.npar.simultaneous(
    Arsenic[Well.type=="Background"], k=1, m=4, r=5,
    rule="k.of.m", lb=0, pi.type="upper")

print(pred.list.1.of.4, conf.cov.sig.digits=7)

pred.list.CA.w.3 <- pred.int.npar.simultaneous(
    Arsenic[Well.type=="Background"], m=3, r=5,
    rule="CA", lb=0, pi.type="upper")

print(pred.list.CA.w.3, conf.cov.sig.digits=7)

pred.list.CA.w.4 <- pred.int.npar.simultaneous(
    Arsenic[Well.type=="Background"], m=4, r=5,
    rule="CA", lb=0, pi.type="upper")

print(pred.list.CA.w.4, conf.cov.sig.digits=7)

pred.list.Modified.CA <- pred.int.npar.simultaneous(
    Arsenic[Well.type=="Background"], r=5,
    rule="Modified.CA", lb=0, pi.type="upper")

print(pred.list.Modified.CA, conf.cov.sig.digits=7)
detach()
```

Parametric vs. Nonparametric Simultaneous Prediction Intervals

As with standard nonparametric prediction intervals, simultaneous nonparametric prediction intervals can easily handle nondetect data. On the other hand, sample size requirements are larger than for simultaneous parametric prediction intervals.

TOLERANCE INTERVALS

A *tolerance interval* for some population is an interval on the real line constructed so as to contain $\beta 100\%$ of the population (i.e., $\beta 100\%$ of all future observations), where $0 < \beta < 1$ (usually β is bigger than 0.5). The quantity $\beta 100\%$ is called the **coverage**. (Note: Do not confuse our use of the symbol β here with the probability of a Type II error. The symbol β is used here to be consistent with previous literature on tolerance intervals.)

As with a prediction interval, the basic idea of a tolerance interval is to assume a particular probability distribution (e.g., normal, lognormal, etc.) for some process generating the data (e.g., quarterly observations of chemical concentrations in groundwater), compute sample statistics from a baseline sample, and then use these sample statistics to construct a tolerance interval, assuming the distribution of the data does not change in the future. For example, if X denotes a random variable from some population, and we know what the population looks like (e.g., N(10, 2)) so we can compute the quantiles of the population, then a $\beta 100\%$ two-sided tolerance interval is given by:

$$\left[x_{1-\beta/2} , \ x_{\beta/2} \right] \tag{6.63}$$

where x_p denotes the pth quantile of the distribution of X. Similarly, a $\beta 100\%$ one-sided upper tolerance interval is given by:

$$\left[-\infty , \ x_\beta \right] \tag{6.64}$$

and a $\beta 100\%$ one-sided lower tolerance interval is given by:

$$\left[x_{1-\beta} , \ \infty \right] \tag{6.65}$$

Note that in the case when the distribution of X is known, a $\beta 100\%$ tolerance interval is exactly the same as a $(1-\alpha)100\%$ prediction interval for $k = 1$ future observation, where $\beta = 1-\alpha$ (see Equations (6.1) to (6.3)).

Usually the true distribution of X is unknown, so the values of the tolerance limits have to be estimated based on estimating the parameters of the distribution of X. In this case, a tolerance interval is a random interval; that is, the lower and/or upper bounds are random variables computed based on sample statistics in the baseline sample.   Given this uncertainty in the bounds, there are two ways to construct tolerance intervals (Guttman, 1970):

- A *β-content* tolerance interval with *confidence level* $(1-\alpha)100\%$ is constructed so that it contains *at least* $\beta100\%$ of the population (i.e., the coverage is at least $\beta100\%$) with probability $(1-\alpha)100\%$.
- A *β-expectation* tolerance interval is constructed so that it contains *on average* $\beta100\%$ of the population (i.e., the average coverage is $\beta100\%$).

A β-expectation tolerance interval with coverage $\beta100\%$ is equivalent to a prediction interval for $k = 1$ future observation with associated confidence level $\beta100\%$. Note that there is no explicit confidence level associated with a β-expectation tolerance interval. If a β-expectation tolerance interval is treated as a β-content tolerance interval, the confidence level associated with this tolerance interval is usually around 50% (e.g., Guttman, 1970, Table 4.2, p. 76). Thus, a β-content tolerance interval with coverage $\beta100\%$ will usually be wider than a β-expectation tolerance interval with the same coverage if the confidence level associated with the β-content tolerance interval is more than 50%.

Relationship between Tolerance Intervals and Confidence Intervals for Percentiles

It can be shown (e.g., Conover, 1980, pp. 119–121) that an upper confidence interval for the pth quantile with confidence level $(1-\alpha)100\%$ is equivalent to an upper β-content tolerance interval with coverage $100p\%$ and confidence level $(1-\alpha)100\%$. Also, a lower confidence interval for the pth quantile with confidence level $(1-\alpha)100\%$ is equivalent to a lower β-content tolerance interval with coverage $100(1-p)\%$ and confidence level $(1-\alpha)100\%$.

Using Tolerance Intervals to Determine Contamination

Prediction and tolerance intervals have long been applied to quality control and life testing problems. In environmental monitoring, USEPA has proposed using tolerance intervals in at least two different ways.

- **Compliance-to-Background Comparisons**. Construct a tolerance interval based on background data, then compare data from a compliance well or site to the tolerance interval. If any compliance data are outside of the tolerance interval, then declare contamination is present.

- **Compliance-to-Fixed Standard Comparisons.** Construct a tolerance interval based on compliance data, then compare the tolerance limit to a fixed standard (e.g., GWPS). If the tolerance limit is greater (less) than the fixed standard, declare contamination is present.

We will discuss the problems with each of these approaches below.

Compliance-to-Background Comparisons

USEPA (1989b, Chapter 5, pp. 20–24) proposes using β-content normal tolerance limits in detection monitoring to compare chemical concentrations at compliance wells with concentrations at background wells. USEPA (1992c, pp. 49–56) gives a confusing discussion, explaining the idea of a β-content tolerance interval on page 49, but then proposing β-expectation normal tolerance limits on subsequent pages and incorrectly associating a confidence level with a β-expectation tolerance interval. USEPA (1994b, pp. 4.8–4.9) suggests a "Hot-Measurement Comparison" using a one-sided upper tolerance interval based on background data to test for contamination at a cleanup site. All of the observations at the cleanup site are compared with the upper tolerance limit, and if any observations fall above the upper tolerance limit, then the site is declared contaminated.

The main problem with this method of deciding whether contamination is present is that the Type I error (the probability of declaring contamination is present when in fact it is not) depends on the sample size for the compliance well or site! For example, suppose we know the background distribution of the chemical of concern, and suppose the distribution at the compliance well or at the cleanup site is exactly the same as the background distribution. Furthermore, suppose we construct a one-sided upper 95% tolerance interval following Equation (6.64) above. Since the distribution of the chemical concentrations is exactly the same at the compliance well or cleanup site as for background, the probability that one observation falls above the tolerance limit is 5%. In fact, if Y denotes the number of observations that fall above the tolerance limit, then Y has a binomial distribution with parameters n_c and $p = 0.05$, where n_c denotes the number of observations taken at the compliance well or cleanup site. Thus, the probability that at least one observation falls above the upper tolerance limit (and thus the probability of a Type I error) is given by:

$$\alpha = \Pr\left(Y \geq 1\right) = 1 - \Pr\left(Y = 0\right)$$

$$\text{(6.66)}$$

$$= 1 - 0.95^{n_c}$$

Figure 6.7 displays the Type I error rate as a function of sample size at the compliance well or cleanup site (n_c). Clearly, this method of determining whether contamination is present encourages the party responsible for the compliance well or cleanup site to take as few observations as possible. Of course in practice we do not know the true background distribution, so we must create β-content or β-expectation tolerance intervals based on sample statistics, but the relationship between the Type I error rate and sample size is similar.

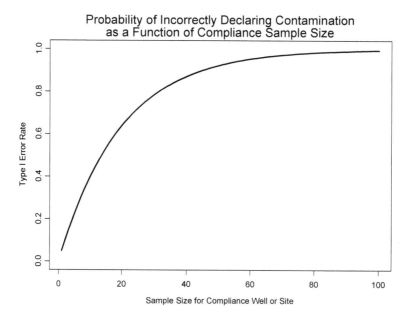

Figure 6.7:   Type I error rate for compliance-to-background comparison (based on a 95% upper tolerance limit) as a function of sample size at the compliance well or cleanup site

USEPA (1992c, p. 51) acknowledges the problem with this method of determining contamination by stating "…note that when testing a large number of samples, the nature of a Tolerance interval practically ensures that a few measurements will be above the upper Tolerance limit, even when no contamination has occurred. In these cases, the offending wells should probably be resampled in order to verify whether or not there is definite evidence of contamination." The problem with this approach is that by resampling we have changed the Type I error rate, but USEPA (1992c) does not tell you how to compute this new Type I error rate nor how to control it. This is why for groundwater monitoring it is much better to use simultaneous prediction intervals because we know how to compute the Type I error.

For comparing a potentially contaminated site with a background or reference site, instead of using a one-sided upper tolerance limit as suggested in USEPA (1994b, pp. 4.8–4.9), a better way to test for potential hot spots is to compute a $(1-\alpha)100\%$ prediction interval for the next k observations based on the background site, setting k equal to the number of observations collected at the potentially contaminated site. This allows you to control the Type I error at a pre-specified level of $\alpha100\%$.

Compliance-to-Fixed Standard Comparisons

USEPA (1989b, Chapter 6, pp. 11–15; 1992c, pp. 49–54) proposes using β-content tolerance limits (assuming a normal distribution) in compliance monitoring to compare concentrations at compliance wells with a fixed standard (e.g., MCL or ACL). This is really the same thing as testing whether the pth percentile of the distribution is less than or equal to the fixed standard (see Examples 5.24 and 5.25 in Chapter 5). To correctly implement this method, data from the compliance well are used to construct a one-sided *lower* tolerance limit. If this lower tolerance limit is greater than the fixed standard, then contamination is declared to be present. Note that both USEPA (1989b, Chapter 6, pp. 11–15) and USEPA (1992c, pp. 52–54) incorrectly use the *upper* tolerance limit.

There is nothing wrong with this approach to testing for contamination, assuming the fixed standard is really a fixed standard and not some statistical estimate (e.g., the background mean). The main problem with this approach is the multiple comparisons problem: if several compliance wells need to be compared to a fixed standard, then the probability of incorrectly declaring contamination is present at at least one of the compliance wells when in fact it is not present at any of the compliance wells can be substantially greater than $1-\beta$ as the number of compliance wells increases. In this case, the Bonferroni method or some other adjustment should probably be used.

Tolerance Intervals for a Normal Distribution

For a normal distribution with mean μ and standard deviation σ, the form of a $\beta100\%$ tolerance interval is exactly the same as for a prediction interval (see Equations (6.9) to (6.11)). The derivation of K is explained below.

The Derivation of K for a β-Expectation Tolerance Interval

As stated earlier, a β-expectation tolerance interval with coverage $\beta100\%$ is equivalent to a prediction interval for $k = 1$ future observation with associated confidence level $\beta100\%$. This is because the probability that any single future observation will fall into this interval is $\beta100\%$, so the distribution of the number of N future observations that will fall into this interval is binomial with parameters $n = N$ and $p = \beta$. Hence the expected *proportion* of future observations that will fall into this interval is $\beta100\%$ and is independent of the value of N. See the previous section Prediction Intervals for a Normal Distribution for information on how these intervals are constructed.

The Derivation of K for a β-Content Tolerance Interval

For a two-sided β-content tolerance interval with coverage $\beta100\%$ and associated confidence level $(1-\alpha)100\%$, the constant K is approximated by:

$$K \approx r \sqrt{\frac{n - 1}{\chi^2_{n-1,\alpha}}} \qquad (6.67)$$

where n denotes the number of background observations, $\chi^2_{v,p}$ denotes the pth quantile of a chi-square distribution with v degrees of freedom, and r is the solution to the equation:

$$\Phi\left(\frac{1}{\sqrt{n}} + r\right) - \Phi\left(\frac{1}{\sqrt{n}} - r\right) = \beta \qquad (6.68)$$

where $\Phi()$ denotes the cumulative distribution function of a standard normal distribution (Wald and Wolfowitz, 1946; Guttman, 1970, p. 59; Gibbons, 1994, p. 85). Wald and Wolfowitz (1946) show that this approximation is quite good, even for values of n as small as 2, provided both β and $1-\alpha$ are greater than 0.95. Furthermore, Ellison (1964) shows that for this approximation, the error in the confidence level is on the order of $1/n$.

For a one-sided β-content tolerance interval with coverage $\beta100\%$ and associated confidence level $(1-\alpha)100\%$, the constant K is given by:

$$K = \frac{t_{n-1,\,z_\beta\sqrt{n},\,1-\alpha}}{\sqrt{n}} \qquad (6.69)$$

where $t_{v,\delta,p}$ denotes the pth quantile of the non-central Student's t-distribution with v degrees of freedom and non-centrality parameter δ, and z_p denotes the pth quantile of a standard normal distribution (Guttman, 1970, pp. 87–89; Owen, 1962, p. 117; Odeh and Owen, 1980; Gibbons, 1994, pp. 85–86).

Example 6.6: Comparing Lower Tolerance Limits to an ACL for Aldicarb

Example 5.24 in Chapter 5 discusses a permit in which an ACL of 50 ppm for aldicarb should not be exceeded more than 5% of the time. Using the data from three groundwater monitoring compliance wells (four monthly samples at each well) shown in Table 5.14, Example 5.24 computed a lower 99% confidence limit for the 95th percentile for the distribution at each of the three compliance wells, yielding 25.3, 25.7, and 25.5 ppm. This is equivalent to computing a lower β-content tolerance limit with coverage 5% and associated confidence level of 99%.

Menu

To produce the one-sided lower β-content tolerance limits with coverage 5% and associated confidence level 99% for each of the compliance wells using the ENVIRONMENTALSTATS for S-PLUS pull-down menu, follow these steps.

1. In the Object Explorer, make sure **epa.89b.aldicarb2.df** is highlighted.
2. On the S-PLUS menu bar, make the following menu choices: **EnvironmentalStats>Estimation>Tolerance Intervals**. This will bring up the Tolerance Interval dialog box.
3. For Data to Use, make sure the **Pre-Defined Data** button is selected.
4. In the Data Set box, make sure **epa.89b.aldicarb2.df** is selected.
5. In the Variable box, select **Aldicarb**.
6. In the Subset Rows with box, type **Well=="1"**.
7. Click on the **Interval** tab.
8. Under the Distribution group, for Type, make sure the **Parametric** button is selected, and in the Distribution box select **Normal**.
9. Under the Tolerance Interval group, for Coverage Type select the **Content** button, for TI Type select **lower**, in the Coverage (%) box type **5**, and in the Confidence Level (%) box type **99**.
10. Click **OK** or **Apply**.

This will produce the one-sided lower β-content tolerance limit for the first well. To produce the tolerance limits for the other two wells, in Step 6 above, type **Well=="2"** or **Well=="3"**.

Command

To produce the one-sided lower β-content tolerance limits with coverage 5% and associated confidence level 99% for each of the compliance wells using the ENVIRONMENTALSTATS for S-PLUS Command or Script Window, type these commands.

```
attach(epa.89b.aldicarb2.df)
```

```
lapply(split(Aldicarb, Well), tol.int.norm,
    coverage=0.05, cov.type="content", ti.type="lower",
    conf.level=0.99)
detach()
```

Tolerance Intervals for a Lognormal Distribution

As with prediction intervals, a tolerance interval for a lognormal distribution is constructed by simply taking the natural logarithm of the observations and constructing a tolerance interval based on the normal distribution, then exponentiating the tolerance limits to produce a tolerance interval on the original scale of the data (Hahn and Meeker, 1991, p. 73). To construct a tolerance interval for a lognormal distribution using the ENVIRONMENTALSTATS for S-PLUS pull-down menu, follow steps similar to those shown in the previous section for a normal distribution, except choose **Lognormal** in the distribution box. To construct a tolerance interval for a lognormal distribution using the ENVIRONMENTALSTATS for S-PLUS Command or Script Window, type commands similar to those shown in the previous section for a normal distribution, except use the function `tol.int.lnorm` instead of `tol.int.norm`.

Example 6.7: Using an Upper Tolerance Limit vs. an Upper Prediction Limit to Determine Contamination at a Cleanup Site

Recall the 1,2,3,4-Tetrachlorobenzene data shown in Table 3.2 in Chapter 3 and displayed in Figures 3.1 and 3.2. In this example, we will compute an upper tolerance limit and an upper prediction limit based on the Reference area data and compare the observations from the Cleanup site to each of these two limits. The one-sided upper 95% β-content tolerance limit with associated confidence level 95% is 1.42 ppb, whereas the one-sided upper 95% prediction limit for the next $k = 77$ observations (there are 77 observations in the Cleanup area) is 2.68 ppb. There are 14 observations in the Cleanup area larger than 1.42 ppb and 7 observations larger than 2.68, so both methods indicate residual contamination in the Cleanup area. The method based on the upper tolerance limit, however, should never be used. We used it here just to show that the prediction limit is larger than the tolerance limit, since the prediction limit controls the Type I error rate at the specified level of 5%.

Menu

To produce the one-sided upper β-content tolerance limit based on the Reference area TcCB data using the ENVIRONMENTALSTATS for S-PLUS pull-down menu, follow these steps.

1. In the Object Explorer, highlight **epa.94b.tccb.df**.
2. On the S-PLUS menu bar, make the following menu choices: **EnvironmentalStats>Estimation>Tolerance Intervals**. This will bring up the Tolerance Interval dialog box.
3. For Data to Use, make sure the **Pre-Defined Data** button is selected.
4. In the Data Set box, make sure **epa.94b.tccb.df** is selected.
5. In the Variable box, select **TcCB**.
6. In the Subset Rows with box, type **Area=="Reference"**.
7. Click on the **Interval** tab.
8. Under the Distribution group, for Type, make sure the **Parametric** button is selected, and in the Distribution box select **Lognormal**.
9. Under the Tolerance Interval group, for Coverage Type select the **Content** button, for TI Type select **upper**, in the Coverage (%) box type **95**, and in the Confidence Level (%) box type **95**.
10. Click **OK** or **Apply**.

To produce the one-sided upper prediction limit based on the Reference area TcCB data using the ENVIRONMENTALSTATS for S-PLUS pull-down menu, follow these steps.

1. In the Object Explorer, highlight the data frame **epa.94b.tccb.df**.
2. On the S-PLUS menu bar, make the following menu choices: **EnvironmentalStats>Estimation>Prediction Intervals**. This will bring up the Prediction Interval dialog box.
3. For Data to Use, make sure the **Pre-Defined Data** button is selected.
4. In the Data Set box, make sure **epa.94b.tccb.df** is selected.
5. In the Variable box, select **TcCB**.
6. In the Subset Rows with box, type **Area=="Reference"**.
7. Click on the **Interval** tab.
8. In the Distribution/Sample Size section, make sure the Type button is set to **Parametric**, that **Lognormal** is selected in the Distribution box.
9. Under the Prediction Interval section, **uncheck** the Simultaneous box, type **77** in the # Future Obs box, in the PI Type box select **upper**, in the PI Method box select **exact**, and in the Confidence Level (%) box type **95**.
10. Click **OK** or **Apply**.

Command

To produce the one-sided upper β-content tolerance limit and the one-sided upper prediction limit for the next $k = 77$ observations based on the Reference area TcCB data using the ENVIRONMENTALSTATS for S-PLUS Command or Script Window, type these commands.

```
attach(epa.94b.tccb.df)
```

```
tol.int.lnorm(TcCB[Area=="Reference"], coverage=0.95,
    cov.type="content", ti.type="upper",
    conf.level=0.95)
pred.int.lnorm(TcCB[Area=="Reference"], k=77,
    method="exact", pi.type="upper", conf.level=0.95)
detach()
```

Example 6.8: Comparing Lower Tolerance Limits to an ACL for Chrysene

Example 5.25 in Chapter 5 discusses a permit in which an ACL of 80 ppb for chrysene should not be exceeded more than 5% of the time. Using the data from five groundwater monitoring compliance wells (four monthly samples at each well) shown in Table 5.16, Example 5.25 computed a lower 99% confidence limit for the 95th percentile for the distribution at each of the five compliance wells, yielding 22.6, 11.1, 51.6, 26.9, and 32.9 ppb. This is equivalent to computing a lower β-content tolerance limit with coverage 5% and associated confidence level of 99%.

Menu

To produce the one-sided lower β-content tolerance limits with coverage 5% and associated confidence level 99% for each of the compliance wells using the ENVIRONMENTALSTATS for S-PLUS pull-down menu, follow these steps.

1. In the Object Explorer, highlight **epa.92c.chrysene.df**.
2. On the S-PLUS menu bar, make the following menu choices: **EnvironmentalStats>Estimation>Tolerance Intervals**. This will bring up the Tolerance Interval dialog box.
3. For Data to Use, make sure the **Pre-Defined Data** button is selected.
4. In the Data Set box, make sure **epa.92c.chrysene.df** is selected.
5. In the Variable box, select **Chrysene**.
6. In the Subset Rows with box, type **Well=="1"**.
7. Click on the **Interval** tab.
8. Under the Distribution group, for Type, make sure the **Parametric** button is selected, and in the Distribution box, select **Lognormal**.
9. Under the Tolerance Interval group, for Coverage Type select the **Content** button, for TI Type select **lower**, in the Coverage (%) box type **5**, and in the Confidence Level (%) box type **99**.
10. Click **OK** or **Apply**.

This will produce the one-sided lower β-content tolerance limit for the first well. To produce the tolerance limits for the other four wells, in Step 6 above, type **Well=="2"**, **Well=="3"**, **Well=="4"**, or **Well=="5"**.

Command

To produce the one-sided lower β-content tolerance limits with coverage 5% and associated confidence level 99% for each of the compliance wells using the ENVIRONMENTALSTATS for S-PLUS Command or Script Window, type these commands.

```
attach(epa.92c.chrysene.df)
lapply(split(Chrysene, Well), tol.int.lnorm,
   coverage=0.05, cov.type="content", ti.type="lower",
   conf.level=0.99)
detach()
```

Tolerance Intervals for a Poisson Distribution

Gibbons (1987b; 1994, pp. 38–40) used the Poisson distribution to model the number of detected compounds per scan of the 32 volatile organic priority pollutants (VOC), and also to model the distribution of chemical concentration (in ppb) when most of the observations are nondetects. USEPA (1992c, pp. 38–40) explains Gibbons's approach and suggests using β-content Poisson tolerance limits to compare chemical concentrations at compliance wells with concentrations at background wells. As stated in Chapter 5 in Example 5.23, it is by no means clear, nor commonly accepted, that modeling concentration data with a Poisson distribution is valid.

β-Expectation Tolerance Intervals for a Poisson Distribution

As stated earlier in this chapter, a β-expectation tolerance interval with coverage β100% is equivalent to a prediction interval for $k = 1$ future observation with associated confidence level β100%. See the previous section Prediction Intervals for a Poisson Distribution for information on how these intervals are constructed.

β-Content Tolerance Intervals for a Poisson Distribution

Zacks (1970) showed that for monotone likelihood ratio (MLR) families of discrete distributions, a uniformly most accurate upper β100% β-content tolerance interval with associated confidence level $(1-\alpha)100\%$ is constructed by finding the upper $(1-\alpha)100\%$ confidence limit for the parameter associated with the distribution, and then computing the βth quantile of the distribution assuming the true value of the parameter is equal to the upper confidence limit. This idea can be extended to one-sided lower and two-sided tolerance limits. It can be shown that all distributions that are one-parameter exponential families have the MLR property, and the Poisson distribution is a one-parameter exponential family, so the method of Zacks (1970) can be applied to a Poisson distribution.

Let X denote a random variable with a Poisson distribution with mean λ, and let $x_{p|\lambda}$ denote the pth quantile of this distribution. That is,

$$\Pr\left(X < x_{p|\lambda}\right) \leq p \leq \Pr\left(X \leq x_{p|\lambda}\right) \qquad (6.70)$$

Note that due to the discrete nature of the Poisson distribution, there are several values of p associated with one particular value of X. For example, for $\lambda = 2$, the value 1 is the pth quantile for any value of p between 0.14 and 0.406.

For a one-sided upper tolerance interval, the first step is to compute the one-sided upper $(1-\alpha)100\%$ confidence limit for λ based on the n observations (see the section Confidence Interval for the Mean of a Poisson Distribution in Chapter 5). Denote this upper confidence limit by UCL. The one-sided upper $\beta100\%$ tolerance limit is then given by:

$$\left[\; 0\;,\; x_{\beta|\lambda=UCL}\;\right] \qquad (6.71)$$

Similarly, for a one-sided lower tolerance interval, the first step is to compute the one-sided lower $(1-\alpha)100\%$ confidence limit for λ, which we will denote by LCL. The one-sided lower $\beta100\%$ tolerance limit is then given by:

$$\left[\; x_{(1-\beta)|\lambda=LCL}\;,\; \infty\;\right] \qquad (6.72)$$

Finally, for a two-sided tolerance interval, the first step is to compute the two-sided $(1-\alpha)100\%$ confidence limits for λ, which we will denote by LCL and UCL. The two-sided $\beta100\%$ tolerance limit is then given by:

$$\left[\; x_{\frac{1-\beta}{2}|\lambda=LCL}\;,\; x_{\frac{1+\beta}{2}|\lambda=UCL}\;\right] \qquad (6.73)$$

Although there are several ways to compute the confidence limits for λ, ENVIRONMENTALSTATS for S-PLUS uses the exact method when computing β-content tolerance limits for a Poisson distribution.

Example 6.9: An Upper Tolerance Interval for Benzene Concentrations

Based on an example in USEPA (1992c, p. 40), Example 5.26 in Chapter 5 discusses creating a one-sided upper 95% confidence interval for the 95th percentile of benzene concentrations, based on observations from six background wells, with the idea that this limit will be used as a threshold value for concentrations observed in downgradient monitoring wells (i.e., if the concentration at a downgradient well exceeds this limit, this indicates there may be contamination in the groundwater). This is the "Hot-Measurement Comparison," and as already discussed above, there are some major problems with this technique.

Creating a one-sided upper 95% confidence interval for the 95th percentile is equivalent to creating a one-sided upper β-content tolerance interval with 95% coverage and an associated 95% confidence level. The data are shown in Table 5.2, and in ENVIRONMENTALSTATS for S-PLUS they are stored in the data frame `epa.92c.benzene1.df`. Nondetect values are reported as "<2." Of the 36 values, 33 are nondetects. Following Gibbons (1987, p. 574), USEPA (1992c, p. 36) assumes that these data can be modeled as having come from a Poisson process. The three detected values represent a count of the number of molecules of benzene observed out of a much larger number of water molecules. The 33 nondetect values are also assumed to represent counts, where the count must be equal to 0 or 1, since they are recorded as "<2" ppb. Following USEPA (1992c), for this example we will set all of the values of the nondetects to 1.

For this data set, the estimated mean is 1.94 and the one-sided upper β-content tolerance limit with coverage 95% and confidence level 95% is 5. As we stated in Example 5.23, these data do not appear to follow a Poisson distribution too well.

Menu

To produce the one-sided upper β-content tolerance interval with coverage 95% and associated confidence level 95% for the benzene concentrations using the ENVIRONMENTALSTATS for S-PLUS pull-down menu, follow these steps.

1. In the Object Explorer, highlight **new.epa.92c.benzene1.df**.
2. On the S-PLUS menu bar, make the following menu choices: **EnvironmentalStats>Estimation>Tolerance Intervals**. This will bring up the Tolerance Interval dialog box.
3. For Data to Use, make sure the **Pre-Defined Data** button is selected.
4. In the Data Set box, make sure **new.epa.92c.benzene1.df** is selected.
5. In the Variable box, select **new.Benzene**.
6. Click on the **Interval** tab.

7. Under the Distribution group, for Type, make sure the **Parametric** button is selected, and in the Distribution box, select **Poisson**.
8. Under the Tolerance Interval group, for Coverage Type select the **Content** button, for TI Type select **upper**, in the Coverage (%) box type **95**, and in the Confidence Level (%) box type **95**.
9. Click **OK** or **Apply**.

Command

To produce the one-sided upper β-content tolerance interval with coverage 95% and associated confidence level 95% for the benzene concentrations using the ENVIRONMENTALSTATS for S-PLUS Command or Script Window, type these commands.

```
attach(new.epa.92c.benzene1.df)
tol.int.pois(new.Benzene, coverage=0.95,
   cov.type="content", ti.type="upper",
   conf.level=0.95)
detach()
```

Nonparametric Tolerance Intervals

So far we have discussed how to construct tolerance intervals assuming the background data come from a normal, lognormal, or Poisson distribution. You can also construct tolerance intervals assuming other kinds of parametric distributions (e.g., Hahn and Meeker, 1991; Nelson, 1982) but we will not discuss these here. You can also construct tolerance intervals without making any assumption about the distribution of the background data, except that the distribution is continuous. These kind of tolerance intervals are called *nonparametric tolerance intervals*.

The form of a nonparametric tolerance interval is exactly the same as for a nonparametric prediction interval (see Equations (6.42) to (6.48)). Let C be a random variable denoting the coverage of a nonparametric tolerance intervals. Wilks (1941) showed that the distribution of C follows a beta distribution with shape parameters $v-u$ and $w+u$ when the unknown distribution is continuous, where u denotes the rank of the lower tolerance limit, v denotes the rank of the upper tolerance limit, and w is defined in Equation (6.44).

Nonparametric β-Expectation Tolerance Intervals

As stated earlier in this chapter, a β-expectation tolerance interval with coverage β100% is equivalent to a prediction interval for $k = m = 1$ future observation with associated confidence level β100% (see the previous section Nonparametric Prediction Intervals). For a β-expectation tolerance interval, the expected coverage is simply the mean of a beta random variable with shape parameters $v-u$ and $w+u$, which is given by:

$$E\left(C\right) = \frac{v - u}{n + 1} \tag{6.74}$$

For example, a one-sided upper tolerance limit using the maximum value results in an expected coverage of $n/(n+1)$.

Nonparametric β-Content Tolerance Intervals

For a β-content tolerance interval, if the coverage $C = \beta$ is specified, then the associated confidence level $(1-\alpha)100\%$ is computed as:

$$1 - \alpha = 1 - I\left(\beta; v - u, w + u\right) \tag{6.75}$$

where $I(x;\ a,\ b)$ denotes the cumulative distribution function of the beta distribution with shape parameters a and b evaluated at x. Similarly, if the confidence level associated with the tolerance interval is specified as $(1-\alpha)100\%$, then the coverage $C = \beta$ is computed as:

$$\beta = x_{\alpha, v - u, w + u} \tag{6.76}$$

where $x_{p,\,a,\,b}$ denotes the pth quantile of the beta distribution with shape parameters a and b.

Example 6.10: Using an Upper Tolerance Limit vs. a Simultaneous Upper Prediction Limit to Determine Contamination at Compliance Wells

Based on an example in USEPA (1992c, pp. 55–56), Example 5.27 in Chapter 5 discusses creating a one-sided upper confidence interval for the 95th percentile of copper concentrations, based on observations from three background wells, with the idea that this limit will be used as a threshold value for concentrations observed in two compliance wells (i.e., if any concentrations at the compliance wells exceed this limit, this indicates there may be contamination in the groundwater). This is the "Hot-Measurement Comparison," and as already discussed above there are some major problems with this technique. Creating a one-sided upper confidence interval for the 95th percentile is equivalent to creating a one-sided upper β-content tolerance interval with 95% coverage. In this example, we will compare the results of using a one-sided upper tolerance limit vs. using a one-sided simultaneous upper prediction limit.

The data are shown in Table 5.19, and in ENVIRONMENTALSTATS for S-PLUS they are stored in the data frame `epa.92c.copper2.df`. Note that 15 out of the 24 observations at the background wells are nondetects recorded as "<5." Observations at the three background wells were taken monthly over a period of 8 months. Observations at the two compliance wells were taken only for the last 4 months of sampling. To implement a simultaneous prediction interval with this data set, we have to decide what rule to use (k-of-m, California, or Modified California) and whether to assume the prediction interval applies only within each month or across all 4 months. As stated earlier in this chapter in the section Simultaneous Prediction Intervals, current regulations prohibit evaluating false positive risks over more than one monitoring event, so our simultaneous prediction interval has to account for $r = 2$ compliance wells and controls the FWFPR within a single month but not over several months. On the other hand, if any resamples are required, the time between resamples may need to be at least a month (depending on the flow-rate of the groundwater). In practice, we would also need to take into account the number of constituents that are being monitored each month (n_c) and adjust the required confidence level of the simultaneous prediction interval accordingly.

Using the maximum value of 9.2 ppb at the background wells as the upper tolerance limit yields a 95% β-content tolerance interval with an associated confidence level of only 71%. If we wanted to create a 95% β-content tolerance interval with associated confidence level 95%, we would need $n = 59$ background observations (see Chapter 8).

Using the maximum value of 9.2 ppb at the background wells as the upper simultaneous prediction limit for the $r = 2$ compliance wells yields prediction intervals with associated confidence levels of 99.9%, 98.9%, and 99.8% for the 1-of-3, California (with $m = 3$), and Modified California plans, respectively. Using these prediction limits, Well 4 in Month 8 indicates possible contamination, but all three plans require resampling, so we cannot conclude anything for the given data set since no data are shown beyond Month 8.

As we stated earlier, the method based on the upper tolerance limit should never be used because we do not know how to control the Type I error rate. Using a simultaneous prediction interval allows us to control the Type I error rate at a specified level.

Menu

To produce the nonparametric one-sided upper β-content tolerance interval with coverage 95% and associated confidence level 71% for the copper concentrations using the ENVIRONMENTALSTATS for S-PLUS pull-down menu, follow these steps.

1. In the Object Explorer, highlight **epa.92c.copper2.df**.
2. On the S-PLUS menu bar, make the following menu choices: **EnvironmentalStats>Estimation>Tolerance Intervals**. This will bring up the Tolerance Interval dialog box.
3. For Data to Use, make sure the **Pre-Defined Data** button is selected.
4. In the Data Set box, make sure **epa.92c.copper2.df** is selected.
5. In the Variable box, select **Copper**.
6. In the Subset Rows with box, type **Well.type=="Background"**.
7. Click on the **Interval** tab.
8. Under the Distribution group, for Type, select the **Nonparametric** button.
9. Under the Tolerance Interval group, for Coverage Type select the **Content** button, for Supply select the **Coverage** button, for TI Type select **upper**, in the Coverage (%) box type **95**, and in the Lower Bound box type **0**.
10. Click **OK** or **Apply**.

To produce the simultaneous one-sided upper prediction intervals for the $r =$ 2 compliance wells, follow these steps.

1. In the Object Explorer, highlight **epa.92c.copper2.df**.
2. On the S-PLUS menu bar, make the following menu choices: **EnvironmentalStats>Estimation>Prediction Intervals**. This will bring up the Prediction Interval dialog box.
3. For Data to Use, make sure the **Pre-Defined Data** button is selected.
4. In the Data Set box, make sure **epa.92c.copper2.df** is selected.
5. In the Variable box, select **Copper**.
6. In the Subset Rows with box, type **Well.type=="Background"**.
7. Under the Results group, in the Conf Level Sig Digits box type **7**.
8. Click on the **Interval** tab.
9. In the Distribution/Sample Size section, set the Type button to **Non-parametric**.
10. Under the Prediction Interval section, **check** the Simultaneous box, in the # Future Obs box type **3**, in the Min # Obs PI Should Contain box type **1**, and in the # Future Sampling Occasions box type **2**. In the Rule box select **k of m**, in the PI Type box select **upper**, and in the Lower Bound box type **0**.
11. Click **OK** or **Apply**.

The above steps produce information for the 1-of-3 plan. To produce information for the California plan with $m = 3$, Step 10 above becomes:

10. Under the Prediction Interval section, **check** the Simultaneous box, in the # Future Obs box type **3**, and in the # Future Sampling Occasions box type **2**. In the Rule box select **CA**, in the PI Type box select **upper**, and in the Lower Bound box type **0**.

Finally, to produce information for the Modified California plan, Step 10 above becomes:

10. Under the Prediction Interval section, **check** the Simultaneous box, and in the # Future Sampling Occasions box type **2**. In the Rule box select **Modified CA**, in the PI Type box select **upper**, and in the Lower Bound box type **0**.

Command

To produce the one-sided upper β-content tolerance interval with coverage 95% and associated confidence level 71% for the copper concentrations using the ENVIRONMENTALSTATS for S-PLUS Command or Script Window, type these commands.

```
attach(epa.92c.copper2.df)

tol.int.npar(Copper[Well.type=="Background"],
    coverage=0.95, cov.type="content", ti.type="upper",
    lb=0)
```

To produce the simultaneous one-sided upper prediction intervals for the $r =$ 2 compliance wells, type these commands.

```
pred.list.1.of.3 <- pred.int.npar.simultaneous(
    Copper[Well.type=="Background"], k=1, m=3, r=2,
    rule="k.of.m", lb=0, pi.type="upper")
print(pred.list.1.of.3, conf.cov.sig.digits=7)

pred.list.CA <- pred.int.npar.simultaneous(
    Copper[Well.type=="Background"], m=3, r=2,
    rule="CA", lb=0, pi.type="upper")
print(pred.list.CA, conf.cov.sig.digits=7)

pred.list.Modified.CA <- pred.int.npar.simultaneous(
    Copper[Well.type=="Background"], r=2,
    rule="Modified.CA", lb=0, pi.type="upper")
print(pred.list.Modified.CA, conf.cov.sig.digits=7)

detach()
```

Parametric vs. Nonparametric Tolerance Intervals

As with nonparametric prediction intervals, nonparametric tolerance intervals can easily handle nondetect data. On the other hand, sample size requirements are larger than for parametric tolerance intervals.

CONTROL CHARTS

Control charts are a graphical and statistical method of assessing the performance of a system over time (Montgomery, 1997; Ryan, 1989). They were developed in the 1920s by Walter Shewhart, and have been employed widely in industry to maintain process control (e.g., manufacturing a part for a car, airplane, or computer to within certain specifications). In the context of groundwater monitoring, they have been suggested as an alternative to prediction or tolerance limits for monitoring constituent concentrations at compliance wells when enough historical data are available at each compliance well to establish reliable background values for each well (USEPA, 1989b, pp. 7-1 to 7-12; USEPA, 1992c, pp. 75–79; Gibbons, 1994, pp. 160–168; ASTM, 1996).

We will explain Shewhart control charts, cumulative summation (CUSUM) charts, and combined Shewhart-CUSUM control charts. All of these procedures assume the observations at a particular compliance well are independent and follow a normal distribution with some constant mean μ and standard deviation σ. Of course, the mean and standard deviation can vary from well to well. USEPA (1989b, p. 7–12) and ASTM (1996) suggest using a minimum of at least $n = 8$ historical observations to estimate the background mean μ and standard deviation σ. If any substantial seasonal variability is present, you will need to take observations over at least 2 years in order to identify the seasonal pattern and remove it from the observations.

Shewhart Control Charts

The basic idea of a Shewhart control chart is to plot the observations over time and compare them to established upper and/or lower control limits that are based on historical data (Montgomery, 1997, Chapter 5; Gibbons, 1994, p. 161; Ryan, 1989, Chapters 5 and 6). For the simplest Shewhart control chart, once a single observation falls outside the control limit(s), this is an indication that the process is "out of control" and needs to be investigated. The forms of a Shewhart upper control limit (UCL) and lower control limit (LCL) for a single observation are:

$$UCL = \mu + L\sigma \qquad\qquad (6.77)$$

$$LCL = \mu - L\sigma \tag{6.78}$$

(Montgomery, 1997, p. 135). In practice, the true mean μ and standard deviation σ are rarely known, but instead must be estimated from historical data. Letting $\overline{x}$ and s denote the sample mean and standard deviation from the historical data, the upper and lower control limits then become:

$$UCL = \overline{x} + Ls \tag{6.79}$$

$$LCL = \overline{x} - Ls \tag{6.80}$$

which are exactly the same forms as a one-sided upper and one-sided lower prediction limit with $K = L$ (see Equations (6.10) and (6.11)). In industrial applications, we usually want the observations to stay within both of the bounds, that is, to stay less than the UCL and greater than the LCL. The constant L is often set to $L = 3$, and then the UCL and LCL are called "3-sigma control limits" (Montgomery, 1997, p. 134). In the case of known values of μ and σ, $L = 3$ corresponds to the 99.87 percentile of the standard normal distribution, so if the process is "in control" then each single observation has a probability of 99.74% of staying within both of the bounds.

For groundwater monitoring, we are almost always interested in only one-sided control limits. Usually, we want the observed values of a constituent at a compliance well to stay below the UCL (which is based on background data). Sometimes, however, we may want the observed values to stay above the LCL (e.g., for pH). USEPA (1989b, p. 7-8; 1992c, p. 78) recommends setting $L = 4.5$. Assuming the mean and standard deviation are estimated with $n = 8$ background observations, using $L = 4.5$ corresponds to a one-sided 95% prediction interval for the next $k = 50$ observations (using the exact method for prediction intervals), or to a one-sided 99.81% prediction interval for the next $k = 1$ observation. Hence, for each sampling occasion, for one particular compliance well and one particular constituent, if you use a Shewhart control chart with $L = 4.5$, the probability of incorrectly declaring contamination when in fact it is not present is only $1-0.9981 = 0.0019$. On the other hand, the constant L is not adjusted for the number of compliance wells (r) nor the number of constituents being monitored (n_c), so for each sampling occasion, the FWFPR is around

$$\text{FWFPR} \approx 1 - 0.9981^{r \times n_c} \tag{6.81}$$

As we stated earlier in this chapter, the number of compliance wells can be as large as 150, and the number of constituents monitored can vary from 5 to 62 or more, so the FWFPR can be very close to 100%.

CUSUM Control Charts

Shewhart control charts are a good device to detect sudden changes in the process being monitored. To detect a gradual trend in the process, Page (1954) proposed using Cumulative Summation (CUSUM) charts (Montgomery, 1997, Chapter 7; Gibbons, 1994, pp. 161–163; Ryan, 1989, p. 105). As its name suggests, a CUSUM chart involves cumulative sums. For the ith future sampling occasion, the ith one-sided upper cumulative sum S_i^+ and one-sided lower cumulative sum S_i^- are given by:

$$S_i^+ = \max\left[0, \ \left(z_i - k\right) + S_{i-1}^+\right] \tag{6.82}$$

$$S_i^- = \max\left[0, \ \left(-z_i - k\right) + S_{i-1}^-\right] \tag{6.83}$$

where $\max[a, b]$ denotes the maximum of a and b,

$$S_0^+ = S_0^- = 0 \tag{6.84}$$

$$z_i = \frac{x_i - \overline{x}}{s} \tag{6.85}$$

and k denotes a positive reference value that must be set by the user and corresponds to half the size of a linear trend (in units of standard deviations) deemed worthy of detecting quickly. USEPA (1989b, p. 7-7) recommends using $k = 1$ (i.e., it is important to detect a trend of two standard deviations quickly). With a CUSUM chart, you declare a process "out of control" when the upper cumulative sums are greater than some pre-specified upper decision bound (also called the decision interval) denoted h, or the lower cumulative sums are less than $-h$. For groundwater monitoring, we usually are only interested in the behavior of the upper cumulative sums. USEPA

(1989b, p. 7-7) recommends setting $h = 5$, while ASTM (1996) recommends setting $h = 4.5$ to make it easy to plot the Shewhart and CUSUM plots on the same graph.

So what are the cumulative sums measuring? If the process is "in control" so that all future observations come from a normal distribution with a fixed mean μ and standard deviation σ, then the quantity z_i in Equation (6.85) is approximately distributed as a N(0, 1) random variable and is "bouncing around" 0, so the quantity $z_i - k$ in Equation (6.82) is "bouncing around" $-k$, and therefore the ith upper cumulative sum S_i^+ will tend to bounce around 0. If the process is "out of control" and the future observation x_i comes from a normal distribution with mean μ_i, where $\mu_i > \mu$, then z_i is bouncing around some positive number approximately equal to $(\mu_i - \mu)/\sigma$, and $z_i - k$ is bouncing around some number approximately equal to $[(\mu_i - \mu)/\sigma] - k$. If the μ_i's are large enough, then eventually the upper cumulative sums will start taking on positive values, start increasing over time, and eventually become larger than the upper decision bound h.

The upper cumulative sums defined in Equation (6.82) are "standardized" upper cumulative sums based on the standardized z-values defined in Equation (6.85). Rather than plotting the standardized upper cumulative sums over time and comparing them to the upper decision bound h, you can plot the unstandardized upper cumulative sums (here denoted U_i^+) and compare them to the unstandardized upper decision bound h_U. The unstandardized upper cumulative sums and decision bound are given by:

$$U_i^+ = \overline{x} + S_i^+\, s \qquad\qquad (6.86)$$

$$h_U = \overline{x} + h\, s \qquad\qquad (6.87)$$

The FWFPR based on using CUSUM charts is rather difficult to compute. As for Shewhart control charts, however, the constants k and h associated with a CUSUM chart are not adjusted for the number of compliance wells (r) nor the number of constituents being monitored (n_c), so for each sampling occasion, the FWFPR can be quite close to 100%.

Combined Shewhart-CUSUM Control Charts

USEPA (1989b, pp. 7-1 to 7-12; 1992c, pp. 75–79), Gibbons (1994, pp. 160–168), and ASTM (1996) all recommend using combined Shewhart-

CUSUM control charts for groundwater monitoring in order to detect both sudden large releases and smaller gradual releases. This requires choosing the control chart parameters L (for the Shewhart control chart) and k and h (for the CUSUM control chart). As already stated, USEPA (1989b, p. 7-7) recommends using $L = 4.5$, $k = 1$, and $h = 5$, based on the recommendations of Lucas (1982) and Starks (1988). Using both Shewhart and CUSUM control charts will increase the FWFPR for each sampling occasion beyond the FWFPR obtained using just one of these kinds of control charts.

Control Charts vs. Prediction Limits

Control charts are sometimes recommended as an alternative to prediction limits because besides providing objective decision rules for determining the presence of outliers and/or trend that may indicate contamination, they allow you to look at the observations at each well over time and get a feel for the data, possibly detecting outliers and/or trends by eye (e.g., USEPA, 1989b, 1992c). Plotting is not restricted to control charts, however. You can just as easily make plots of the observations over time for each well and add a line showing a prediction limit instead of a control limit.

More importantly, unlike simultaneous prediction limits, confidence levels associated with CUSUM and combined Shewhart-CUSUM control charts are hard to determine, and EPA guidelines do not account for adjusting the control chart parameters to account for the number of compliance wells and the number of constituents monitored during each sampling event (Gibbons, 1994, p. 164; ASTM, 1996, p. 12; Davis, 1998c). ASTM (1996) suggests several possible modifications to control charts by allowing resampling and updating background data in order to attempt to control the FWFPR on each monitoring occasion. Davis (1998c) performed simulation studies to determine the number of monitoring occasions (run lengths) that would occur before contamination was mistakenly declared at at least one compliance well for at least one constituent when in fact no contamination had occurred at any of the wells for any of the constituents. He found that control charts yield average run lengths (ARLs) much smaller than average run lengths associated with simultaneous prediction limits. Although some of the modified procedures suggested in ASTM (1996) yield average run lengths comparable to simultaneous prediction limits, the power of these procedures to detect contamination is not adequate. We therefore recommend using simultaneous prediction limits instead of control charts.

Example 6.11: Shewhart and CUSUM Control Charts for Nickel Concentrations

Table 6.6 displays nickel concentrations (ppb) in groundwater sampled at a compliance well over two separate 8-month periods in 1995 and 1996. In ENVIRONMENTALSTATS for S-PLUS, these data are stored in the data frame epa.98.nickel2.df. The first year (1995) is considered the baseline period, and the second year is considered the compliance period. The probability plot for the baseline data indicates there is no evidence that these data deviate grossly from a normal distribution. The sample mean and standard deviation based on the baseline data are 25.1 and 11.5 ppb.

Month	Baseline Period (1995)	Compliance Period (1996)
1	32.8	19.0
2	15.2	34.5
3	13.5	17.8
4	39.6	23.6
5	37.1	34.8
6	10.4	28.8
7	31.9	43.7
8	20.6	81.8

Table 6.6   Nickel concentrations (ppb) at a compliance well

Figure 6.8 displays the plot of the observations from the compliance period (1996) vs. time, along with both the upper and lower Shewhart control limits based on the baseline data with $L = 4.5$. In this figure, the eighth month is above the upper control limit and indicates an "out of control" condition.

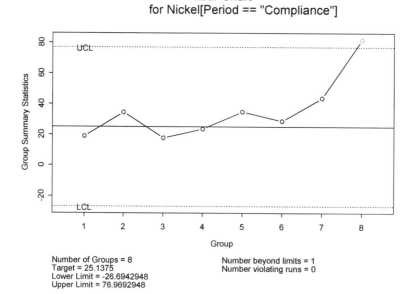

xbar Chart
for Nickel[Period == "Compliance"]

Number of Groups = 8
Target = 25.1375
Lower Limit = -26.6942948
Upper Limit = 76.9692948

Number beyond limits = 1
Number violating runs = 0

Figure 6.8   Shewhart control chart for the compliance period nickel data, based on using the baseline period data to establish background

Figure 6.9 displays the plot of the standardized upper and lower cumulative sums vs. time, along with the standardized CUSUM upper and lower decision bounds (h and $-h$, here denoted UDB and LDB) based on the baseline data with $k = 1$ and $h = 5$. None of the cumulative sums indicates an out of control condition.

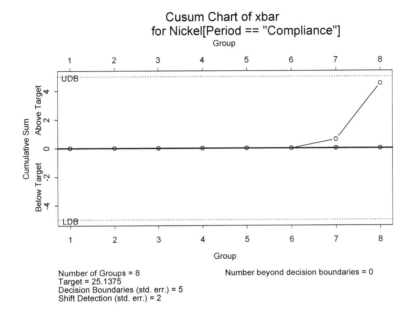

Cusum Chart of xbar
for Nickel[Period == "Compliance"]

Number of Groups = 8   Number beyond decision boundaries = 0
Target = 25.1375
Decision Boundaries (std. err.) = 5
Shift Detection (std. err.) = 2

Figure 6.9   Standardized CUSUM control chart for the compliance period nickel data

Menu

In the current version of S-PLUS 2000 you cannot create the Shewhart or CUSUM control charts for the nickel data using the S-PLUS menu because the menu version of control charts requires more than one observation per month.

Command

To create the Shewhart control chart for the nickel data using the S-PLUS Command or Script Window, type these commands.

```
attach(epa.98.nickel2.df)
qcc.baseline <- qcc(Nickel[Period=="Baseline"],
   type="xbar",
   std.dev=sd(Nickel[Period=="Baseline"]),
   labels=Month[Period=="Baseline"])
```

```
shewhart(qcc.baseline,
  newdata=Nickel[Period=="Compliance"], nsigmas=4.5,
  labels=Month[Period=="Compliance"])
```

Note that in the call to the function `shewhart()`, the argument `nsigma` corresponds to the constant L.

To create the CUSUM chart, type these commands.

```
cusum(qcc.baseline,
  newdata=Nickel[Period=="Compliance"],
  decision.int=5, se.shift=2,
  labels=Month[Period=="Compliance"])
detach()
```

Note that in the call to the function `cusum()`, the argument `decision.int` corresponds to the upper decision bound h, and the argument `se.shift` corresponds to $2k$ (that is, twice the value of the reference value k).

SUMMARY

- Any activity that requires comparing new values to "background" or "standard" values creates a decision problem: if the new values greatly exceed the background or standard value, is this evidence of a true difference (i.e., is there contamination)?
- Any rule you make to choose between deciding "Contamination" vs. "No Contamination" will result in instances when you make the correct decision and in instances when you make a mistake.
- Assuming contamination is not present, declaring contamination is present when it is not is called a *Type I error* or *false-positive*, and declaring contamination is absent when it is really present is called a *Type II error* or *false-negative*.
- One way to come up with an objective decision rule to decide whether there is contamination or not is to base it on some kind of statistical interval or test.
- You can use S-Plus and EnvironmentalStats for S-Plus to create prediction intervals, simultaneous prediction intervals, tolerance intervals, and control charts.
- For monitoring several chemical constituents in groundwater at several compliance wells, simultaneous prediction intervals are the best way to control the *facility-wide false positive rate (FWFPR)* and at the same time maintain adequate power to detect contamination.

- Using a tolerance interval for a compliance-to-background comparison is inappropriate because the Type I error rate increases with the sample size at the compliance well or site.
- Control charts do not perform as well as simultaneous prediction limits in terms of controlling the false positive and false negative rates over time.

EXERCISES

6.1. Consider the copper data shown in Table 3.14 of Exercise 3.7 in Chapter 3. In ENVIRONMENTALSTATS for S-PLUS these data are stored in the data frame `epa.92c.copper1.df`.

    a. Create a normal probability plot by combining the data from both background wells to determine whether these data may be adequately modeled with a normal distribution.

    b. Assume there is no substantial spatial or temporal variability in the copper concentrations. Also, assume the data at the compliance well will be tested monthly, and use the background well data to create a one-sided upper 95% prediction interval for the next $k = 6$ future observations, and compare the data from the compliance well to the upper prediction limit. Is there any evidence of contamination based on this method?

    c. Plot the compliance well data vs. time, and add a horizontal line at the upper prediction limit to the plot.

    d. Now assume there are $r = 10$ compliance wells and $n_c = 20$ constituents being monitored on each sampling occasion. Create a one-sided upper 95% simultaneous prediction interval based on the 1-of-3 rule, and compare the data from the compliance well to the upper prediction limit.

    e. Add a horizontal line at the upper simultaneous prediction limit to the plot you created in part **c** above.

    f. Compute a one-sided upper 95% β-content tolerance interval with associated confidence level 95% based on the background data, and compare the data from the compliance well with the upper tolerance limit.

    g. Add a horizontal line at the upper tolerance limit to the plot you created in part **c** above.

    h. Explain the difference between the upper prediction limit, the upper simultaneous prediction limit, and the upper tolerance limit. What does each limit assume?

6.2. Repeat parts **b-h** of Exercise **6.1** using nonparametric prediction and tolerance intervals. For part **b**, does the confidence level of the prediction interval achieve 95%? For part **d**, does the confidence level of the simultaneous prediction interval achieve $1-(0.05/20) \times 100\% = 99.75\%$? For part **f**, if you set the coverage at 95% does the associated confidence level achieve 95%? What do you need to do to attain the desired confidence levels?

6.3. Using the copper data, create control limits for a combined Shewhart-CUSUM control chart using the background data, then plot the compliance well data vs. time along with the control limits. Is there any evidence of contamination based on this method?

6.4. Table 6.7 below displays concentrations of trichloroethylene (ppb) in groundwater from three background wells and one compliance well. Sampling was done monthly for 6 months, with observations taken at the background wells for all 6 months and observations taken at the compliance well the last 4 months. In ENVIRONMENTALSTATS for S-PLUS these data are stored in the data frame tce.df.

a.   Assume there is no substantial spatial or temporal variability in the trichloroethylene concentrations. Use the background well data to create a one-sided upper nonparametric prediction interval for the next $k = 4$ future observations, and compare the data from the compliance well to the upper prediction limit. Is there any evidence of contamination based on this method?

b.   What are the confidence level and Type I error rate associated with the prediction interval you computed in part **a**? If you wanted to increase the confidence level associated with the prediction interval, what would you have to do?

c.   Now assume there are $r = 20$ compliance wells and $n_c = 30$ constituents being monitored on each sampling occasion. Create one-sided upper simultaneous prediction intervals based on the 1-of-3 rule, the California rule with $m = 3$, and the Modified California rule. Compare the data from the compliance well to these upper prediction limits.

d.   What are the confidence levels and FWFPRs associated with the simultaneous prediction intervals you created in part **c**? If you wanted to achieve a FWFPR of 5% for each sampling occasion, what would the confidence levels of the prediction intervals have to be?

| Month | Background Wells | | | Compliance Well |
	Well 1	Well 2	Well 3	Well 4
1	<5	7	<5	
2	<5	6.5	<5	
3	8	<5	10.5	7.5
4	<5	6	<5	<5
5	9	12	<5	8
6	10	<5	9	14

Table 6.7   Trichloroethylene concentrations (ppb) in groundwater at background and compliance wells

6.5.   Compute a one-sided upper β-content nonparametric tolerance interval for trichloroethylene concentrations using the data from the background wells in Table 6.7.

   a.   If the coverage of this tolerance interval is set to 95%, what is the associated confidence level?

   b.   What would you have to do to increase the associated confidence level of the tolerance interval?

   c.   Why is it a bad idea to use tolerance intervals for compliance-to-background comparisons?

7   HYPOTHESIS TESTS

Comparing Groups to Standards
and One Another

In Chapters 2 and 6, we introduced the idea of the hypothesis testing framework. In Chapter 6 we discussed three tools you can use to make an objective decision about whether contamination is present or not: prediction intervals, tolerance intervals, and control charts. In this chapter, we provide a full discussion of the statistical hypothesis testing framework, discuss the relationship between confidence intervals and formal hypothesis tests, and discuss hypothesis tests to make inferences about a single population and compare two or more populations.

THE HYPOTHESIS TESTING FRAMEWORK

We introduced the hypothesis testing framework back in Chapter 2. Our first example involved deciding whether to wear a jacket or not, and our decision depended on our belief about whether it would rain that day (Table 2.1). Our second example involved deciding whether a site or well is contaminated or not (Table 2.2). Table 7.1 below reproduces Table 2.2. In this case, the null hypothesis is that no contamination is present.

Your Decision	Reality	
	No Contamination	Contamination
Contamination	Mistake: Type I Error (Probability = α)	Correct Decision (Probability = $1-\beta$)
No Contamination	Correct Decision	Mistake: Type II Error (Probability = β)

Table 7.1   Hypothesis testing framework for deciding on the presence of contamination in the environment when the null hypothesis is "no contamination"

In Step 5 of the DQO process (see Chapter 2), you usually link the principal study question you defined in Step 2 with some population parameter such as the mean, median, 95th percentile, etc. For example, if the study question is "Is the concentration of 1,2,3,4-tetrachlorobenzen (TcCB) in the soil at a Cleanup site significantly above background levels?" then you may decide to reformulate this question as "Is the average concentration of TcCB in the soil at the Cleanup site greater than the average concentration of TcCB in the soil at a Reference site?"

A *hypothesis test* or *significance test* is a formal mathematical mechanism for objectively making a decision in the face of uncertainty, and is usually used to answer a question about the value of a population parameter. A *two-sided hypothesis test* about a population parameter θ (theta) is used to test the null hypothesis

$$H_0 : \theta = \theta_0 \tag{7.1}$$

versus the two-sided alternative hypothesis

$$H_a : \theta \neq \theta_0 \tag{7.2}$$

where H_0 (pronounced "H-naught") denotes the null hypothesis that the true value of θ is equal to some specified value θ_0 (theta-naught). A *lower one-sided hypothesis test* is used to test the null hypothesis

$$H_0 : \theta \geq \theta_0 \tag{7.3}$$

versus the lower one-sided alternative hypothesis

$$H_a : \theta < \theta_0 \tag{7.4}$$

and an *upper one-sided hypothesis test* is used to test the null hypothesis

$$H_0 : \theta \leq \theta_0 \tag{7.5}$$

versus the upper one-sided alternative hypothesis

$$H_a : \theta > \theta_0 \tag{7.6}$$

In environmental monitoring, we are almost always concerned only with one-sided hypotheses, such as "The average concentration of TcCB in the soil at the Cleanup site is less than or equal to 2 ppb," or "The average value of pH at the compliance well is greater than or equal to 7." We rarely consider two-sided hypothesis tests.

Hypotheses tests are usually based on a **test statistic**, say T, which is computed from a random sample from the population. If T is "too extreme," then we decide to reject the null hypothesis in favor of the alternative hypothesis. For example, suppose we are interested in determining whether the true mean of a distribution is less than or equal to some hypothesized value μ_0 vs. the alternative that the true mean is bigger than μ_0. This is simply the one-sided upper hypothesis given in Equations (7.5) and (7.6) with θ replaced by μ. We can use Student's t-statistic, which we will discuss in more detail later in this chapter, to test this hypothesis. Student's t-statistic is a scaled version of the sample mean minus the hypothesized mean:

$$t = \frac{\bar{x} - \mu_0}{s/\sqrt{n}} \qquad (7.7)$$

Since the sample mean is an unbiased estimator of the true mean μ, if the true mean is equal to μ_0, then the sample mean is "bouncing around" μ_0 and the t-statistic is "bouncing around" 0. The distribution of the t-statistic under the null hypothesis is shown in Figure 5.9 in Chapter 5 for sample sizes of $n = 2$, 5, and ∞. On the other hand, if the true mean μ is larger than μ_0, then the sample mean is bouncing around μ, the numerator of the t-statistic is bouncing around $\mu-\mu_0$, and the t-statistic is bouncing around some positive number. So if the t-statistic is "large" we will probably reject the null hypothesis in favor of the alternative hypothesis.

Parametric vs. Nonparametric Tests

For a **parametric test**, the test statistic T is usually some estimator of θ (possibly shifted by subtracting a number and scaled by dividing by a number), and the distribution of T under the null hypothesis depends on the distribution of the population (e.g., normal, lognormal, Poisson, etc.). For a **nonparametric** or **distribution-free test**, T is usually based on the ranks of the data in the random sample, and the distribution of T under the null hypothesis does not depend on the distribution of the population.

For example, for a two-sample t-test (see below), the test statistic is a scaled version of the difference between the two sample means, and both populations are assumed to be normally distributed. For the Wilcoxon rank sum test (see below), the test statistic is the sum of the ranks in the first sample, and the distribution of this statistic under the null hypothesis does not depend on the distribution of the two populations.

Type I and Type II Errors (Significance Level and Power)

As stated above, a hypothesis test involves using a test statistic computed from data collected from an experiment to make a decision. A test statistic is a random quantity (e.g., some expression involving the sample mean); if you repeat the experiment or get new observations, you will often get a different value for the test statistic. Because you are making your decision based on the value of a random quantity, you will sometimes make the "wrong" choice. Table 7.2 below illustrates the general hypothesis testing framework; it is simply a generalization of Table 7.1 above.

	Reality	
Your Decision	H_0 True	H_0 False
Reject H_0	Mistake: Type I Error (Probability = α)	Correct Decision (Probability = $1-\beta$)
Do Not Reject H_0	Correct Decision	Mistake: Type II Error (Probability = β)

Table 7.2   The framework of a hypothesis test

As we explained in Chapter 2, statisticians call the two kinds of mistakes you can make a *Type I error* and a *Type II error*. Of course, in the real world, once you make a decision, you take an action (e.g., clean up the site or do nothing), and you hope to find out eventually whether the decision you made was the correct decision. For a specific hypothesis test, the probability of making a Type I error is usually denoted with the Greek letter α (alpha) and is called the *significance-level* or *α-level* of the test. This probability is also called the *false positive rate*. The probability of making a Type II error is usually denoted with the Greek letter β (beta). This probability is also called the *false negative rate*. The probability $1-\beta$ denotes the probability of correctly deciding to reject the null hypothesis when in fact it is false. This probability is called the *power* of the hypothesis test.

Note that in Table 7.2 above the phrase "Do Not Reject H_0" is used instead of "Decide H_0 is True." This is because in the framework of hypothesis testing, you assume the null hypothesis is true unless you have enough evidence to reject it. If you end up not rejecting H_0 because of the value of your test statistic, you may have unknowingly committed a Type II error; i.e., the alternative hypothesis is really true, but you did not have enough evidence to reject H_0.

A critical aspect of any hypothesis test is deciding on the acceptable values for the probabilities of making a Type I and Type II error. This is part of Step 6 of the DQO process. **The choice of values for α and β is a subjective policy decision**. The possible choices for α and β are limited by the sample size of the experiment (usually denoted n), the variability inherent in the data (usually denoted σ), and the magnitude of the difference between the null and alternative hypothesis (usually denoted δ or Δ). Conventional choices for α are 1%, 5%, and 10%, but these choices should be made in the context of balancing the cost of a Type I and Type II error. For most hypothesis tests, there is a well-defined relationship between α, β, n and the scaled difference δ/σ (see Chapter 8). A very important fact is that for a specified sample size, if you reduce the Type I error, then you increase the Type II error, and vice-versa.

P-Values

When you perform a hypothesis test, you usually compute a quantity called the p-value. The *p-value* is the probability of seeing a test statistic as extreme or more extreme than the one you observed, assuming the null hypothesis is true. Thus, if the p-value is less than or equal to the specified value of α (the Type I error level), you reject the null hypothesis, and if the p-value is greater than α, you do not reject the null hypothesis. For hypothesis tests where the test statistic has a continuous distribution, under the null hypothesis (i.e., if H_0 is true), the p-value is uniformly distributed between 0 and 1. When the test statistic has a discrete distribution, the p-value can take on only a discrete number of values.

To get an idea of the relationship between p-values and Type I errors, consider the following example. Suppose your friend is a magician and she has a fair coin (i.e., the probability of a "head" and the probability of a "tail" are both 50%) and a coin with heads on both sides (so the probability of a head is 100% and the probability of a tail is 0%). She takes one of these coins out of her pocket, begins to flip it several times, and tells you the outcome after each flip. You have to decide which coin she is flipping. Of course if a flip comes up tails, then you automatically know she is flipping the fair coin. If the coin keeps coming up heads, however, how many flips in a row coming up heads will you let go by before you decide to say the coin is the two-headed coin?

Table 7.3 displays the probability of seeing various numbers of heads in a row under the null hypothesis that your friend is flipping the fair coin. (Of course, under the alternative hypothesis, all flips will result in a head and the probability of seeing any number of heads in a row is 100%.) Suppose you decide you will make your decision after seeing the results of five flips, and if you see $T = 5$ heads in a row then you will reject the null hypothesis and say your friend is flipping the two-headed coin. If you observe five heads in a row, then the p-value associated with this outcome is 0.0312; that is, there is a probability of 3.12% of getting five heads in a row when you flip a fair coin five times. Therefore, the Type I error rate associated with your decision rule is 0.0312. If you want to create a decision rule for which the Type I error rate is no greater than 1%, then you will have to wait until you see the outcome of seven flips. If your decision rule is to reject the null hypothesis after seeing $T = 7$ heads in a row, then the actual Type I error rate is 0.78%. If you do see seven heads in a row, the p-value is 0.0078.

# Heads in a Row (T)	Probability (%)
1	50
2	25
3	12.5
4	6.25
5	3.12
6	1.56
7	0.78
8	0.39
9	0.20
10	0.10

Table 7.3   The probability of seeing T heads in a row in T flips of a fair coin

For this example, the power associated with your decision rule is the probability of correctly deciding your friend is flipping the two-headed coin when in fact that is the one she is flipping. This is a special example in which the power is equal to 100%, because if your friend really is flipping the two headed coin then you will always see a head on each flip and no matter what value of T you choose for the cut-off, you will always see T heads in a row. Usually, however, there is an inverse relationship between the Type I and Type II error, so that the smaller you set the Type I error, the smaller the power of the test (see Chapter 8).

Relationship between Hypothesis Tests and Confidence Intervals

Consider the null hypothesis shown in Equation (7.1), where θ is some population parameter of interest (e.g., mean, proportion, 95th percentile, etc.). There is a one-to-one relationship between hypothesis tests concerning θ and a confidence interval for this parameter. A $(1-\alpha)100\%$ confidence interval for θ consists of all possible values of θ that are associated with not rejecting the null hypothesis at significance level α. Thus, if you know how to create a confidence interval for a parameter, you can perform a hypothesis test for that parameter, and vice-versa. Table 7.4 shows the explicit relationship between hypothesis tests and confidence intervals.

Whenever you report the results of a hypothesis test, you should almost always report the corresponding confidence interval as well. This is because if you have a small sample size, you may not have much power to uncover the fact that the null hypothesis is not true, even if there is a huge difference between the postulated value of θ (e.g., $\theta_0 \leq 5$ ppb) and the true value of θ

(e.g., $\theta = 20$ ppb). On the other hand, if you have a large sample size, you may be very likely to detect a small difference between the postulated value of θ (e.g., $\theta_0 \leq 5$ ppb) and the true value of θ (e.g., $\theta = 6$ ppb), but this difference may not really be important to detect.   Confidence intervals help you sort out the important distinction between a *statistically significantly difference* and a *scientifically meaningful difference*.

Test Type	Alternative Hypothesis	Corresponding Confidence Interval	Rejection Rule Based on CI
Two-sided	$\theta \neq \theta_0$	Two-sided [LCL, UCL]	$LCL > \theta_0$ or $UCL < \theta_0$
Lower	$\theta < \theta_0$	Upper [$-\infty$, UCL]	$UCL < \theta_0$
Upper	$\theta > \theta_0$	Lower [LCL, ∞]	$LCL > \theta_0$

Table 7.4   Relationship between hypothesis tests and confidence intervals

OVERVIEW OF UNIVARIATE HYPOTHESIS TESTS

Table 7.5 summarizes the kinds of univariate hypothesis tests that we will talk about in this chapter. In Chapter 9 we will talk about hypothesis tests for regression models. We will not discuss hypothesis tests for multivariate observations. A good introduction to multivariate statistical analysis is Johnson and Wichern (1998).

One-Sample	Two-Samples	Multiple Samples
Goodness-of-Fit		
Proportion	Proportions	Proportions
Location	Locations	Locations
Variability	Variability	Variability

Table 7.5   Summary of the kinds of hypothesis tests discussed in this chapter

GOODNESS-OF-FIT TESTS

Most commonly used parametric statistical tests assume the observations in the random sample(s) come from a normal population. So how do you know whether this assumption is valid? We saw in Chapter 3 how to make a visual assessment of this assumption using Q-Q plots. Another way to verify this assumption is with a goodness-of-fit test, which lets you specify what kind of distribution you think the data come from and then compute a test statistic and a p-value.

A goodness-of-fit test may be used to test the null hypothesis that the data come from a specific distribution, such as "the data come from a normal distribution with mean 10 and standard deviation 2," or to test the more general null hypothesis that the data come from a particular family of distributions, such as "the data come from a lognormal distribution." Goodness-of-fit tests are mostly used to test the latter kind of hypothesis, since in practice we rarely know or want to specify the parameters of the distribution.

In practice, goodness-of-fit tests may be of limited use for very large or very small sample sizes. Almost any goodness-of-fit test will reject the null hypothesis of the specified distribution if the number of observations is very large, since "real" data are never distributed according to any theoretical distribution (Conover, 1980, p. 367). On the other hand, with only a very small number of observations, no test will be able to determine whether the observations appear to come from the hypothesized distribution or some other totally different looking distribution.

Tests for Normality

Two commonly used tests to test the null hypothesis that the observations come from a normal distribution are the Shapiro-Wilk test (Shapiro and Wilk, 1965), and the Shapiro-Francia test (Shapiro and Francia, 1972). The Shapiro-Wilk test is more powerful at detecting short-tailed (platykurtic) and skewed distributions, and less powerful against symmetric, moderately long-tailed (leptokurtic) distributions. Conversely, the Shapiro-Francia test is more powerful against symmetric long-tailed distributions and less powerful against short-tailed distributions (Royston, 1992b; 1993). These tests are considered to be two of the very best tests of normality available (D'Agostino, 1986b, p. 406).

The Shapiro-Wilk Test

The Shapiro-Wilk test statistic can be written as:

$$W = \left[\sum_{i=1}^{n} a_i x_{(i)} \right]^2 \Big/ \sum_{i=1}^{n} (x_i - \bar{x})^2 \qquad (7.8)$$

where $x_{(i)}$ denotes the ith ordered observation, a_i is the ith element of the $n \times 1$ vector $\underline{a}$, and the vector $\underline{a}$ is defined by:

$$\underline{a}^T = \underline{m}^T \mathbf{V}^{-1} \Big/ \sqrt{\underline{m}^T \mathbf{V}^{-1} \mathbf{V}^{-1} \underline{m}} \qquad (7.9)$$

where T denotes the transpose operator, and $\underline{m}$ is the vector of expected values and V is the variance-covariance matrix of the order statistics of a random sample of size n from a standard normal distribution. That is, the values of $\underline{a}$ are the expected values of the standard normal order statistics weighted by their variance-covariance matrix, and normalized so that $\underline{a}^T \underline{a} = 1$. It can be shown that the W-statistic in Equation (7.8) is the same as the square of the sample correlation coefficient between the vectors $\underline{a}$ and $\underline{x}_{()}$:

$$ W = \left\{ r \left[\underline{a}, \underline{x}_{(\ )} \right] \right\}^2 \tag{7.10} $$

where

$$ r\left(\underline{x}, \underline{y} \right) = \frac{\displaystyle\sum_{i=1}^{n} (x_i - \bar{x})(y_i - \bar{y})}{\sqrt{\displaystyle\sum_{i=1}^{n} (x_i - \bar{x})^2 \sum_{i=1}^{n} (y_i - \bar{y})^2}} \tag{7.11} $$

(see Chapter 9 for an explanation of the sample correlation coefficient).

Small values of W yield small p-values and indicate the null hypothesis of normality is probably not true. Royston (1992a) presents an approximation for the coefficients $\underline{a}$ necessary to compute the Shapiro-Wilk W-statistic, and also a transformation of the W-statistic that has approximately a standard normal distribution under the null hypothesis. Both of these approximations are used in ENVIRONMENTALSTATS for S-PLUS.

The Shapiro-Francia Test

Shapiro and Francia (1972) introduced a modification of the W-test that depends only on the expected values of the order statistics $(\underline{m})$ and not on the variance-covariance matrix (V):

$$ W' = \left[\sum_{i=1}^{n} b_i x_{(i)} \right]^2 \bigg/ \sum_{i=1}^{n} (x_i - \bar{x})^2 \tag{7.12} $$

where b_i is the ith element of the vector $\underline{b}$ defined as:

$$\underline{b} = \underline{m}/\sqrt{\underline{m}^T \underline{m}} \qquad (7.13)$$

Several authors, including Ryan and Joiner (1973), Filliben (1975), and Weisberg and Bingham (1975), note that the W'-statistic is intuitively appealing because it is the squared sample correlation coefficient associated with a normal probability plot. That is, it is the squared correlation between the ordered sample values $\underline{x}_{()}$ and the expected normal order statistics $\underline{m}$:

$$W' = \left\{ r\left[\underline{b}, \underline{x}_{()}\right] \right\}^2 = \left\{ r\left[\underline{m}, \underline{x}_{()}\right] \right\}^2 \qquad (7.14)$$

Weisberg and Bingham (1975) introduced an approximation of the Shapiro-Francia W'-statistic that is easier to compute. They suggested using Blom scores (Blom, 1958, pp. 68–75; see Chapter 3) to approximate the elements of $\underline{m}$:

$$\tilde{W}' = \frac{\left[\sum_{i=1}^{n} c_i x_{(i)}\right]^2}{\sum_{i=1}^{n}(x_i - \bar{x})^2} = \left\{ r\left[\underline{c}, \underline{x}_{()}\right] \right\}^2 \qquad (7.15)$$

where c_i is the ith element of the vector $\underline{c}$ defined by:

$$\underline{c} = \tilde{\underline{m}}/\sqrt{\tilde{\underline{m}}^T \tilde{\underline{m}}} \qquad (7.16)$$

$$\tilde{m}_i = \Phi^{-1}\left(\frac{i - 3/8}{n + 1/4}\right) \qquad (7.17)$$

and Φ denotes the standard normal cdf. That is, the values of the elements of $\underline{m}$ in Equation (7.13) are replaced with their estimates based on the usual plotting positions for a normal distribution (see Chapter 3).

Filliben (1975) proposed the probability plot correlation coefficient (PPCC) test that is essentially the same test as the test of Weisberg and Bingham (1975), but Filliben used different plotting positions. Looney and Gulledge (1985) investigated the characteristics of Filliben's PPCC test using various plotting position formulas and concluded that the PPCC test based on Blom plotting positions performs slightly better than tests based on other plotting positions. The Weisberg and Bingham (1975) approximation to the Shapiro-Francia W'-statistic is the square of Filliben's PPCC test statistic based on Blom plotting positions. Royston (1992c) provides a method for computing p-values associated with the Weisberg-Bingham approximation to the Shapiro-Francia W'-statistic, and this method is implemented in ENVIRONMENTALSTATS for S-PLUS.

The Shapiro-Wilk and Shapiro-Francia tests can be used to test whether observations appear to come from a normal distribution, or the transformed observations (e.g., Box-Cox transformed) come from a normal distribution. Hence, these tests can test whether the data appear to come from a normal, lognormal, or three-parameter lognormal distribution for example, as well as a zero-modified normal or zero-modified lognormal distribution.

Example 7.1: *Testing the Normality of the Reference Area TcCB Data*

In Chapter 3 we saw that the Reference area TcCB data appear to come from a lognormal distribution based on histograms (Figures 3.1, 3.2, 3.10, and 3.11), an empirical cdf plot (Figure 3.16), normal Q-Q plots (Figures 3.18 and 3.19), Tukey mean-difference Q-Q plots (Figures 3.21 and 3.22), and a plot of the PPCC vs. λ for a variety of Box-Cox transformations (Figure 3.24). Here we will formally test whether the Reference area TcCB data appear to come from a normal or lognormal distribution.

Assumed Distribution	Shapiro-Wilk (W)	Shapiro-Francia (W')
Normal	0.918 (p=0.003)	0.923 (p=0.006)
Lognormal	0.979 (p=0.55)	0.987 (p=0.78)

Table 7.6   Results of tests for normality and lognormality for the Reference area TcCB data

Table 7.6 lists the results of these two tests. The second and third columns show the test statistics with the p-values in parentheses. The p-values clearly indicate that we should not assume the Reference area TcCB data come from a normal distribution, but the assumption of a lognormal distri-

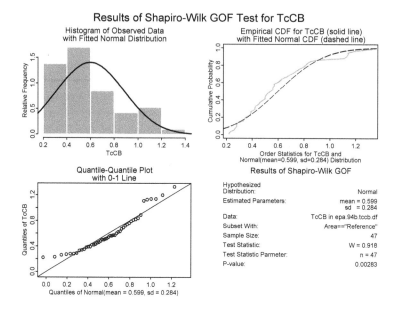

Figure 7.1   Companion plots for the Shapiro-Wilk test for normality for the Reference area TcCB data

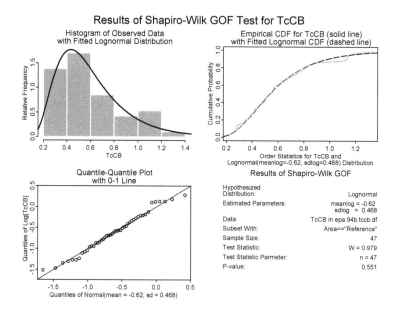

Figure 7.2   Companion plots for the Shapiro-Wilk test for lognormality for the Reference area TcCB data

bution appears to be adequate. Figure 7.1 and Figure 7.2 show companion plots for the results of the Shapiro-Wilk tests for normality and lognormality, respectively. These plots include the observed distribution overlaid with the fitted distribution, the observed and fitted cdf, the normal Q-Q plot, and the results of the hypothesis test.

Menu

To perform the Shapiro-Wilk test for normality using the ENVIRONMENTALSTATS for S-PLUS pull-down menu, follow these steps.

1. In the Object Explorer, find and highlight the data frame **epa.94b.tccb.df**.
2. On the S-PLUS menu bar, make the following menu choices: **EnvironmentalStats>Hypothesis Tests>GOF Tests>One Sample>Shapiro-Wilk**. This will bring up the Shapiro-Wilk GOF Test dialog box.
3. For Data to Use, select **Pre-Defined Data**. For Data Set make sure **epa.94b.tccb.df** is selected, for Variable select **TcCB**, and in the Subset Rows with box type **Area=="Reference"**. In the Distribution box select **Normal**.
4. Click on the **Plotting** tab. Under Plotting Information, in the Significant Digits box type **3**, then click **OK** or **Apply**.

To perform the Shapiro-Wilk test for lognormality, repeat the above steps, but in Step 3 select **Lognormal** in the Distribution box. To perform the Shapiro-Francia tests, repeat the above procedure, but in Step 2 make the menu selection **EnvironmentalStats>Hypothesis Tests>GOF Tests>One Sample>Shapiro-Francia**.

Command

To perform the Shapiro-Wilk tests using the ENVIRONMENTALSTATS for S-PLUS Command or Script Window, type these commands.

```
attach(epa.94b.tccb.df)
TcCB.ref <- TcCB[Area=="Reference"]
sw.list.norm <- sw.gof(TcCB.ref)
sw.list.norm
sw.list.lnorm <- sw.gof(TcCB.ref, dist="lnorm")
sw.list.lnorm
```

To plot the results of these tests as shown in Figure 7.1 and Figure 7.2, type these commands.

```
plot.gof.summary(sw.list.norm, digits=3)
```

```
plot.gof.summary(sw.list.lnorm, digits=3)
```

To perform the Shapiro-Francia tests, type these commands.

```
sf.list.norm <- sf.gof(TcCB.ref)
sf.list.norm
sf.list.lnorm <- sf.gof(TcCB.ref, dist="lnorm")
sf.list.lnorm
```

To plot the results of the Shapiro-Francia tests, type these commands.

```
plot.gof.summary(sf.list.norm, digits=3)
plot.gof.summary(sf.list.lnorm, digits=3)
detach()
```

Testing Several Groups for Normality

Current regulations for monitoring groundwater at hazardous and solid waste sites usually require performing statistical analyses even when there are only small sample sizes at each monitoring well. As we noted above, goodness-of-fit tests are not very useful with small sample sizes; there is simply not enough information to determine whether the data appear to come from the hypothesized distribution or not. Gibbons (1994, p. 228) suggests pooling the measures from several upgradient wells to establish "background." Due to spatial variability, the wells may have different means and variances, yet you would like to test the assumption of a normal distribution for the chemical concentration at each of the upgradient wells.

Wilk and Shapiro (1968) suggest two different test statistics for the problem of testing the normality of K separate groups, using the results of the Shapiro-Wilk test applied to random samples from each of the K groups. Both test statistics are functions of the K p-values that result from performing the test on each of the K samples. Under the null hypothesis that all K samples come from normal distributions, the p-values represent a random sample from a uniform distribution on the interval [0,1]. Since these two test statistics are based solely on the p-values, they are really meta-analysis statistics (Fisher and van Belle, 1993, p. 893), and can be applied to the problem of combining the results from K independent hypothesis tests, where the hypothesis tests are not necessarily goodness-of-fit tests.

The test based on normal scores uses the test statistic G given by

$$G = \frac{1}{\sqrt{K}} \sum_{i=1}^{n} G_i \qquad (7.18)$$

where

$$G_i = \Phi^{-1}(p_i) \qquad (7.19)$$

Φ denotes the cdf of the standard normal distribution, and p_i denotes the p-value from the goodness-of-fit test for the ith group. Under the null hypothesis that all K samples come from normal distributions, each normal score G_i represents a random variable from a standard normal distribution, and so the distribution of G also follows a standard normal distribution. Under the alternative hypothesis that at least one of the K groups does not come from the hypothesized distribution, the p-values for the groups that do not come from the hypothesized distribution will tend to be small, so the corresponding normal scores for those p-values will be small, and G will tend to be small. Hence, the null hypothesis is rejected if G is "too small," so the p-value for this test statistic is given by:

$$p = \Phi(G) \qquad (7.20)$$

Wilk and Shapiro (1968) suggest creating normal probability plots of the normal scores in Equation (7.19) as an additional aid in determining whether all of the groups come from the same kind of distribution.

The test based on chi-squared scores uses the test statistic C given by

$$C = \sum_{i=1}^{n} C_i \qquad (7.21)$$

where

$$C_i = F_2^{-1}(1 - p_i) = -2\log(p_i) \qquad (7.22)$$

and F_v denotes the cdf of a chi-squared distribution with v degrees of freedom. Under the null hypothesis that all K samples come from normal distributions, each chi-squared score C_i represents a random variable from a chi-squared distribution with 2 degrees of freedom, and so the distribution of C follows a chi-squared distribution with $2K$ degrees of freedom. Under the alternative hypothesis that at least one of the K groups does not come from the hypothesized distribution, the p-values for the groups that do not come from the hypothesized distribution will tend to be small, so the corresponding chi-squared scores for those p-values will be large, and C will tend to be large. Hence, the null hypothesis is rejected if C is "too large," so the p-value for this test statistic is given by:

$$p = 1 - F_{2K}(C) \qquad (7.23)$$

Just as for the normal scores statistic, you can look at chi-squared probability plots of the chi-squared scores in Equation (7.22) as an additional aid in determining whether all of the groups come from the same kind of distribution.

Wilk and Shapiro (1968) note that the probability plots based on the normal scores in Equation (7.19) and the chi-square scores in Equation (7.22) are equivalent. The two overall test statistics G and C, however, are different measures. Based on simulations, Wilk and Shapiro (1968) found that G and C perform similarly for highly skewed non-normal alternatives, although C is more sensitive to symmetric long-tailed alternatives and G is more sensitive to finite range alternatives.

Example 7.2: Group Test of Normality for Arsenic Concentrations

The guidance document USEPA (1992c, p. 21) contains observations of arsenic concentrations (ppm) collected over 4 months at six monitoring wells. These data are displayed in Table 7.7 and are stored in the data frame epa.92c.arsenic1.df in ENVIRONMENTALSTATS for S-PLUS. Figure 7.3 displays the observations for each well, and Figure 7.4 displays the log-transformed observations. Here we will use the Shapiro-Wilk group test to test the null hypotheses that the observations at each well represent a sample from some kind of normal or lognormal distribution, but the population means and/or variances may differ between wells.

Month	Well 1	Well 2	Well 3	Well 4	Well 5	Well 6
1	22.9	2.0	2.0	7.84	24.9	0.34
2	3.09	1.25	109.4	9.3	1.3	4.78
3	35.7	7.8	4.5	25.9	0.75	2.85
4	4.18	52	2.5	2.0	27	1.2

Table 7.7   Arsenic concentrations (ppm) from groundwater monitoring wells (USEPA, 1992c, p. 21)

Null Hypothesis	Normal Scores (G)	Chi-Squared Scores (C)
Normal	−2.62 (p=0.004)	30.2 (p=0.003)
Lognormal	−0.59 (p=0.28)	14.1 (p=0.29)

Table 7.8   Results of Shapiro-Wilk group tests for the arsenic data

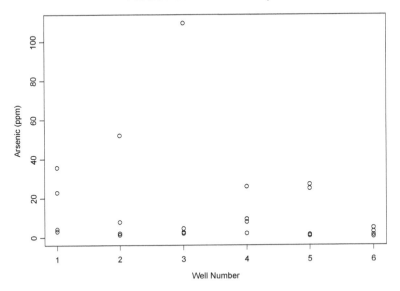

Figure 7.3   Arsenic concentrations (ppm) by well

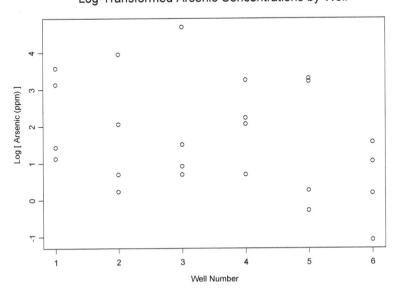

Figure 7.4   Log-transformed arsenic concentrations by well

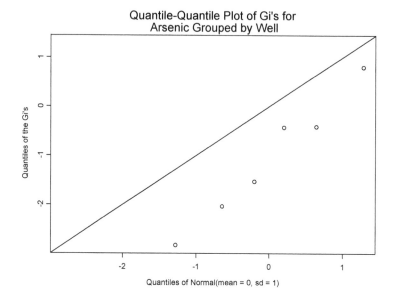

Figure 7.5　Normal Q-Q plot of the normal scores used in the group Shapiro-Wilk goodness-of-fit test for normality of the arsenic concentrations

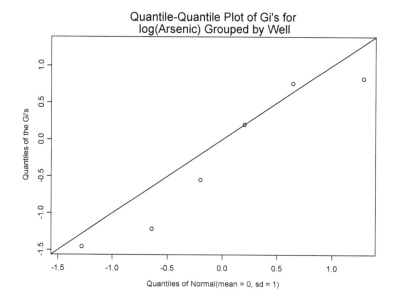

Figure 7.6　Normal Q-Q plot of the normal scores used in the group Shapiro-Wilk goodness-of-fit test for lognormality of the arsenic concentrations

Table 7.8 displays the results of the normal scores and chi-squared scores tests for the null hypotheses of normality and lognormality. Figure 7.5 and Figure 7.6 display the normal Q-Q plots of the normal scores for each of the hypotheses. There is clear evidence that the concentrations do not come from normal distributions, but the assumption of lognormal distributions appears to be adequate.

Menu

To compute the Shapiro-Wilk group test statistics and create the companion probability plot using the ENVIRONMENTALSTATS for S-PLUS pull-down menu, follow these steps.

1. In the Object Explorer, find and highlight the data frame **epa.92c.arsenic1.df**.
2. On the S-PLUS menu bar, make the following menu choices: **EnvironmentalStats>Hypothesis Tests>GOF Tests>Group> Shapiro-Wilk**. This will bring up the Shapiro-Wilk Group GOF Test dialog box.
3. In the Data Set box, make sure **epa.92c.arsenic1.df** is selected. In the Variable box select **Arsenic**, and in the Grouping Variable box select **Well**. In the Distribution box, make sure **Normal** is selected.
4. Click on the **Plotting** tab. Under Plot Type select the **Gi Q-Q** button, then click **OK** or **Apply**.

This creates the results for the test of a normal distribution. To create the results for the test of a lognormal distribution, in Step 3 select **Lognormal** in the Distribution box.

Command

To compute the Shapiro-Wilk group test statistics and create the companion probability plot using the ENVIRONMENTALSTATS for S-PLUS Command or Script Window, type these commands.

```
attach(epa.92c.arsenic1.df)
sw.list <- sw.group.gof(Arsenic, Well)
sw.list
plot(sw.list, plot.type="Gi Q-Q")
sw.log.list <- sw.group.gof(Arsenic, Well,
    dist="lnorm")
sw.log.list
plot(sw.log.list, plot.type="Gi Q-Q")
detach()
```

Tests for Other Kinds of Distributions

Three other commonly used goodness-of-fit tests are the Kolmogorov-Smirnov goodness-of-fit test (Zar, 1999, p. 475), the chi-square goodness-of-fit test (Zar, 1999, p. 462), and the probability plot correlation coefficient (PPCC) goodness-of-fit test (Filliben, 1975; Vogel, 1986). All of these tests are available in S-PLUS and/or ENVIRONMENTALSTATS for S-PLUS (see the help files). Here, we will give a brief explanation of how these tests work. You can look at the help files to get more detailed information and see examples.

The Kolmogorov-Smirnov goodness-of-fit test lets you test whether a set of observations comes from a specified theoretical distribution based on looking at the maximum vertical distance between the empirical cdf for the observations and the cdf for the hypothesized theoretical distribution. The test also lets you test whether two sets of data appear to come from the same distribution by looking at the distance between the two empirical cdfs.

The chi-square test requires you to group the observations into bins (just as you have to do to create a histogram). Once the observations are grouped, you can count the observed number of observations in a group and compare this number with the "expected" number of observations in that group based on the hypothesized distribution (for example, for a normal distribution, you do not expect to see too many observations in a group that is more than two standard deviations away from the mean). The chi-square statistic is based on adding up the squares of the differences between the observed and expected number of observations.

The probability plot correlation coefficient (PPCC) test is based on the sample correlation coefficient for the Q-Q plot. The sampling distribution of this test statistic does not depend on the distribution parameters, only on the type of distribution (e.g., normal, lognormal, extreme value, etc.) For tests other than testing for the normal distribution, the sampling distribution of this test statistic must be simulated. Vogel (1986) provides a table of empirical percentage points for the distribution of the PPCC test statistic based on the null hypothesis that the data come from an extreme value distribution.

Outlier Tests

Tests for outliers are closely related to goodness-of-fit tests. An *outlier* can be defined as an observation that is "far away" from the rest of the observations. Outliers can occur in one-dimensional space (one variable) or multi-dimensional space (several variables). Outliers are valuable sources of information and should never be discarded without thorough investigation. There are at least three possible causes for an outlier to show up in a set of observations:

- The outlier is actually not a valid value, but rather the result of a measurement or coding error.
- The outlier is associated with a different population than the population that the rest of the data were drawn from. If this is the case, you need to figure out how you ended up sampling from a different population. For example, maybe a different laboratory analysis was used on the physical sample associated with the outlier.
- The outlier is a valid value from the same population as the other observations, and you have simply come up with a "rare event" in your sample.

Each of the above three possibilities needs to be investigated to determine how to handle outliers. A data set that does not conform to a pre-conceived distribution (e.g., normal) is no reason for deleting or ignoring outliers. In fact, the hole in the ozone layer over Antarctica that was discovered in the mid-1980s might have been discovered sooner, but the software used to collect and manage the Nimbus-7 satellite ozone observations included an algorithm to flag low-level "outliers" in the data (Stolarski et al., 1986). In cases where there is no reason to delete outliers and a transformation does not seem to force the data to conform to the normality assumption of standard parametric estimation and testing, nonparametric and robust methods should be used.

An outlier test is used to determine whether one or more observations in a data set really come from a different distribution than the rest of the observations. Several outlier tests have been proposed over the years, and Barnett and Lewis (1995) present a thorough discussion of these tests. Gibbons (1994, chapter 13) and Gilbert (1987, chapter 15) also discuss some of these tests. One simple way to test for outliers is to perform a goodness-of-fit test with and without the suspected outliers.

TEST OF A SINGLE PROPORTION

One of the simplest hypothesis tests concerns the value of the proportion p from a binomial distribution (i.e., the probability of "success"). The two-sided and one-sided hypotheses about the proportion are given in Equations (7.1) to (7.6) above, with θ replaced by p. We assume that our observations x_1, x_2, ..., x_n are independent observations from a $B(1, p)$ distribution. Then the test statistic for the *binomial test* is simply the number of observed "successes":

$$x = \sum_{i=1}^{n} x_i \qquad (7.24)$$

Under the null hypothesis that $p = p_0$, the test statistic x has a $B(n, p_0)$ distribution. For a two-sided alternative hypothesis (Equation (7.2)), the p-value is computed as:

$$\text{p-value} = \Pr\left(X_{n, p_0} \leq n - m\right) + \Pr\left(X_{n, p_0} \geq m\right) \quad (7.25)$$

where $X_{n, p}$ denotes a random variable that follows a $B(n, p)$ distribution and m is defined by:

$$m = \max\left(x, n - x\right) \quad (7.26)$$

For a one-sided lower alternative hypothesis (Equation (7.4)), the p-value is computed as:

$$\text{p-value} = \Pr\left(X_{n, p_0} \leq x\right) \quad (7.27)$$

and for a one-sided upper alternative hypothesis (Equation (7.6)), the p-value is computed as:

$$\text{p-value} = \Pr\left(X_{n, p_0} \geq x\right) \quad (7.28)$$

Example 7.3: Comparing Ozone Concentrations to a Standard

Figure 7.7 below displays daily maximum ozone concentrations (ppb) at ground level recorded between May 1 and September 30, 1974 at a site in Yonkers, New York. In S-PLUS these data are stored in a list called ozone, which has two components called yonkers and stamford. These data are discussed in Chambers et al. (1983) and presented in an appendix (p. 346). They are also discussed in Cleveland (1993, p. 148). Although there should be 153 observations, there were a total of 21 days for which no measurements were available either at the Yonkers site and/or the Stamford site, so only 132 values are stored in the list ozone. Chambers et al. (1983, p. 346) present the original time series, including the days with missing values at each site. In ENVIRONMENTALSTATS for S-PLUS, the full data set, including missing values, is stored in the multivariate calendar time series ozone.orig.cts.

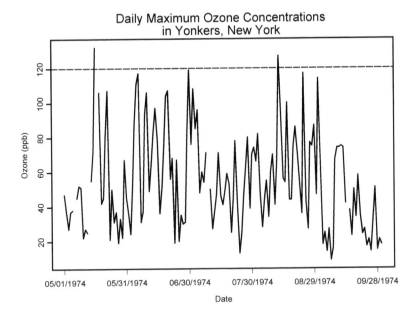

Figure 7.7   Daily maximum ozone concentrations (ppb) in Yonkers, New York
between May 1 and September 30, 1974

When these ozone measurements were taken, the U.S. Federal standard
for ozone required that the concentration should not exceed 120 ppb more
than 1 day per year at any particular station. This was a deterministic air
quality standard that ignored the inherent variability in air quality data. As
noted in Chapter 1, however, the ozone standard has changed to account for
inherent variability in the data. For this example, we will test the null hypothesis that the probability p that the daily maximum ozone exceeds 120
ppb is less than or equal to 1/365 (0.27%) vs. the one-sided upper alternative
that p is greater than 1/365. This is the same as testing the null hypothesis
that the expected number of days exceeding 120 ppb within a 1-year period
is 1. Ott (1995, Chapter 4) discusses a slightly different model for ozone exceedance.

There were 2 days out of 148 days with recorded data for which the
ozone concentration was above 120 ppb. This yields an estimated probability of exceedance of 2/148 = 1.35%. Under the null hypothesis that the true
probability of exceeding 120 ppb on any given day is 1/365, the probability
of seeing at least 2 days out of 148 over 120 ppb is 6.3%. Therefore, the p-
value is 0.063. Figure 7.8 displays the distribution of the number of ex-
ceedances out of 148 days under the null hypothesis. The sum of the areas
of the histogram bars for all numbers greater than or equal to 2 is equal to the

p-value. Depending on what we choose for our Type I error rate (e.g., 10%, 5%, 1%, etc.), we may or may not reject the null hypothesis that $p = 1/365 = 0.0027$. The lower 95% confidence interval for p (see Chapter 5) is [0.0024, 1], which includes $p = 1/365$, whereas the lower 90% confidence interval for p is [0.0036, 1], which does not include $p = 1/365$.

We need to add several caveats to this example. First of all, our "sample" of days does not span the whole year, yet there is probably important seasonal variability in the ozone concentrations. Second, our sample of days does not span more than 1 year, and there may be important year-to-year variability as well. Third, the binomial test assumes our observations are independent of one another. This is probably not a valid assumption with time series data collected close together in time. We will revisit this data set in Chapter 11 when we discuss analyzing time series data.

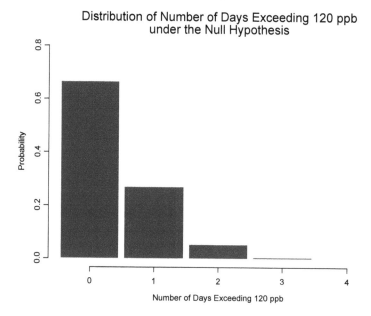

Distribution of Number of Days Exceeding 120 ppb
under the Null Hypothesis

Number of Days Exceeding 120 ppb

Figure 7.8   Distribution of number of days out of 148 exceeding 120 ppb under the null hypothesis that the true probability of exceeding 120 ppb on any given day is 1/365

Menu

To perform the binomial test using the S-Plus pull-down menu, follow these steps.

1. On the S-PLUS menu bar, make the following menu choices: **Statistics>Compare Samples>Counts and Proportions>Binomial Test**. This will bring up the Exact Binomial Test dialog box.
2. Under the Data group, in the No. of Successes box type **2**, and in the No. of Trials box type **148**.
3. Under the Test Hypotheses group, in the Hypothesized Proportion box type **1/365**, and in the Alternative Hypothesis box select **greater**.
4. Click on **OK** or **Apply**.

Command

To perform the binomial test using the S-PLUS Command or Script Window, type this command.

```
binom.test(x=
    sum(ozone.orig.cts[,"yonkers"] > 120, na.rm=T),
    n=sum(!is.na(ozone.orig.cts[,"yonkers"])), p=1/365,
    alternative="greater")
```

TESTS OF LOCATION

A frequent question in environmental statistics is "Is the concentration of chemical X greater than Y units?" For example, in groundwater assessment (compliance) monitoring at hazardous and solid waste sites, the concentration of a chemical in the groundwater at a downgradient well must be compared to a groundwater protection standard (GWPS). If the concentration is "above" the GWPS, then the site enters corrective action monitoring. As another example, soil screening at a Superfund site involves comparing the concentration of a chemical in the soil with a pre-determined soil screening level (SSL). If the concentration is "above" the SSL, then further investigation and possible remedial action is required. Determining what it means for the chemical concentration to be "above" a GWPS or an SSL is a policy decision: the average of the distribution of the chemical concentration must be above the GWPS or SSL, or the 95th percentile must be above the GWPS or SSL, or something else. Often, the first interpretation is used.

Table 7.9 displays six kinds of tests for location, along with the parameter the test is concerned with, the statistic the test is based on, and the assumptions of the test. The term "IID" stands for independent and identically distributed (i.e., independent observations from the same distribution). Note that for tests that assume a normal or symmetric distribution, the median is the same as the mean. We will discuss each of these tests below.

Test Name	Test on	Based on	Assumptions
Student's t-Test	Mean (Median)	Estimated Mean	IID observations from a normal distribution
Fisher's Randomization Test	Mean (Median)	Permuting Signs of Observations	IID observations from a symmetric distribution
Wilcoxon Signed Rank Test	Mean (Median)	Permuting Signs of Ranks	IID observations from a symmetric distribution
Chen's Modified t-Test	Mean	Estimated Mean and Skew	IID observations from a skewed distribution
Sign Test	Median	Signs of Observations	IID observations
Bootstrap Confidence Intervals	Any Kind of Location (Mean, Median, etc.)	Bootstrap Confidence Intervals	IID observations

Table 7.9   Six tests for location

Student's t-Test

One of the most commonly used statistical tests is Student's t-test to test whether the mean of a distribution is equal to a specific value. The two-sided and one-sided hypotheses you can test are given in Equations (7.1) to (7.6), with θ replaced by μ. Student's t-statistic is simply the difference between the sample mean and hypothesized mean, scaled by the estimated standard deviation of the sample mean:

$$t = \frac{\bar{x} - \mu_0}{s/\sqrt{n}} \qquad (7.29)$$

In this formula, $\bar{x}$ denotes the sample mean and s denotes the sample standard deviation (see Equations (5.7), (5.50), and (5.76) in Chapter 5). The "degrees of freedom" associated with this statistic are $n-1$, since we have to estimate the population standard deviation with s.

In order to use Student's t-test, the following assumptions must hold:

1. The observations all come from the same normal distribution.
2. The observations are independent of one another.

Under the null hypothesis that the true mean is equal to μ_0, the t-statistic in Equation (7.29) bounces around 0 and follows a Student's t-distribution with $n-1$ degrees of freedom (Johnson et al., 1995, p. 362; Fisher and van Belle,

1993, p. 143; Sheskin, 1997, Chapter 2; Zar, 1999, pp. 92–93). The first requirement is not as important as the second one. Several authors have investigated the behavior of Student's t-test under various deviations from normality and found that the test is fairly robust (i.e., the true Type I error rate is close to the assumed Type I error rate) to departures from normality (Zar, 1999, p. 95). On the other hand, positive or negative correlation between the observations can greatly affect the Type I error rate (Millard et al., 1985).

For a two-sided alternative hypothesis (Equation (7.2)), the p-value is computed as:

$$p = \Pr\left(t_{n-1} > |t|\right) \tag{7.30}$$

where t_v denotes a random variable that follows Student's t-distribution with v degrees of freedom. For a one-sided lower alternative hypothesis (Equation (7.4)), the p-value is computed as:

$$p = \Pr\left(t_{n-1} < t\right) \tag{7.31}$$

and for a one-sided upper alternative hypothesis (Equation (7.6)), the p-value is computed as:

$$p = \Pr\left(t_{n-1} > t\right) \tag{7.32}$$

Figure 7.9 illustrates how the two-sided p-value is computed based on a sample size of $n = 10$ when the observed value of the statistic is $t = 2$. In this figure, the p-value is equal to the combined area of the two shaded regions. If we were testing the one-sided upper hypothesis (Equations (7.5) and (7.6)), then the p-value would be equal to the area of the shaded region on the right-hand side.

Confidence Interval for the Mean Based on Student's t-Test

Based on the relationship between tests of hypothesis and confidence intervals (see Table 7.4), we can construct a confidence interval for the population mean based on Student's t-statistic. The formulas for the two-sided and one-sided confidence intervals were presented in Chapter 5 (Equations (5.78) to (5.80)).

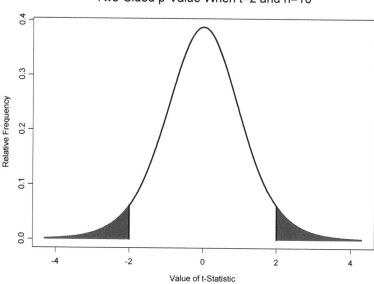

Figure 7.9   Computation of the two-sided p-value (area of the shaded regions) for the one-sample t-test based on a test statistic of $t = 2$ and a sample size of $n = 10$

Example 7.4: Testing Whether the Mean Aldicarb Concentration is above the MCL

The guidance document USEPA (1989b, p. 6-4) contains observations of aldicarb at three compliance wells. These data are displayed in Table 7.10 below and are stored in the data frame `epa.89b.aldicarb1.df` in ENVIRONMENTALSTATS for S-PLUS.

Month	Well 1	Well 2	Well 3
Jan	19.9	23.7	5.6
Feb	29.6	21.9	3.3
Mar	18.7	26.9	2.3
Apr	24.2	26.1	6.9

Table 7.10   Aldicarb concentrations (ppb) at three compliance wells, from USEPA (1989b, p. 6-4)

The Maximum Contaminant Level (MCL) has been set at 7 ppb. For each well, we will test the null hypothesis that the average aldicarb concentration is less than or equal to the MCL of 7 ppb versus the alternative hypothesis that the average concentration is bigger than 7 ppb.

For each of the three wells, Table 7.11 displays the sample means, sample standard deviations, t-statistics and their associated p-values, and the one-sided lower 99% confidence intervals for the true average concentration. We see that the average concentrations at Wells 1 and 2 are significantly greater than 7 ppb ($p < 0.01$), but the average concentration at Well 3 is not significantly different from 7 ppb.

Well	Sample Mean	Sample SD	t (p-value)	Lower 99% CI
1	23.1	4.9	6.5 (0.004)	[11.9 , ∞]
2	24.6	2.3	15.4 (0.0003)	[19.5 , ∞]
3	4.5	2.1	−2.4 (0.95)	[−0.25, ∞]

Table 7.11   Results of testing whether the true average aldicarb concentration is above the MCL of 7 ppb using Student's t-test

Menu

To test the null hypothesis that the true average aldicarb concentration at Well 1 is less than or equal to 7 ppb using Student's t-test and the S-PLUS pull down menu, we have to first create a new data frame that splits the observations at each well into separate columns (we will call it new.epa.89b.aldicarb1.df). To create this new data frame, follow these steps.

1. In the Object Explorer find and highlight the data frame **epa.89b.aldicarb1.df**.
2. On the S-PLUS menu bar, make the following menu choices: **Data>Restructure>Unstack**. This will bring up the Unstack Columns dialog box.
3. Under the From group, in the Data Set box, make sure **epa.89b.aldicarb1.df** is selected, and in the Columns box select **Aldicarb**.
4. Under the Row Grouping group, in the Type box make sure **Group Column** is selected and in the Group Column box select **Well**.
5. Under the To group, in the Data Set box type **new.epa.89b.aldicarb1.df**.
6. Click on **OK**. A data window entitled new.epa.89b.aldicarb1.df should appear.
7. In the data window, right click on the first column and select Properties. In the Name box, type **Well.1.Aldicarb**, then click **OK**. Repeat this for the second and third columns, renaming them **Well.2.Aldicarb** and **Well.3.Aldicarb**, respectively.
8. Exit the data window. Do not save the changes to a file.

To test the null hypothesis that the true average aldicarb concentration at Well 1 is less than or equal to 7 ppb, follow these steps.

1. In the Object Explorer find and highlight the data frame **new.epa.89b.aldicarb1.df**.
2. On the S-PLUS menu bar, make the following menu choices: **Statistics>Compare Samples>One Sample>t Test**. This will bring up the One-sample t Test dialog box.
3. In the Data Set box, make sure **new.epa.89b.aldicarb1.df** is selected, and in the Variable box select **Well.1.Aldicarb**.
4. In the Mean Under Null Hypothesis box, type **7** and in the Alternative Hypothesis box select **greater**.
5. In the Confidence Level box, type **0.99**, then click **OK** or **Apply**.

To test the null hypothesis for Wells 2 and 3, repeat the above steps, but in Step 3 select **Well.2.Aldicarb** or **Well.3.Aldicarb** in the Variable box.

Command

To test the null hypothesis that the true average aldicarb concentration at a well is less than or equal to 7 ppb using Student's t-test and the S-PLUS Command or Script Window, type these commands.

```
attach(epa.89b.aldicarb1.df)
t.test(Aldicarb[Well=="1"], alternative="greater",
   mu=7, conf.level=0.99)
detach()
```

This will produce the results for Well 1. To produce the results for Wells 2 and 3, replace `Well=="1"` with `Well=="2"` or `Well=="3"`.

Testing the Mean of a Lognormal Distribution

In Chapter 5 in the section Confidence Interval for the Mean of a Lognormal Distribution we explained that you *cannot* produce a confidence interval for the mean of a lognormal distribution by computing a confidence interval for the mean based on the log-transformed observations and then exponentiating the confidence limits (in fact, this produces a confidence interval for the *median* of the lognormal distribution). Similarly, suppose we want to test a hypothesis about the mean of a lognormal distribution, for example the null hypothesis and upper one-sided alternative given in Equations (7.5) and (7.6), where θ denotes the mean of the lognormal distribution. You might be tempted to perform a t-test by computing the sample mean of the log-transformed observations and comparing this sample mean with $\log(\theta_0)$. The results of this t-test, however, apply to the hypothesis that the *median* of the lognormal distribution is less than or equal to θ_0, not the mean. To perform a hypothesis test concerning the mean of a lognormal distribution, you must create a confidence interval for the mean (using one of the methods described in Chapter 5) and reject or not reject the null hypothesis based on the relationship between confidence intervals and tests of hypotheses (Table 7.4).

Fisher's Randomization (Permutation) Test

In 1935, R.A. Fisher introduced the idea of a ***randomization test*** (Manly, 1997, p. 91; Efron and Tibshirani, 1993, Chapter 15), which is based on trying to answer the question: "Did the observed pattern happen by chance, or does the pattern indicate the null hypothesis is not true?" A randomization test works by simply enumerating all of the possible outcomes under the null hypothesis, then seeing where the observed outcome fits in. A randomization test is also called a ***permutation test***, because it involves permuting the observations during the enumeration procedure (Manly, 1997, p. 3).

In the past, randomization tests have not been used as extensively as they are now because of the "large" computing resources needed to enumerate all of the possible outcomes, especially for large sample sizes. The advent of more powerful desktop PCs and software has allowed randomization tests to become much easier to perform. Depending on the sample size, however, it may still be too time consuming to enumerate all possible outcomes. In this case, the randomization test can still be performed by sampling from the randomization distribution, and comparing the observed outcome to this sampled permutation distribution.

For a one-sample location test, Fisher proposed using the test statistic

$$T = \sum_{i=1}^{n} y_i \qquad (7.33)$$

where

$$y_i = x_i - \mu_0 \qquad (7.34)$$

and μ_0 denotes the hypothesized value of the population mean (Manly, 1997, p. 96). The test assumes all of the observations come from the same distribution that is symmetric about the true population mean (hence the mean is the same as the median for this distribution). Under the null hypothesis that the true mean is μ_0, the y_i's are equally likely to be positive or negative. Therefore, the permutation distribution of the test statistic T consists of enumerating all possible ways of permuting the signs of the y_i's and computing the resulting sums. For n observations, there are 2^n possible permutations of the signs, because each observation can either be positive or negative.

For a one-sided upper alternative hypothesis (Equation (7.6)), the p-value is computed as the proportion of sums in the permutation distribution that are greater than or equal to the observed sum T. For a one-sided lower alternative hypothesis (Equation (7.4)), the p-value is computed as the proportion of sums in the permutation distribution that are less than or equal to the observed sum T. For a two-sided alternative hypothesis (Equation (7.2)), the p-value is computed by using the permutation distribution of the absolute value of T (i.e., $|T|$) and computing the proportion of values in this permutation distribution that are greater than or equal to the observed value of $|T|$.

Confidence Intervals Based on Permutation Tests

Based on the relationship between hypothesis tests and confidence intervals (see Table 7.4), it is possible to construct a two-sided or one-sided $(1-\alpha)100\%$ confidence interval for the mean μ based on the one-sample permutation test by finding the value(s) of μ_0 that correspond to obtaining a p-value of α (Manly, 1997, pp. 17–19, 97). A confidence interval based on the bootstrap however (see Chapter 5), will yield a similar type of confidence interval (Efron and Tibshirani, 1993, p. 214).

Example 7.5: Testing Whether the Mean Aldicarb Concentration is above the MCL Using Fisher's Randomization Test

In this example we will repeat the analysis in Example 7.4 but we will use Fisher's randomization test instead of Student's t-test. In this case, we have $\mu_0 = 7$ ppb. Table 7.12 lists the 16 possible permutations of the signs of the y_i's for Well 1, along with the corresponding sums. The observed sum of the y_i's for Well 1 is $T = 64.4$, and the p-value for the one-sided upper alternative that the true mean is greater than 7 ppb is p = 1/16 = 0.0625. Figure 7.10 displays the permutation distribution of the sums of the y_i's for Well 1. For Wells 2 and 3, the observed sums are $T = 70.6$ and $T = -9.9$, and the corresponding p-values are p = 0.0625 and p = 1.

Menu

To perform Fisher's one-sample randomization test for the aldicarb data using the ENVIRONMENTALSTATS for S-Plus pull-down menu, follow these steps.

1. In the Object Explorer find and highlight the data frame **epa.89b.aldicarb1.df**.

Permutation of Signs for (Observations – 7)				Sum
-12.9	-22.6	-11.7	-17.2	-64.4
12.9	-22.6	-11.7	-17.2	-38.6
-12.9	22.6	-11.7	-17.2	-19.2
12.9	22.6	-11.7	-17.2	6.6
-12.9	-22.6	11.7	-17.2	-41
12.9	-22.6	11.7	-17.2	-15.2
-12.9	22.6	11.7	-17.2	4.2
12.9	22.6	11.7	-17.2	30
-12.9	-22.6	-11.7	17.2	-30
12.9	-22.6	-11.7	17.2	-4.2
-12.9	22.6	-11.7	17.2	15.2
12.9	22.6	-11.7	17.2	41
-12.9	-22.6	11.7	17.2	-6.6
12.9	-22.6	11.7	17.2	19.2
-12.9	22.6	11.7	17.2	38.6
12.9	22.6	11.7	17.2	64.4

Table 7.12   All possible permutations of the signs of the y_i's for Well 1 aldicarb data

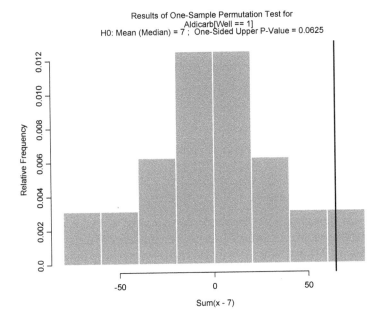

Figure 7.10   Permutation distribution of the sum of the y_i's for the Well 1 aldicarb data

2. On the S-PLUS menu bar, make the following menu choices: **EnvironmentalStats>Hypothesis Tests>Compare Samples>One Sample>Permutation Test**. This will bring up the One-Sample Permutation Test dialog box.
3. For Data to Use make sure the **Pre-Defined Data** button is selected, in the Data Set box make sure **epa.89b.aldicarb1.df** is selected, in the Variable box select **Aldicarb**, and in the Subset Rows with box type **Well=="1"**.
4. In the Mean Under Null Hypothesis box, type **7** and in the Alternative Hypothesis box select **greater**. For Test Method select the **Exact** button, then click **OK** or **Apply**.

This will produce the results for Well 1. To test the null hypothesis for Wells 2 and 3, repeat the above steps, but in Step 3 in the Subset Rows with box type **Well=="2"** or **Well=="3"**.

Command

To perform Fisher's one-sample randomization test for the aldicarb data using the ENVIRONMENTALSTATS for S-PLUS Command or Script Window, type these commands.

```
attach(epa.89b.aldicarb1.df)
Well.1.list <-
   one.sample.permutation.test(Aldicarb[Well== "1"],
   alt="greater", mu=7, exact=T)
plot(Well.1.list)
detach()
```

This will produce the results for Well 1. To produce the results for Wells 2 and 3, replace `Well=="1"` with `Well=="2"` or `Well=="3"`.

Wilcoxon Signed Rank Test

The Wilcoxon signed rank test is exactly the same test as Fisher's one-sample permutation test, except instead of permuting the signs of the y_i's defined in Equation (7.34), the Wilcoxon signed rank test permutes the signed ranks of the absolute values of the y_i's (Conover, 1980, p. 280; Fisher and van Belle, 1993, p. 310; Hollander and Wolfe, 1999, p. 36; Lehmann, 1975, p. 123; Sheskin, 1997, p. 83; Zar, 1999, p. 165). That is, the test statistic T is given by:

$$T = \sum_{i=1}^{n} R_i = \sum_{i=1}^{n} sign(y_i) \, Rank(|y_i|) \qquad (7.35)$$

where

$$sign(x) = \begin{cases} 1, & x > 0 \\ 0, & x = 0 \\ -1, & x < 0 \end{cases} \qquad (7.36)$$

As for the general one-sample permutation test, the Wilcoxon signed rank test assumes all of the observations come from the same distribution that is symmetric about the true population mean (hence the mean is the same as the median for this distribution). Under the null hypothesis that the true mean is μ_0, the y_i's are equally likely to be positive or negative, hence the sum of the signed ranks (T) bounces around 0. The permutation distribution of the test statistic T consists of enumerating all possible ways of permuting the signed ranks and computing the resulting sums.

If we let R_+ denote the sum of the positive signed ranks and R_- denote the sum of the negative signed ranks, then it can be shown that

$$R_+ + R_- = \frac{n(n+1)}{2} \qquad (7.37)$$

Most textbooks and software packages describe the Wilcoxon signed rank test in terms of one of these two statistics, rather than the statistic T defined in Equation (7.35). S-PLUS bases the test on R_+.

Because the test statistic R_+ is based on the ranks of the observations, the permutation distribution of R_+ does not have to be computed for each new set of observations. Instead, the permutation distribution can be easily tabulated for distinct sample sizes. In S-PLUS, the permutation distribution is tabulated for sample sizes up to $n = 25$. If ties are present, then the permutation distribution depends on the number of tied groups (e) and the number of ties within each group. In this case, S-PLUS uses a normal approximation to the permutation distribution of R_+ (see below).

Wilcoxon suggested that observations that yield y_i's equal to 0 should be discarded before computing the test statistic and permutation distribution, but Pratt suggested leaving these observations in. A study by Conover showed that each method is more powerful then the other in different situations (Conover, 1980, p. 288). S-PLUS does not discard observations that yield y_i's equal to 0. If these are present, a normal approximation to the permutation distribution is used (see below).

For "large" values of n (e.g., $n > 25$), a normal approximation to the distribution of R_+ (or R_-) can be used to determine the p-value. The following quantity is approximately distributed as a standard normal random variable under the null hypothesis:

$$z = \frac{R_+ - \mu_{R_+}}{\sigma_{R_+}} \tag{7.38}$$

where

$$\mu_{R_+} = \frac{n(n+1) - n_0(n_0+1)}{4} \tag{7.39}$$

$$\sigma_{R_+}^2 = \frac{n(n+1)(2n+1) - n_0(n_0+1)(2n_0+1)}{24} \tag{7.40}$$

$$-\frac{1}{48}\sum_{j=1}^{e}\left(t_j^3 - t_j\right)$$

where n_0 denotes the number of differences equal to 0 (i.e., $y_i = 0$), e denotes the number of groups of ties, and t_j denotes the number of ties in the jth group of ties (Lehmann, 1975, p. 130; Zar, 1999, p. 167). A continuity correction may be used (Sheskin, 1997, p. 90; Zar, 1999, p. 168).

Confidence Interval for the Mean (Median) Based on the Signed Rank Test

Based on the relationship between hypothesis tests and confidence intervals (see Table 7.4), we can construct a confidence interval for the population mean (median) based on the signed rank test. Conover (1980, pp. 288–290), Holland and Wolfe (1973, pp. 33–38), and Lehmann (1973, pp. 175–185) describe the Hodges-Lehmann estimate of the population median and this method of constructing confidence intervals for the median.

Example 7.6: Testing Whether the Mean Aldicarb Concentration is above the MCL Using the Wilcoxon Signed Rank Test

In this example we will repeat the analysis in Example 7.4 and Example 7.5, but we will use the Wilcoxon signed rank test. We want to determine whether the mean aldicarb concentration is greater than the MCL of 7 ppb. Table 7.13 displays the statistics and p-values (in parentheses) based on Student's t-test, Fisher's permutation test, and the Wilcoxon signed rank test. In this case, the Wilcoxon signed rank test yields essentially the same results as the permutation test.

Test	Well 1	Well 2	Well 3
Student's t	6.5 (0.004)	15.4 (0.0003)	-2.4 (0.95)
Permutation	64.4 (0.0625)	70.6 (0.0625)	-9.9 (1)
Signed Rank	10 (0.0625)	10 (0.0625)	4 (1)

Table 7.13   Statistics and p-values for testing whether the true average aldicarb concentration is above the MCL of 7 ppb

Menu

To perform the Wilcoxon signed rank test for the aldicarb data using the S-PLUS pull-down menu, follow these steps.

1. In the Object Explorer, find the data frame **new.epa.89b.aldicarb1.df** you created earlier and highlight it.
2. On the S-PLUS menu bar, make the following menu choices: **Statistics>Compare Samples>One Sample>Wilcoxon Signed Rank Test**. This will bring up the One-sample Wilcoxon Test dialog box.
3. In the Data Set box, make sure **new.epa.89b.aldicarb1.df** is selected, and in the Variable box select **Well.1.Aldicarb**.
4. In the Mean Under Null Hypothesis box, type **7** and in the Alternative Hypothesis box select **greater**, then click **OK** or **Apply**.

To test the null hypothesis for Wells 2 and 3, repeat the above steps, but in Step 3 select **Well.2.Aldicarb** or **Well.3.Aldicarb** in the Variable box.

Command

To perform the Wilcoxon signed rank test for the aldicarb data using the S-PLUS Command or Script Window, type these commands.

```
attach(epa.89b.aldicarb1.df)
wilcox.test(Aldicarb[Well== "1"],
    alternative="greater", mu=7)
detach()
```

This will produce the results for Well 1. To produce the results for Wells 2 and 3, replace `Well=="1"` with `Well=="2"` or `Well=="3"`.

Chen's Modified t-Test

Student's t-test, Fisher's one-sample permutation test, and the Wilcoxon signed rank test all assume that the underlying distribution is symmetric about its mean. Chen (1995b) developed a modified t-statistic for performing a one-sided test of hypothesis on the mean of a skewed distribution. Her test can be applied to the upper one-sided hypothesis (Equations (7.5) and (7.6) with θ replaced by μ) in the case of a positively skewed distribution, or to the lower one-sided hypothesis (Equations (7.3) and (7.4) with θ replaced by μ) in the case of a negatively skewed distribution. Since environmental data are usually positively skewed, her test would usually be applied to the case of testing the one-sided upper hypothesis.

The test based on Student's t-statistic of Equation (7.29) is fairly robust to departures from normality in terms of maintaining Type I error and power, provided that the sample size is sufficiently large. In the case when the underlying distribution is positively skewed and the sample size is small, the sampling distribution of the t-statistic under the null hypothesis does not follow a Student's t-distribution, but is instead negatively skewed. For the test against the one-sided upper alternative in Equation (7.6), this leads to a Type I error smaller than the one assumed and a loss of power (Chen, 1995b, p. 767). Similarly, in the case when the underlying distribution is negatively skewed and the sample size is small, the sampling distribution of the t-statistic is positively skewed. For the test against the lower alternative in Equation (7.4), this also leads to a Type I error smaller than the one assumed and a loss of power.

In order to overcome these problems, Chen (1995b) proposed the following modified t-statistic that takes into account the skew of the underlying distribution:

$$t_2 = t + a\left(1 + 2t^2\right) + 4a^2\left(t + 2t^3\right) \qquad (7.41)$$

where

$$a = \frac{\sqrt{\hat{\beta}_1}}{6\sqrt{n}} = \frac{\hat{\mu}_3/\hat{\sigma}^3}{6\sqrt{n}} \qquad (7.42)$$

$$\hat{\mu}_3 = \frac{n}{(n-1)(n-2)} \sum_{i=1}^{n} (x_i - \bar{x})^3 \qquad (7.43)$$

$$\hat{\sigma}^3 = s^3 = \left[\frac{1}{n-1} \sum_{i=1}^{n} (x_i - \bar{x})^2 \right]^{3/2} \qquad (7.44)$$

Note that the numerator in Equation (7.42) is an estimate of the skew of the underlying distribution and is based on unbiased estimators of central moments (see Chapter 5). For a positively skewed distribution, Chen's modified t-test rejects the null hypothesis in favor of the upper one-sided alternative if the modified t-statistic in Equation (7.41) is too large. For a negatively skewed distribution, Chen's modified t-test rejects the null hypothesis in favor of the lower one-sided alternative if the modified t-statistic is too small.

Chen (1995b) performed a simulation study in which she compared her modified t-statistic to a critical value based on the normal distribution (z-value), a critical value based on Student's t-distribution (t-value), and the average of the critical z-value and t-value. Based on the simulation study, Chen (1995b) suggests using either the z-value or average of the z-value and t-value when n (the sample size) is small (e.g., $n \leq 10$) or α (the Type I error) is small (e.g., $\alpha \leq 0.01$), and using either the t-value or the average of the z-value and t-value when $n \geq 20$ or $\alpha \geq 0.05$. ENVIRONMENTALSTATS for S-PLUS returns three different p-values: one based on the normal distribution, one based on Student's t-distribution, and one based on the average of these two p-values. This last p-value should roughly correspond to a p-value based on the distribution of the average of a normal and Student's t random variable.

Confidence Interval for the Mean Based on Chen's t-Test

Because of the relationship between hypothesis tests and confidence intervals (see Table 7.4), you can compute a one-sided confidence interval for the true mean μ based on Chen's modified t-test. Unfortunately, the test of the upper one-sided hypothesis (Equations (7.5) and (7.6) with θ replaced by μ) for a positively skewed distribution yields a lower one-sided confidence interval, but we are often interested in obtaining a one-sided upper confidence interval.

Example 7.7: Testing Whether the Mean Chromium Concentration is below an SSL

In Example 5.19 in Chapter 5 we introduced a set of 15 observations of chromium concentrations (mg/kg) (see Table 5.5). Here, we will use Chen's modified t-test to test the null hypothesis that the average chromium concentration is less than or equal to 100 mg/kg vs. the alternative that it is greater than 100 mg/kg. The estimated mean, standard deviation, and skew are 175 mg/kg, 319 mg/kg, and 3.6, respectively. The modified t-statistic is equal to 1.56, the p-value based on the normal distribution is 0.06, and the lower 95% confidence interval is [96.9, ∞).

Menu

To perform Chen's modified t-test for the chromium data using the ENVIRONMENTALSTATS for S-PLUS pull-down menu, follow these steps.

1. In the Object Explorer, find and highlight the vector **epa.92d.chromium.vec**.
2. On the S-PLUS menu bar, make the following menu choices: **EnvironmentalStats>Hypothesis Tests>Compare Samples>One Sample>t-Test for Skewed Data**. This will bring up the One-Sample t-Test for Skewed Data dialog box.
3. For Data to Use select the **Pre-Defined data** button, and in the Data Set box and Variable box make sure **epa.92d.chromium.vec** is selected.
4. In the Mean Under Null Hypothesis box, type **100**, in the Alternative Hypothesis box select **greater**, in the Confidence Level (%) box type **95**, and in the CI Method box select **z**, then click **OK** or **Apply**.

Command

To perform Chen's modified t-test for the chromium data using the ENVIRONMENTALSTATS for S-PLUS Command or Script Window, type this command.

```
chen.t.test(epa.92d.chromium.vec, mu=100)
```

Sign Test

The *sign test* (Conover, 1980, p. 122; Fisher and van Belle, 1993, p. 308; Hollander and Wolfe, 1999, p. 60; Lehmann, 1975, p. 120; Sheskin, 1997, p. 125; Zar, 1999, p. 110) tests the null hypothesis that the *median* of the distribution, here denoted θ, is equal to some specified value θ_0. This test only requires that the observations are independent and that they all come from one or more distributions (not necessarily the same ones) that all have the same population median. The test statistic T is simply the number of observations that are greater than θ_0. Under the null hypothesis, the distribution of T is a binomial random variable with parameters $n = n$ and $p = 0.5$. Usually, however, cases for which the observations are equal to θ_0 are discarded, so the distribution of T is taken to be binomial with parameters $n = r$ and $p = 0.5$, where r denotes the number of observations not equal to θ_0.

For a two-sided alternative hypothesis (Equation (7.2)), the p-value is computed as:

$$p = \Pr\left(X_{r,0.5} \leq r - m\right) + \Pr\left(X_{r,0.5} \geq m\right) \quad (7.45)$$

where $X_{r,p}$ denotes a random variable that follows a B(r, p) distribution and m is defined by:

$$m = \max\left(T, r - T\right) \quad (7.46)$$

For a one-sided lower alternative hypothesis (Equation (7.4)), the p-value is computed as:

$$p = \Pr\left(X_{m,0.5} \leq T\right) \quad (7.47)$$

and for a one-sided upper alternative hypothesis (Equation (7.6)), the p-value is computed as:

$$p = \Pr\left(X_{m,0.5} \geq T\right) \quad (7.48)$$

It is obvious that the sign test is simply a special case of the binomial test with $p_0 = 0.5$ (see the earlier section Test of a Single Proportion).

Confidence Interval for the Median Based on the Sign Test

Based on the relationship between hypothesis tests and confidence intervals (see Table 7.4), we can construct a confidence interval for the population median based on the sign test (e.g., Hollander and Wolfe, 1999, p. 72; Lehmann, 1975, p. 182). It turns out that this is equivalent to using the formulas for a nonparametric confidence intervals for the 0.5 quantile given in Chapter 5 (Equations (5.154), (5.157), and (5.159)).

Example 7.8: : Testing Whether the Median Chromium Concentration is below an SSL

Again consider the chromium data of Example 5.19 in Chapter 5 and Example 7.7 above. Here, we will use the sign test to test the null hypothesis that the *median* chromium concentration is less than or equal to 100 mg/kg vs. the alternative that it is greater than 100 mg/kg. The estimated median is 110 mg/kg. There are 8 out of 15 observations greater than 100 mg/kg, the p-value is equal to 0.5, and the lower 94% confidence limit is 41 mg/kg. The confidence limit for the median is based on the nonparametric method of constructing confidence intervals for quantiles discussed in Chapter 5.

Menu

To perform the sign test for the chromium data using the ENVIRONMENTALSTATS for S-PLUS pull-down menu, follow these steps.

1. In the Object Explorer, find and highlight the vector **epa.92d.chromium.vec**.
2. On the S-PLUS menu bar, make the following menu choices: **EnvironmentalStats>Hypothesis Tests>Compare Samples>One Sample>Sign Test**. This will bring up the Sign Test dialog box.
3. For Data to Use select the **Pre-Defined data** button, and in the Data Set box and Variable box make sure **epa.92d.chromium.vec** is selected.
4. In the Median Under Null Hypothesis box, type **100**, in the Alternative Hypothesis box select **greater**, and in the Confidence Level (%) box type **95**, then click **OK** or **Apply**.

Command

To perform the sign test for the chromium data using the ENVIRONMENTALSTATS for S-PLUS Command or Script Window, type this command.

```
sign.test(epa.92d.chromium.vec, mu=100,
    alternative="greater")
```

Tests Based on Bootstrapped Confidence Intervals

Because of the relationship between hypothesis tests and confidence intervals (see Table 7.4), you can construct a bootstrap confidence interval for the mean or median or any other population quantity to test a null hypothesis concerning that quantity (Efron and Tibshirani, 1993, p. 214). Unlike Student's t-test, Fisher's randomization test, and the Wilcoxon signed rank test, tests based on bootstrapped confidence intervals do not require the assumption that the underlying population is symmetric. Also, unlike Chen's modified t-test for skewed distributions, you can obviously produce any kind of confidence interval (two-sided, lower, or upper).

Example 7.9: Testing Whether the Mean Chromium Concentration is below an SSL Using a Bootstrap Confidence Interval

Again consider the chromium data of Example 5.19 in Chapter 5 and Example 7.7 above. Here, we will use a bootstrap confidence interval for the mean to test the null hypothesis that the mean chromium concentration is less than or equal to 100 mg/kg vs. the alternative that it is greater than 100 mg/kg. The estimated mean is 175 mg/kg. Figure 5.12 displays the bootstrap distribution of the sample mean for the chromium data. Table 7.14 displays the 90%, 95%, and 99% lower and upper confidence bounds for the population mean based on the BCa bootstrap distribution. The one-sided upper test of hypothesis is based on the one-sided lower confidence limit. Since the 90% lower confidence limit is 98.8 and is less than 100 mg/kg, the p-value for the test of hypothesis is greater than 0.1.

Limit Type	90%	95%	99%
Lower	98.8	90.8	77.8
Upper	348	418	508

Table 7.14   One-sided lower and upper BCa bootstrap confidence limits for mean chromium concentration

Menu

To produce the confidence limits shown in Table 7.14 using the S-PLUS pull-down menu, follow these steps.

1. In the Object Explorer, highlight **epa.92d.chromium.vec**.
2. On the S-PLUS menu bar, make the following menu choices: **Statistics>Resample>Bootstrap**. This will bring up the Bootstrap Inference dialog box.
3. In the Data Set box, make sure **epa.92d.chromium.vec** is selected. In the Expression box, type **mean(epa.92d.chromium.vec)**.
4. Click on the **Options** tab. In the Number of Resamples box, type **2000**. In the Random Number Seed box, type **47**. (Note: This is done just so that your results match the ones in this example. Leaving this box blank or explicitly supplying a different seed may yield very slightly different confidence limits.)
5. Click on the **Results** tab. In the Percentiles Levels box, type **c(0.01, 0.05, 0.1, 0.9, 0.95, 0.99)**. Click **OK** or **Apply**.

Command

To produce the confidence limits shown in Table 7.14 using the S-PLUS Command or Script Window, type these commands.

```
chromium.boot.list <-
   bootstrap(data=epa.92d.chromium.vec,
   statistic=mean(epa.92d.chromium.vec), B=2000,
   seed=47)

summary(chromium.boot.list,
   probs=c(1, 5, 10, 90, 95, 99)/100))
```

To produce the plot of the bootstrap distribution for the mean of chromium shown in Figure 5.12, type this command.

```
plot(chromium.boot.list)
```

Comparing the Performance of Tests of Location

Kempthorne and Doerfler (1969) compared the performance of Student's t-test, the one-sample permutation test, the Wilcoxon signed rank test, and the sign test. The study involved simulating observations from eight different distributions, and the sample sizes considered ranged from $n = 3$ to $n = 8$. They concluded that the one-sample permutation test performed the best overall, that the Wilcoxon signed rank test should only be used for computational convenience (a moot matter in the age of statistical software), and that the sign test should never be used. It is important to note, however, that all but one of the distributions used in the study was symmetric. When the underlying distribution is highly positively skewed and the sample size is small, the true Type I error rates of the t-test, one-sample permutation test, and Wilcoxon signed rank test are highly inflated relative to the assumed Type I error rates (which was the motivation for Chen to develop her modified t-test), whereas the Type I error rates for Chen's modified t-test and the sign test are maintained at their assumed level.

For large sample sizes, Conover (1980, p. 291) and Hollander and Wolfe (1999, pp. 104–105) present the asymptotic relative efficiencies e (i.e., the relative powers for a fixed sample size) of the t-test, Wilcoxon signed rank test, and sign test compared to each other, assuming a normal, uniform, or double exponential distribution. The Wilcoxon signed rank test is almost as powerful as the t-test under the normal distribution ($e = 3/\pi = 0.955$), as powerful as the t-test for the uniform distribution ($e = 1$), and more powerful than the t-test for the double exponential distributions ($e = 1.5$). The sign test is less powerful than the t-test for the normal and uniform distributions ($e = 2/\pi = 0.637$ and $e = 1/3 = 0.333$), but more powerful for the double exponential distribution ($e = 2$). Unfortunately, these results have little use in environmental statistics, since the underlying distribution is usually positively skewed.

The sign test concerns the median, while Chen's modified t-test concerns the mean. The test based on bootstrap confidence limits can be used to test the mean, median, or any other location parameter.

TESTS ON PERCENTILES

The section Estimates and Confidence Intervals for Distribution Quantiles (Percentiles) in Chapter 5 discusses how to create confidence intervals for quantiles for a normal, lognormal, or Poisson distribution, as well as a nonparametric method to create confidence intervals under any kind of continuous distribution. Based on the relationship between hypothesis tests and confidence intervals (see Table 7.4), you can perform a test of hypothesis concerning a particular percentile based on any of these methods for constructing confidence intervals.

Example 7.10 Testing a Hypothesis about the 95th Percentile of Aldicarb Concentrations

In Example 5.24 in Chapter 5, we discussed data on aldicarb concentrations from three groundwater monitoring compliance wells. In that example, the permit established an ACL of 50 ppm that could not be exceeded more than 5% of the time. Therefore, our null hypothesis is:

$$H_0 : x_{0.95} \leq 50 \qquad\qquad (7.49)$$

and our alternative hypothesis is:

$$H_a : x_{0.95} > 50 \qquad\qquad (7.50)$$

where x_p denotes the pth percentile of the distribution. In this example, we will assume that the distribution of concentrations at each of the three compliance wells is normal. The last row of Table 5.15 displays the one-sided lower 99% confidence limits for the 95th percentile. Since all of these confidence limits are less than 50 ppm, we know that the p-values associated with each of the hypothesis tests are all bigger than 0.01. For any particular well, if we want to know whether the p-value is larger than some specified value α^*, we need to construct a lower $(1-\alpha^*)100\%$ confidence limit for the 95th percentile. If this lower confidence limit is less than 50 ppm, then we know the p-value is larger than α^*.

TESTS ON VARIABILITY

Just as you can perform tests of hypothesis on measures of location (mean, median, percentile, etc.), you can do the same thing for measures of spread or variability. Usually, we are interested in estimating variability only because we want to quantify the uncertainty of our estimated location or percentile. Sometimes, however, we are interested in estimating variability and quantifying the uncertainty in our estimate of variability (for example, for performing a sensitivity analysis for power or sample size calculations), or testing whether the population variability is equal to a certain value. There are at least two possible methods of performing a hypothesis test on variability:

- Perform a hypothesis test for the population variance based on the chi-squared statistic, assuming the underlying population is normal.
- Perform a hypothesis test for any kind of measure of spread assuming any kind of underlying distribution based on a bootstrap confidence interval.

The procedure for the second method was already explained in the context of testing a hypothesis on the location of a distribution in the section Tests Based on Bootstrapped Confidence Intervals. In the rest of this section we will explain the first method.

Test on Variance Based on the Chi-Square Statistic

The two-sided and one-sided hypotheses on the population variance that you can test are given in Equations (7.1) to (7.6), with θ replaced by σ^2. In Chapter 5 we discussed creating a confidence interval for the population variance of a normal distribution, and we noted in Equation 5.81 that the quantity

$$\chi^2 = \frac{(n-1)\, s^2}{\sigma^2} \tag{7.51}$$

(where s^2 denotes the sample variance) follows a chi-square distribution with $n-1$ degrees of freedom, assuming the underlying observations come from a normal distribution. Thus, under the null hypothesis

$$H_0 \; : \; \sigma^2 \; = \; \sigma_0^2 \tag{7.52}$$

the quantity

$$\chi^2 = \frac{(n-1)\,s^2}{\sigma_0^2} \qquad (7.53)$$

follows a chi-square distribution with $n-1$ degrees of freedom.

For a one-sided upper alternative (Equation (7.6)), you reject the null hypothesis if χ^2 is "too large," and the p-value is computed as:

$$p_U = \text{Pr}\left(\chi^2_{n-1} \geq \chi^2\right) \qquad (7.54)$$

where χ^2_{ν} denotes a random variable that follows the chi-square distribution with ν degrees of freedom (Fisher and van Belle, 1993, pp. 115–119; Sheskin, 1997, Chapter 3; Zar, 1999, p. 112). For a one-sided lower alternative (Equation (7.4)), you reject the null hypothesis if χ^2 is "too small," and the p-value is computed as:

$$p_L = \text{Pr}\left(\chi^2_{n-1} \leq \chi^2\right) \qquad (7.55)$$

For a two-sided alternative hypothesis (Equation (7.2)), you reject the null hypothesis if χ^2 is "too small" or "too large," and the p-value is computed as:

$$p = 2\,\text{min}\left(p_L,\, p_U\right) \qquad (7.56)$$

As we stated earlier, Student's t-test is fairly robust to departures from normality (i.e., the Type I error rate is maintained), as long as the sample size is reasonably "large." The chi-square test on the population variance, however, is extremely sensitive to departures from normality. For example, if the underlying population is skewed, the actual Type I error rate will be larger than assumed.

Example 7.11: Testing the Variance of Log-Transformed TcCB Concentrations

We saw in Example 5.16 that the estimated mean and standard deviation for the log-transformed Reference area TcCB data are about –0.62 and 0.47

log(ppb). The estimated variance is 0.22. Suppose we want to test the null hypothesis that the variance of the population of all log-transformed observations is no greater than 0.2 log(ppb). In this case, the chi-square statistic is 50.4 with $47-1 = 46$ degrees of freedom, and the p-value is 0.30, so we would not reject the null hypothesis.

Menu

To perform the chi-square test using the ENVIRONMENTALSTATS for S-PLUS pull-down menu, follow these steps.

1. In the Object Explorer highlight **new.epa.94b.tccb.df**.
2. On the S-PLUS menu bar, make the following menu choices:
3. **EnvironmentalStats>Hypothesis Tests>Compare Samples>One Sample>Chi-Square Test on Variance**. This will bring up the One-Sample Chi-Square Test on Variance dialog box.
4. For Data to Use, make sure the **Pre-Defined Data** button is selected.
5. In the Data Set box, make sure **new.epa.94b.tccb.df** is selected.
6. In the Variable box, select **log.TcCB**.
7. In the Subset Rows with box, type **Area=="Reference"**.
8. In the Hypotheses section, in the Variance Under Null Hypothesis box type **0.2**, and in the Alternative Hypothesis box select **greater**.
9. Click **OK** or **Apply**.

Command

To perform the chi-square test using the ENVIRONMENTALSTATS for S-PLUS Command or Script Window, type these commands.

```
attach(epa.94b.tccb.df)
var.test(log(TcCB[Area=="Reference"]),
    alternative="greater", sigma.squared=0.2)
detach()
```

COMPARING LOCATIONS BETWEEN TWO GROUPS: THE SPECIAL CASE OF PAIRED DIFFERENCES

As we discussed in Chapter 1, one of the tenets of the scientific method is that ideally subjects in the "control group" are identical to subjects in the "treatment group" except for the fact that the subjects in the "treatment group" are exposed to some supposed causal mechanism. A frequently used device to try to approach this ideal is the concept of blocking, in which observations are taken in pairs. For example, if we want to determine whether a new anti-wrinkle cream is more effective than Brand X, we could perform an experiment in which each subject applies the new cream to either his or her right hand and applies Brand X to the other hand. When we use pairing, we can determine whether an effect exists by looking at the differences in the paired observations (e.g., right-hand vs. left-hand) and performing a one-sample test of location on these paired differences.

Example 7.12: Testing for a Difference in Ozone Concentrations between Two Cities

Consider the ozone concentrations measured in Stamford, Connecticut, and Yonkers, New York that we introduced in Example 7.3. Figure 7.11 displays a histogram of the daily differences in ozone concentrations at these two sites on the days for which measurements were available at both sites.

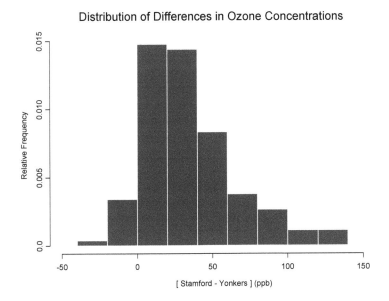

Figure 7.11   Relative frequency histogram of daily differences in ozone concentrations between Stamford and Yonkers

It is fairly obvious from this figure that if we test the null hypothesis that the average difference in concentrations between the two sites is 0 ppb vs. the alternative that it is greater than 0, we will get a significant result. The observed distribution of the paired differences looks slightly positively skewed, but this may have just been a chance occurrence based on the particular time period we are looking at. Table 7.15 displays the results of various tests. All of the tests yield a p-value that is essentially 0. What is more informative, as usual, is a confidence interval for the mean difference (note, however, that the sign test and the associated confidence interval concern the median, not the mean).

Test	Statistic (one-sided p-value)	95% Lower CI for Mean of Differences
Student's Paired t	13.0 (0)	[31, ∞]
Permutation	4,616 (0)	
Signed Ranks	9.6 (0)	
Chen's Modified Paired t	22.2 (0)	[31, ∞]
Paired Sign	122 (0)	[23, ∞]
Bootstrap CI		[31, ∞]

Table 7.15   Results of testing whether the mean difference in ozone concentration between Stamford and Yonkers is different from 0

We should note here that we are taking the differences between two time series, so we should not automatically assume that the differences represent independent observations (although taking differences usually does decrease the magnitude of serial correlation compared to the magnitude of serial correlation in the original time series). We will talk more about these data in Chapter 11 when we discuss dealing with time series data in detail.

Menu

To perform the above paired sample tests using the pull-down menu, find the list **ozone** in the Object Explorer and highlight it. Follow the menu selections listed below. The Data Set should be **ozone**, the first variable should be **stamford**, the second variable should be **yonkers**, and the Alternative Hypothesis should be **greater**.

- *Student's Paired t-Test.* Make the following menu choices: **Statistics>Compare Samples>Two Samples>t Test**. Under Type of t Test, select the **Paired t** button.
- *Permutation Test.* Make the following menu choices: **EnvironmentalStats>Hypothesis Tests>Compare Samples>Two Samples>Permutation Tests>Locations**. Under Type of Test, select the **Paired** button.
- *Wilcoxon Signed Rank Test.* Make the following menu choices: **Statistics>Compare Samples>Two Samples>Wilcoxon Rank Test**. Under Type of Rank Test, select the **Signed Rank** button.
- *Chen's Modified Paired t-Test.* Make the following menu choices: **EnvironmentalStats>Hypothesis Tests>Compare Samples>Two Samples>Paired t-Test for Skewed Data**.

- *Paired Sign Test.* Make the following menu choices: **EnvironmentalStats>Hypothesis Tests>Compare Samples>Two Samples>Paired Sign Test.**
- *Test on Bootstrapped Confidence Interval.* Make the following menu choices: **Statistics>Resample>Bootstrap.** In the Expression box, type **mean(stamford-yonkers).**

Command

To perform the above paired sample tests using the Command or Script Window, type these commands.

```
attach(ozone)
t.test(stamford, yonkers, alternative="greater",
   paired=T)
perm.test.list <-
   two.sample.permutation.test.location(stamford,
   yonkers, alternative="greater", paired=T)
plot(perm.test.list)
wilcox.test(stamford, yonkers, alternative="greater",
   paired=T)
chen.t.test(stamford, yonkers, alternative="greater",
   paired=T)
sign.test(stamford, yonkers, alternative="greater",
   paired=T)
boot.list <- bootstrap(data=ozone,
   statistic=mean(stamford-yonkers), B=2000, seed=47)
summary(boot.list)
plot(boot.list, xlab="Stamford - Yonkers (ppb)",
   main=paste("Bootstrap Distribution of the ",
   "Average Paired Differences\n",
   "for the ozone Data", sep=""))
detach()
```

COMPARING LOCATIONS BETWEEN TWO GROUPS

In the section Tests of Location, we discussed comparing chemical concentrations to a fixed standard. Another frequent question in environmental statistics is "Is the concentration of chemical X in Area A greater than the concentration of chemical X in Area B?," where Area B is usually a "background" or "reference" area, and Area A is a potentially contaminated well or site. For example, in groundwater detection monitoring at hazardous and solid waste sites, the concentration of a chemical in the groundwater at a downgradient well must be compared to "background." If the concentration is "above" the background then the site enters assessment monitoring. As another example, soil cleanup at a Superfund site may involve comparing the concentration of a chemical in the soil at a "cleaned up" site with the concentration at a "background" site. If the concentration at the "cleaned up" site is "greater" than the background concentration, then further investigation and remedial action may be required. Determining what it means for the chemical concentration to be "greater" than background is a policy decision: you may want to compare averages, medians, 95th percentiles, etc.

Table 7.16 displays six kinds of tests to compare "location" between two groups, along with the parameter the test is concerned with, the statistic the test is based on, and the assumptions of the test. We will discuss each of these tests below. For all of these tests, we will assume there are n_1 observations from group 1 and n_2 observations from group 2. We will denote the observations from group 1 by

$$x_{11}, x_{12}, \ldots, x_{1n_1} \tag{7.57}$$

and the observations from group 2 by

$$x_{21}, x_{22}, \ldots, x_{2n_2} \tag{7.58}$$

For a two-sample test for a difference in locations, a two-sided test tests

$$H_0 : \delta = \delta_0 \quad \text{vs.} \quad H_a : \delta \neq \delta_0 \tag{7.59}$$

where δ denotes the true difference in the measures of location for the two groups:

$$\delta = \theta_1 - \theta_2 \tag{7.60}$$

The quantity δ_0 denotes the hypothesized difference in the measures of location (usually, $\delta_0 = 0$), θ_1 denotes the measure of location for group 1, and θ_2 denotes the measure of location for group 2 (e.g., population mean, population median, etc.). Similarly, the one-sided lower test tests

$$H_0 : \delta \geq \delta_0 \quad \text{vs.} \quad H_a : \delta < \delta_0 \qquad (7.61)$$

and the one-sided upper test tests

$$H_0 : \delta \leq \delta_0 \quad \text{vs.} \quad H_a : \delta > \delta_0 \qquad (7.62)$$

Test Name	Test on Difference between ...	Based on	Assumptions
Student's t-Test	Means (Medians)	Estimated Means for Each Group	IID observations from a normal distribution, with a possibly shifted mean for one group
Fisher's Randomization Test	Means or Medians	Permuting Group Assignments of the Observations	IID observations, with possibly shifted mean or median for one group
Wilcoxon Rank Sum Test	Medians (Means)	Permuting the Ranks of the Observations	IID observations, with possibly shifted median for one group
Linear Rank Test	Medians (Means)	Permuting a Function of the Ranks of the Observations	IID observations, with possibly shifted median for one group
Quantile Test	Quantiles	Permuting the Ranks of the Observations in the Tail	IID observations, with possibly shifted upper (or lower) tail for one group
Bootstrap Confidence Intervals	Any Kind of Location	Bootstrap Confidence Intervals	IID observations within each group. Distributions of the two groups do not have to be the same

Table 7.16   Six kinds of tests to compare differences in locations between two groups

Student's Two-Sample t-Test

One of the most commonly used statistical tests to compare two groups is Student's two-sample t-test to compare the mean of group 1 (μ_1) with the mean of group 2 (μ_2). The two-sided and one-sided hypotheses you can test are given in Equations (7.59) to (7.62), with θ replaced by μ in Equation (7.60). Intuitively, the test statistic should be based on the differences in the sample means. If we assume the underlying distributions for the two groups are normal, and that the two distributions have the same standard deviation σ, then under the null hypothesis the quantity

$$z = \frac{(\bar{x}_1 - \bar{x}_2) - \delta_0}{\sigma_{\bar{x}_1 - \bar{x}_2}} \qquad (7.63)$$

where

$$\bar{x}_1 = \frac{1}{n_1} \sum_{i=1}^{n_1} x_{1i} \qquad (7.64)$$

$$\bar{x}_2 = \frac{1}{n_2} \sum_{i=1}^{n_2} x_{2i} \qquad (7.65)$$

$$\sigma_{\bar{x}_1 - \bar{x}_2}^2 = \frac{\sigma^2}{n_1} + \frac{\sigma^2}{n_2} = \sigma^2 \left(\frac{1}{n_1} + \frac{1}{n_2} \right) \qquad (7.66)$$

follows a standard normal distribution. The quantity in Equation (7.66) is the variance of the difference in the sample means, and the denominator in Equation (7.63) is the standard deviation of the difference in the sample means (regardless of whether the underlying distributions are normal).

The statistic z in Equation (7.63) is of little use in practice because we usually do not know the value of the population standard deviation σ. Student's two-sample t-statistic simply replaces the true standard deviation σ with an estimate of the standard deviation:

$$t = \frac{(\bar{x}_1 - \bar{x}_2) - \delta_0}{\hat{\sigma}_{\bar{x}_1 - \bar{x}_2}} \qquad (7.67)$$

where

$$\hat{\sigma}^2_{\bar{x}_1 - \bar{x}_2} = \hat{\sigma}^2 \left(\frac{1}{n_1} + \frac{1}{n_2} \right) \tag{7.68}$$

$$\hat{\sigma}^2 = s_p^2 = \frac{(n_1 - 1)\, s_1^2 + (n_2 - 1)\, s_2^2}{(n_1 - 1) + (n_2 - 1)} \tag{7.69}$$

$$s_1^2 = \frac{1}{n_1 - 1} \sum_{i=1}^{n_1} (x_{1i} - \bar{x}_1)^2 \tag{7.70}$$

$$s_2^2 = \frac{1}{n_2 - 1} \sum_{i=1}^{n_2} (x_{2i} - \bar{x}_2)^2 \tag{7.71}$$

The degrees of freedom associated with the t-statistic are:

$$\nu = n_1 + n_2 - 2 \tag{7.72}$$

Note that the pooled estimate of variance in Equation (7.69) is simply a weighted average of the two individual estimates of variance for each group, weighted by the degrees of freedom associated with each group. The p-values associated with Student's two-sample t-test are computed as shown in Equations (7.30) to (7.32), with the $n-1$ degrees of freedom replaced by n_1+n_2-2 degrees of freedom.

In order to use Student's two-sample t-test, the following assumptions must hold:

1. The observations are normally distributed.
2. The means may differ between the two groups, but the variances are the same, therefore the two distributions have the exact same shape.
3. The observations are independent of one another.

These assumptions can be written mathematically as follows:

$$X_{1i} = \mu_1 + \varepsilon_{1i} \, , \quad i = 1, 2, \dots, n_1$$

$$(7.73)$$

$$X_{2j} = \mu_2 + \varepsilon_{2j} \, , \quad j = 1, 2, \dots, n_2$$

where the error terms ε_{1i} and ε_{2i} come from a normal distribution with mean 0 and standard deviation σ. You can verify assumption 1 by creating a normal Q-Q plot for each group (see Chapter 3), and performing a goodness-of-fit test. You can verify assumption 2 by creating a two-sample Q-Q plot: plot the quantiles for group 2 vs. the quantiles for group 1. As we stated in Chapter 3, if the two groups have the same shape and dispersion, and differ only by a shift in location, then the two-sample Q-Q plot should fall on a line parallel to the 0-1 line. On the other hand, if the two groups differ greatly in variances, then the Q-Q plot should fall on a line with a slope that is different from 1. You could also perform a hypothesis test to test whether the two variances are the same before implementing the t-test (see the section Comparing Variability between Several Groups below), but this test is much more sensitive to nonnormality than the t-test is, so this two-step approach is not recommended (Zar, 1999, p. 129).

Under the null hypothesis that the true difference in the means is equal to δ_0, the t-statistic in Equation (7.67) bounces around 0 and follows a Student's t-distribution with n_1+n_2-2 degrees of freedom (Fisher and van Belle, 1993, p. 154; Sheskin, chapter 8; Zar, 1999, p. 125). The above assumptions are listed in increasing order of importance. Several authors have investigated the behavior of Student's two-sample t-test under various deviations from these assumptions and found that the test is fairly robust (i.e., the true Type I error rate is close to the assumed Type I error rate) to departures from normality or when the variances of the two underlying distributions are not equal, especially for moderate to large sample sizes with approximately equal sample sizes in each group (Zar, 1999, p. 127). On the other hand, positive or negative correlation between the observations can greatly affect the Type I error rate (Millard et al., 1985). Helsel and Hirsch (1992, p. 128) note that the t-test is adversely affected by outliers and cannot be easily applied to data with censored observations.

Modification for Differences in Variances

The case where the population variances differ between the two groups is called the "Behrens-Fisher problem." As we already stated, the t-test is fairly robust to modest deviations from this assumption, but if there is a gross difference in the variances (e.g., a difference of four-fold is serious; Johnson and Wichern, 1998, p. 311), then the Type I error rate will be affected. In this case, a modified from of the two-sample t-test can be used.

This modified form is sometimes called Welch's t-test (Zar, 1999, p. 128) and sometimes called Satterthwaite's t-test (USEPA, 1996a, p. 3.3-2).

If we know the population standard deviation for group 1, σ_1, and the population standard deviation for group 2, σ_2, then the formula for the variance of the difference in the sample means (Equation (7.66)) becomes

$$\sigma^2_{\bar{x}_1 - \bar{x}_2} = \frac{\sigma_1^2}{n_1} + \frac{\sigma_2^2}{n_2} \qquad (7.74)$$

Since we do not know the values of σ_1 and σ_2, the form of the t-statistic is the same as before (Equation (7.67)), except the estimate of the variance of the difference in the sample means (Equation (7.68)) is replaced by:

$$\hat{\sigma}^2_{\bar{x}_1 - \bar{x}_2} = \frac{s_1^2}{n_1} + \frac{s_2^2}{n_2} \qquad (7.75)$$

and the associated degrees of freedom are given by:

$$\nu = \frac{\left(\dfrac{s_1^2}{n_1} + \dfrac{s_2^2}{n_2} \right)^2}{\dfrac{s_1^4 / n_1^2}{n_1 - 1} + \dfrac{s_2^4 / n_2^2}{n_2 - 1}} \qquad (7.76)$$

It is important to decide whether it makes sense to focus on a possible difference in location between the two groups if you already know there is a difference in the variances. If the difference in variances is due to the fact that both underlying populations are skewed, then you may wish to consider performing the test on the log-transformed observations.

Comparing the Means of Log-Transformed Data

Environmental data are often positively skewed, so in these cases we know the normality assumption of the t-test is not satisfied. If we are willing to assume the observations from each group come from lognormal distributions, then we could perform the two-sample t-test on the log-transformed observations, but we need to be clear about what we are testing. Recall from Chapter 4 that if μ is the mean of the distribution of the log-transformed observations, then e^μ is the *median* of the distribution of the untransformed observations. Since, the null hypothesis

$$H_0 : \mu_1 - \mu_2 = \delta_0 \tag{7.77}$$

is equivalent to the null hypothesis

$$H_0 : e^{\mu_1 - \mu_2} = e^{\delta_0} \tag{7.78}$$

and this null hypothesis is equivalent to the null hypothesis

$$H_0 : \frac{e^{\mu_1}}{e^{\mu_2}} = e^{\delta_0} \tag{7.79}$$

the two-sample t-test performed on the log-transformed observations is testing a hypothesis about the **ratio of the medians**. Note that the null hypothesis of equal means for the distributions based on the log-transformed observations ($H_0: \mu_1-\mu_2 = 0$) corresponds to the null hypothesis of equal medians for the lognormal distributions ($H_0: e^{\mu_1} = e^{\mu_2}$).

On the other hand, the t-test requires that the variances of the underlying distributions of the log-transformed observations are the same, which implies that the coefficient of variation is the same for the two underlying lognormal distributions (see Equation (4.33) of Chapter 4). If you are willing to assume the two lognormal distributions have the same coefficient of variation, then the two-sample t-test performed on the log-transformed observations is testing a hypothesis about the **ratio of the means**. To show this, let θ_1 and θ_2 denote the means of the lognormal distributions for groups 1 and 2. Then

$$\frac{\theta_1}{\theta_2} = \frac{e^{\mu_1 + \sigma^2/2}}{e^{\mu_2 + \sigma^2/2}} = \frac{e^{\mu_1}}{e^{\mu_2}} \tag{7.80}$$

where σ^2 denotes the common variance for the distributions of the log-transformed observations (see Equation (4.32) of Chapter 4). In this case, the null of hypothesis of equal means for the distributions of the log-transformed observations is the same as the null hypothesis of equal means for the lognormal distributions.

Confidence Interval for the Difference in Means Based on Student's t-Test

Based on the relationship between hypothesis tests and confidence intervals (see Table 7.4), we can construct confidence intervals for the difference between the means ($\delta = \mu_1 - \mu_2$) based on Student's t-statistic. The formulas for these intervals are very similar to the formulas for the one-sample case. The two-sided $(1-\alpha)100\%$ confidence interval is given by:

$$\left[\left(\bar{x}_1 - \bar{x}_2 \right) - t_{\nu, 1-\alpha/2}\ \hat{\sigma}_{\bar{x}_1 - \bar{x}_2} \ , \right.$$

$$\left. \left(\bar{x}_1 - \bar{x}_2 \right) + t_{\nu, 1-\alpha/2}\ \hat{\sigma}_{\bar{x}_1 - \bar{x}_2} \right] \tag{7.81}$$

where the quantity $\hat{\sigma}_{\bar{x}_1 - \bar{x}_2}$ and the degrees of freedom ν depend on whether you are willing to assume equal variances between groups. The one-sided upper $(1-\alpha)100\%$ confidence interval is given by:

$$\left[-\infty,\ \left(\bar{x}_1 - \bar{x}_2 \right) + t_{\nu, 1-\alpha}\ \hat{\sigma}_{\bar{x}_1 - \bar{x}_2} \right] \tag{7.82}$$

and the one-sided lower $(1-\alpha)100\%$ confidence interval is given by:

$$\left[\left(\bar{x}_1 - \bar{x}_2 \right) - t_{\nu, 1-\alpha}\ \hat{\sigma}_{\bar{x}_1 - \bar{x}_2} \ ,\ \infty \right] \tag{7.83}$$

If you are computing confidence intervals based on log-transformed observations, then you can exponentiate the confidence limits to create confidence limits for the ratio of the medians, and in the case of a constant coefficient of variation, confidence limits for the ratio of the means.

Example 7.13: Comparing Sulfate Concentrations between a Background and Compliance Well

Table 7.17 lists sulfate concentrations (ppm) observed in groundwater at a background and downgradient well. In ENVIRONMENTALSTATS for S-PLUS these data are stored in the data frame `sulfate.df`. We will use Student's two-sample t-test to test the null hypothesis that the average sulfate concentration at the two wells is the same vs. the alternative hypothesis that the average sulfate concentration at the downgradient well is larger than the average concentration at the upgradient well.

Year	Month	Background	Downgradient
1995	Jan	560	
	Apr	530	
	Jul	570	600
	Oct	490	590
1996	Jan	510	590
	Apr	550	630
	Jul	550	610
	Oct	530	630

Table 7.17   Sulfate concentrations (ppm) at a background and downgradient well

The estimated background mean and standard deviation are 536.3 and 26.7 ppm, and the estimated downgradient mean and standard deviation are 608.3 and 18.3 ppm. The t-statistic is 5.66 with 12 degrees of freedom, and the one-sided p-value is 5×10^{-5}. The 99% one-sided lower confidence interval for the difference between the two means is [37.9, ∞]. Thus, if you decide on a 1% significance level, then you would reject the null hypothesis that the average concentration of sulfate is the same at the two wells in favor of the alternative hypothesis that the average concentration of sulfate is larger at the downgradient well.

In general, for monitoring groundwater at hazardous and solid waste sites, you should *not* use t-tests or ANOVA (see below) to test for differences between background and downgradient wells. Instead you should use simultaneous prediction limits (see Chapter 6).

Menu

To perform Student's two-sample t-test on the data shown in Table 7.17 using the S-PLUS pull-down menu, follow these steps.

1. In the Object Explorer, find and highlight **sulfate.df**.
2. On the S-PLUS menu bar, make the following menu choices: **Statistics>Compare Samples>Two Samples>t Test**. This will bring up the Two-sample t Test dialog box.
3. For Data Set, make sure **sulfate.df** is selected. For Variable 1 select **Sulfate**, for Variable 2 select **Well.type**, and **check** the box that says Variable 2 is a Grouping Variable.
4. Under Alternative Hypothesis select **less**, and for Confidence Level type **0.99**. Click **OK** or **Apply**.

Command

To perform Student's two-sample t-test on the data shown in Table 7.17 using the S-PLUS Command or Script Window, type these commands.

```
attach(sulfate.df)
```

```
t.test(Sulfate[Well.type=="Downgradient"],
    Sulfate[Well.type=="Background"],
    alternative="greater", conf.level = 0.99)
detach()
```

Example 7.14: Testing the Difference in Means between Areas for the Log-Transformed TcCB Data

We saw in Chapter 3 that the TcCB concentration data looks positively skewed for both the Cleanup and Reference areas. Figures 3.9, 3.10, 3.13, and 3.20 show the strip plots, histograms, boxplots, and two-sample Q-Q plot comparing the distributions of the log-transformed TcCB concentrations for the Cleanup and Reference areas. All of these figures indicate that while there may or may not be a difference in means between the two areas, there is certainly a difference in variances. Thus, if we perform a two-sample t-test on the log-transformed data, we are testing a hypothesis only about the ratio of the medians, not the means.

Here we will test the null hypothesis that the ratio of the median concentration in the Cleanup area to the median concentration in the Reference are is 1 vs. that alternative that the ratio is greater than 1 (i.e., the median concentration in the Cleanup area is greater than the median concentration in the Reference area). As we saw in Chapter 3, the estimated medians for the original TcCB data are 0.43 ppb in the Cleanup area and 0.54 ppb in the Reference area. For the log-transformed data, the Cleanup area estimated mean and standard deviation are –0.55 and 1.36 ppb and the Reference area estimated mean and standard deviation are –0.62 and 0.47 ppb. The resulting t-statistic is 0.35 with 122 degrees of freedom and an associated p-value of 0.36. The lower 99% confidence interval for the difference of the means is [–0.41, ∞]. If we do not assume the variances are the same, then the t-statistic is 0.43 with 102 degrees of freedom and an associated p-value of 0.34, and the lower 99% confidence interval for the difference of the means is [–0.33, ∞]. Thus, we would not reject the null hypothesis and would conclude that the median concentration of TcCB is the same in both areas.

Menu

To perform Student's two-sample t-test on the log-transformed TcCB data using the S-PLUS pull-down menu, follow these steps.

1. In the Object Explorer, find and highlight **new.epa.94b.tccb.df**.
2. On the S-PLUS menu bar, make the following menu choices: **Statistics>Compare Samples>Two Samples>t Test**. This will bring up the Two-sample t Test dialog box.

3. For Data Set, make sure **new.epa.94b.tccb.df** is selected. For Variable 1 select **log.TcCB**, for Variable 2 select **Area**, and **check** the box that says Variable 2 is a Grouping Variable.
4. Under Alternative Hypothesis select **greater**, and for Confidence Level type **0.99**. Click **OK** or **Apply**.

This will produce the results for the test that assumes equal variances. To perform the test that does not assumed equal variances, in Step 3 above **uncheck** the box that says Assume Equal Variances.

Command

To perform Student's two-sample t-test on the TcCB data using the S-PLUS Command or Script Window, type these commands.

```
attach(epa.94b.tccb.df)
t.test(log(TcCB[Area=="Cleanup"]),
   log(TcCB[Area=="Reference"]),
   alternative="greater", conf.level = 0.99)
detach()
```

This will produce the results for the test that assumes equal variances. To perform the test that does not assumed equal variances, include the argument `var.equal=F` in the call to `t.test`.

Two-Sample Permutation Tests

The two sample permutation test is based on trying to answer the question, "Did the observed difference in means or medians (or any other measure of location) happen by chance, or does the observed difference indicate that the null hypothesis is not true?" Under the null hypothesis, the underlying distributions for each group are the same, therefore it should make no difference which group an observation gets assigned to. The two-sample permutation test works by simply enumerating all possible permutations of group assignments, and for each permutation computing the difference between the measures of location for each group (Manly, 1997, p. 97; Efron and Tibshirani, 1993, p. 202). The measure of location for a group could be the mean, median, or any other measure you want to use. For example, if the observations from Group 1 are 3 and 5, and the observations from Group 2 are 4, 6, and 7, then there are 10 different ways of splitting these five observations into one group of size 2 and another group of size 3. Table 7.18 lists all of the possible group assignments, along with the differences in the group means.

In this example, the observed group assignments and difference in means are shown in the second row of Table 7.18. For a one-sided upper alternative (Equation (7.62)), the p-value is computed as the proportion of times that the differences of the means in the permutation distribution are greater than or equal to the observed difference in means. For a one-sided lower alternative hypothesis (Equation (7.61)), the p-value is computed as the proportion of times that the differences in the means in the permutation distribution are less than or equal to the observed difference in the means. For a two-sided alternative hypothesis (Equation (7.59)), the p-value is computed as the proportion of times the absolute values of the differences in the means in the permutation distribution are greater than the absolute value of the observed difference in the means. For this simple example, the one-sided upper, one-sided lower, and two-sided p-values are 0.9, 0.2 and 0.4, respectively.

Group 1	Group 2	Mean 1 – Mean 2
3,4	5,6,7	-2.5
3,5	4,6,7	-1.67
3,6	4,5,7	-0.83
3,7	4,5,6	0
4,5	3,6,7	-0.83
4,6	3,5,7	0
4,7	3,5,6	0.83
5,6	3,4,7	0.83
5,7	3,4,6	1.67
6,7	3,4,5	2.5

Table 7.18   All possible group assignments and corresponding differences in means

In this simple example, we assumed the hypothesized differences in the means under the null hypothesis was $\delta_0 = 0$. If we had hypothesized a different value for δ_0, then we would have had to subtract this value from each of the observations in Group 1 *before* permuting the group assignments to compute the permutation distribution of the differences of the means. As in the case of the one-sample permutation test, if the sample sizes for the groups become too large to compute all possible permutations of the group assignments, the permutation test can still be performed by sampling from the permutation distribution and comparing the observed difference in locations to the sampled permutation distribution of the difference in locations.

Unlike the two-sample Student's t-test, we do not have to worry about the normality assumption when we use a permutation test. The permutation test still assumes, however, that under the null hypothesis, the distributions of the observations from each group are exactly the same, and under the alternative hypothesis there is simply a shift in location (that is, the whole distribution of group 1 is shifted by some constant relative to the distribution of group 2). Mathematically, this can be written as follows:

Probability Density Functions for a Location Shift

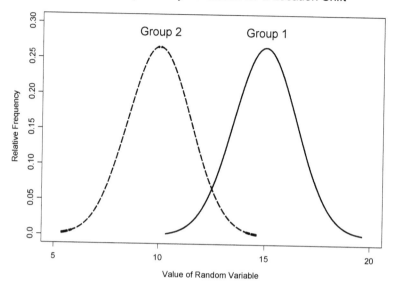

Cumulative Distribution Functions for a Location Shift

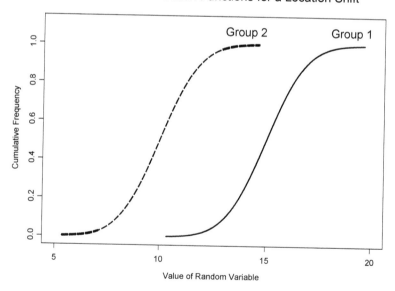

Figure 7.12   Probability density functions and cumulative distribution functions for groups 1 and 2, where the distribution of group 1 is shifted by 5 units relative to distribution of group 2

$$F_1\left(t\right) \;=\; F_2\left(t - \delta\right)\;,\quad -\infty < t < \infty \qquad\qquad \textbf{(7.84)}$$

where F_1 and F_2 denote the cumulative distribution functions for group 1 and group 2, respectively. (If $\delta > 0$, this implies that the observations in group 1 tend to be larger than the observations in group 2, and if $\delta < 0$, this implies that the observations in group 1 tend to be smaller than the observations in group 2.) Figure 7.12 displays this kind of location shift for $\delta = 5$. Thus, the shape and spread (variance) of the two distributions should be the same whether the null hypothesis is true or not. Therefore, the type I error rate for a permutation test can be affected by differences in variances between the two groups.

Confidence Intervals Based on the Two-Sample Permutation Test

It is possible to construct a confidence interval for the difference between the two means or medians (or any other measure of location) based on the two-sample permutation test by finding the value(s) of δ_0 that correspond to obtaining a p-value of α (Manly, 1997, pp. 17–19, 98). As we stated earlier when we were talking about the one-sample permutation test for location, however, a confidence interval based on the bootstrap will yield a similar type of confidence interval (Efron and Tibshirani, 1993, p. 214).

Example 7.15: Testing the Difference between Areas for the TcCB Data

In this example we will use the two-sample permutation test to test for a difference in means and a difference in medians between the Cleanup and Reference area for the TcCB data. The observed mean concentrations of TcCB in the Cleanup and Reference areas are 3.9 and 0.6 ppb, respectively, yielding a difference of 3.3 ppb. Figure 7.13 displays the permutation distribution of the difference in the means based on sampling the permutation distribution 5,000 times. The one-sided upper p-value for the alternative hypothesis that the mean concentration in the Cleanup area is greater than the mean concentration in the Reference area is $p = 0.021$.

The observed median concentrations of TcCB in the Cleanup and Reference areas are 0.43 and 0.54 ppb, respectively, yielding a difference of –0.11 ppb. Figure 7.14 displays the permutation distribution of the difference in the medians based on sampling the permutation distribution 5,000 times. The one-sided upper p-value for the alternative hypothesis that the median concentration in the Cleanup area is greater than the median concentration in the Reference area is $p = 0.93$.

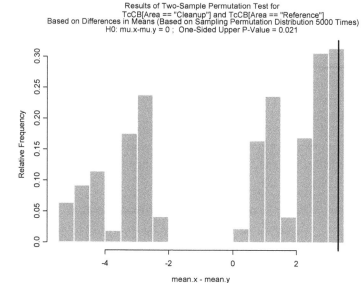

Figure 7.13 Results of two-sample permutation test for difference in the means for the TcCB data

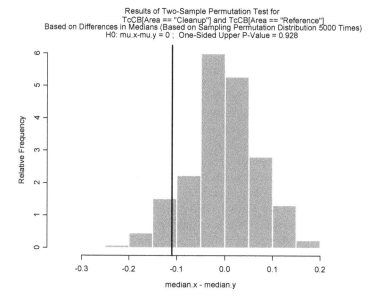

Figure 7.14 Results of the two-sample permutation test for difference in the medians for the TcCB data

Menu

To perform the two-sample permutation tests on the TcCB data using the ENVIRONMENTALSTATS for S-PLUS pull-down menu, follow these steps.

1. In the Object Explorer, find and highlight **epa.94b.tccb.df**.
2. On the S-PLUS menu bar, make the following menu choices: **EnvironmentalStats>Hypothesis Tests>Compare Samples>Two Samples>Permutation Tests>Locations**. This will bring up the Two-Sample Permutation Test for Locations dialog box.
3. For Data to Use, make sure **Pre-Defined Data** is selected. For Data Set, make sure **epa.94b.tccb.df** is selected. For Variable 1 select **TcCB**, for Variable 2 select **Area**, and **check** the box that says Variable 2 is a Grouping Variable.
4. For Type of Test make sure **Two-Sample** is selected, and for Test Parameter make sure **Mean** is selected.
5. Under Alternative Hypothesis select **greater**, for Test Method make sure **Sample Perm Dist** is selected, and for Set Seed with type **47**. Click **OK** or **Apply**.

This will produce the results for the test of the difference in means. To perform the test for the difference in medians, in Step 4 above for Test Parameter select **Median**.

Command

To perform the two-sample permutation tests on the TcCB data using the ENVIRONMENTALSTATS for S-PLUS Command or Script Window, type these commands.

```
attach(epa.94b.tccb.df)
mean.list <- two.sample.permutation.test.location(
    TcCB[Area=="Cleanup"], TcCB[Area=="Reference"],
    alternative="greater", seed=47)
plot(mean.list)
median.list <- two.sample.permutation.test.location(
    TcCB[Area=="Cleanup"], TcCB[Area=="Reference"],
    fcn="median", alternative="greater", seed=47)
plot(median.list)
detach()
```

The call to the argument `seed` simply lets you reproduce the figures as shown here. If you use the default value of `seed=.Random.seed`, you may get slightly different looking figures.

Wilcoxon Rank Sum Test

The Wilcoxon rank sum test (also called the Mann-Whitney test) is exactly the same test as the two-sample permutation test for the difference in means, but it is based on the ranks of the observations instead of the observations themselves (Conover, 1980, p. 215; Fisher and van Belle, 1993, p. 315; Helsel and Hirsch, 1992, p. 118; Hollander and Wolfe, 1999, p. 106; Lehmann, 1975, p. 5; Sheskin, 1997, p. 181; Zar, 1999, p. 146). Instead of looking at the permutation distribution of the difference between the average ranks for each group, the equivalent test statistic that is used is the sum of the ranks for the first group:

$$W = \sum_{i=1}^{n_1} R_i = \sum_{i=1}^{n_1} Rank\left(x_{1i}\right) \tag{7.85}$$

As for the general two-sample permutation test, the Wilcoxon rank sum test assumes that under the null hypothesis, the distributions of the observations from each group are exactly the same, and under the alternative hypothesis there is simply a shift in location (that is, the whole distribution of group 2 is shifted by some constant relative to the distribution of group 1; see Equation (7.84)). While at first this may appear to imply that the shape and spread (variance) of the two distributions should be the same whether the null hypothesis is true or not, the requirement is not quite so strict. Since the Wilcoxon rank sum test is based on the ranks of the observations, and ranks are preserved under monotonic transformations, the requirements for the Wilcoxon rank sum test are really that there must be some monotonic transformation of the observations (e.g., no transformation, log transformation, square-root transformation, etc.) such that the distributions of the transformed observations are the same in shape and spread, but may differ by location. For example, if the distribution of group 1 is lognormal with mean $\theta = 5$ and coefficient of variation $\tau = 1$, and the distribution of group 2 is lognormal with mean $\theta = 10$ and coefficient of variation $\tau = 1$, then these two distributions obviously do not have the same standard deviation (the standard deviation is 5 for group 1 and 10 for group 2). The distributions of the log-transformed observations, however, are both normal with the same standard deviation (see Equation (4.35) in Chapter 4). Thus, the Wilcoxon rank sum test could be applied in this case.

Some references suggest the Wilcoxon rank sum test is applicable even when the variances are different because the Type I error is not adversely affected (e.g., USEPA, 1996a, p. 3.3-2; Zar, 1999, p. 129). Strictly speaking, however, the Type I error level is not maintained when the two populations differ in dispersion or shape, unless there is some monotonic transformation that yields two distributions with the same dispersion and shape (Hollander and Wolfe, 1999, p. 120). On the other hand, unlike the t-test or the two-sample permutation test based on differences in the means, the Wilcoxon rank sum test is unaffected by gross "outliers" because it is based on the ranks instead of the actual observations.

Because the Wilcoxon rank sum test is based on the ranks of the observations, the permutation distribution of W does not have to be computed for each new set of observations, but instead can be easily tabulated for fixed sample sizes. In S-PLUS, the permutation distribution is tabulated for sample sizes up to 49 in either or both groups. If ties are present, the permutation distribution depends on the number of tied groups (e) and the number of ties within each group. In this case, S-PLUS uses a normal approximation to the permutation distribution of W.

For "large" sample sizes (e.g., $n_1 \geq 10$ and $n_2 \geq 10$), a normal approximation to the distribution of W can be used to determine the p-value (Lehmann, 1975, p. 13). The following quantity is approximately distributed as a standard normal random variable under the null hypothesis:

$$z = \frac{W - \mu_W}{\sigma_W} \tag{7.86}$$

where

$$\mu_W = \frac{n_1 N}{2} \tag{7.87}$$

$$\sigma_W^2 = \frac{n_1 n_2 (N + 1)}{12} - \frac{n_1 n_2 \sum_{j=1}^{e} \left(t_j^3 - t_j\right)}{12N (N - 1)} \tag{7.88}$$

$$N = n_1 + n_2 \tag{7.89}$$

The quantity e denotes the number of groups of ties, and t_j denotes the number of ties in the jth group of ties (Lehmann, 1975, p. 20). A continuity correction may be used (Sheskin, 1997, p. 187; Zar, 1999, p. 151).

Confidence Interval for the Difference in Medians (Means) Based on the Wilcoxon Rank Sum Test

Based on the relationship between hypothesis tests and confidence intervals (see Table 7.4), we can construct a confidence interval for the difference between the population medians (means) of the two groups based on the Wilcoxon rank sum test. Conover (1980, pp. 223–225), Helsel and Hirsch (1992, pp. 132–134), Holland and Wolfe (1973, p. 78), and Lehmann (1973, pp. 81–95) describe the Hodges-Lehmann estimate of the difference in the population medians and this method of constructing a confidence interval for this difference.

Example 7.16: Testing for a Difference between Areas for the TcCB Data

In this example we will repeat the comparison we made in Example 7.15, but we will use the Wilcoxon rank sum test. The permutation test based on the difference in the medians yielded a one-sided upper p-value of p = 0.93 for the alternative hypothesis that the median concentration in the Cleanup area is greater than the median concentration in the Reference area. Not surprisingly, the Wilcoxon rank sum test yields a similar p-value of p = 0.88 based on the normal approximation with the continuity correction ($z=-1.17$).

Menu

To perform the Wilcoxon rank sum test on the TcCB data using the S-PLUS pull-down menu, follow these steps.

1. In the Object Explorer, find and highlight **epa.94b.tccb.df**.
2. On the S-PLUS menu bar, make the following menu choices: **Statistics>Compare Samples>Two Samples>Wilcoxon Rank Test**. This will bring up the Two-sample Wilcoxon Test dialog box.
3. For Data Set, make sure **epa.94b.tccb.df** is selected. For Variable 1 select **TcCB**, for Variable 2 select **Area**, and **check** the box that says Variable 2 is a Grouping Variable. Under Alternative Hypothesis select **greater**, then click **OK** or **Apply**.

Command

To perform the Wilcoxon rank sum test on the TcCB data using the S-PLUS Command or Script Window, type these commands.

```
attach(epa.94b.tccb.df)
```

```
wilcox.test(TcCB[Area=="Cleanup"],
   TcCB[Area=="Reference"], alternative="greater")
detach()
```

Linear Rank Tests for Shift in Location

The Wilcoxon rank sum test is an example of a linear rank test. A linear rank test can be written as follows:

$$L = \sum_{i=1}^{n_1} a\left(R_{1i}\right) \qquad (7.90)$$

where $a()$ is some function and is called the *score function*. A linear rank test is based on the sum of the scores for group 1. For the Wilcoxon rank sum test, the function $a()$ is simply the identity function. Other functions may work better at detecting a small shift in location, depending on the shape of the underlying distributions. See the ENVIRONMENTALSTATS for S-PLUS help file for `two.sample.linear.rank.test` for more information.

Quantile Test for Shift in the Tail

The tests we have considered so far (Student's t-test, two-sample permutation tests for a difference in location, the Wilcoxon rank sum test, and linear rank tests for shifts in location) are all designed to detect a shift in the whole distribution of group 1 relative to the distribution of group 2 (see Equation (7.84) and Figure 7.12). Sometimes, we may be interested in detecting a difference between the two distributions where only a portion of the distribution of group 1 is shifted relative to the distribution of group 2. The mathematical notation for this kind of shift is

$$F_1\left(t\right) = \left(1 - \varepsilon\right) F_2\left(t\right) + \varepsilon F_3\left(t\right) , \quad -\infty < t < \infty \qquad (7.91)$$

where ε denotes a fraction between 0 and 1. In the statistical literature, the distribution of group 1 is sometimes called a "contaminated" distribution, because it is the same as the distribution of group 2, except it is partially contaminated with another distribution. If the distribution of group 1 is partially shifted to the right of the distribution of group 2, F_3 denotes a cdf such that

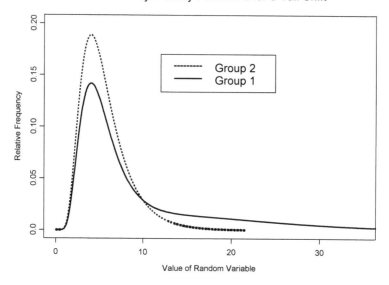

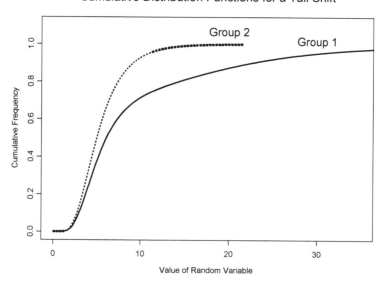

Figure 7.15   Probability density functions and cumulative distribution functions for groups 1 and 2, where 25% of the distribution of group 1 has a median that is shifted by 15 units relative to the median of the distribution of group 2

$$F_3\left(t\right) \leq F_2\left(t\right)\ ,\quad -\infty < t < \infty \qquad (7.92)$$

with a strict inequality for at least one value of t. If the distribution of group 1 is partially shifted to the left of the distribution of group 2, F_3 denotes a cdf such that

$$F_3\left(t\right) \geq F_2\left(t\right)\ ,\quad -\infty < t < \infty \qquad (7.93)$$

with a strict inequality for at least one value of t. Figure 7.15 illustrates the case where 25% of the distribution of group 1 has a median that is shifted by 15 units to the right of the median of the distribution of group 2.

The quantile test is a two-sample rank test to detect a shift in a proportion of one population relative to another population (Johnson et al., 1987). Under the null hypothesis, the two distributions are the same. If the alternative hypothesis is that the distribution of group 1 is partially shifted to the right of the distribution of group 2, the test combines the observations, ranks them, and computes k, which is the number of observations from group 1 out of the r largest observations. The test rejects the null hypothesis if k is too large. The p-value is computed as

$$p = \sum_{i=k}^{r} \binom{N-r}{n_1 - i}\binom{r}{i}\Big/\binom{N}{n_1} \qquad (7.94)$$

where N is defined in Equation (7.89). The value of r can determined by choosing a target quantile q. The value of r is smallest rank such that

$$\frac{r}{N+1} > q \qquad (7.95)$$

Note that because of the discrete nature of ranks, any quantile q' such that

$$\frac{r}{N+1} > q' \geq \frac{r-1}{N+1} \qquad (7.96)$$

will yield the same value for r as the quantile q does. Alternatively, you can choose a value for r, and the bounds on an associated quantile are then given by Equation (7.96).

The optimal choice of q or r (i.e., the choice that yields the largest power) depends on the true underlying distribution of group 2 (with cdf F_2), the contaminating distribution (with cdf F_3), and the mixing proportion ε. Johnson et al. (1987) performed a simulation study and showed that the quantile test performs better than the Wilcoxon rank sum test and the normal scores test (a linear rank test) under the alternative of a mixed normal distribution with a shift of at least 2 standard deviations in the contaminating distribution. USEPA (1994b, pp. 7.17–7.21) showed that when the mixing proportion ε is small and the shift is large, the quantile test is more powerful than the Wilcoxon rank sum test, and when ε is large and the shift is small the Wilcoxon rank sum test is more powerful than the quantile test. Both USEPA (1994b) and USEPA (1996a) recommend employing both the Wilcoxon test to test for a location shift and the quantile test to test for a shift in the tail.

Example 7.17: Testing for a Shift in the Tail of TcCB Concentrations

Following Example 7.5 on pages 7.23 to 7.24 of USEPA (1994b), we will perform the quantile test for the TcCB data. There are 77 observations from the Cleanup area and 47 observations from the Reference Area for a total of $N = 124$ observations. The target rank is set to $r = 9$, resulting in a target quantile upper bound of $q = 0.928$. Of the 9 largest observations, all 9 are from the Cleanup area, yielding a p-value of 0.01. Hence, we would probably reject the null hypothesis that the distribution of concentrations of TcCB is the same in both areas in favor of the alternative hypothesis that the distribution for the Cleanup area is partially shifted to the right compared to the distribution for the Reference area. (Note that in Example 7.16, the Wilcoxon rank sum test did not detect a difference between the Cleanup and Reference area concentrations.)

Menu

To perform the quantile test on the TcCB data using the ENVIRONMENTALSTATS for S-PLUS pull-down menu, follow these steps.

1. In the Object Explorer, find and highlight **epa.94b.tccb.df**.
2. On the S-PLUS menu bar, make the following menu choices: **EnvironmentalStats>Hypothesis Tests>Compare Samples>Two Samples>Quantile Test**. This will bring up the Two-Sample Quantile Test dialog box.
3. For Data to Use, make sure **Pre-Defined Data** is selected. For Data Set, make sure **epa.94b.tccb.df** is selected. For Variable 1 select

TcCB, for Variable 2 select **Area**, and **check** the box that says Variable 2 is a Grouping Variable.

4. Under Alternative Hypothesis select **greater**. For Specify Target select the **Rank** button, and for Target Rank type **9**, then click **OK** or **Apply**.

Command

To perform the quantile test on the TcCB data using the ENVIRONMENTALSTATS for S-PLUS Command or Script Window, type these commands.

```
attach(epa.94b.tccb.df)
quantile.test(TcCB[Area=="Cleanup"],
  TcCB[Area=="Reference"], alternative="greater",
  target.r=9)
detach()
```

Tests Based on Bootstrap Confidence Intervals

Because of the relationship between hypothesis tests and confidence intervals (see Table 7.4), you can construct a bootstrap confidence interval for the difference in the means, medians, or any other measure of location to test a null hypothesis concerning this difference (Efron and Tibshirani, 1993, p. 214). Unlike the Student's t-test, two-sample permutation test, Wilcoxon rank sum test, or quantile test, tests based on bootstrapped confidence intervals do not require the assumption that under the null hypothesis the underlying populations are exactly the same in shape and dispersion.

Permutation Tests vs. Tests Based on Bootstrap Confidence Intervals

In Chapter 5 we constructed a two-sided 95% bootstrap confidence interval for the difference in the mean concentration of TcCB in the Cleanup and Reference areas (see Figure 5.14): [0.75, 13]. We computed this confidence interval by sampling *with* replacement from the observed Cleanup area data and sampling *with* replacement from the observed Reference area data. On the other hand, the two-sample permutation test involves permuting all group assignments. If we perform this test by sampling from the permutation distribution, then we sample all observations *without* replacement, but vary their group assignment.

Example 7.18: Testing for a Difference between Areas for the TcCB Data

In this example we will repeat the comparisons we made in Example 7.15, Example 7.16, and Example 7.17, but we will base our tests on bootstrap confidence intervals. For all of our tests, the null hypothesis is that the parameter is the same in the Cleanup and Reference areas, and the alternative hypothesis is that the parameter is larger in the Cleanup area. Table 7.19 shows one-sided lower confidence limits for the difference between the means, medians, and 90th percentiles for the Cleanup and Reference area TcCB data. For the difference between the medians, all of the lower confidence limits are less than 0, so the corresponding p-value is larger than 10%. On the other hand, for the difference between the means and the difference between the 90th percentiles, the 99% lower confidence limits do not contain 0, so the p-values are less than 1%.

Parameter	90%	95%	99%
Mean	1.36	0.96	0.55
Median	-0.21	-0.24	-0.29
90th Percentile	0.46	0.35	0.19

Table 7.19   One-sided lower BCa confidence limits for the difference between the Cleanup and Reference areas for three selected parameters

Menu

To obtain the confidence intervals shown in Table 7.19 using the S-PLUS pull-down menu, follow the steps shown in the section Nonparametric Confidence Intervals for the Difference between Two Means or Two Medians in Chapter 5. For the difference in the 90th percentiles, in Step 4, in the Expression box, type **quantile(TcCB[Area=="Cleanup"], prob=0.9) - quantile(TcCB[Area=="Reference"], prob=0.9)**. Also, after Step 8, click on the **Results** tab, and in the Percentile Levels box, type **c(0.1, 0.05, 0.01)**.

Command

To obtain the confidence intervals shown in Table 7.19 using the S-PLUS Command or Script Window, follow the commands shown in the section Nonparametric Confidence Intervals for the Difference between Two Means or Two Medians in Chapter 5. For the difference in the 90th percentiles, set the value of the `statistic` argument to

```
statistic=quantile(TcCB[Area=="Cleanup"], prob=0.9) -
    quantile(TcCB[Area=="Reference"], prob=0.9)
```

Also, in the call to the `summary` function, include the argument

```
probs = c(0.1, 0.05, 0.01)
```

COMPARING TWO PROPORTIONS

Sometimes in environmental data analysis we are interested in determining whether two probabilities or rates or proportions differ from each other. For example, we may ask the question: "Does exposure to pesticide X increase the risk of developing cancer Y?," where cancer Y may be liver cancer, stomach cancer, or some other kind of cancer. One way environmental scientists attempt to answer this kind of question is by conducting experiments on rodents in which one group (the "treatment" or "exposed" group) is exposed to the pesticide and the other group (the control group) is not. The incidence of cancer Y in the exposed group is compared with the incidence of cancer Y in the control group. We will talk more about extrapolating results from experiments involving rodents to consequences in humans (and the associated difficulties) in Chapter 13 on Risk Assessment.

In this case, we assume the observations from each group shown in Equations (7.57) and (7.58) come from a binomial distribution with size parameter $n = 1$. That is, the observations can only take on two possible values: 0 (failure) or 1 (success). In our rodent example, a "success" is the occurrence of cancer. We are interested in whether the probability of "success" in group 1 (p_1) is the same as the probability of "success" in group 2 (p_2). The kinds of hypotheses we can test are shown in Equations (7.59), (7.61), and (7.62), where the parameter δ denotes the differences in the two probabilities:

$$\delta = p_1 - p_2 \qquad (7.97)$$

There are at least two ways of performing a hypothesis test to compare two proportions:

- Use Fisher's exact test.
- Create a statistic based on the difference between the sample proportions, and compute the p-value based on a normal approximation.

We will explain how to perform each procedure in the rest of this section.

Fisher's Exact (Permutation) Test

In Equations (5.28), (5.30), (5.34), and (5.58) in Chapter 5 we showed that when the observations are from a B(1, p) distribution, the sample mean is an estimate of p. Fisher's exact test is simply a permutation test for the difference between two means from two different groups as described in the section Two-Sample Permutation Tests, where the underlying populations are binomial with size parameter $n = 1$, but possibly different values of p. The null hypothesis is that the two proportions, p_1 and p_2, are the same (i.e.,

$\delta = 0$). Fisher's exact test is usually described in terms of testing hypotheses concerning a 2×2 contingency table (Fisher and van Bell, 1993, p. 185; Hollander and Wolfe, 1999, p. 473; Sheskin, 1997, p. 221; Zar, 1999, p. 543). The probabilities associated with the permutation distribution can be computed by using the hypergeometric distribution (see Chapter 4).

Example 7.19: Testing Whether Exposure to Ethylene Thiourea Increases the Incidence of Thyroid Tumors in Rats

Rodricks (1992, p. 133) presents data from an experiment in which different groups of rats were exposed to various concentration levels of ethylene thiourea (ETU), a decomposition product of a certain class of fungicides that can be found in treated foods. In the group exposed to a dietary level of 250 ppm of ETU, 16 out of 69 rats (23%) developed thyroid tumors, whereas in the control group (no exposure to ETU) only 2 out of 72 (3%) rats developed thyroid tumors. If we use Fisher's exact test to test the null hypothesis that the proportion of rats exposed to 250 ppm of ETU who will develop thyroid tumors over their lifetime is no greater than the proportion of rats not exposed to ETU who will develop tumors, we get a one-sided upper p-value of 0.0002. Therefore, we conclude that the true underlying rate of tumor incidence in the exposed group is greater than in the control group. Figure 7.16 displays the permutation distribution of the difference between the sample proportions, as well as the observed difference between the sample proportions.

Menu

To perform Fisher's exact test to compare the incidence of thyroid tumors between the exposed and unexposed groups of rats using the ENVIRONMENTALSTATS for S-PLUS pull-down menu, follow these steps.

1.   On the S-PLUS menu bar, make the following menu choices: **EnvironmentalStats>Hypothesis Tests>Compare Samples>Two Samples>Permutation Tests>Proportions**. This will bring up the Two-Sample Permutation Test for Proportions dialog box.
2.   For Data to Use, select **Expression**. For Data Form, select **# Succ and Trials**. For # Successes, type **c(16,2)**. For # Trials type **c(69,72)**. Under Alternative Hypothesis select **greater**.
3.   Click **OK** or **Apply**.

Note that you can perform Fisher's exact test using the S-PLUS pull down menu, but only two-sided p-values are computed.

Command

To perform Fisher's exact test using the ENVIRONMENTALSTATS for S-PLUS Command or Script Window, type these commands.

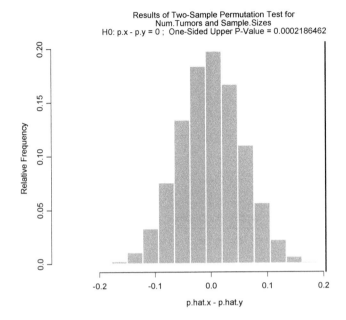

Figure 7.16   Results of the two-sample permutation test (Fisher's exact test) for the difference in the proportion of rats who develop thyroid tumors in the exposed and unexposed groups

```
Num.Tumors <- c(16, 2)
Sample.Sizes <- c(69, 72)
prop.list <- two.sample.permutation.test.proportion(
   x=Num.Tumors, y=Sample.Sizes,
   x.and.y="Number Successes and Trials",
   alternative="greater")
prop.list
plot(prop.list)
```

Note that you can perform Fisher's exact test using the S-PLUS function `fisher.test`, but only two-sided p-values are computed.

Approximate Test Based on the Normal Approximation to the Binomial Distribution

Another way to test whether two population proportions are equal (i.e., $\delta = 0$) is to use a statistic based on the normal approximation to the binomial distribution:

$$z = \frac{\hat{p}_1 - \hat{p}_2}{\hat{\sigma}_{\hat{p}_1 - \hat{p}_2}} \qquad (7.98)$$

where

$$\hat{\sigma}^2_{\hat{p}_1 - \hat{p}_2} = \bar{p}\,(1 - \bar{p})\left(\frac{1}{n_1} + \frac{1}{n_2}\right) \qquad (7.99)$$

$$\bar{p} = \frac{n_1\hat{p}_1 + n_2\hat{p}_2}{n_1 + n_2} \qquad (7.100)$$

$$\hat{p}_i = \frac{1}{n_i}\sum_{j=1}^{n_i} x_{ij} \,, \quad i = 1, 2 \qquad (7.101)$$

(Fisher and van Belle, 1993, p. 187; Hollander and Wolfe, 1999, p. 459; Sheskin, 1997, p. 226; Zar, 1999, p. 556). Under the null hypothesis that the two proportions are equal, the statistic z is approximately distributed as a standard normal random variable. We can also test the more general hypothesis $\delta = \delta_0$ (where $\delta_0 \neq 0$) with the following statistic:

$$z = \frac{(\hat{p}_1 - \hat{p}_2) - \delta_0}{\hat{\sigma}_{\hat{p}_1 - \hat{p}_2}} \qquad (7.102)$$

where

$$\hat{\sigma}^2_{\hat{p}_1 - \hat{p}_2} = \frac{\hat{p}_1\,(1 - \hat{p}_1)}{n_1} + \frac{\hat{p}_2\,(1 - \hat{p}_2)}{n_2} \qquad (7.103)$$

(Sheskin, 1997, p. 229). Note the similarity between Equations (7.102) and (7.67) and Equations (7.103) and (7.68).

A correction for continuity is usually suggested (e.g., Sheskin, 1997, p. 228; Zar, 1999, p. 556), in which case the quantity

$$\frac{1}{2}\left(\frac{1}{n_1} + \frac{1}{n_2}\right) \tag{7.104}$$

is either subtracted from or added to the numerator, depending on whether it is positive or negative, respectively. The square of the z-statistic in Equation (7.98) is equivalent to the chi-square statistic that is computed to test independence in a 2×2 contingency table (Fisher and van Belle, 1993, p. 188; Sheskin, 1997, p. 226; Zar, 1999, p. 556).

An approximate two-sided $(1-\alpha)100\%$ confidence interval for $\delta = p_1 - p_2$, including the continuity correction, is given by:

$$\left[\left(\hat{p}_1 - \hat{p}_2\right) - z_{1-\alpha/2}\hat{\sigma}_{\hat{p}_1 - \hat{p}_2} - \frac{1}{2}\left(\frac{1}{n_1} + \frac{1}{n_2}\right), \right.$$

$$\tag{7.105}$$

$$\left. \left(\hat{p}_1 - \hat{p}_2\right) + z_{1-\alpha/2}\hat{\sigma}_{\hat{p}_1 - \hat{p}_2} + \frac{1}{2}\left(\frac{1}{n_1} + \frac{1}{n_2}\right) \right]$$

(Sheskin, 1997, p. 231; Zar, 1996, p. 556). For the confidence interval, Zar (1999, p. 556) recommends using the formula in Equation (7.99) to compute the standard error of the difference of the proportions, while Fisher and van Belle (1993, p. 188) and Sheskin (1997, p. 231) recommend using the formula in Equation (7.103). The latter makes more sense because for the confidence interval we are not assuming that the two proportions are equal.

Example 7.20: Comparing Incidence Rates of Thyroid Tumors in Rats

In this example, we will again compare the incidence of thyroid tumors in rats exposed and not exposed to ETU as we did in Example 7.19, but here we will perform the approximate test. As before, the estimated incidence rates are 23% and 3% for the exposed and unexposed groups. The one-sided upper p-value is 0.0004 and a one-sided lower 95% confidence interval for the difference between the two incidence rates is [0.10, 1]. Hence, as before, we conclude that the incidence rate is higher in the exposed group than in the unexposed group.

Menu

To perform the approximate test to compare the incidence of thyroid tumors between the exposed and unexposed groups of rats using the S-PLUS pull-down menu, follow these steps.

1. On the S-PLUS menu bar, make the following menu choices: **Statistics>Compare Samples>Two Samples>Counts and Proportions>Proportions Parameters**. This will bring up the Proportions Test dialog box.
2. Make the Data Set box **blank**. For Success Variable, type **c(16,2)**. For Trials Variable, type **c(69,72)**. Under Alternative Hypothesis select **greater**, then click **OK** or **Apply**.

Command

To perform the approximate test using the S-PLUS Command or Script Window, type these commands.

```
Num.Tumors <- c(16, 2)
Sample.Sizes <- c(69, 72)
prop.test(Num.Tumors, Sample.Sizes,
    alternative="greater")
```

COMPARING VARIANCES BETWEEN TWO GROUPS

In this chapter so far, we have concentrated mostly on tests concerning locations and proportions. Sometimes, we are interested in whether the variability is the same between two groups. For example, if we send splits from field samples to two separate labs for analysis, we are not only interested in whether the labs generally agree on the estimated concentration in each physical sample, but also how variable the results are within each lab.

If we are willing to assume the observations from both groups come from normal distributions, we can use the following statistic to test the null hypothesis that the population variances of the two groups are the same:

$$F = \frac{s_1^2}{s_2^2} \qquad\qquad (7.106)$$

where s_1^2 and s_2^2 denote the sample variances from groups 1 and 2, respectively. Under the null hypothesis $H_0 : \sigma_1^2 = \sigma_2^2$, the statistic F follows an

F-distribution with n_1-1 and n_2-1 degrees of freedom (Fisher and van Belle, 1993, p. 155; Sheskin, 1997, p. 161; Zar, 1999, p. 137). Note that this statistic is bounded below by 0, since the sample variances are always greater than 0. Figure 4.8 displays an F-distribution with 5 and 10 degrees of freedom.

For the one-sided upper alternative hypothesis $H_a : \sigma_1^2 > \sigma_2^2$, the p-value is computed as:

$$p = \Pr\left(F_{n_1-1,\,n_2-1} > F\right) \qquad (7.107)$$

where F_{v_1,v_2} denotes a random variable that follows an F-distribution with v_1 and v_2 degrees of freedom. For the one-sided lower alternative hypothesis $H_a : \sigma_1^2 < \sigma_2^2$, the p-value is computed as:

$$p = \Pr\left(F_{n_1-1,\,n_2-1} < F\right) \qquad (7.108)$$

For the two-sided alternative hypothesis $H_a : \sigma_1^2 \neq \sigma_2^2$, the p-value is twice the p-value shown in Equation (7.107) when F is larger than the 50th percentile of the $F_{n_1-1,\,n_2-1}$ distribution, otherwise it is twice the p-value shown in Equation (7.108).

A lower one-sided $(1-\alpha)100\%$ confidence interval for the variance ratio σ_1^2 / σ_2^2 is given by:

$$\left[\frac{s_1^2 / s_2^2}{F_{n_1-1,\,n_2-1,\,1-\alpha}}\,,\ \infty\right) \qquad (7.109)$$

where $F_{v_1,v_2,p}$ denotes the pth quantile of the F-distribution with v_1 and v_2 degrees of freedom. An upper one-sided $(1-\alpha)100\%$ confidence interval for the variance ratio is given by:

$$\left(0 , \frac{s_1^2/s_2^2}{F_{n_1-1,n_2-1,\alpha}} \right] \qquad \text{(7.110)}$$

and a two-sided $(1-\alpha)100\%$ confidence interval for the variance ratio is given by:

$$\left[\frac{s_1^2/s_2^2}{F_{n_1-1,n_2-1,1-\alpha/2}} , \frac{s_1^2/s_2^2}{F_{n_1-1,n_2-1,\alpha/2}} \right] \qquad \text{(7.111)}$$

In the section Student's Two-Sample t-Test we pointed out that in order to use Student's two-sample t-test we must assume the two groups are normally distributed and have the same population variances. We also pointed out that the two-sample t-test is fairly robust to departures from these assumptions, especially for moderate to large sample sizes with approximately equal sample sizes in each group. Some references suggest performing the F-test to make sure the two variances are the same before performing the two-sample t-test. Just like the chi-square test on a single variance, however, the F-test for comparing variances, is extremely sensitive to the normality assumption, while the t-test is not. Levene's test (discussed later in this chapter) is more robust to deviations from normality.

Example 7.21: Comparing Variability in Manganese Concentrations between Two Wells

The guidance document USEPA (1996a, p. 4.5-2) contains manganese concentrations at two different monitoring wells, which are displayed in Table 7.20 below and plotted in Figure 7.17. The estimated sample variances are 9,076 and 12,125 (ppm^2), and the F-statistic is 0.75 with 3 and 4 degrees of freedom. The two-sided p-value is 0.85, therefore we do not reject the null hypothesis that the variability of manganese concentrations at the two wells is the same. A two-sided 95% confidence interval for the variance ratio is [0.08, 11.3].

Well	Manganese Concentrations (ppm)				
1	50	73	244	202	
2	272	171	32	250	53

Table 7.20 Manganese concentrations at two different monitoring wells (USEPA, 1996a, p. 4.5-2)

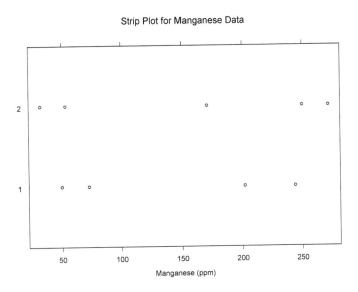

Figure 7.17   Strip plot of manganese concentrations at two different monitoring wells
(USEPA, 1996a, p. 4.5-2)

Menu

To perform the F-test for equal variances for the manganese concentrations at the two monitoring wells using the ENVIRONMENTALSTATS for S-PLUS pull-down menu, follow these steps.

1. In the Object Explorer, find and highlight **epa.96a.manganese.df**.
2. On the S-PLUS menu bar, make the following menu choices: **EnvironmentalStats>Hypothesis Tests>Compare Samples>Two Samples>F Test for Equal Variances**. This will bring up the Two-Sample F Test for Equal Variances dialog box.
3. For Data to Use, make sure **Pre-Defined Data** is selected. For Data Set, make sure **epa.96a.manganese.df** is selected. For Variable 1 select **Manganese**, for Variable 2 select **Well**, and **check** the box that says Variable 2 is a Grouping Variable. Click **OK** or **Apply**.

Command

To perform the F-test for equal variances using the S-PLUS Command or Script Window, type these commands.

```
attach(epa.96a.manganese.df)
Manganese.1 <- Manganese[Well=="1"]
Manganese.2 <- Manganese[Well=="2"]
```

```
var.test(Manganese.1, Manganese.2)
detach()
```

THE MULTIPLE COMPARISONS PROBLEM

So far we have discussed tests for one group (goodness-of-fit, test on a proportion, test on a location or percentile, and test on variance) and two groups (comparing two locations, proportions, or variances). We have not yet considered the problem of testing the equality of several locations, proportions, or variances. Obviously, one way to do this is to perform all pairwise tests. For example, if we want to test whether the average concentration of lead in the soil is the same at sites A, B, and C, we could perform three two-sample t-tests to compare A with B, A with C, and B with C. A problem with this approach is that if we set the Type I error rate to 5% for each of the three tests, the probability that we incorrectly conclude there is a difference between at least two of the sites when in fact there is no difference between any of the sites is bigger than 5%.

To see how this can happen, consider a simple example with a hypothesis test based on flipping a coin. If the coin lands "heads," reject the null hypothesis, and if the coin lands "tails" do not reject the null hypothesis. Assuming the coin is fair (i.e., the probability of a head is ½), the probability of making a Type I error is ½. Now suppose you perform two hypothesis tests. For each test, the Type I error rate is ½, but the probability of incorrectly rejecting at least one of the null hypotheses when in fact both are true is the probability of getting at least one head out of two tosses of the coin, which is ¾.

For comparing three or more locations, we can initially avoid the problem of performing all pairwise comparisons by using parametric or nonparametric one-way analysis of variance (ANOVA). Similarly, for comparing three or more proportions we can use the extension of Fisher's exact test or the chi-square test, and for comparing three or more variances we can use Bartlett's or Levene's test. All of these methods are discussed later in this chapter. But if we reject the null hypothesis that all the means (or medians, proportions, variances, etc.) are equal, we would usually like to know which ones are different from one another and which are the same. In this case, we are back to the problem of pairwise comparisons.

Experiment-Wise Type I Error Rate

If we perform k hypothesis tests, and the null hypothesis is true for all k tests, the probability of incorrectly rejecting the null hypothesis for at least one of these tests is called the **experiment-wise Type I error rate**, here denoted α^*. The Type I error rate α associated with an individual hypothesis test is sometimes called the **comparison-wise Type I error rate** (Milliken and

Johnson, 1992, p. 30). In the special case when each of the tests is performed with a comparison-wise Type I error rate of α and the outcomes of the tests are independent of one another, the experiment-wise Type I error rate α^* is easily computed because the number of false positives we will observe is a binomial random variable with size parameter k and probability parameter α (i.e., a $B(k, \alpha)$ random variable):

$$\alpha^* = 1 - (1 - \alpha)^k \qquad (7.112)$$

Thus, the more hypothesis tests you perform, the greater the experiment-wise Type I error rate. Figure 7.18 displays the experiment-wise Type I error rate as a function of the number of hypothesis tests for various values of the comparison-wise Type I error rate α.

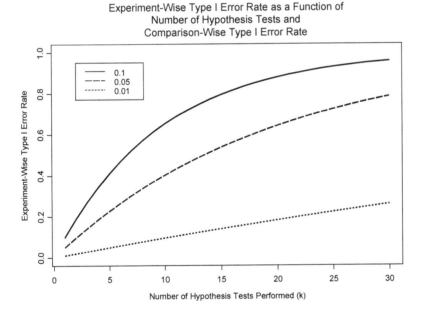

Figure 7.18   Experiment-wise Type I error rate as a function of number of tests performed and comparison-wise Type I error rate, assuming the test outcomes are independent

In general, the outcomes of the hypothesis tests you are interested in are usually not independent for at least two reasons:

- If you perform pairwise comparisons, these are not independent of one another because some of the same data are used in each comparison. For example, to compare site A to B and A to C, the same data from site A are used in both comparisons.

- Inherent temporal or spatial correlation may cause the results of several hypothesis tests to be correlated.

Often, the best we can do is put an upper bound on the experiment-wise Type I error rate, which is simply the sum of the individual Type I error rates for each test. This follows from a fundamental law of probability, which states that for two events E_1 and E_2, the probability that at least one of these events occurs is less than or equal to the sum of the probabilities of each event:

$$\Pr \left(E_1 \cup E_2 \right) \leq \Pr \left(E_1 \right) + \Pr \left(E_2 \right) \tag{7.113}$$

Multiple Comparisons and the Bonferroni Inequality

The term *multiple comparisons* (also called *simultaneous statistical inference*) refers to the situation where several hypothesis tests are performed simultaneously to answer the same question or several related questions (Miller, 1981a). We have already discussed one example in environmental statistics extensively in Chapter 6 in the context of groundwater monitoring programs at a hazardous or solid waste sites. In this context, the experiment-wise Type I error rate is also called the facility-wide false positive rate (FWFPR). Another example is contained in USEPA (1994b), which suggests performing three different hypothesis tests (the Wilcoxon rank sum test, the quantile test, and a "hot-measurement" test) to compare soil chemical concentrations at a cleanup site with those in a reference area.

As part of the DQO process (Chapter 2), it is important to be able to set the overall Type I error rate and power for a set of hypothesis tests in order to balance the costs of these two kinds of mistakes. Several statistical methods have been proposed to deal with the problem of multiple comparisons (Miller, 1981a). Probably the most common technique relies on the **Bonferroni inequality** which, although theoretically crude, actually works quite well in many circumstances. The Bonferroni procedure is based on the fundamental result shown in Equation (7.113) and simply sets $\alpha = \alpha^*/k$ for each individual test, where k is the number of hypothesis tests being performed. Other parametric and nonparametric univariate techniques are also available, as well as techniques that rely on multivariate analysis.

One of the major problems with standard multiple comparisons procedures is the reduction in the power of the tests after controlling the experiment-wise Type I error rate. For example, if we want to perform 10 hypothesis tests and maintain an experiment-wise Type I error rate of 0.05, the Bonferroni procedure sets the comparison-wise (individual) α-levels to $\alpha =$

0.005. With such small individual α-levels, any individual test will have a low probability of detecting a difference (change, contamination, etc.) unless the difference is very large. As we already discussed in Chapter 6, in the context of groundwater monitoring, simultaneous prediction intervals with re-sampling take care of the multiple comparisons problem while at the same time yield adequate power for detecting contamination.

COMPARING LOCATIONS BETWEEN SEVERAL GROUPS

Sometimes we are interested in comparing observations from several different groups to determine whether all of the groups could be considered to come from the same population. For example, we may be interested in characterizing the "background" concentration of lead, and we may have initially stratified our sampling area into three different sub-areas, and now we want to determine whether the distributions of lead concentrations in each of the three sub-areas are the same.

One-way Analysis of Variance (ANOVA) is a commonly used statistical method to compare two or more groups simultaneously. Parametric one-way ANOVA is an extension of the two-sample t-test. It simultaneously compares the sample means from each group, and assumes all of the groups follow a normal distribution with the same variance, but possibly different population means. Nonparametric one-way ANOVA (also called the Kruskall-Wallis test) is an extension of the Wilcoxon rank sum test. It is the same as parametric ANOVA except it is based on the ranks of the observations (it compares the averages of the ranks from each group simultaneously), and assumes all of the groups follow the same distribution (not necessarily normal) with the same amount of dispersion but possibly different medians. Both parametric and nonparametric ANOVA are based on the idea that if the amount of variability between the sample means or medians is large relative to the amount of variability within each group, then this is good evidence that the population means or medians are not the same. Hence the name "Analysis of Variance."

USEPA (1989b, p. 5-5) recommends using ANOVA to simultaneously compare concentrations from several downgradient wells with background well concentrations for groundwater detection monitoring. There are three main problems with this strategy. First, ANOVA is used to determine whether there are any differences between any of the groups, whereas in groundwater detection monitoring we are only interested in whether a compliance well differs from background, not whether it differs from another compliance well. Second, the ANOVA may indicate a statistically significant difference between monitoring wells which is simply a result of natural spatial variability rather than any influence from the hazardous or solid waste facility. Third, because several constituents are usually monitored, you need to perform an ANOVA for each constituent, and you must therefore somehow control the facility-wide false positive rate (FWFPR). Both Gibbons (1994, p. 258) and Davis (1994) recommend against using ANOVA for groundwater detection monitoring (although Davis, 1994 recommends using ANOVA to estimate spatial and temporal components of variability; see Chapter 6). For groundwater monitoring at hazardous waste sites, using simultaneous prediction intervals with re-sampling, as explained in Chapter 6, is a much better approach.

Parametric ANOVA

Suppose there are k groups and we want to compare the averages among the k groups. Following the notation in Equations (7.57) and (7.58), we will let

$$x_{i1}, x_{i2}, \dots, x_{in_i} \tag{7.114}$$

denote the n_i observations from group i ($i = 1, 2, \dots, k$). Also, we will let N denote the total sample size. That is,

$$N = \sum_{i=1}^{k} n_i \tag{7.115}$$

Parametric ANOVA assumes the observations all come from normal distributions, with the same variance for each group but possibly different means. Mathematically, these assumptions can be written as follows:

$$X_{ij} = \mu_i + \varepsilon_{ij},$$

$$i = 1, 2, \dots k \; ; \; j = 1, 2, \dots, n_i \tag{7.116}$$

where the error terms ε_{ij} come from a normal distribution with mean 0 and standard deviation σ. We want to compare the k groups by testing the null hypothesis

$$H_0 : \mu_1 = \mu_2 = \dots = \mu_k \tag{7.117}$$

against the alternative hypothesis that at least one of the group means differs from the others. As we stated above, parametric one-way ANOVA is an extension of the two-sample t-test. It simultaneously compares the sample means from each group, and assumes all of the groups follow a normal distribution with the same variance, but possibly different population means.

To test the null hypothesis in Equation (7.117), we can use the following statistic:

$$F = \frac{\hat{\sigma}^2_{between}}{\hat{\sigma}^2_{within}} \qquad (7.118)$$

where

$$\hat{\sigma}^2_{between} = \frac{\sum\limits_{i=1}^{k} n_i \left(\overline{x}_i - \overline{x} \right)^2}{k - 1}$$

$$\qquad (7.119)$$

$$= \frac{SS_{between}}{DF_{between}} = MS_{between}$$

$$\hat{\sigma}^2_{within} = \frac{\sum\limits_{i=1}^{k} \left(n_i - 1 \right) s_i^2}{\sum\limits_{i=1}^{k} \left(n_i - 1 \right)} = \frac{\sum\limits_{i=1}^{k} \sum\limits_{j=1}^{n_i} \left(x_{ij} - \overline{x}_i \right)^2}{N - k}$$

$$\qquad (7.120)$$

$$= \frac{SS_{within}}{DF_{within}} = MS_{within}$$

$$\overline{x}_i = \frac{1}{n_i} \sum\limits_{j=1}^{n_i} x_{ij} \quad , \quad i = 1, 2, \ldots, k \qquad (7.121)$$

$$\overline{X} = \frac{\displaystyle\sum_{i=1}^{k} n_i \overline{x}_i}{\displaystyle\sum_{i=1}^{k} n_i} = \frac{\displaystyle\sum_{i=1}^{k} \sum_{j=1}^{n_i} x_{ij}}{N} \qquad (7.122)$$

In the above equations, the notation SS stands for "Sum of Squares", DF stands for "Degrees of Freedom," and MS stands for "Mean Square." These are commonly used terms in analysis of variance (Helsel and Hirsch, 1992, p. 164; Fisher and van Belle, p. 427; Sheskin, 1997, pp. 336–338; Zar, 1999, pp. 178–189)

Looking at Equation (7.120), you can see that the denominator of the F-statistic in Equation (7.118) is simply the pooled estimate of variance we used for two groups (shown in Equation (7.69)) extended to k groups. This is a valid estimate of the population variance common to all groups, no matter whether the null hypothesis of equal means is true or not. If the null hypothesis is true, then the numerator of the F-statistic in Equation (7.118) is also a valid estimate of the population variance. If, however, the null hypothesis is not true, then the numerator is bouncing around something larger than the population variance and the F-statistic will tend to be larger than 1.

Under the null hypothesis that all the means are equal, the F-statistic in Equation (7.118) follows an F-distribution with $k-1$ and $N-k$ degrees of freedom. The p-value associated with the test is computed as:

$$p = \Pr\left(F_{k-1,N-k} > F\right) \qquad (7.123)$$

Note that there is no one-sided vs. two-sided p-value associated with this test. The p-value is inherently "two-sided" since the alternative hypothesis is that at least one of the means is different from the rest. In fact, for the case when $k = 2$ groups, this test is equivalent to the two-sample t-test for testing the two-sided alternative shown in Equation (7.59) with $\delta = \mu_1-\mu_2$ and $\delta_0 = 0$. As with the t-test, parametric ANOVA is fairly robust to departures from normality and unequal variances, especially if the sample sizes for each group are equal or nearly equal (Zar, 1999, p. 185).

Example 7.22: Comparing Fecal Coliform Counts between Seasons

Lin and Evans (1980) reported fecal coliform measures (organisms per 100 ml) from the Illinois River taken between 1971 and 1976. Helsel and Hirsch (1992, p. 162) present a small subset of these data, which is shown in Table 7.21 below.

Summer	Fall	Winter	Spring
100	65	28	22
220	120	58	53
300	210	120	110
430	280	230	140
640	500	310	320
1600	1100	500	1300

Table 7.21   Selected fecal coliform counts (organisms per 100 ml) from the Illinois River (from Lin and Evans, 1980; as presented in Helsel and Hirsch, 1992, p. 162)

Figure 7.19 displays a strip plot of these data along with the means and CVs for each season, and the mean for all of the seasons combined. In this example, we would like to determine whether the mean fecal coliform count is the same between seasons. Except for the Winter season, the data do not look normally distributed, but instead positively skewed. (Of course since the observations are counts and therefore discrete, they cannot truly come from a normal distribution, which is continuous. Since we are interested in comparing means using parametric ANOVA, however, we are mainly interested in making sure the observations in each group are approximately normal and that the variances are similar.) The Shapiro-Francia test for normality (equivalent to the PPCC test) yields individual p-values of 0.04, 0.06, 0.6, and 0.005 for each season, and an overall group p-value of 0.003. Thus, we probably do not want to assume these data can be modeled with a normal distribution.

Figure 7.20 displays a strip plot of the log-transformed data, along with the means and standard deviations for each season (based on the log-transformed data), and the mean for all of the seasons combined. These data are more symmetric, and the Shapiro-Francia test for normality yields individual p-values of 1, 1, 0.97, and 0.99 for each season, and an overall group p-value of 1 (of course, with such small sample sizes we can only detect gross deviations from normality). In this example, we will use one-way ANOVA to test whether the mean concentrations of the log-transformed data are significantly different. Based on our discussion in the section Comparing the Means of Log-Transformed Data, this is equivalent to testing whether the mean fecal coliform count is the same between all seasons, assuming we are willing to assume the coefficient of variation is the same for all four seasons.

Looking at Figure 7.20, we can see that the variability between seasonal means is small compared to the variability with each season. The one-way ANOVA performed on the log-transformed observations yields an F-statistic of 1.12 and a p-value of 0.36. Therefore, we do not have sufficient evidence to reject the null hypothesis that the average log-transformed fecal coliform count differs between seasons, and thus we conclude that the average fecal coliform count does not differ between seasons either.

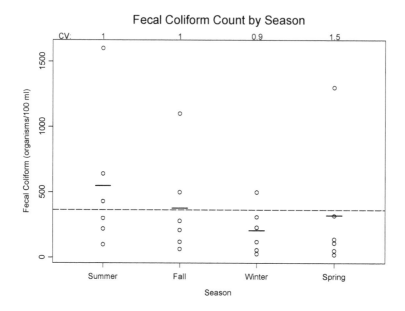

Figure 7.19   Strip plot of the fecal coliform data, with group means (solid lines) and overall mean (dashed line)

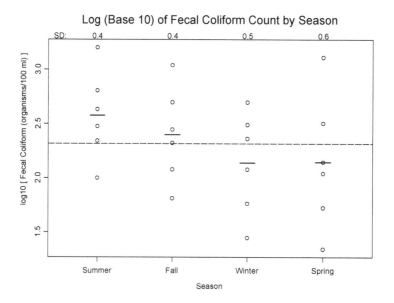

Figure 7.20   Strip plot of the log-transformed fecal coliform data, with group means (solid lines) and overall mean (dashed line)

Menu

To perform the one-way ANOVA on the log-transformed data using the S-PLUS pull-down menu, follow these steps.

1.   In the Object Explorer, find and highlight the data frame **lin.and.evans.80.df**.
2.   On the S-PLUS menu bar, make the following menu choices: **Statistics>Compare Samples>k Samples>One-way ANOVA**. This will bring up the One-way Analysis of Variance dialog box.
3.   In the Data Set box **lin.and.evans.80.df** should be selected. For Variable type **log10(Fecal.Coliform)**. For Grouping Variable select **Season**. Click **OK** or **Apply**.

Command

To perform the one-way ANOVA on the log-transformed data using the S-PLUS Command or Script Window, type these commands.

```
attach(lin.and.evans.80.df)
aov.list <- aov(log10(Fecal.Coliform) ~ Season,
    data=lin.and.evans.80.df)
summary(aov.list)
detach()
```

Post-Test Multiple Comparisons

If we had rejected the null hypothesis of equal means in the last example, we could have used one of several possible multiple comparisons procedures to determine which means differed from each other. These procedures are explained in standard statistics textbooks that discuss ANOVA, including Fisher and van Belle (1993, Chapter 12), Helsel and Hirsch (1992, pp. 195–202), Milliken and Johnson (1992, Chapter 3), Sheskin (1997, pp. 339–369), and Zar (1999, Chapter 11). See the S-PLUS documentation for instructions on how to implement these procedures.

K-Sample Permutation Tests and Nonparametric ANOVA

In the section Two-Sample Permutation Tests we explained how to perform a two-sample permutation test to compare two means. This idea can be extended to comparing k samples (Manly, 1997, Chapter 7). In the section Wilcoxon Rank Sum Test we pointed out that the Wilcoxon rank sum test is equivalent to a two-sample permutation test for the difference between the means based on the ranks. Similarly, the Kruskal-Wallis test is the nonparametric analogue of one-way parametric ANOVA and is based on the ranks of the observations (Conover, 1980, p. 229; Fisher and van Belle, 1993, p. 430;

Helsel and Hirsch, 1992, p. 159; Hollander and Wolfe, 1999, p. 190; Lehmann, 1975, p. 204; Sheskin, 1997, Chapter 17; Zar, 1999, p. 195). The assumptions of the Kruskal-Wallis test are the same as for parametric one-way ANOVA except that the underlying distributions for each group do not have to be normal.

Based on the notation in Equation (7.114), we will let r_{ij} denote the rank of the jth observation in the ith group ($i = 1, 2, ..., k$; $j = 1, 2, ..., n_i$), where the ranking is based on combining the observations from all of the k groups. The Kruskal-Wallis test statistic H is given by:

$$H = \frac{12}{N(N+1)} \sum_{i=1}^{k} n_i \left(\bar{r}_i - \bar{r} \right)^2 \qquad (7.124)$$

where

$$\bar{r}_i = \frac{1}{n_i} \sum_{j=1}^{n_i} r_{ij} \ , \quad i = 1, 2, ..., k \qquad (7.125)$$

$$\bar{r} = \frac{1}{N} \sum_{i=1}^{k} \sum_{j=1}^{n_i} r_{ij} = \frac{N+1}{2} \qquad (7.126)$$

Under the null hypothesis that the medians for all of the groups are the same, the averages of the ranks for each group should be similar so H should be "small." We reject the null hypothesis if H is too "large."

For "large" sample sizes (e.g., $n_i > 4$), the p-value can be computed by approximating the distribution of the statistic H with a chi-square distribution with $k-1$ degrees of freedom. In the case of ties in the ranks, the statistic H is replaced with H':

$$H' = \frac{H}{1 - \dfrac{\sum_{g=1}^{m} t_g}{N^3 - N}} \qquad (7.127)$$

where m denotes the number of groups of ties, and t_g denotes the number of ties in the gth group of ties.

Example 7.23: Comparing Median Fecal Coliform Counts between Seasons

Here we will repeat Example 7.22, except we will use the Kruskal-Wallis test to compare the median fecal coliform count between seasons. The observed seasonal medians are 365, 245, 175, and 125 organisms/100 ml. The Kruskal-Wallis statistic is 2.69, which yields a p-value of 0.44 based on the chi-square distribution with 3 degrees of freedom. Hence we conclude we do not have enough evidence to reject the null hypothesis that the median fecal coliform count is the same between seasons.

Menu

To perform the Kruskal-Wallis test using the S-PLUS pull-down menu, follow these steps.

1. In the Object Explorer, find and highlight the data frame **lin.and.evans.80.df**.
2. On the S-PLUS menu bar, make the following menu choices: **Statistics>Compare Samples>k Samples>Kruskal-Wallis Rank Test**. This will bring up the Kruskal-Wallis Rank Sum Test dialog box.
3. In the Data Set box **lin.and.evans.80.df** should be selected. For Variable select **Fecal.Coliform**. For Grouping Variable select **Season**. Click **OK** or **Apply**.

Command

To perform the Kruskal-Wallis test using the S-PLUS Command or Script Window, type these commands.

```
attach(lin.and.evans.80.df)
kruskal.test(Fecal.Coliform, Season)
detach()
```

COMPARING PROPORTIONS BETWEEN SEVERAL GROUPS

Fisher's exact test and the chi-square test for 2×2 contingency tables can be extended to tests on $2 \times k$ contingency tables to compare k proportions simultaneously (Conover, 1980, Chapter 4; Fisher and van Belle, 1993, Chapter 7; Helsel and Hirsch, 1992, Chapter 14; Sheskin, 1997, pp. 209–255; Zar, 1999, Chapter 23). See the S-PLUS documentation for information on how to perform these tests with S-PLUS.

COMPARING VARIABILITY BETWEEN SEVERAL GROUPS

Sometimes, we are interested in whether the variability is the same between k groups. For example, if we send splits from field samples to k separate labs for analysis, we are not only interested in whether the labs generally agree on the estimated concentration in each physical sample, but also how variable the results are within each lab. In this section we will discuss two tests you can use to test whether the variances of k groups are the same (also called a test of *homogeneity of variance*): Bartlett's test and Levene's test.

We will use a more general model than the one that we used for parametric ANOVA: we will assume the observations all come from normal distributions, with possibly different means *and* possibly different variances for each group. This is the same model as Equation (7.116), except that now the error terms ε_{ij} are assumed to come from a normal distribution with mean 0 and standard deviation σ_i ($i = 1, 2, ..., k$). We are interested in testing the null hypothesis

$$H_0 : \sigma_1^2 = \sigma_2^2 = \cdots = \sigma_k^2 \qquad (7.128)$$

versus the alternative hypothesis that at least one of the group variances is different from the others.

In the section Parametric ANOVA we pointed out that ANOVA is fairly robust to departures from normality and homogeneity of variance, especially for moderate to large sample sizes with approximately equal sample sizes in each group. For this reason, Zar (1999, p. 204) does not recommend that you perform a test of homogeneity of variance before performing a parametric ANOVA.

Bartlett's Test

Bartlett's test of the null hypothesis (7.128) is based on comparing the logarithm of the pooled estimate of variance with the sum of the logarithms of the estimated variances for each group. The test statistic is given by:

$$B = (N - k) \log \left(\hat{\sigma}_{within}^2 \right) - \sum_{i=1}^{k} \log \left(s_i^2 \right) \qquad (7.129)$$

Under the null hypothesis of equal variances, the statistic B is approximately distributed as a chi-square random variable with $k-1$ degrees of freedom (Zar, 1999, p. 202; USEPA, 1989b, p. 4-17). The p-value is thus computed as:

$$p = \Pr\left(\chi^2_{k-1} > B\right) \tag{7.130}$$

where χ^2_v denotes a chi-square random variable with v degrees of freedom.

A more accurate approximation to the chi-square distribution is obtained by dividing the statistic B by a correction factor:

$$B' = \frac{B}{C} \tag{7.131}$$

where

$$C = 1 + \frac{1}{3(k-1)}\left[\sum_{i=1}^{k}(n_i - 1) - \frac{1}{N-k}\right] \tag{7.132}$$

(Milliken and Johnson, 1992, p. 19; Snedecor and Cochran, 1989, p. 251; Zar, 1999, pp. 202–203). Unlike Levene's test (see the next section) Bartlett's test is very sensitive to non-normality (Milliken and Johnson, 1992, p. 19; Zar, 1999, pp. 185, 204).

Example 7.24: Testing Homogeneity of Variance in Arsenic Concentrations between Wells Using Bartlett's Test

Consider the arsenic concentrations of Example 7.2. In that example we showed that the observations at each well data do not appear to come from normal populations, but a lognormal model seems to be adequate. Here, we will perform Bartlett's test on both the untransformed and log-transformed observations. For the untransformed observations, Bartlett's test yields a value of 21.1 (with correction) on 5 degrees of freedom and an associated p-value of 0.0008. Because these observations are highly skewed, Bartlett's test is not recommended for the untransformed data. For the log-transformed data, Bartlett's test yields a value of 1.8 on 5 degrees of freedom and an associated p-value of 0.88.

Menu

To perform Bartlett's test on the untransformed and log-transformed observations using the ENVIRONMENTALSTATS for S-PLUS pull-down menu, follow these steps.

1. In the Object Explorer, find and highlight the data frame **epa.92c.arsenic1.df**.
2. On the S-PLUS menu bar, make the following menu choices: **EnvironmentalStats>Hypothesis Tests>Compare Samples>k Samples>Homogeneous Variances**. This will bring up the Test for Homogeneous Variances dialog box.
3. Under Data to Use, select **Pre-Defined Data**. In the Data Set box **epa.92c.arsenic1.df** should be selected. For Response Variable select **Arsenic**. For Grouping Variable select **Well**. For Test select **Bartlett**, and make sure the Correct box is **checked**. Click **OK** or **Apply**.

This will perform the test on the untransformed observations. To perform the test on the log-transformed observations, modify Step 3 above as follows.

3. Under Data to Use, select **Expression**. In the Response Variable box type **log(epa.92c.arsenic1.df$Arsenic)**. In the Grouping Variable box type **epa.92c.arsenic1.df$Well**. For Test select **Bartlett**, and make sure the Correct box is **checked**. Click **OK** or **Apply**.

Command

To perform Bartlett's test on the untransformed and log-transformed observations using the ENVIRONMENTALSTATS for S-PLUS Command or Script Window, type these commands.

```
attach(epa.92c.arsenic1.df)
var.group.test(Arsenic, groups=Well, test="Bartlett")
var.group.test(log(Arsenic), groups=Well,
  test="Bartlett")
detach()
```

Levene's Test

Levene's test of the null hypothesis (7.128) is based on computing absolute deviations from the group mean within each group, then comparing these absolute deviations via a standard one-way analysis of variance (Milliken and Johnson, 1992, p. 19; Snedecor and Cochran, 1989, p. 252;

USEPA, 1992c, p. 23). The within-group absolute deviations are computed as:

$$z_{ij} = \left| x_{ij} - \bar{x}_i \right|,$$

(7.133)

$$i = 1, 2, \ldots, k; \quad j = 1, 2, \ldots, n_i$$

Levene's test is almost as powerful as Bartlett's test when the underlying distributions are normal, and unlike Bartlett's test it tends to maintain the assumed α-level when the underlying distributions are not normal (Snedecor and Cochran, 1989, p. 252; Milliken and Johnson, 1992, p. 22; Conover et al., 1981). Thus, Levene's test is generally recommended over Bartlett's test.

Example 7.25: Testing Homogeneity of Variance in Arsenic Concentrations between Wells Using Levene's Test

We will repeat Example 7.24, but we will use Levene's test instead of Bartlett's test. For the untransformed observations, Levene's test yields an F-statistic of 4.6 with 5 and 18 degrees of freedom and an associated p-value of 0.007. For the log-transformed data, Levene's test yields an F-statistic of 1.2 with 5 and 18 degrees of freedom and an associated p-value of 0.36. Thus, we conclude that for the untransformed observations, the population variances are in fact different between the wells. If we want to make inferences based on the log-transformed observations, then we do not have enough evidence to reject the null hypothesis that the population variances (for the log-transformed observations) are the same between wells.

Menu

To perform Levene's test on the untransformed and log-transformed observations using the ENVIRONMENTALSTATS for S-PLUS pull-down menu, follow the same steps as above for performing Bartlett's test, except in Step 3 for Test select **Levene**.

Command

To perform Levene's test on the untransformed and log-transformed observations using the ENVIRONMENTALSTATS for S-PLUS Command or Script Window, use the same commands as above for performing Bartlett's test, except set the `test` argument to `var.group.test` to `test="Levene"`.

SUMMARY

- In Step 5 of the DQO process (see Chapter 2), you usually link the principal study question you defined in Step 2 with some population parameter such as the mean, median, 95th percentile, etc.

- A *hypothesis test* or *significance test* is a formal mathematical mechanism for objectively making a decision in the face of uncertainty, and is usually used to answer a question about the value of a population parameter.

- Hypotheses tests are usually based on a *test statistic*, say T, which is computed from a random sample from the population. A test statistic is a random quantity

- Because you make your decision based on the value of a random quantity, you will sometimes make the "wrong" choice. Statisticians call the two kinds of mistakes you can make a *Type I error* (rejecting the null hypothesis when it is true) and a *Type II error* (not rejecting the null hypothesis when it is false). See Table 7.2.

- The probability of making a Type I error is usually denoted with the Greek letter α (alpha) and is called the *significance-level* or *α-level* of the test. The probability of making a Type II error is usually denoted with the Greek letter β (beta). The probability $1-\beta$ denotes the probability of correctly deciding to reject the null hypothesis when in fact it is false. This probability is called the *power* of the hypothesis test.

- **The choice of values for α and β is a subjective, policy decision**. The possible choices for α and β are limited by the sample size of the experiment, the variability inherent in the data, and the magnitude of the difference between the null and alternative hypothesis. If you reduce the Type I error, then you increase the Type II error, and vice-versa.

- The *p-value* is the probability of seeing a test statistic as extreme or more extreme than the one you observed, assuming the null hypothesis is true. The p-value is *not* the probability that the null hypothesis is true. If the test statistic has a continuous distribution, then the p-value is uniformly distributed between 0 and 1 if the null hypothesis is true. If the null hypothesis is not true, then the p-value will tend to be "small."

- A $(1-\alpha)100\%$ confidence interval for θ consists of all possible values of θ that are associated with not rejecting the null hypothesis at significance level α. Thus, if you know how to create a confidence interval for a parameter, you can perform a hypothesis test for that parameter, and vice-versa (see Table 7.4).

- In this chapter we presented hypothesis tests concerning proportions, locations, variability, and goodness-of-fit to a distribution.

EXERCISES

7.1. Consider the arsenic concentration data of Exercise 3.2 shown in Table 3.13 (stored in the data frame `arsenic3.df` in ENVIRONMENTALSTATS for S-PLUS). Perform a Shapiro-Wilk goodness-of-fit test for normality on the data from the Background well.

7.2. We considered the S-PLUS built-in data set `halibut` in Exercise 3.4. Perform a Shapiro-Wilk goodness-of-fit test for normality on both catch-per-unit-effort (CPUE) and biomass.

7.3. Consider the nickel concentration data of Exercise 3.1 shown in Table 3.12 (stored in the data frame `epa.92c.nickel1.df` in ENVIRONMENTALSTATS for S-PLUS).

    a. Plot the observations by well.
    b. Perform a Shapiro-Wilk group test for normality and log-normality on these data, grouping the observations by well.
    c. Plot the results of the goodness-of-fit tests.

7.4. Consider the S-PLUS data set `rain.nyc1`, which contains annual total New York City precipitation (inches) from 1869 to 1957. In Exercise 3.6 we looked at several Box-Cox transformations and computed the value of the PPCC. Perform the PPCC goodness-of-fit test for normality for the original data and the transformed data, using the transformation you chose as the "best" in Exercise 3.6.

7.5. The built-in data set `quakes.bay` in S-PLUS contains the location, time, and magnitude of earthquakes in the San Francisco Bay Area from 1962 to 1981.

    a. Plot a histogram of these data.
    b. Try fitting a 3-parameter lognormal distribution to these data. Plot the results of the Shapiro-Wilk goodness-of-fit test (observed with fitted distribution, observed with fitted cdf, normal Q-Q plot, etc.).
    c. Why does the 3-parameter lognormal model not work here?

7.6. In Example 7.3 we constructed 90% and 95% lower confidence interval for the probability p of daily maximum ozone exceeding 120 ppb. Use ENVIRONMENTALSTATS for S-PLUS to construct these confidence intervals.

7.7.   As an example of the robustness of the t-test to departures from normality, compare the results of the t-test and Chen's modified t-test using the data set `rain.nyc1` to test the null hypothesis that the annual total rainfall in New York City is less than or equal to 40 inches vs. the alternative that it is greater than 40 inches. (Plot the data to get an idea of what it looks like.)

7.8.   As an example of how skewness can affect the results of the t-test, compare the results of the t-test and Chen's modified t-test using the TcCB observations from the Cleanup area to test the null hypothesis that the average TcCB concentration in the Cleanup area is less than or equal to 1 ppb vs. the alternative that it is greater than 1 ppb. (Plot the data to get an idea of what it looks like.)

7.9.   For the one-sample Wilcoxon signed rank test, tabulate the distribution of R_+ for $n = 2, 3$, and 4, assuming there are no ties.

7.10. Use the aldicarb data of Example 7.5 and Example 7.6 to show that the Wilcoxon signed rank test is really the same thing as the one-sample permutation test based on the signed ranks. That is, for each well, create a vector that contains the signed ranks of the quantities (Aldicarb-7), then perform a one-sample permutation test for the null hypothesis that the population mean of these quantities is 0.

7.11. In Example 7.13 we use the two-sample t-test to compare sulfate concentrations at a background and downgradient well. The resulting t-statistic is 5.66 with 12 degrees of freedom.

>   **a.**   Plot the pdf of a t-distribution with 12 degrees of freedom and add a vertical line at $x = 5.66$.
>   **b.**   Explain what part of this plot represents the p-value for the test of the null hypothesis that the average sulfate concentrations at the two wells are the same against the alternative hypothesis that the average concentration of sulfate at the downgradient well is larger than the average concentration at the background well.

7.12. Consider the copper concentrations shown in Table 3.14 in Exercise 3.7 and stored in the data frame `epa.92c.copper1.df` in ENVIRONMENTALSTATS for S-PLUS. Use the t-test and Wilcoxon rank sum test to compare the data from the two background wells.

7.13. Consider the Wilcoxon rank sum test to compare the locations of two groups. Show that if you know the sum of the ranks of the observations in the first group, you can figure out the difference between the average ranks for the two groups. This shows that the sum of the ranks for the first group is an equivalent statistic to the difference between the average ranks.

7.14. In Example 7.19 and Example 7.20 we compared the incidence of thyroid tumors in rats exposed and not exposed to ETU. In the control group, 2 out of 72 (3%) rats developed thyroid tumors, and in the group exposed to a dietary level of 250 ppm of ETU, 16 out of 69 rats (23%) developed thyroid tumors. Suppose in the exposed group only 6 out of 69 (9%) developed thyroid tumors. Is this difference of 3% vs. 9% statistically significant? Use both Fisher's exact test and the approximate test.

7.15. Consider the nickel concentration data of Exercise 3.1 shown in Table 3.12 (stored in the data frame `epa.92c.nickel1.df` in ENVIRONMENTALSTATS for S-PLUS). Perform a test for variance homogeneity (between wells) for both the original and log-transformed observations using both Bartlett's and Levene's test.

7.16. Using the nickel data from the previous exercise, perform a parametric and nonparametric one-way ANOVA to compare the concentrations between wells. Use both the original and log-transformed data. Do the results differ much for the original vs. log-transformed data?

8 DESIGNING A SAMPLING PROGRAM, PART II

Sample Size and Power

In Chapters 5 to 7 we explained various statistical methods for estimating parameters and performing hypothesis tests, for example, to decide whether soil at a Superfund site has been adequately cleaned up. In all of these chapters we concentrated on specifying the confidence level for a confidence interval, or specifying the Type I (false positive) error rate for a hypothesis test (e.g., the probability of declaring a site is contaminated when in fact it is clean; see Tables 7.1 and 7.2 in Chapter 7). In this chapter, we will talk about how you can determine the sample size you will need for a study. If the study involves estimating a parameter, you will need to decide how wide you are willing to have your confidence intervals. If the study involves testing a hypothesis, you will need to decide how you want to balance the Type I and Type II error rates based on detecting a specified amount of change. The tools we talk about in this chapter are used in Steps 6 and 7 of the DQO process (see Chapter 2) to specify limits on the decision error rates and in the iterative step of optimizing the design.

DESIGNS BASED ON CONFIDENCE INTERVALS

In Chapters 5 and 7 we discussed various ways to estimate parameters and create confidence intervals for the parameters. In this section, we will discuss how you can investigate the trade-off between sample size, confidence level, and the half-width of the confidence interval.

Confidence Interval for the Mean of a Normal Distribution

Equation (5.78) displays the formula for constructing a two-sided $(1-\alpha)100\%$ confidence interval for the population mean, assuming the observations come from a normal distribution. The half-width of this confidence interval (denoted h) is given by:

$$h = t_{n-1, 1-\alpha/2} \frac{s}{\sqrt{n}} \tag{8.1}$$

which depends on the sample size (n), the confidence level ($1-\alpha$), and the estimated standard deviation (s). Equation (8.1) can be rewritten to show the required sample size as a function of the half-width:

$$
n = \left(t_{n-1,1-\alpha/2} \, \frac{s}{h} \right)^2
\tag{8.2}
$$

(Zar, 1999, p. 105). Because the quantity n appears in both sides of this equation, the equation must be solved iteratively.

You can use formulas (8.1) and (8.2) to help you decide how many samples you need to take for a given precision (half-width) of the confidence interval. To apply these formulas, you will need to assume a value for the estimated standard deviation s. The value you choose for s is usually based on a pilot study, a literature search, or "expert judgment." It is usually a good idea to use some kind of upper bound on s to ensure an adequate sample size.

Example 8.1: Design for Estimating the Mean of Sulfate Concentrations

In Example 7.13 we estimated the mean and standard deviation of sulfate concentrations (ppm) at a background well as 536.3 and 26.7, respectively, based on a sample size of $n = 8$ quarterly samples. A two-sided 95% confidence interval for this mean is [514, 559], which has a half-width of 22.5 ppm. Assuming a standard deviation of about $s = 30$, if we had taken only four observations, the half-width of the confidence interval would have been about 48 ppm. Also, if we want the confidence interval to have a half-width of 10 ppm, we would need to take $n = 38$ observations. Figure 8.1 displays the half-width of the confidence interval as a function of the sample size for various confidence levels.

Menu

To compute the half-width of the confidence interval assuming a sample size of $n = 4$ using the ENVIRONMENTALSTATS for S-PLUS pull-down menu, follow these steps.

1. On the S-PLUS menu bar, make the following menu choices: **EnvironmentalStats>Sample Size and Power>Confidence Intervals>Normal Mean>Compute**. This will bring up the Normal CI Half-Width and Sample Size dialog box.
2. For Compute, select **CI Half-Width**, for Sample Type select **One-Sample**, for Confidence Level(s) (%) type **95**, for n type **4**, and for Standard Deviation(s) type **30**. Click **OK** or **Apply**.

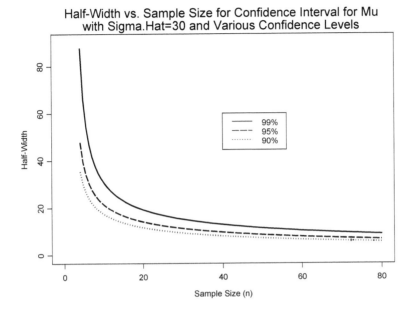

Figure 8.1   The half-width of the confidence interval for the mean of background sulfate concentrations (ppm) as a function of sample size and confidence level

To compute the required sample size for a specified half-width of 10, follow the same steps as above, except for Compute select **Sample Size**, and for Half-Width type **10**.

To create the plot shown in Figure 8.1, follow these steps.

1. On the S-PLUS menu bar, make the following menu choices: **EnvironmentalStats>Sample Size and Power>Confidence Intervals>Normal Mean>Plot**. This will bring up the Plot Normal CI Design dialog box.
2. For X Variable select **Sample Size**, for Y Variable select **Half-Width**, for Minimum X type **4**, for Maximum X type **80**, for Sample Type select **One-Sample**, for Confidence Level (%) type **99**, and for Standard Deviation type **30**.
3. Click on the **Plotting** tab. In the X-Axis Limits box type **c(0, 80)**. In the Y-Axis Limits box type **c(0, 90)**. Click **Apply**.
4. Click on the **Model** tab. For Confidence Level (%) type **95**. Click on the **Plotting** tab, **check** the Add to Current Plot box, and for Line Type type **4**. Click **Apply**.
5. Click on the **Model** tab. For Confidence Level (%) type **90**. Click on the **Plotting** tab, and for Line Type type **8**. Click **OK**.

Command

To compute the half-width of the confidence interval assuming a sample size of $n = 4$ using the ENVIRONMENTALSTATS for S-PLUS Command or Script Window, type this command.

```
ci.norm.half.width(n.or.n1=4, sigma.hat=30)
```

To compute the required sample size for a specified half-width of 10, type this command.

```
ci.norm.n(half.width=10, sigma.hat=30)
```

To create the plot shown in Figure 8.1, type these commands.

```
plot.ci.norm.design(sigma.hat=30, range.x.var=c(4,
   80), conf=0.99, xlim = c(0, 80), ylim = c(0, 90),
   main="Half-Width vs. Sample Size for Confidence
   Interval for Mu\nWith Sigma.Hat=30 and Various
   Confidence Levels")
plot.ci.norm.design(sigma.hat=30, range.x.var=c(4,
   80), conf=0.95, plot.lty=4, add=T)
plot.ci.norm.design(sigma.hat=0.5, range.x.var=c(4,
   80), conf=0.90, plot.lty=2, add=T)
legend(40, 60, c("99%", "95%", "90%"), lty=c(1, 4, 2),
   lwd=1.5)
```

Confidence Interval for the Difference between Two Means of Normal Distributions

Equation (7.81) displays the formula for constructing a two-sided $(1-\alpha)100\%$ confidence interval for the difference between two population means, assuming the observations from each group come from normal distributions. Assuming equal variances in each group, the half-width of this confidence interval is given by:

$$h = t_{n_1 + n_2 - 2, 1 - \alpha/2} \; s_p \sqrt{\frac{1}{n_1} + \frac{1}{n_2}} \qquad (8.3)$$

Assuming equal sample sizes in each group (i.e., $n_1 = n_2 = n$), Equation (8.3) can be rewritten to show the required sample size in each group as a function of the half-width:

$$n = 2\left(t_{2n-2,1-\alpha/2}\,\frac{s_p}{h}\right)^2 \qquad (8.4)$$

(Zar, 1999, p. 131). As in the one-sample case, you must supply a value for the pooled estimate of standard deviation s_p.

If the sample size for group 2 (n_2) is constrained to a certain value, the required sample size for group 1 can be computed as:

$$n_1 = \frac{n\,n_2}{2n_2 - n} \qquad (8.5)$$

(Zar, 1999, p. 131). If $2n_1-n$ is less than or equal to 0, then either n_2 must be increased, the half-width h must be increased, and/or the confidence level $(1-\alpha)100\%$ must be decreased.

Example 8.2: Design for Estimating the Difference in Sulfate Concentrations between a Background and Compliance Well

In Example 7.13 we computed the sample means and standard deviations of sulfate concentrations in ground water at a background and compliance well (sampled quarterly) based on sample sizes of 8 and 6, respectively. The background well mean and standard deviation were 536 and 26.7 ppm, and the compliance well mean and standard deviation were 608 and 18.3 ppm. The standard deviation based on the pooled estimate of variance is 23.6 ppm. The 95% confidence interval for the difference between the two means (compliance – background) is [44.3, 99.8] which has a half-width of 27.8 ppm. Assuming a standard deviation of about 25 ppm, if we had taken 10 observations at each well, the half-width would have been about 23.5 ppm. Also, if we want the confidence interval to have a half-width of 10 ppm, we would need to take $n_1 = n_2 = 50$ observations (12.5 years of quarterly sampling) at each well.

Menu

To compute the half-width of the confidence interval assuming a sample size of $n_1 = n_2 = 10$ at each well using the ENVIRONMENTALSTATS for S-PLUS pull-down menu, follow these steps.

1. On the S-PLUS menu bar, make the following menu choices: **EnvironmentalStats>Sample Size and Power>Confidence Intervals>Normal Mean>Compute**. This will bring up the Normal CI Half-Width and Sample Size dialog box.

2. For Compute, select **CI Half-Width**, for Sample Type select **Two-Sample**, for Confidence Level(s) (%) type **95**, for n1 type **10**, for n2 type **10**, and for Standard Deviation(s) type **25**.   Click **OK** or **Apply**.

To compute the required sample size for a specified half-width of 10 ppm, follow the same steps as above, except for Compute select **Sample Size**, and for Half-Width type **10**.

Command

To compute the half-width of the confidence interval assuming a sample size of $n_1 = n_2 = 10$ at each well using the ENVIRONMENTALSTATS for S-PLUS Command or Script Window, type this command.

```
ci.norm.half.width(n.or.n1=10, n2=10, sigma.hat=25)
```

To compute the required sample size for a specified half-width of 10 ppm, type this command.

```
ci.norm.n(half.width=10, sigma.hat=25,
   sample.type="two.sample")
```

Confidence Interval for a Binomial Proportion

Equations (5.92) to (5.98) display formulas used to construct a two-sided $(1-\alpha)100\%$ confidence interval for a binomial proportion. Equation (5.101) displays the formula based on the normal approximation, and Equation (5.103) displays the formula based on the normal approximation and continuity correction. Using any three of these methods, you can compute the half-width of the confidence interval for a given sample size, or compute the required sample size given a specific half-width, as long as you specify an assumed value for the estimated proportion. For example, using the normal approximation with the continuity correction (Equation (5.103)), the formula for the half-width (h) given the sample size (n) is:

$$h = z_{1-\alpha/2}\sqrt{\frac{\hat{p}\hat{q}}{n}} + \frac{1}{2n} \tag{8.6}$$

and the formula for the sample size given a specific half-width is:

$$n = \left[\sqrt{z_{1-\alpha/2}^2\,\hat{p}\hat{q} + 2h} - z_{1-\alpha/2}\sqrt{\hat{p}\hat{q}}\right]^{-2} \tag{8.7}$$

where

$$\hat{q} = 1 - \hat{p} \qquad\qquad (8.8)$$

Example 8.3: Design for Estimating the Probability of a Nondetect

In Example 5.22 we computed a confidence interval for the probability of observing a nondetect value for a benzene concentration in groundwater. Of $n = 36$ past observations, 33 of these (about 92%) were nondetects, and the two-sided 95% confidence interval for the binomial proportion is [0.78, 0.98]. The half-width of this interval is 0.1, or 10 percentage points. Assuming an estimated proportion of 90%, if we had taken only $n = 10$ observations, the half-width of the confidence interval would have been about 0.22 (22 percentage points). Also, if we want the confidence interval to have a half-width of 0.02 (two percentage points), we would need to take $n = 914$ observations. Figure 8.2 displays the half-width of the confidence interval as a function of the sample size for various confidence levels.

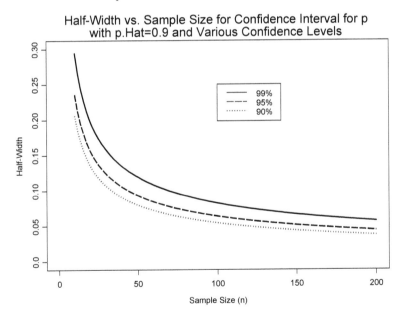

Figure 8.2 The half-width of the confidence interval for the probability of a nondetect as a function of sample size and confidence level, assuming an estimated nondetect proportion of about 90%

Menu

To compute the half-width of the confidence interval assuming a sample size of $n = 10$ using the ENVIRONMENTALSTATS for S-PLUS pull-down menu, follow these steps.

1. On the S-PLUS menu bar, make the following menu choices: **EnvironmentalStats>Sample Size and Power>Confidence Intervals>Binomial Proportion>Compute**. This will bring up the Binomial CI Half-Width and Sample Size dialog box.
2. For Compute, select **CI Half-Width**, for Sample Type select **One-Sample**, for Confidence Level(s) (%) type **95**, for n type **10**, and for Proportion type **0.9**. Click **OK** or **Apply**.

To compute the required sample size for a specified half-width of 0.02, follow the same steps as above, except for Compute select **Sample Size**, and for Half-Width type **0.02**.

To create the plot shown in Figure 8.2, follow these steps.

1. On the S-PLUS menu bar, make the following menu choices: **EnvironmentalStats>Sample Size and Power>Confidence Intervals>Binomial Proportion>Plot**. This will bring up the Plot Binomial CI Design dialog box.
2. For X Variable select **Sample Size**, for Y Variable select **Half-Width**, for Minimum X type **10**, for Maximum X type **200**, for Sample Type select **One-Sample**, for Confidence Level (%) type **99**, and for Proportion type **0.9**.
3. Click on the **Plotting** tab. In the X-Axis Limits box type c(0, 200). In the Y-Axis Limits box type **c(0, 0.3)**. Click **Apply**.
4. Click on the **Model** tab. For Confidence Level (%) type **95**. Click on the **Plotting** tab, **check** the Add to Current Plot box, and for Line Type type **4**. Click **Apply**.
5. Click on the **Model** tab. For Confidence Level (%) type **90**. Click on the **Plotting** tab, and for Line Type type **8**. Click **OK**.

Command

To compute the half-width of the confidence interval assuming a sample size of $n = 10$ using the ENVIRONMENTALSTATS for S-PLUS Command or Script Window, type this command.

```
ci.binom.half.width(n.or.n1=10, p.hat=0.9, approx=F)
```

To compute the required sample size for a specified half-width of 0.02, type this command.

```
ci.binom.n(half.width=0.02, p.hat=0.9)
```

To create the plot shown in Figure 8.2, type these commands.

```
plot.ci.binom.design(p.hat=0.9, range.x.var=c(10,
   200), conf=0.99, xlim=c(0, 200), ylim=c(0, 0.3),
   main="Half-Width vs. Sample Size for Confidence
   Interval for p\nWith p.Hat=0.9 and Various
   Confidence Levels")
plot.ci.binom.design(p.hat=0.9, range.x.var=c(10,
   200), conf=0.95, plot.lty=4, add=T)
plot.ci.binom.design(p.hat=0.9, range.x.var=c(10,
   200), conf=0.90, plot.lty=2, add=T)
legend(100, 0.25, c("99%", "95%", "90%"), lty=c(1, 4,
   2), lwd=1.5)
```

Confidence Interval for the Difference between Two Binomial Proportions

For the two sample case, Equation (7.105) displays the formula used to construct a two-sided $(1-\alpha)100\%$ confidence interval for the difference between two binomial proportions. You can compute the half-width of the confidence interval for given sample sizes, or compute the required sample size given a specific half-width, as long as you specify assumed values for both of the estimated proportions. For example, the formula for the half-width (h) given the sample sizes (n_1 and n_2) is:

$$h = z_{1-\alpha/2}\sqrt{\frac{\hat{p}_1\hat{q}_1}{n_1} + \frac{\hat{p}_2\hat{q}_2}{n_2} + \frac{1}{2}\left(\frac{1}{n_1} + \frac{1}{n_2}\right)} \qquad (8.9)$$

and the formula for the sample size for group 1 (n_1) given a specific half-width and ratio $r = n_2/n_1$ is:

$$n_1 = \left(1 + \frac{1}{r}\right)^2 \left\{\sqrt{z_{1-\alpha/2}^2\left(\hat{p}_1\hat{q}_1 + \frac{\hat{p}_2\hat{q}_2}{r}\right) + 2\left(1 + \frac{1}{r}\right)h}\right.$$

$$\left. - z_{1-\alpha/2}\sqrt{\hat{p}_1\hat{q}_1 + \frac{\hat{p}_2\hat{q}_2}{r}}\right\}^{-2} \qquad (8.10)$$

Example 8.4: Design for Estimating the Difference between Incidence Rates of Thyroid Tumors in Rats

In Examples 7.19 and 7.20 we compared the incidence of thyroid tumors in rats exposed and not exposed to ethylene thiourea (ETU). In the exposed group, 16 out of 69 rats (23%) developed thyroid tumors, whereas in the un-exposed group only 2 out of 72 (3%) developed thyroid tumors.

The two-sided 95% confidence interval for the difference between the incidence rates (exposed – unexposed) is [0.08, 0.32], which has a half-width of 0.12, or 12 percentage points. Assuming estimated incidence rates of about 20% and 5%, if we had used only $n_1 = n_2 = 20$ rats in each group, the half-width of the confidence interval would have been about 0.25 (25 percentage points). Also, if we want the confidence interval to have a half-width of 0.02 (two percentage points), we would need to use $n_1 = n_2 = 2,092$ rats in each group.

Menu

To compute the half-width of the confidence interval assuming a sample size of $n_1 = n_2 = 20$ rats in each group using the ENVIRONMENTALSTATS for S-PLUS pull-down menu, follow these steps.

1. On the S-PLUS menu bar, make the following menu choices: **EnvironmentalStats>Sample Size and Power>Confidence Intervals>Binomial Proportion>Compute**. This will bring up the Binomial CI Half-Width and Sample Size dialog box.
2. For Compute, select **CI Half-Width**, for Sample Type select **Two-Sample**, for Confidence Level(s) (%) type **95**, for n1 type **20**, for n2 type **20**, for Proportion 1 type **0.2**, and for Proportion 2 type **0.05**. Click **OK** or **Apply**.

To compute the required sample size for a specified half-width of 0.02, follow the same steps as above, except for Compute select **Sample Size**, and for Half-Width type **0.02**.

Command

To compute the half-width of the confidence interval assuming a sample size of $n_1 = n_2 = 20$ rats in each group using the ENVIRONMENTALSTATS for S-PLUS Command or Script Window, type this command.

```
ci.binom.half.width(n.or.n1=20, p.hat.or.p1.hat=0.2,
    n2=20, p2.hat=0.05)
```

To compute the required sample sizes for a specified half-width of 0.02, type this command.

```
ci.binom.n(half.width=0.02, p.hat.or.p1.hat=0.2,
   p2.hat=0.05)
```

DESIGNS BASED ON NONPARAMETRIC CONFIDENCE, PREDICTION, AND TOLERANCE INTERVALS

In Chapter 5 we discussed nonparametric confidence intervals, and in Chapter 6 we discussed nonparametric prediction and tolerance intervals. In this section we will discuss how you can investigate the trade-off between sample size and confidence levels. In general, there is a heavy price to pay for not making an assumption about the parent distribution of the observations: required sample sizes for a specified confidence level are usually very large!

Nonparametric Confidence Interval for a Quantile

In Chapter 5 we discussed how to construct nonparametric confidence intervals for a quantile (percentile) based on using order statistics. Equations (5.154) to (5.160) show the formulas for the two-sided and one-sided confidence intervals and their associated confidence levels. Based on these formulas, you can compute the required sample size for a specific confidence level, or the associated confidence level for a specific sample size. Table 5.18 displays sample sizes and associated confidence levels for a one-sided upper nonparametric confidence interval for the 95th percentile based on using the maximum value (the largest order statistic) as the upper confidence limit.

Example 8.5: Design for an Upper Bound on the 95th Percentile

In Example 5.27 we computed a nonparametric upper confidence bound for the 95th percentile of copper concentration in groundwater at three background monitoring wells. Out of $n = 24$ observations, the largest one was 9.2 ppb. The confidence level associated with the one-sided upper confidence interval [0, 9.2] was only 71%. If we had taken only $n = 12$ observations (four at each well), the associated confidence level would have been 46%. As we noted in Example 6.10, if we had wanted a confidence level of at least 95% we would have had to have taken $n = 59$ observations (about 20 at each well).

Menu

To compute the confidence level of the one-sided upper confidence interval for the 95th percentile assuming a sample size of $n = 12$ using the ENVIRONMENTALSTATS for S-PLUS pull-down menu, follow these steps.

1. On the S-PLUS menu bar, make the following menu choices: **EnvironmentalStats>Sample Size and Power>Confidence Intervals>Nonparametric>Compute**. This will bring up the Nonparametric CI Confidence Level and Sample Size dialog box.
2. For Compute select **Confidence Level**, and for n type **12**. Click **OK** or **Apply**.

To compute the required sample sizes for a specified confidence level of 95%, follow the same steps as above, except for Compute select **Sample Size**, and for Confidence Level (%) type **95**.

Command

To compute the confidence level of the one-sided upper confidence interval for the 95th percentile assuming a sample size of $n = 12$ using the ENVIRONMENTALSTATS for S-PLUS Command or Script Window, type this command.

```
ci.npar.conf.level(n=12, p=0.95, ci.type="upper")
```

To compute the required sample size for a specified confidence level of 95%, type this command.

```
ci.npar.n(p=0.95, ci.type="upper")
```

Nonparametric Prediction Intervals

In Chapter 6 we discussed how to construct nonparametric prediction intervals based on using order statistics. Equations (6.42) to (6.48) show the formulas for the two-sided and one-sided prediction intervals for the next k out of m observations. Equations (6.49) to (6.53) show the formulas for the confidence level associated with the nonparametric prediction interval. These formulas depend on k, m, which order statistics are used (usually the smallest and/or largest values are used), and the sample size (n). Based on these formulas, you can compute the required sample size for a specific confidence level, or the associated confidence level for a specific sample size. Table 8.1 shows the required sample size for a two-sided prediction interval for the next m future observations ($k = m$) for various values of m and required confidence levels, assuming we are using the minimum and maximum values as the prediction limits.

Example 8.6: Design to Detect TcCB Contamination at the Cleanup Site

In Example 6.7 we compared using a tolerance interval and a prediction interval to determine whether there is evidence of residual TcCB contamination in the Cleanup area based on using the Reference area to calculate

Confidence Level (%)	# Future Observations (m)	Required Sample Size(n)
90	1	19
	5	93
	10	186
95	1	39
	5	193
	10	386

Table 8.1   Required sample sizes for a two-sided nonparametric prediction interval for the next m observations

"background." In that example, we assumed the Reference area data come from a lognormal distribution. We also noted that the tolerance interval approach should never be used. Here we will not make any assumption about the distribution of TcCB in the Reference area and we will determine how many observations we need to take in the Reference area to create a one-sided upper prediction limit to encompass the next $k = m = 77$ observations (which we will take from the Cleanup area).

If we want an associated confidence level of 95%, we will need to take n = 1,463 observations! Recall that if we are willing to make an assumption about the distribution of TcCB in the Reference area (e.g., lognormal), then there is no requirement on sample size in order to construct a 95% level prediction interval for the next 77 observations (although an increasing sample size will decrease the magnitude of the upper prediction limit).

Menu

To compute the required sample size to construct a one-sided upper nonparametric prediction limit for the next $k = m = 77$ observations with a specified confidence level of 95% using the ENVIRONMENTALSTATS for S-PLUS pull-down menu, follow these steps.

1. On the S-PLUS menu bar, make the following menu choices: **EnvironmentalStats>Sample Size and Power>Prediction Intervals>Nonparametric>Compute**. This will bring up the Nonparametric PI Confidence Level and Sample Size dialog box.
2. For Compute select **Sample Size**, for Confidence Level (%) type **95**, for # Future Obs type **77**, for Min # Obs PI Should Contain type **77**. Click **OK** or **Apply**.

Command

To compute the required sample size to construct a one-sided upper nonparametric prediction limit for the next $k = m = 77$ observations with a specified confidence level of 95% using the ENVIRONMENTALSTATS for S-PLUS Command or Script Window, type this command.

```
pred.int.npar.n(m=77, pi.type="upper")
```

Nonparametric Tolerance Intervals

In Chapter 6 we discussed how to construct nonparametric tolerance intervals based on using order statistics. The form of a nonparametric tolerance interval is exactly the same as for a nonparametric prediction interval (see Equations (6.42) to (6.48)). For a β-content tolerance interval, Equation (6.75) shows the confidence level for a specified coverage, and Equation (6.76) show the coverage for a specified confidence level. These formulas depend on which order statistics are used (usually the smallest and/or largest values are used), and the sample size (n). Based on these formulas, you can compute the required sample size for a specific confidence level and coverage, the associated confidence level for a specific sample size and coverage, or the associated coverage for a specific sample size and confidence level. Table 8.2 shows the required sample size for a two-sided tolerance interval for various required confidence levels and coverages, assuming we are using the minimum and maximum values as the prediction limits.

Confidence Level (%)	Coverage (%)	Required Sample Size(n)
90	80	18
	90	38
	95	77
95	80	22
	90	46
	95	93

Table 8.2   Required sample size for a two-sided nonparametric tolerance interval

Example 8.7: Design for an Upper Bound on the 95th Percentile

Example 8.7: Design for an Upper Bound on the 95th Percentile

In Example 8.5 we discussed required sample sizes for a one-sided upper nonparametric confidence interval for the 95th percentile. We noted that to achieve a confidence level of at least 95% would require $n = 59$ observations. Based on the relationship between confidence intervals for percentiles and tolerance intervals, this is the same thing as saying that to construct a one-sided upper tolerance interval with at least 95% coverage and a 95% confidence level requires a sample size of $n = 59$.

Menu

To compute the required sample size for a specified confidence level of 95% and coverage of 95% using the ENVIRONMENTALSTATS for S-PLUS pull-down menu, follow these steps.

1. On the S-PLUS menu bar, make the following menu choices: **EnvironmentalStats>Sample Size and Power>Tolerance Intervals>Nonparametric>Compute**. This will bring up the Nonparametric TI Confidence Level and Sample Size dialog box.
2. For Compute select **Sample Size**, for Coverage (%) type **95**, and for Confidence Level (%) type **95**. Click **OK** or **Apply**.

Command

To compute the required sample size for a specified confidence level of 95% and coverage of 95% using the ENVIRONMENTALSTATS for S-PLUS Command or Script Window, type this command.

```
tol.int.npar.n(coverage=0.95, ti.type="upper",
   conf.level=0.95)
```

DESIGNS BASED ON HYPOTHESIS TESTS

In Chapter 7 we discussed various ways to perform hypothesis tests on population parameters (e.g., means, medians, proportions, etc.). We also discussed the two kinds of mistakes you can make when you perform a hypothesis test: Type I and Type II errors. In Chapters 6 and 7 we concentrated on maintaining the Type I error rate (significance level) at a pre-specified value. In this section, we will discuss how you can investigate the trade-off between sample size, Type I error, Type II error or power, and amount of deviation from the null hypothesis.

Testing the Mean of a Normal Distribution

In Equation (7.29) we introduced Student's t-statistic to test the null hypothesis $H_0: \mu = \mu_0$ assuming the observations come from a normal (Gaussian) distribution:

$$t = \frac{\bar{x} - \mu_0}{s/\sqrt{n}} \tag{8.11}$$

When the null hypothesis is true, the sample mean $\bar{x}$ is bouncing around the true mean μ_0, so the t-statistic is bouncing around 0 (it follows a t-distribution with $n-1$ degrees of freedom). Figure 7.9 displays the distribution of this statistic for a sample size of $n = 10$ when the null hypothesis is true.

Now suppose the null hypothesis is false and that the true mean μ is some value different from μ_0. We can rewrite the t-statistic as follows:

$$t = \frac{\bar{x} - \mu_0}{s/\sqrt{n}} = \frac{\bar{x} - \mu}{s/\sqrt{n}} + \frac{\mu - \mu_0}{s/\sqrt{n}}$$

$$= \frac{\bar{x} - \mu}{s/\sqrt{n}} + \sqrt{n} \, \frac{\delta}{s}$$

(8.12)

where

$$\delta = \mu - \mu_0 \qquad\qquad (8.13)$$

denotes the amount of deviation from the null hypothesis. Looking at the second line of Equation (8.12), we see that under the alternative hypothesis Student's t-statistic is the sum of a random variable that follows Student's t-distribution (with $n-1$ degrees of freedom) and a second random variable (because s is a random variable) that is bouncing around some positive or negative value, depending on whether δ is positive or negative.

As a concrete example, suppose we are interested in testing the null hypothesis H_0: $\mu = 20$ against the one-sided alternative hypothesis H_0: $\mu > 20$ using a significance level of 5%. Our test statistic would be computed as:

$$t = \frac{\bar{x} - 20}{s/\sqrt{10}} \qquad\qquad (8.14)$$

and we would reject the null hypothesis if $t > t_{9, .95} = 1.83$. Figure 8.3 shows the distribution of the observations under the null hypothesis $\mu = 20$ and the particular alternative hypotheses $\mu = 21$ assuming a standard deviation of $\sigma = 1$. Figure 8.4 shows the distribution of the t-statistic under the null and alternative hypotheses, along with the critical value 1.83. When the null hypothesis is true, the area under the distribution curve of the t-statistic to the right of the critical value of 1.83 (cross-hatched area) is the *Type I error rate*, which we set equal to 5%. When the alternative hypothesis is true and $\mu = 21$, the area under the distribution curve of the t-statistic to the right of the critical value of 1.83 (single-hatched and cross-hatched area) is the *power* of the test, which in this case is about 90%.

In general, when the null hypothesis is false, Student's t-statistic follows a *non-central t-distribution* with non-centrality parameter given by:

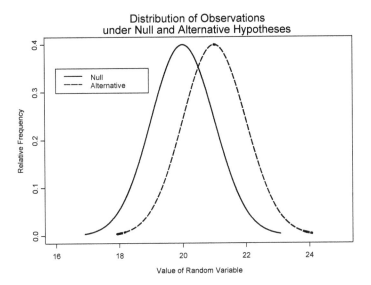

Figure 8.3   The distribution of the original observations under the null ($\mu = 20$) and alternative ($\mu = 21$) hypotheses, assuming a population standard deviation of $\sigma = 1$

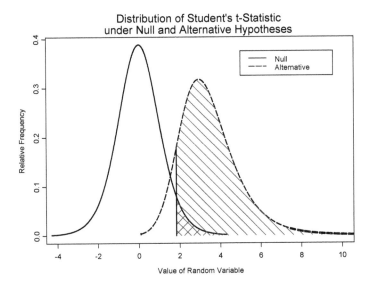

Figure 8.4   The distribution of Student's t-statistic under the null ($\mu = 20$) and alternative ($\mu = 21$) hypotheses, assuming a population standard deviation of $\sigma = 1$

$$\Delta \;=\; \sqrt{n}\; \frac{\left(\mu \,-\, \mu_0\right)}{\sigma} \;=\; \sqrt{n}\; \frac{\delta}{\sigma} \tag{8.15}$$

(Johnson et al., 1995, pp. 508–510). For the one-sided upper alternative hypothesis H_0: $\mu > \mu_0$, the power of this test is given by:

$$Power \;=\; Pr\left[t \;\geq\; t_{n-1,1-\alpha}\right]$$

$$\tag{8.16}$$

$$= \; 1 \,-\, G\left(t_{n-1,1-\alpha},\, n \,-\, 1,\, \Delta\right)$$

where $G(x,\; v,\; \Delta)$ denotes the cumulative distribution function of the non-central t-distribution with v degrees of freedom and non-centrality parameter Δ evaluated at x. For the one-sided lower alternative hypothesis H_0: $\mu < \mu_0$, the power of this test is given by:

$$Power \;=\; Pr\left[t \;\leq\; t_{n-1,\alpha}\right]$$

$$\tag{8.17}$$

$$= \; G\left(t_{n-1,\alpha},\, n \,-\, 1,\, \Delta\right)$$

Finally, for the two-sided alternative H_0: $\mu \neq \mu_0$, the power of this test is given by:

$$Power \;=\; Pr\left[t \;\leq\; t_{n-1,\alpha/2}\right] + Pr\left[t \;\geq\; t_{n-1,1-\alpha/2}\right]$$

$$= \; G\left(t_{n-1,\alpha/2},\, n \,-\, 1,\, \Delta\right) + \tag{8.18}$$

$$1 \,-\, G\left(t_{n-1,1-\alpha/2},\, n \,-\, 1,\, \Delta\right)$$

Looking at Equations (8.15) to (8.18), we see that the power of the one-sample t-test depends on the sample size (n), the Type I error rate (α), and the relative deviation from the null hypothesis (δ/σ). Often, δ is called the *minimal detectable difference (MDD)*, and δ/σ is called the **scaled minimal detectable difference**.

Example 8.8: Design for Testing Whether the Mean Aldicarb Concentration is above the MCL

In Example 7.4 we tested whether the mean aldicarb concentration in groundwater at three different monitoring wells was above a Maximum Contaminant Level (MCL) of 7 ppb. The estimated standard deviations at each of the three wells were 4.9, 2.3, and 2.1 ppb, respectively. Assuming a population standard deviation of about 5 ppb, for $n = 4$ observations we have a power of only 14% to detect an average aldicarb concentration that is 5 ppb above the MCL (i.e., a scaled MDD of 1) if we set the significance level to 1%. Alternatively, we would need a sample size of at least $n = 16$ to detect an average aldicarb concentration that is 5 ppb above the MCL if we set the significance level to 1% and desire a power of 90%. Figure 8.5 plots power as a function of sample size for a significance level of 1%, assuming a scaled minimal detectable difference of 1. Figure 8.6 plots the scaled minimal detectable difference as a function of sample size for a significance level of 1%, assuming a power of 90%.

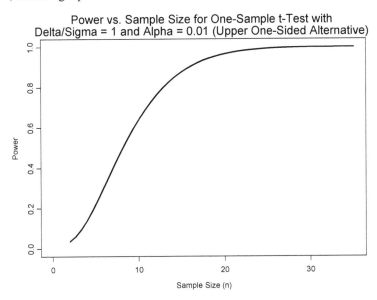

Figure 8.5   Power vs. sample size for a significance level of 1%, assuming a scaled minimal detectable difference of $\delta/\sigma = 1$

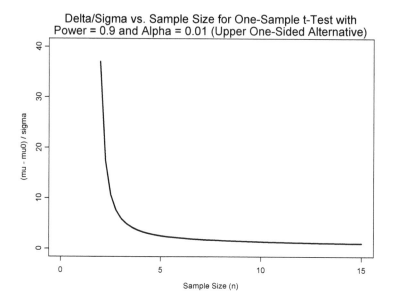

Figure 8.6   Scaled minimal detectable difference vs. sample size for a significance level of 1%, assuming a power of 90%

Menu

To compute the power to detect a scaled minimal detectable difference of 1 assuming a sample size of $n = 4$ and a significance level of 1% using the ENVIRONMENTALSTATS for S-PLUS pull-down menu, follow these steps.

1. On the S-PLUS menu bar, make the following menu choices: **EnvironmentalStats>Sample Size and Power>Hypothesis Tests>Normal Mean>One- and Two-Samples>Compute**. This will bring up the Normal Power and Sample Size dialog box.
2. For Compute select **Power**, for Sample Type select **One-Sample**, for Alpha(s) (%) type **1**, for n type **4**, for Standard Deviation(s) type **5**, for Null Mean type **7**, for Alternative Mean type **12**, and for Alternative select **"greater"**. Click **OK** or **Apply**.

To compute the required sample size to detect a scaled minimal detectable difference of 1 assuming a significance level of 1% and a required power of 90%, repeat Step 1 above, and replace Step 2 with the following:

1. For Compute select **Sample Size**, for Sample Type select **One-Sample**, for Alpha(s) (%) type **1**, for Power(s) (%) type **90**, for Standard Deviation(s) type **5**, for Null Mean type **7**, for Alternative

Mean type **12**, and for Alternative select **"greater"**. Click **OK** or **Apply**.

To produce the plot of power vs. sample size shown in Figure 8.5 follow these steps.

1. On the S-PLUS menu bar, make the following menu choices: **EnvironmentalStats>Sample Size and Power>Hypothesis Tests>Normal Mean>One- and Two-Samples>Plot**. This will bring up the Plot T-Test Design dialog box.
2. For X Variable select **Sample Size**, for Y Variable select **Power**, for Minimum X type **2**, for Maximum X type **35**, for Sample Type select **One-Sample**, **check** the Use Exact Algorithm box, for Alpha (%) type **1**, for Standard Deviation type **5**, for Null Mean type **7**, for Alternative Mean type **12**, and for Alternative select **"greater"**.
3. Click on the **Plotting** tab. For X-Axis Limits type **c(0,35)**, and for Y-Axis Limits type **c(0,1)**. Click **OK** or **Apply**.

To produce the plot of the scaled minimal detectable difference vs. sample size shown in Figure 8.6, repeat Step 1 above and replace Steps 2 and 3 with the following:

2. For X Variable select **Sample Size**, for Y Variable select **Delta/Sigma**, for Minimum X type **2**, for Maximum X type **15**, for Sample Type select **One-Sample**, **check** the Use Exact Algorithm box, for Alpha (%) type **1**, for Power (%) type **90**, and for Alternative select **"greater"**.
3. Click on the **Plotting** tab. For X-Axis Limits type **c(0,15)**, and for Y-Axis Limits type **c(0,40)**. Click **OK** or **Apply**.

Command

To compute the power to detect a scaled minimal detectable difference of 1 assuming a sample size of $n = 4$ and a significance level of 1% using the ENVIRONMENTALSTATS for S-PLUS Command or Script Window, type this command.

```
t.test.power(n.or.n1=4, delta.over.sigma=1,
   alpha=0.01, alternative="greater", approx=F)
```

To compute the required sample size to detect a scaled minimal detectable difference of 1 assuming a significance level of 1% and a required power of 90%, type this command.

```
t.test.n(delta.over.sigma=1, alpha=0.01, power=0.9,
   alternative="greater", approx=F)
```

To produce the plot of power vs. sample size shown in Figure 8.5 type this command.

```
plot.t.test.design(alpha=0.01, delta.over.sigma=1,
    range.x.var=c(2, 35), xlim=c(0, 35), ylim=c(0, 1),
    alternative="greater", approx=F)
```

To produce the plot of the scaled minimal detectable difference vs. sample size shown in Figure 8.6 type this command.

```
plot.t.test.design(y.var="delta.over.sigma",
    alpha=0.01, power=0.9, range.x.var=c(2, 15),
    xlim=c(0, 15), ylim=c(0, 40),
    alternative="greater", approx=F)
```

Testing the Difference between Two Means of Normal Distributions

In Equations (7.67) to (7.72) we introduced Student's two-sample t-statistic to test the null hypothesis H_0: $\mu_1-\mu_2 = \delta = \delta_0$ assuming the observations in groups 1 and 2 both come from normal distributions with the same standard deviation σ and possibly different means. Usually, $\delta_0 = 0$, and we will assume that is the case here. When the null hypothesis is true, the difference in the sample means $\bar{x}_1 - \bar{x}_2$ is bouncing around the difference between the population means δ_0 so the t-statistic is bouncing around 0 (it follows a t-distribution with n_1+n_2-2 degrees of freedom). Now suppose the null hypothesis is false and that the difference between the population means δ is some value different from δ_0. In this case, the Student's two-sample t-statistic follows a non-central t-distribution with non-centrality parameter given by:

$$\Delta = \sqrt{\frac{n_1 \, n_2}{n_1 + n_2}} \left(\frac{\delta}{\sigma} \right) \tag{8.19}$$

(Johnson et al., 1995, pp. 508–510). The formulas for the power of the two-sample t-test for the one-sided upper, one-sided lower, and two-sided alternative hypotheses are the same as those for the one-sample t-test shown in Equations (8.16) to (8.18), except the formula for Δ is given in Equation (8.19) rather than Equation (8.15), and the degrees of freedom are n_1+n_2-2 instead of $n-1$.

Example 8.9: Design for Comparing Sulfate Concentrations between a Background and Compliance Well

In Example 7.13 we compared the average sulfate concentration in ground water from a background and downgradient well based on $n_1 = 6$ quarterly observations at the downgradient well and $n_2 = 8$ quarterly observations at the background well. The estimated means were 608.3 and 536.3 ppm, yielding a difference of 72.1 ppm. The estimated standard deviations were 18.3 and 26.7 ppm, yielding a pooled standard deviation estimate of 23.6 ppm.

Assuming a population standard deviation of about 25 ppm for both the background and downgradient wells, if we took only four observations at each well, we would have a power of only 27% to detect an average sulfate concentration at the downgradient well that is 12.5 ppm above the average concentration at the background well (i.e., a scaled MDD of 0.5) if we set the significance level to 10%. Alternatively, we would need sample sizes of at least $n_1 = n_2 = 69$ (over 17 years of quarterly monitoring) to detect an increase in sulfate concentration at the downgradient well that is 12.5 ppb above the average at the background well if we set the significance level to 10% and desire a power of 95%. Finally, if we specify a significance level of 10%, a power of 95%, and sample sizes of $n_1 = n_2 = 16$ (4 years of quarterly monitoring), then the smallest increase in sulfate concentration we can detect at the downgradient well is 26.2 ppm (a scaled MDD of 1.05).

Menu

To compute the power to detect a scaled minimal detectable difference of 0.5 assuming a sample size of $n_1 = n_2 = 4$ and a significance level of 10% using the ENVIRONMENTALSTATS for S-PLUS pull-down menu, follow these steps.

1. On the S-PLUS menu bar, make the following menu choices: **EnvironmentalStats>Sample Size and Power>Hypothesis Tests>Normal Mean>One- and Two-Samples>Compute**. This will bring up the Normal Power and Sample Size dialog box.

2. For Compute select **Power**, for Sample Type select **Two-Sample**, for Alpha(s) (%) type **10**, for n1 type **4**, for n2 type **4**, for Standard Deviation(s) type **25**, for Mean 2 type **535**, for Mean 1 type **547.5**, and for Alternative select **"greater"**. Click **OK** or **Apply**.

To compute the required sample sizes to detect a scaled minimal detectable difference of 0.5 assuming a significance level of 10% and a required power of 95%, repeat Step 1 above, and replace Step 2 with the following:

2. For Compute select **Sample Size**, for Sample Type select **Two-Sample**, for Alpha(s) (%) type **10**, for Power(s) (%) type **95**, for

Standard Deviation(s) type **25**, for Mean 2 type **535**, for Mean 1 type **547.5**, and for Alternative select **"greater"**. Click **OK** or **Apply**.

To compute the minimal detectable difference assuming a significance level of 10%, a power of 95%, and sample sizes of $n_1 = n_2 = 16$, repeat Step 1 above, and replace Step 2 with the following:

2. For Compute select **Min. Difference**, for Sample Type select **Two-Sample**, for Alpha(s) (%) type **10**, for Power(s) (%) type **95**, for n1 type **16**, for n2 type **16**, for Standard Deviation(s) type **25**, for Mean 2 type **535**, and for Alternative select **"greater"**. Click **OK** or **Apply**.

Command

To compute the power to detect a scaled minimal detectable difference of 0.5 assuming a sample size of $n_1 = n_2 = 4$ and a significance level of 10% using the ENVIRONMENTALSTATS for S-PLUS Command or Script Window, type this command.

```
t.test.power(n.or.n1=4, n2=4, delta.over.sigma=0.5,
    alpha=0.1, alternative="greater", approx=F)
```

To compute the required sample sizes to detect a scaled minimal detectable difference of 0.5 assuming a significance level of 10% and a required power of 95%, type this command.

```
t.test.n(delta.over.sigma=0.5, alpha=0.1, power=0.95,
    sample.type="two.sample", alternative="greater",
    approx=F)
```

To compute the scaled minimal detectable difference assuming a significance level of 10%, a power of 95%, and sample sizes of $n_1 = n_2 = 16$, type this command.

```
t.test.scaled.mdd(n.or.n1=16, n2=16, alpha=0.1,
    power=0.95, alternative="greater", approx=F)
```

Testing the Mean of a Lognormal Distribution

We explained in Chapter 7 that to perform a hypothesis test on the mean of a lognormal distribution (denoted θ) you have to create a confidence interval for the mean and base the test on the relationship between hypothesis tests and confidence intervals (Table 7.4). In particular, performing a one-sample t-test on the log-transformed observations involves testing a hypothesis about the *median* of the distribution, not the mean.

Although you cannot use Student's t-test based on the log-transformed observations to test a hypothesis about the mean of a lognormal distribution (θ), you *can* use the t-distribution to estimate the power of a test on θ that is based on confidence intervals or Chen's modified t-test if you are willing to assume the population coefficient of variation τ stays constant for all possible values of θ you are interested in, and you are willing to postulate possible values for τ.

First, let us rewrite the null and alternative hypotheses as follows. The null hypothesis $H_0: \theta = \theta_0$ is equivalent to:

$$H_0 : \frac{\theta}{\theta_0} = 1 \tag{8.20}$$

The three possible alternative hypotheses are the upper one-sided alternative

$$H_a : \frac{\theta}{\theta_0} > 1 \tag{8.21}$$

the lower one-sided alternative

$$H_a : \frac{\theta}{\theta_0} < 1 \tag{8.22}$$

and the two-sided alternative

$$H_a : \frac{\theta}{\theta_0} \neq 1 \tag{8.23}$$

Let μ and σ denote the population mean and standard deviation of the log-transformed observations (see Equations (4.32) to (4.35) for the relationship between θ, τ, μ, and σ). By Equation (4.35), if the coefficient of variation τ is constant, then so is the standard deviation of the log-transformed observations σ. Hence, by Equation (4.32), the ratio of the true mean to the hypothesized mean can be written as:

$$R = \frac{\theta}{\theta_0} = \frac{e^{\mu + \sigma^2/2}}{e^{\mu_0 + \sigma^2/2}} = \frac{e^{\mu}}{e^{\mu_0}} = e^{\mu - \mu_0} \qquad (8.24)$$

which only involves the difference $\mu - \mu_0$. Thus, for given values of R and τ, the power of the test of the null hypothesis (8.20) against any of the alternatives (8.21) to (8.23) can be computed based on the power of a one-sample t-test with the scaled minimal detectable difference defined as follows:

$$\frac{\delta}{\sigma} = \frac{\log(R)}{\sqrt{\log(\tau^2 + 1)}} \qquad (8.25)$$

You can also compute the required sample size for a specified power, scaled MDD, and significance level, or compute the scaled MDD for a specified sample size, power, and significance level.

Example 8.10: Design for Comparing Soil Contaminant Concentrations to Soil Screening Levels

The guidance document **Soil Screening Guidance: Technical Background Document** (USEPA, 1996c, pp. 81–110) presents a strategy for screening soil for contamination. The strategy involves taking a number of soil samples in an exposure area (EA; e.g., a ¼-acre lot), determining the contaminant concentration in each physical sample, and comparing the average concentration to a soil screening level (SSL).   If the true mean concentration θ were known, the decision rule is:

*   If $\theta > $ SSL, investigate the site further for possible remedial action.
*   If $\theta \le $ SSL, "walk away" from the site; no further investigation is necessary.

To put the decision rule in the context of a hypothesis test, there are two possible null hypotheses:

* H_0: $\theta > $ SSL (the site is contaminated)
* H_0: $\theta \le $ SSL (the site is clean)

USEPA (1996c) uses the first null hypothesis when describing a strategy involving something called the Max test, and it uses the second null hypothesis when describing a strategy involving Chen's modified t-test.

For the present example, we will assume we will be using Chen's modified t-statistic or a test based on a confidence interval for a lognormal mean. USEPA (1996c, p. 87) suggests the following error rates:

- Pr(Walk Away | $\theta > 2$ SSL) = 5%
- Pr(Investigate | $\theta \le$ SSL/2) = 20%

Figure 8.7 displays the design performance goal diagram based on these error rates.

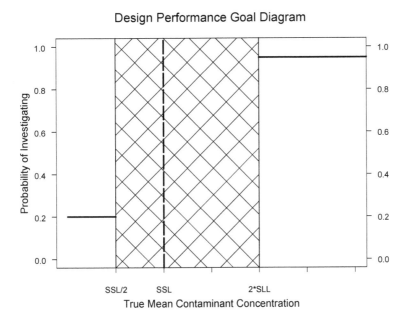

Design Performance Goal Diagram

Figure 8.7   Design performance goal diagram for the soil screening study

Because the error rates are defined in terms of SSL/2 and 2 SSL, we will have to use the following null hypothesis and associated Type I and Type II error rates:

- H_0: $\theta \le \theta_0$=SSL/2 (the site is clean)
- α = Pr(Type I Error) = Pr(Investigate | $\theta \le \theta_0$) = 20%
- β = Pr(Type II Error) = Pr(Walk Away | $\theta > 4\,\theta_0$) = 5%

Figure 8.8 illustrates one possible scenario in which SSL = 10 and the distribution of contamination follows a lognormal distribution with a coefficient of variation of 2. In this case, SSL/2 = 5 and 2 × SSL = 20. Note that the Type II error rate (or equivalently, the power) is defined for a ratio of $R = 4$, because (2 × SSL)/(SSL/2) = 4.

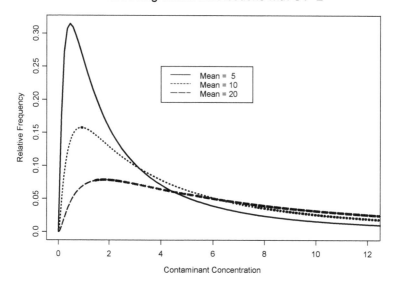

Figure 8.8   Possible distribution of soil contamination (SSL = 10)

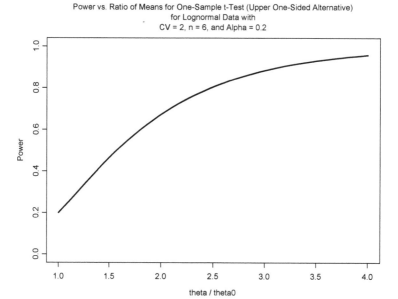

Figure 8.9   Power vs. $R(\theta/\theta_0)$ assuming CV = 2, $n = 6$, and a Type I error rate of 20%

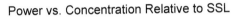

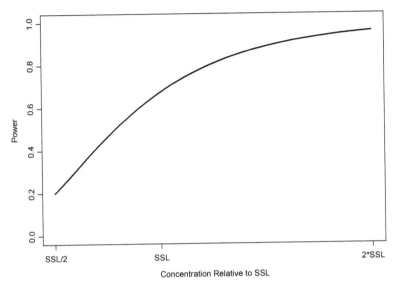

Figure 8.10   Power vs. concentration relative to SSL, assuming CV = 2, n = 6, and a Type I error rate of 20%

Based on the null hypothesis and the specified Type I and Type II error rates and value for R, we can determine the required sample size for a specified coefficient of variation. If we assume CV = 2, then the required sample is $n = 6$. Figure 8.9 plots the power of the test vs. the ratio $R = \theta/\theta_0$ assuming CV = 2 and a sample size of $n = 6$. Comparing this figure with Figure 8.7, you can see that we have met the design performance goal. Figure 8.10 plots the same thing as Figure 8.9, but re-labels the x-axis in terms of the SSL so it is easier to compare it with Figure 8.7.

Note that we have performed power and sample size calculations for only one exposure area. In practice, you will probably have to deal with multiple exposure areas and/or multiple chemicals, and therefore you will need to worry about the multiple comparisons problem. USEPA (1996c, p. 129) notes that "Guidance for performing multiple hypothesis tests is beyond the scope of the current document. Obtain the advice of a statistician ...."

Menu

To determine the required sample size assuming CV = 2, a specified Type I error rate of 20%, and a specified power of 95% when $R = 4$ using the ENVIRONMENTALSTATS for S-PLUS pull-down menu, follow these steps.

1. On the S-Plus menu bar, make the following menu choices: **EnvironmentalStats>Sample Size and Power>Hypothesis Tests>Lognormal Mean>Compute**. This will bring up the Lognormal Power and Sample Size dialog box.
2. For Compute select **Sample Size**, for Sample Type select **One-Sample**, for Alpha(s) (%) type **20**, for Power(s) (%) type **95**, for Coefficient(s) of Variation type **2**, for Null Mean type **1/2**, for Alt. Mean type **2**, and for Alternative select **"greater"**. Click **OK** or **Apply**.

To create the power curve shown in Figure 8.9, follow these steps.

1. On the S-Plus menu bar, make the following menu choices: **EnvironmentalStats>Sample Size and Power>Hypothesis Tests>Lognormal Mean>Plot**. This will bring up the Plot T-Test Design for Lognormal Data dialog box.
2. For X Variable select **Ratio of Means**, for Y Variable select **Power**, for Minimum X type **1**, for Maximum X type **4**, for Sample Type select **One-Sample**, **check** the Use Exact Algorithm box, for Alpha (%) type **20**, for n type **6**, for Coefficient of Variation type **2**, and for Alternative select **"greater"**.
3. Click on the **Plotting** tab. For Y-Axis Limits type **c(0,1)**. Click **OK** or **Apply**.

Command

To determine the required sample size assuming CV = 2, a specified Type I error rate of 20%, and a specified power of 95% when $R = 4$ using the EnvironmentalStats for S-Plus Command or Script Window, type this command.

```
t.test.lnorm.alt.n(ratio.of.means=4, cv=2, alpha=0.2,
    power=0.95, alternative="greater", approx=F)
```

To create the power curve shown in Figure 8.9, type this command.

```
plot.t.test.lnorm.alt.design(x.var="ratio.of.means",
    y.var="power", range.x.var=c(1, 4), cv=2,
    n.or.n1=6, alpha=0.2, alternative="greater",
    approx=F, ylim=c(0,1))
```

To create the power curve shown in Figure 8.10, type this command.

```
plot.t.test.lnorm.alt.design(x.var="ratio.of.means",
    y.var="power", range.x.var=c(1, 4), cv=2,
    n.or.n1=6, alpha=0.2, alternative="greater",
    approx=F, ylim=c(0, 1), xaxt="n",
```

```
   xlab="Concentration Relative to SSL",
   main="Power vs. Concentration Relative to SSL")
axis(side=1, at=c(1, 2, 4),
   labels=c("SSL/2", "SSL", "2*SSL"))
```

Testing the Difference between Two Means of Lognormal Distributions

In Chapter 7 we explained that to test the difference between two means of lognormal distributions, you can perform the two-sample t-test on the log-transformed observations, as long as you are willing to assume the coefficient of variation τ is the same for both groups (see Equations (7.77) to (7.80)). In this case, the ratio of the means is given by:

$$R = \frac{\theta_1}{\theta_2} = \frac{e^{\mu_1 + \sigma^2/2}}{e^{\mu_2 + \sigma^2/2}} = \frac{e^{\mu_1}}{e^{\mu_2}} = e^{\mu_1 - \mu_2} \qquad (8.26)$$

which only involves the difference $\mu_1 - \mu_2$. Thus, for given values of R and τ, the power of the test of the null hypothesis of equal means against any one-sided or two-sided alternative can be computed based on the power of a two-sample t-test with the scaled minimal detectable difference defined as in Equation (8.25) and R defined as in Equation (8.26). You can also compute the required sample sizes for a specified power, scaled MDD, and significance level, or compute the scaled MDD for specified sample sizes, power, and significance level.

Example 8.11: Design for Comparing TcCB Concentrations in a Reference Area with Concentrations in a Cleanup Area

In Example 7.14 we compared the means of the log-transformed TcCB concentrations in the Reference and Cleanup areas. The sample mean and coefficient of variation for the (untransformed) Reference area data are 0.6 ppb and 0.47, whereas the sample mean and coefficient of variation for the Cleanup area are 3.9 ppb and 5.1. Obviously, in this case the assumption of equal coefficients of variation is suspect. Suppose we want to design a sampling program for another Reference and Cleanup area. Here we will assume the CV is about 1, and we want to detect a two-fold difference between the Cleanup and Reference area with 95% power, assuming a 1% Type I error rate. If we take equal samples in each area, we will need to take a total of $n_1 = n_2 = 47$ samples in each area.

Menu

To compute the required sample sizes assuming CV = 1, a significance level of 1%, and a power of 95% when *R* = 2 using the ENVIRONMENTALSTATS for S-PLUS pull-down menu, follow these steps.

1. On the S-PLUS menu bar, make the following menu choices: **EnvironmentalStats>Sample Size and Power>Hypothesis Tests>Lognormal Mean>Compute**. This will bring up the Lognormal Power and Sample Size dialog box.
2. For Compute select **Sample Size**, for Sample Type select **Two-Sample**, for Alpha(s) (%) type **1**, for Power(s) (%) type **95**, for Coefficient(s) of Variation type **1**, for Mean 2 type **1**, for Mean 2 type **2**, and for Alternative select **"greater"**. Click **OK** or **Apply**.

Command

To compute the required sample sizes assuming CV = 1, a significance level of 1%, and a power of 95% when *R* = 2 using the ENVIRONMENTALSTATS for S-PLUS Command or Script Window, type this command.

```
t.test.lnorm.alt.n(ratio.of.means=2, cv=1, alpha=0.01,
   power=0.95, sample.type="two.sample",
   alternative="greater", approx=F)
```

Testing the Difference between Several Means of Normal Distributions

In Equations (7.118) to (7.122) we introduced the F-statistic to test the null hypothesis H_0: $\mu_1 = \mu_2 = ... = \mu_k$ assuming the observations in each of k groups come from normal distributions with the same standard deviation σ and possibly different means. When the null hypothesis is true, the F-statistic in Equation (7.118) is bouncing around 1 and has an F-distribution with $k-1$ and $N-k$ degrees of freedom. Now suppose the null hypothesis is false and that at least one of the means is different from the other $k-1$ means. In this case, the F-statistic follows a non-central F-distribution with non-centrality parameter given by:

$$\Delta = \frac{\sum_{i=1}^{k} n_i(\mu_i - \bar{\mu})^2}{\sigma^2} \tag{8.27}$$

where

$$\bar{\mu} = \sum_{i=1}^{k} \mu_i \qquad (8.28)$$

(Scheffé, 1959, pp. 38–39,62–65). The power of the F-test is given by:

$$Power = \Pr\left[F \geq F_{k-1,N-k,1-\alpha}\right] \qquad (8.29)$$

$$= 1 - H\left(F_{k-1,N-k,1-\alpha}, \, k-1, \, N-k, \, \Delta\right)$$

where $H(x, \, \nu_1, \, \nu_2, \, \Delta)$ denotes the cumulative distribution function of the non-central F-distribution with ν_1 and ν_2 degrees of freedom and non-centrality parameter Δ evaluated at x.

Example 8.12: Design for Comparing Fecal Coliform Counts between Seasons

In Example 7.22 we used analysis of variance to compare fecal coliform counts (organisms per 100 ml) in the Illinois River between seasons, based on 6 years of quarterly samples (i.e., $n_i = 6$ for each of the $k = 4$ seasons). In that example, the log-transformed counts appeared to follow the normality and equal variance assumptions better than the original count data. The observed means for the log-transformed count data for summer, fall, winter, and spring are 2.6, 2.4, 2.1, and 2.1, and the observed standard deviations are 0.4, 0.4, 0.5, and 0.6.

Suppose we want to design another sampling program. If we assume the true population means of the log-transformed observations are 2.5, 2.5, 2.0, and 2.0, assume a population standard deviation of about 0.6, use a significance level of 5%, and desire a power of 95% to detect a difference, we will need to take $n_i = 26$ samples in each season (i.e., we will need observations from 26 years of quarterly sampling). If we only use 6 years of sampling, then we will only achieve a power of about 31%.

Menu

To compute the required sample sizes using the ENVIRONMENTALSTATS for S-PLUS pull-down menu, follow these steps.

1. On the S-PLUS menu bar, make the following menu choices: **EnvironmentalStats>Sample Size and Power>Hypothesis**

Tests>Normal Mean>k Samples>Compute. This will bring up the ANOVA Power and Sample Size dialog box.

2. For Compute select **Sample Size**, for Alpha (%) type **5**, for Power (%) type **95**, for Standard Deviation type **0.6**, and for Means type **c(2.5,2.5,2,2)**. Click **OK** or **Apply**.

To compute the power assuming only 6 years of sampling, repeat Step 1 above and replace Step 2 with the following:

2. For Compute select **Power**, for Alpha (%) type **5**, for Sample Size(s) type **rep(6,4)**, for Standard Deviation type **0.6**, and for Means type **c(2.5,2.5,2,2)**. Click **OK** or **Apply**.

Command

To compute the required sample sizes using the ENVIRONMENTALSTATS for S-PLUS Command or Script Window, type this command.

```
aov.n(mu.vec=c(2.5, 2.5, 2, 2), sigma=0.6)
```

To compute the power assuming only 6 years of sampling, type this command.

```
aov.power(n.vec=rep(6,4), mu.vec=c(2.5, 2.5, 2, 2),
   sigma=0.6)
```

Testing a Binomial Proportion

As we noted in Chapter 7, the test on the proportion p from a binomial distribution, where p denotes the probability of "success" in one trial, is based on the observed number of successes x out of n trials (see Equations (7.24) to (7.28)). Note that x is one observed value of the random variable X, where X follows a B(n, p) distribution. We will write the null hypothesis as

$$H_0 : p = \pi_0 \tag{8.30}$$

and the three possible alternative hypotheses are the upper one-sided alternative

$$H_a : p > \pi_0 \tag{8.31}$$

the lower one-sided alternative

$$H_a : p < \pi_0 \qquad (8.32)$$

and the two-sided alternative

$$H_a : p \neq \pi_0 \qquad (8.33)$$

For the upper one-sided alternative, the test rejects the null hypothesis (8.30) in favor of the alternative hypothesis (8.31) at level-α if $x > q_{1-\alpha}$, where

$$\Pr\left(X > q_{1-\alpha} \mid p = \pi_0\right) = \alpha \qquad (8.34)$$

(Conover, 1980, pp. 95–99; Zar, 1999, pp. 533–535). Because of the discrete nature of the binomial distribution, it is usually not possible to find a number $q_{1-\alpha}$ that satisfies Equation (8.34) for a user-specified value of α. Thus, given a user-specified value of α, the exact test uses a critical upper value $q_{1-\alpha*}$ such that

$$\Pr\left(X > q_{1-\alpha*} \mid p = \pi_0\right) = \alpha* \leq \alpha \qquad (8.35)$$

so the true significance level of the test is given by $\alpha*$ rather than α.

When the probability of success is equal to some value π that is different from π_0, the power of the upper one-sided test is given by:

$$Power = \Pr\left(X > q_{1-\alpha*} \mid p = \pi_0 + \delta\right)$$

$$\qquad (8.36)$$

$$= 1 - F\left(q_{1-\alpha*}, \ n, \ \pi_0 + \delta\right)$$

where

$$\delta = \pi - \pi_0 \qquad (8.37)$$

denotes the minimal detectable difference (MDD) and $F(x, \ n, \ p)$ denotes the cumulative distribution function of a $B(n, p)$ random variable evaluated at x.

For the lower one-sided alternative, the test rejects the null hypothesis (8.30) in favor of the alternative hypothesis (8.32) at level-α if $x \le q_{\alpha*}$, where

$$\Pr\left(X \le q_{\alpha*} \mid p = \pi_0\right) = \alpha^* \le \alpha \qquad (8.38)$$

When the probability of success is equal to some value π that is different from π_0, the power of the lower one-sided test is given by:

$$Power = \Pr\left(X \le q_{\alpha*} \mid p = \pi_0 + \delta\right)$$

$$= F\left(q_{\alpha*}, \ n, \ \pi_0 + \delta\right)$$

$$(8.39)$$

Finally, for the two-sided alternative, the test rejects the null hypothesis (8.30) in favor of the alternative hypothesis (8.33) at level-α if $x \le q_{\alpha*/2}$ or $x > q_{1-\alpha*/2}$, where these quantities are defined in Equations (8.35) and (8.38) with α^* replaced with $\alpha^*/2$. When the probability of success is equal to some value π that is different from π_0, the power of the two-sided test is given by:

$$Power = \Pr\left(X \le q_{\alpha*/2} \mid p = \pi_0 + \delta\right) +$$
$$\Pr\left(X > q_{1-\alpha*/2} \mid p = \pi_0 + \delta\right)$$

$$(8.40)$$

$$= F\left(q_{\alpha*/2}, \ n, \ \pi_0 + \delta\right) +$$
$$1 - F\left(q_{1-\alpha*/2}, \ n, \ \pi_0 + \delta\right)$$

Equations (8.36), (8.39), and (8.40) can also be used to determine the required sample size for a specified power, MDD, value of π_0, and significance level, or to determine the MDD for a specified power, sample size, value of π_0, and significance level.

Example 8.13: Design for Testing Whether the Probability of Exceedance is above a Standard

In Example 7.3 we tested a hypothesis about the probability p that the daily maximum ozone exceeds a standard of 120 ppb. The null hypothesis was that this probability is less than or equal to 1/365 (.00274) (which would result in an average of 1 day per year in which the standard is exceeded). Suppose we want to set the significance level to 5% and monitor for 1, 3, 6, or 12 months. Table 8.3 displays the power of the one-sided test to detect a probability greater than 1/365 given that the true probability is 10 times as large ($10/365 = 0.0274$). The table also shows that the true significance level is far less than the desired 5% for all of the sample sizes considered here. This problem is due to the fact that the value of p under the null hypothesis is very far from 0.5. For values of p closer to 0.5, you should be able to attain a significance level close to the one you desire for moderate to large sample sizes.

Sample Size (n)	Significance Level (%)	Power
30	0.3	20
90	2.6	71
180	1.4	87
365	1.9	99

Table 8.3   Sample sizes, significance levels, and powers for the binomial test of the probability of daily maximum ozone exceeding 120 ppb

Menu

To determine the power to detect a true daily probability of exceedance of 10/365 given a desired significance level of 5% and sample sizes of n = 30, 90, 180, and 365 days using the ENVIRONMENTALSTATS for S-PLUS pull-down menu, follow these steps.

1. On the S-PLUS menu bar, make the following menu choices: **EnvironmentalStats>Sample Size and Power>Hypothesis Tests>Binomial Proportion>Compute**. This will bring up the Binomial Power and Sample Size dialog box.
2. For Compute select **Power**, for Sample Type select **One-Sample**, for Alpha(s) (%) type **5**, for n type **c(30,90,180,365)**, for Null Proportion type **1/365**, and for Alt. Proportion type **10/365**, and for Alternative select **"greater"**. Click **OK** or **Apply**.

Command

To determine the power to detect a true daily probability of exceedance of 0.0274 given a desired significance level of 5% and sample sizes of n =

30, 90, 180, and 365 days using the ENVIRONMENTALSTATS for S-PLUS Command or Script Window, type this command.

```
prop.test.power(n.or.n1=c(30, 90, 180, 365),
    p.or.p1=10/365, p0.or.p2=1/365,
    alternative="greater", approx=F)
```

Testing for the Difference between Two Binomial Proportions

In Chapter 7 we discussed two ways to test for a difference between two binomial proportions: use Fisher's exact test or use an approximate test based on the normal approximation to the binomial distribution (see Equations (7.98) to (7.104)). We will write the null hypothesis as

$$H_0 : p_1 = p_2 \tag{8.41}$$

and the three possible alternative hypotheses are the upper one-sided alternative

$$H_a : p_1 > p_2 \tag{8.42}$$

the lower one-sided alternative

$$H_a : p_1 < p_2 \tag{8.43}$$

and the two-sided alternative

$$H_a : p_1 \neq p_2 \tag{8.44}$$

For the approximate test, the test statistic z defined in Equation (7.98) is approximately distributed as a standard normal random variable under the null hypothesis.

For the upper one-sided alternative, the approximate test rejects the null hypothesis (8.41) in favor of the alternative hypothesis (8.42) at level-α if $z \geq z_{1-\alpha}$. When the two proportions are not equal, the power of the upper one-sided test (without using the continuity correction) is given by:

$$Power = \Pr\left(z \geq z_{1-\alpha}\right)$$

$$= \Pr\left[\frac{\hat{p}_1 - \hat{p}_2}{\hat{\sigma}_{\hat{p}_1 - \hat{p}_2}} \geq z_{1-\alpha}\right]$$

$$= \Pr\left[\frac{(\hat{p}_1 - \hat{p}_2) - \delta}{\sigma_{\hat{p}_1 - \hat{p}_2}} \geq z_{1-\alpha}\frac{\hat{\sigma}_{\hat{p}_1 - \hat{p}_2} - \delta}{\sigma_{\hat{p}_1 - \hat{p}_2}}\right]$$

$$= \Pr\left[Z \geq z_{1-\alpha}\frac{\hat{\sigma}_{\hat{p}_1 - \hat{p}_2} - \delta}{\sigma_{\hat{p}_1 - \hat{p}_2}}\right]$$

$$= 1 - \Phi\left[z_{1-\alpha}\frac{\hat{\sigma}_{\hat{p}_1 - \hat{p}_2} - \delta}{\sigma_{\hat{p}_1 - \hat{p}_2}}\right] \tag{8.45}$$

where Z denotes a standard normal random variable, $\Phi(x)$ denotes the cumulative distribution function of the standard normal distribution evaluated at x, and

$$\delta = p_1 - p_2 \tag{8.46}$$

$$\sigma_{\hat{p}_1 - \hat{p}_2} = \sqrt{\frac{p_1(1 - p_1)}{n_1} + \frac{p_2(1 - p_2)}{n_2}} \tag{8.47}$$

For the lower one-sided alternative, the test rejects the null hypothesis (8.41) in favor of the alternative hypothesis (8.43) at level-α if $z \leq z_\alpha$. When the two proportions are not equal, the power of the lower one-sided test (without using the continuity correction) is given by:

$$Power = \Pr\left(z \leq z_{\alpha}\right)$$

(8.48)

$$= \Phi\left[z_{\alpha}\frac{\hat{\sigma}_{\hat{p}_1-\hat{p}_2} - \delta}{\sigma_{\hat{p}_1-\hat{p}_2}}\right]$$

Finally, for the two-sided alternative, the test rejects the null hypothesis (8.41) in favor of the alternative hypothesis (8.44) at level-α if $z \leq z_{\alpha/2}$ or $z \geq z_{1-\alpha/2}$. When the two proportions are not equal, the power of the two-sided test (without using the continuity correction) is given by:

$$Power = \Pr\left(z \leq z_{\alpha/2}\right) + \Pr\left(z \geq z_{1-\alpha/2}\right)$$

$$= \Phi\left[z_{\alpha/2}\frac{\hat{\sigma}_{\hat{p}_1-\hat{p}_2} - \delta}{\sigma_{\hat{p}_1-\hat{p}_2}}\right] +$$

(8.49)

$$1 - \Phi\left[z_{1-\alpha/2}\frac{\hat{\sigma}_{\hat{p}_1-\hat{p}_2} - \delta}{\sigma_{\hat{p}_1-\hat{p}_2}}\right]$$

Looking at Equations (8.45) to (8.49), we see that the power of the two-sample z-test for binomial proportions depends on the sample sizes (n_1 and n_2), the Type I error rate (α), the deviation from the null hypothesis (δ), and the true proportions p_1 and p_2. These equations can be modified to incorporate the continuity correction.

Example 8.14: Design for Comparing the Incidence Rates of Thyroid Tumors in Rats

In Examples 7.19 and 7.20 we compared the incidence of thyroid tumors in rats exposed and not exposed to ethylene thiourea (ETU). In the exposed group, 16 out of 69 rats (23%) developed thyroid tumors, whereas in the unexposed group only 2 out of 72 (3%) developed thyroid tumors. Suppose we are going to design another study with $n_1 = n_2 = 60$ rats in each group. If we assume the baseline tumor incidence rate in the unexposed group is about 5%, and we want to use a significance level of 5%, the power to detect an incidence rate of 20% in the exposed group (an increase of 15 percentage points) is about 72% (based on using the continuity correction). Also, if we desire a power of 95%, the smallest increase in incidence rate we can detect with this power is about 24 percentage points (i.e., an incidence rate of 29% in the exposed group). Finally, if we want to be able to detect an increase of 15 percentage points with 95% power, we would need to use sample sizes of $n_1 = n_2 = 116$ rats in each group.

Figure 8.11 plots the minimal detectable difference vs. sample size (for each group) assuming a baseline incidence rate of 5%, a significance level of 5%, and a power of 95%. Note that if we want to be able to detect an increase in incidence of 10 percentage points (0.1 on the y-axis; i.e., an incidence rate of 15% in the exposed group), we would need a sample size of over 200 animals in each group.

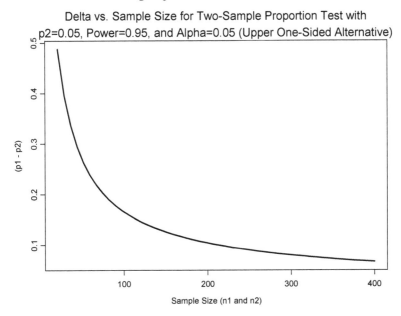

Delta vs. Sample Size for Two-Sample Proportion Test with p2=0.05, Power=0.95, and Alpha=0.05 (Upper One-Sided Alternative)

Figure 8.11   Minimal detectable difference between incidence rates vs. sample size (for each group), assuming a 5% baseline incidence rate, 5% significance level, and 95% power (using continuity correction)

Menu

To compute the power to detect an incidence rate of 20% in the exposed group assuming an incidence rate of 5% in the unexposed group, a significance level of 5%, and sample sizes of 60 rats in each group using the ENVIRONMENTALSTATS for S-PLUS pull-down menu, follow these steps.

1. On the S-PLUS menu bar, make the following menu choices: **EnvironmentalStats>Sample   Size   and   Power>Hypothesis**

Tests>Binomial Proportion>Compute. This will bring up the Binomial Power and Sample Size dialog box.

2. For Compute select **Power**, for Sample Type select **Two-Sample**, for Alpha(s) (%) type **5**, for n1 type **60**, for n2 type **60**, for Proportion 2 type **0.05**, for Proportion 1 type **0.2**, and for Alternative select **"greater"**. Click **OK** or **Apply**.

To compute the minimal detectable difference assuming an incidence rate of 5% in the unexposed group, a significance level of 5%, a power of 95%, and sample sizes of 60 rats in each group, repeat Step 1 above and replace Step 2 with the following.

2. For Compute select **Min. Difference**, for Sample Type select **Two-Sample**, for Alpha(s) (%) type **5**, for Power(s) (%) type **95**, for n1 type **60**, for n2 type **60**, for Proportion 2 type **0.05**, and for Alternative select **"greater"**. Click **OK** or **Apply**.

To compute the required sample size in each group in order to detect a difference of 15 percentage points assuming an incidence rate of 5% in the unexposed group, a significance level of 5%, and a power of 95%, repeat Step 1 above and replace Step 2 with the following.

2. For Compute select **Sample Size**, for Sample Type select **Two-Sample**, for Alpha(s) (%) type **5**, for Power(s) (%) type **95**, for Ratio (n2/n1) type **1**, for Proportion 2 type **0.05**, for Proportion 1 type **0.2**, and for Alternative select **"greater"**. Click **OK** or **Apply**.

To create the plot of MDD vs. sample size (for each group) shown in Figure 8.11, follow these steps.

1. On the S-PLUS menu bar, make the following menu choices: **EnvironmentalStats>Sample Size and Power>Hypothesis Tests>Binomial Proportion>Plot**. This will bring up the Plot Binomial Test Design dialog box.

2. For X Variable select **Sample Size**, for Y Variable select **Delta**, for Minimum X type **20**, for Maximum X type **400**, for Sample Type select **Two-Sample**, for Alpha (%) type **5**, for Power (%) type **95**, for Proportion 2 type **0.05**, and for Alternative select **"greater"**. Click **OK** or **Apply**.

Command

To compute the power to detect an incidence rate of 20% in the exposed group assuming an incidence rate of 5% in the unexposed group, a significance level of 5%, and sample sizes of 60 rats in each group using the ENVIRONMENTALSTATS for S-PLUS Command or Scrip Window, type this command.

```
prop.test.power(n.or.n1=60, p.or.p1=0.2, n2=60,
   p0.or.p2=0.05, alpha=0.05,
   sample.type="two.sample", alternative="greater")
```

To compute the minimal detectable difference assuming an incidence rate of 5% in the unexposed group, a significance level of 5%, a power of 95%, and sample sizes of 60 rats in each group, type this command.

```
prop.test.mdd(n.or.n1=60, n2=60, p0.or.p2=0.05,
   alpha=0.05, power=0.95, sample.type="two.sample",
   alternative="greater")
```

To compute the required sample size in each group in order to detect a difference of 15 percentage points assuming an incidence rate of 5% in the unexposed group, a significance level of 5%, and a power of 95%, type this command.

```
prop.test.n(p.or.p1=0.2, p0.or.p2=0.05, alpha=0.05,
   power=0.95, sample.type="two.sample",
   alternative="greater")
```

To create the plot of MDD vs. sample size (for each group) shown in Figure 8.11, type this command.

```
plot.prop.test.design(x.var="n", y.var="delta",
   alpha=0.05, power=0.95, p0.or.p2=0.05,
   sample.type="two.sample", alternative="greater")
```

Testing for Contamination Using Parametric Prediction Intervals

In Chapter 6 we discussed prediction intervals for normal, lognormal, and Poisson distributions, and in Examples 6.1, 6.2, and 6.7 we showed how you can construct a prediction interval for the next k "future" observations (or means or sums) based on background data and then compare these k "future" observations (or means or sums) at a downgradient well or from a cleanup site with the prediction interval to decide whether there is evidence of contamination. (If any of the k observations fall outside the prediction interval, then this is evidence of contamination.) In this section, we will discuss how to compute power and sample sizes for this kind of test assuming a normal or lognormal distribution.

Formulas for prediction intervals based on a normal distribution are shown in Equations (6.9) to (6.11). Davis and McNichols (1987) show that the power of the test based on a prediction interval for k future observations, assuming a normal distribution, is given by:

$$Power = 1 - p \qquad\qquad (8.50)$$

where p is defined in Equation (6.54), and in this case $m = k$ and $r = 1$. The power of this test depends on the background sample size (n), the number of future observations (k), the Type I error rate (α) (because the constant K depends on α, as well as n and k), and the scaled minimal detectable difference (δ/σ). (In the case when $k = 1$, this test is equivalent to a two-sample t-test with $n_1 = n$ and $n_2 = k = 1$.) This equation for power can also be used to determine the required sample size for a specified power, significance level, scaled MDD, and number of future observations.

Example 8.15: EPA Reference Power Curve for Groundwater Monitoring

In Chapter 6 we gave examples of using prediction intervals and simultaneous prediction intervals to monitor groundwater at hazardous and solid waste sites. At the 1991 US Environmental Protection Agency Statistics Conference, EPA introduced the concept of the "EPA Reference Power Curve" for groundwater monitoring (Davis and McNichols, 1994b, 1999; USEPA, 1992c, p. 65). Any test procedure (e.g., a test based on simultaneous prediction limits) whose power curve (power vs. scaled MDD) compares adequately with the reference curve should be acceptable under regulation. In 1991, the reference power curve was deemed to be based on a parametric (normal) prediction interval for a single well and a single constituent assuming a confidence level of 99% and a background sample size of $n = 8$. The guidance document USEPA (1992c, p. 65) displays a reference power curve based on $n = 16$ background observations. Davis and McNichols (1999) note that some states require monitoring after only $n = 4$ background observations have been collected. Figure 8.12 displays the EPA reference power curve based on a background sample size of $n = 8$ observations.

Menu

To compute the reference power curve using the ENVIRONMENTALSTATS for S-PLUS pull-down menu, follow these steps.

1. On the S-PLUS menu bar, make the following menu choices: **EnvironmentalStats>Sample Size and Power>Hypothesis Tests>Prediction Intervals>Normal>Plot**. This will bring up the Plot Normal PI Test Power Curve dialog box.
2. For Confidence Level (%) type **99**, for n type **8**, for # Future Obs type **1**, and for PI Type select **"upper"**. Click **OK** or **Apply**.

Command

To compute the reference power curve using the ENVIRONMENTALSTATS for S-PLUS Command or Script Window, type this command.

```
plot.pred.int.norm.test.power.curve(n=8, k=1,
   conf.level=0.99)
```

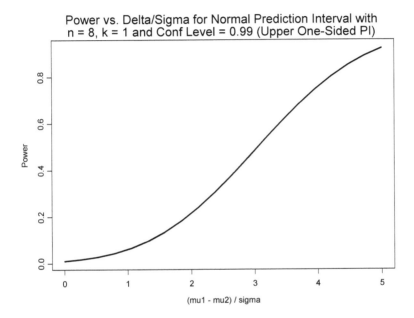

Figure 8.12   U.S. EPA Reference Power Curve based on $n = 8$ background observations and a 99% prediction interval for $k = 1$ future observation

Example 8.16: Design for Comparing TcCB Concentrations in a Reference Area with Concentrations in a Cleanup Area

In Example 6.7 we created a prediction interval for the next $k = 77$ observations of TcCB concentrations in the Cleanup area based on using the $n = 47$ observations from the Reference area. The sample mean and coefficient of variation for the (untransformed) Reference area data are 0.6 ppb and 0.47, whereas the sample mean and coefficient of variation for the Cleanup area are 3.9 ppb and 5.1. Obviously, in this case the assumption of equal coefficients of variation is suspect.

Suppose we want to design a sampling program for another Reference and Cleanup area. Here we will assume the CV is about 1, and we want to detect a 2-fold difference between the Cleanup and Reference area using a

5% Type I error rate. Setting $n = 47$ and $k = 77$ yields a power of only 37%. Compare this with the results in Example 8.11 in which we found we needed only 47 observations in each area to achieve a power of 95% using the two-sample t-test. The t-test is more powerful than the test based on the prediction interval in this case because we are looking for a shift in the mean, whereas the test based on the prediction interval is more powerful for detecting a shift in the tail of the distribution (e.g., due to residual contamination).

Menu

To compute the power assuming CV = 1, a significance level of 1%, $n = 47$, $k = 77$, and $R = 2$ using the ENVIRONMENTALSTATS for S-PLUS pull-down menu, follow these steps.

1. On the S-Plus menu bar, make the following menu choices: **EnvironmentalStats>Sample Size and Power>Hypothesis Tests>Prediction Intervals>Lognormal>Compute**. This will bring up the Lognormal PI Test Power and Sample Size dialog box.
2. For Compute select **Power**, for Confidence Level (%) type **95**, for n type **47**, for # Future Obs type **77**, for Coefficient(s) of Variation type **1**, for Null Mean type **1**, for Alt. Mean type **2**, and for PI Type select **"upper"**. Click **OK** or **Apply**.

Command

To compute the power assuming CV = 1, a significance level of 5%, $n = 47$, $k = 77$, and $R = 2$ using the ENVIRONMENTALSTATS for S-PLUS Command or Script Window, type this command.

```
pred.int.lnorm.alt.test.power(n=47, k=77,
    ratio.of.means=2, cv=1, pi.type="upper",
    conf.level=0.95)
```

Testing for Contamination Using Parametric Simultaneous Prediction Intervals

In Chapter 6 we discussed simultaneous prediction intervals as a method to control the facility-wide false positive rate (FWFPR) of mistakenly declaring contamination is present when it is not when monitoring groundwater at several wells (r) for several constituents (n_c). The formulas for simultaneous prediction intervals based on a normal distribution are exactly the same as for standard prediction intervals (Equations (6.9) to (6.11)), except that K is computed differently. For the k-of-m rule, Equation (6.54) is used to compute K by setting p equal to the desired confidence level and assuming $\delta/\sigma = 0$. For the California and Modified California rules, Equations (6.56) and (6.57) are used, respectively. Once K has been computed, the power of the test based on a simultaneous prediction interval can be computed using Equation (8.50), where p is defined in Equation (6.54), (6.56), or (6.57), depending on which rule is being used, and δ/σ can be set to some arbitrary value.

Equations (6.54), (6.56), and (6.57) assume that the mean on all r future sampling occasions (e.g., at all r wells) is shifted by the same amount δ. In Example 8.15 we noted that the U.S. EPA reference power curve is based on assuming contamination is present at only one well. We can modify the power calculation to specify that the mean is shifted for only a subset of the r future occasions (wells). In this case, the value of K is computed as before based on assuming that $\delta/\sigma = 0$ and that there are r future sampling occasions, but to compute the power the value of p is then computed using this value of K and assuming $r = r_{shifted}$, where $r_{shifted} \le r$.

Davis and McNichols (1999) display power curves based on both parametric and nonparametric simultaneous prediction intervals for each of the three rules (k-of-m, California, and Modified California). Power curves for the nonparametric prediction intervals were derived via Monte Carlo simulation. When the assumptions of the parametric simultaneous prediction intervals are met, the power of the nonparametric simultaneous prediction intervals is only slightly less than the power of its parametric counterpart.

For the k-of-m rule, Davis and McNichols (1987) give tables with "optimal" choices of k (in terms of best power for a given overall confidence level) for selected values of m, r, and n. They found that the optimal ratios of k to m (i.e., k/m) are generally small, in the range of 15 to 50%. Davis (1998a) states that if the FWFPR is kept constant, then the California rule offers little increased power compared to the k-of-m rule, and can actually decrease the power of detecting contamination. The Modified California Rule has been proposed as a compromise between a 1-of-m rule and the California rule. For a given FWFPR, the Modified California rule achieves better power than the California rule, and still requires at least as many observations in bounds as out of bounds, unlike a 1-of-m rule.

As we noted in Chapter 6, Davis and McNichols (1999) caution that statistically significant spatial variability is almost ubiquitous in groundwater monitoring data for which the proportion of nondetects is low enough that you can measure spatial variability fairly well. In such cases, standard normal simultaneous prediction intervals are not valid since they assume the population standard deviation σ is the same at all monitoring wells. In Chapter 6 we noted that in such cases Davis and McNichols (1999) recommend using well-specific simultaneous prediction intervals (i.e., $r = 1$ for each well) and using the Bonferroni inequality to set the per-constituent and well significance level to $\alpha/(r \times n_c)$.

***Example 8.17: Design for Comparing Constituent Concentrations
between Background and Compliance Wells***

In Example 6.4 we computed simultaneous upper prediction limits for arsenic concentrations based on the $n = 12$ observations at the background well shown in Table 6.2 assuming various numbers of compliance wells (r). In this example, we will assume there are $r = 10$ compliance wells and 7 other constituents being monitored, for a total of $n_c = 8$ constituents. We want the FWFPR set to 5%, so using the Bonferroni inequality we set the per-constituent significance level to $\alpha = 0.05/8 = 0.00625$, which means we desire a confidence level of $(1-0.00625)100\% = 99.375\%$. Figure 8.13 shows the power curve for this design for detecting contamination for a single constituent for a single well based on the 1-of-3 rule, along with the EPA reference power curve that was displayed in Figure 8.12.

Note that initially the power curve based on the simultaneous prediction interval lies below the EPA reference power curve. This will usually happen if you want to control the overall Type I error over all r monitoring wells and all n_c constituents being tested, because the EPA reference power curve is based on testing only one constituent at one well. In Figure 8.13, the power curve crosses the EPA reference power curve at a scaled MDD of about 2.

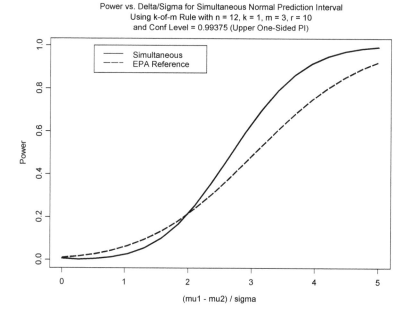

Figure 8.13   EPA reference power curve and power curve for the test based on a normal simultaneous prediction interval using the 1-of-3 rule, assuming $r = 10$ compliance wells, $n_c = 8$ constituents, and a FWFPR of 5%

Menu

To compute the power curve for the test based on the simultaneous prediction interval shown in Figure 8.13 using the ENVIRONMENTALSTATS for S-PLUS pull-down menu, follow these steps.

1. On the S-PLUS menu bar, make the following menu choices: **EnvironmentalStats>Sample Size and Power>Hypothesis Tests>Prediction Intervals>Normal>Plot**. This will bring up the Plot Normal PI Test Power Curve dialog box.

2. For Confidence Level (%) type **99.375**, for n type **12**, **check** the Simultaneous box, for # Future Obs type **3**, for Min # Obs PI Should Contain type **1**, for # Future Sampling Occasions type **10**, for Rule select **k.of.m**, for # Shifted type **1**, and for PI Type select **"upper"**. Click **Apply**.

To add the EPA reference power curve to the plot, make the following changes to the open dialog box.

3. **Uncheck** the Simultaneous box, for Confidence Level (%) type **99**, for n type **8**, for # Future Obs type **1**, and for PI Type select **"upper"**. Click on the **Plotting** tab, **check** the Add to Current Plot box, and for Line Type type **4**. Click **OK** or **Apply**.

Command

To compute the power curve for the test based on the simultaneous prediction interval shown in Figure 8.13 using the ENVIRONMENTALSTATS for S-PLUS Command or Script Window, type this command.

```
plot.pred.int.norm.simultaneous.test.power.curve(n=12,
    k=1, m=3, r=10, rule="k.of.m", pi.type="upper",
    conf.level=1-0.05/8, r.shifted=1, n.points=20)
```

To add the EPA reference power curve and legend to the plot, type these commands.

```
plot.pred.int.norm.test.power.curve(n=8, k=1,
    conf.level=0.99, n.points=20, add=T, plot.lty=4)
legend(0.5, 1, c("Simultaneous", "EPA Reference"),
    lty=c(1, 4), lwd=1.5)
```

Testing for Contamination Using Nonparametric Simultaneous Prediction Intervals

Earlier in this chapter we discussed how to compute sample size and confidence levels for nonparametric prediction intervals. In this section we will extend that discussion to nonparametric simultaneous prediction intervals.

In Chapter 6 we discussed how to construct nonparametric simultaneous prediction intervals based on using order statistics. Equations (6.60) to (6.62) show the formulas for the confidence level associated with a one-sided upper nonparametric simultaneous prediction interval for the k-of-m, California, and Modified California rule, respectively. These formulas depend on k, m, the number of future observations (r), which order statistic is used (v; usually the largest or second to largest value is used), and the sample size (n). Based on these formulas, you can compute the required sample size for a specific confidence level, or the associated confidence level for a specific sample size.

Example 8.18: Design for Comparing Constituent Concentrations between Background and Compliance Wells, Part II

Here we will follow Example 8.17 and assume there are $r = 10$ compliance wells and $n_c = 8$ constituents. We want the FWFPR set to 5%, so as before using the Bonferroni inequality we set the per-constituent significance level to $\alpha = 0.05/8 = 0.00625$, which means we desire a confidence level of $(1-0.00625)100\% = 99.375\%$. Table 8.4 shows required sample sizes for various values of m for each of the three kinds of rules, assuming that we are using the maximum value as the upper prediction limit.

Rule	# Future Observations (m)	Required Sample Size(n)
1-of-m	1	1,590
	2	55
	3	20
CA	3	78
	4	95
Modified CA	4	28

Table 8.4   Required sample sizes for a one-sided upper nonparametric simultaneous prediction interval based on the maximum value, assuming $r = 10$ compliance wells, $n_c = 8$ constituents, and a FWFPR of 5%

Menu

To compute the required sample sizes shown in Table 8.4 for the 1-of-m rule using the ENVIRONMENTALSTATS for S-PLUS pull-down menu, follow these steps.

1. On the S-PLUS menu bar, make the following menu choices: **EnvironmentalStats>Sample Size and Power>Prediction Intervals>Nonparametric>Compute**. This will bring up the Nonparametric PI Confidence Level and Sample Size dialog box.

2. For Compute select **Sample Size**, for Confidence Level (%) type **99.375, check** the Simultaneous box, for # Future Obs type **1:3**, for Min # Obs PI Should Contain type **1**, for # Future Sampling Occasions type **10**, for Rule select **k.of.m**, and for PI Type select **"upper"**. Click **OK** or **Apply**.

To compute the required sample sizes for the California rule, follow Steps 1 and 2 above, except for Rule select **CA** and for # Future Obs type **3:4**. To compute the required sample size for the Modified California rule, follow Steps 1 and 2 above, except for Rule select **Modified CA**.

Command

To compute the required sample sizes shown in Table 8.4 for the 1-of-*m* rule using the ENVIRONMENTALSTATS for S-PLUS Command or Script Window, type this command.

```
pred.int.npar.simultaneous.n(k=1, m=1:3, r=10,
    rule="k.of.m", pi.type="upper",
    conf.level=1-0.05/8)
```

To compute the required sample sizes for the California rule, type this command.

```
pred.int.npar.simultaneous.n(m=3:4, r=10, rule="CA",
    pi.type="upper", conf.level=1-0.05/8)
```

To compute the required sample size for the Modified California rule, type this command.

```
pred.int.npar.simultaneous.n(r=10, rule="Modified.CA",
    pi.type="upper", conf.level=1-0.05/8)
```

OPTIMIZING A DESIGN BASED ON COST CONSIDERATIONS

So far in all of our discussions of sampling design, we have avoided the issue of cost. Quite often, the available sample size for a study is entirely determined by budget. This situation, however, does not excuse the people responsible for the study from considering the effectiveness of the design! If the budget only allows for a total of 10 observations at a background and assessment site, and the power of detecting the kind of difference you are interested in detecting is less than adequate (e.g., 50% or less), you should think about increasing the budget, figuring out ways to decrease the inherent variability, or simply not do the study in the first place. In the two subsections below, we will discuss examples of how to incorporate costs into the sampling design process.

Trading off Cost and Sample Size to Minimizing the Variance of the Estimated Mean for a Stratified Random Sampling Design

In Chapter 2 we discussed stratified random sampling designs. Equation (2.5) shows the formula for the estimator of the mean over all strata, and Equation (2.9) shows the formula for the variance of this estimator. Cochran (1977, pp. 96–99) and Gilbert (1987, pp. 50–52) show how to minimize this variance if you know the variability within each stratum and the cost of taking a sample within each stratum. Cochran assumes the following cost function:

$$C = c_0 + \sum_{h=1}^{L} c_h n_h \qquad (8.51)$$

where C denotes the total cost, c_0 denotes a fixed overhead cost, and c_h denotes the cost per sampling unit within the hth stratum. In this, case the optimal number of samples to take in the hth stratum is given by:

$$n_h = n \; \frac{W_h \sigma_h / \sqrt{c_h}}{\sum\limits_{h=1}^{L} W_h \sigma_h / \sqrt{c_h}} \qquad (8.52)$$

Note that if the cost of sampling is the same in each stratum and the variance is the same in each stratum, then Equation (8.52) yields proportional allocation. In general, however, proportional allocation is not optimal. The formula above shows that you should take more samples in the hth stratum if it is larger than other strata, has more variability than other strata, and/or sampling is cheaper than in other strata (Cochran, 1977, p. 98).

In order to use Equation (8.52), besides knowing the variability and cost per sampling unit within each stratum (which usually requires a pilot study), you need to have a pre-specified value of n, the total sample size. If the cost is fixed, then n is given by:

$$n = \frac{(C - c_0) \sum\limits_{h=1}^{L} \left(W_h \sigma_h / \sqrt{c_h} \right)}{\sum\limits_{h=1}^{L} W_h \sigma_h \sqrt{c_h}} \qquad (8.53)$$

and if the variance of the stratified mean (Equation 2.9 in Chapter 2) is fixed at V, then n is given by:

$$n = \frac{\left(\displaystyle\sum_{h=1}^{L} W_h \sigma_h \sqrt{c_h}\right)\left[\displaystyle\sum_{h=1}^{L} \left(W_h \sigma_h / \sqrt{c_h}\right)\right]}{V + \dfrac{1}{N} \displaystyle\sum_{h=1}^{L} W_h \sigma_h^2} \qquad (8.54)$$

Trading off Cost and Sample Size to Maximize the Power of Detecting a Change

Millard and Lettenmaier (1986) discuss optimal sampling designs in the context of detecting change. They use a standard analysis of variance model that allows for sampling at multiple stations both before and after some event (e.g., operation of a power plant) that may cause a change in the level of the response variable (e.g., density of benthic organisms). The model also allows for seasonal and other auxiliary effects. The sampling cost is assumed to be the sum of a fixed initial cost plus components for the number of sampling occasions, the number of stations sampled per occasion, and the number of replicates taken at each station on each sampling occasion. They show how to optimize the design by maximizing the power to detect a specified change given a fixed cost, or minimizing the cost given a specified power to detect a specified change.

SUMMARY

- Steps 6 and 7 of the DQO process (see Chapter 2) require specification of acceptable limits on decision error rates (i.e., probabilities of Type I and II errors) and optimization of the sampling design. The statistical methods discussed in this chapter allow you to compute decision error rates as well as sample size requirements for confidence intervals and hypothesis tests.
- Parametric confidence intervals based on the normal distribution depend on sample size(s), the estimated standard deviation, and the confidence level. You can use formulas presented in this chapter to help you decide how many samples you need to take for a given precision (half-width) of the confidence interval.
- To apply the formulas for parametric confidence intervals, you need to assume a value for the estimated standard deviation s. The value you choose for s is usually based on a pilot study, a literature search,

or "expert judgment." It is usually a good idea to use some kind of upper bound on s to ensure an adequate sample size.

- Parametric hypothesis tests on a mean or difference between two means involve four quantities: sample size(s), Type I error, Type II error (or equivalently, power), and the scaled minimal detectable difference. You can use the formulas presented in this chapter to determine one of these quantities given the other three.

- Nonparametric confidence, prediction, and tolerance intervals are based on order statistics from the sample. You can use the formulas presented in this chapter to determine confidence levels for a specified sample size or vice-versa.

- In the context of confidence, prediction, and tolerance intervals, in general, there is a heavy price to pay for not making an assumption about the parent distribution of the observations: required sample sizes for specified confidence levels usually are very large. On the other hand, nonparametric hypothesis tests are often almost as powerful as their parametric counterparts when the normal distribution assumption is adequate, and more powerful when the normal distribution assumption is not met.

- For groundwater monitoring at hazardous and solid waste sites, the U.S. EPA has promulgated the concept of the "EPA Reference Power Curve," which in 1991 was deemed to be based on a parametric (normal) prediction interval for a single well and a single constituent assuming a confidence level of 99% and a background sample size of $n = 8$.

- We noted in Chapter 6 that simultaneous prediction limits are the preferred method of handling the problem of testing for contamination from multiple constituents at multiple wells on each sampling occasion. The formulas presented in this chapter allow you to determine the power of using parametric simultaneous prediction intervals, given a specified overall Type I error rate (FWFPR). The power of nonparametric prediction intervals is usually only slightly less than that of parametric prediction intervals.

- Given an equation to model the cost of a sampling design, you can optimize the design to yield the minimum variance of the mean or maximum power to detect change for a fixed cost, or minimize the cost for a specified variance of the mean or power to detect a change.

EXERCISES

8.1. Using the built-in data set `halibut` in S-PLUS, estimate the mean catch per unit effort (CPUE) and create a 95% confidence interval. Assuming a standard deviation of about 50, what is the required sample size for a confidence interval with a half-width of about 20? 10? 5?

8.2. Consider the copper data shown in Table 3.14 of Exercise 3.7 in Chapter 3. Combine the data from Wells 1 and 2, and compute the pooled estimate of standard deviation allowing for different means for the background and compliance well data.

    a. Based on this estimate of standard deviation, how many samples are required at the background and compliance wells in order to achieve a confidence interval half-width of 5 ppb?

    b. Repeat part **a** above, but assume the background well sample size if fixed at $n_2 = 12$.

8.3. Sometimes when the results of an opinion poll are reported, the results are qualified by a statement something like the following: "53% of those polled said they were willing to pay higher taxes to enforce stricter environmental standards. These results are accurate to within three percentage points." Assuming this statement means that the 95% confidence interval for the estimated proportion has a half-width of three percentage points, determine how many people must have been polled. Assume the estimated proportion is about 50%.

8.4. In Examples 7.19 and 7.20 we tested whether there was a significant difference in the incidence of thyroid tumors between rats exposed and not exposed to ethylene thiourea (ETU). The observed incidence was 16/69 (23%) in the exposed group and 2/72 (3%) in the unexposed group. Suppose we want to create a confidence interval for the difference in the incidence rates. How many rats would have to be in each group if we want the half-width of the confidence interval to be 0.05? Assume the incidence rates in the exposed and unexposed groups are about 20% and 5%, respectively.

8.5. Consider Example 8.9 again and the design calculations for a two-sample t-test to compare sulfate concentrations at an upgradient and downgradient well.

    a. Create a plot of power vs. sample size assuming a significance level of 5% and a scaled MDD of 1.

    b. Create a plot of power vs. scaled MDD assuming $n_1 = n_2 = 16$ and a significance level of 5%. Add the EPA reference power curve to this plot.

8.6. Using the halibut data set and without assuming a particular probability distribution, estimate the 95th percentile of the exploitable biomass and try to create a 99% one-sided upper confidence interval. How many observations are required to compute such a confidence interval for the 95th percentile?

8.7. In the United States, there are about 2.3 million deaths per year, and of these deaths approximately 23% are attributed to cancer (see www.census.gov/statab/www). When deciding on acceptable levels for contaminants in the environment, the U.S. EPA often uses a standard that exposure to the chemical should not increase a person's lifetime risk of dying from cancer by more than 1 in 1 million (1×10^{-6}). Using the null hypothesis that the probability of dying from cancer is $H_0: p = 0.23$, what is the power of detecting a minimal detectable difference of $\delta = 1 \times 10^{-6}$ for a one-sided upper alternative hypothesis, assuming a significance level of 5% and sample size of $n = 2{,}300{,}000$?

8.8. Using the information in Problem 8.7, determine a two-sided 95% confidence interval for the true probability of dying from cancer assuming a sample size of $n = 2{,}300{,}000$ and an observed number of deaths from cancer of $x = 529{,}000$ (23% of 2,300,000). What is the half-width of this confidence interval, in terms of percentage points?

8.9. Consider Exercise 8.5 again. Now assume there are $r = 20$ monitoring wells and $n_c = 10$ constituents being monitored. Plot power vs. scaled MDD based on using simultaneous prediction intervals, assuming $n = 16$ and that you are controlling the FWFPR at 5% by using the Bonferroni inequality. Add the EPA reference power curve.

9   LINEAR MODELS

Correlation, Regression, Calibration and Detection Limits, and Dose-Response Models

Often in environmental data analysis we are interested in the relationship between two or more variables. For example, hydrologists often use models to predict sediment load or chemical concentration in a river or stream as a function of the flow. Air quality analysts model ozone or particulate concentration as a function of other meteorological factors such as humidity, radiation, wind speed, temperature, etc. In Chapter 3 we discussed graphs for looking at the relationships between two or more variables. In this chapter we will discuss ways to model these relationships. We will postpone discussion of the special case of looking at the relationship between a variable and how it behaves over time until Chapter 11 on Time Series Analysis.

COVARIANCE AND CORRELATION

The *covariance* between two random variables X and Y is defined as:

$$\sigma_{x,y} = Cov\left(X, Y\right) = E\left[\left(X - \mu_x\right)\left(Y - \mu_y\right)\right] \quad (9.1)$$

where μ_x and μ_y denote the means of X and Y, respectively. The covariance measures how X and Y deviate from their average values as a pair. Figure 9.1 shows a picture indicating the sign (positive or negative) of the cross-product $(X-\mu_x)(Y-\mu_y)$ for each quadrant, where the quadrants are defined by the intersection of the lines $x = \mu_x$ and $y = \mu_y$. If Y tends to be above its average when X is above its average, and Y tends to be below its average when X is below its average, then the covariance is positive. Conversely, if Y tends to be below its average when X is above its average, and Y tends to be above its average when X is below its average, then the covariance is negative. If knowing that X is above or below its average gives us no information on the odds that Y is above or below its average, then the covariance is 0.

It can be shown that

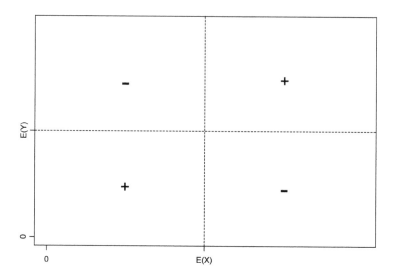

Figure 9.1   Sign of the cross-product $(X-\mu_x)\ (Y-\mu_y)$

$$E\left[\left(X\ -\ \mu_x\right)\left(Y\ -\ \mu_y\right)\right]\ =\ E\left(XY\right)\ -\ \mu_x\mu_y \qquad (9.2)$$

Also, when X and Y are independent, it can be shown that

$$E\left(XY\right)\ =\ E\left(X\right)E\left(Y\right)\ =\ \mu_x\mu_y \qquad (9.3)$$

so when X and Y are independent the covariance is 0. Note that the covariance between a random variable X and itself is the same as the variance of X. That is:

$$\begin{aligned} \sigma_{X,X}\ &=\ Cov\left(X,\ X\right)\ =\ E\left[\left(X\ -\ \mu_x\right)^2\right] \\ &=\ Var\left(X\right)\ =\ \sigma_X^2 \end{aligned} \qquad (9.4)$$

The covariance between X and Y can range from $-\infty$ to ∞, depending on the variances of X and Y. The **correlation** between X and Y, usually denoted with the Greek letter ρ (rho), is the covariance between X and Y scaled by the product of their respective standard deviations:

$$\rho = Cor\left(X, Y\right) = \frac{Cov\left(X, Y\right)}{\sqrt{Var\left(X\right)Var\left(Y\right)}} = \frac{\sigma_{x,y}}{\sigma_x \sigma_y} \quad \text{(9.5)}$$

(Draper and Smith, 1998). The correlation between two random variables is unitless and always lies between -1 and 1. Note that the correlation between a variable and itself is 1.

Figure 9.2 shows four examples of relationships between two random variables X and Y with varying correlations. Each of these graphs was constructed by taking a random sample of 20 (X, Y) pairs from a bivariate population with a known correlation.

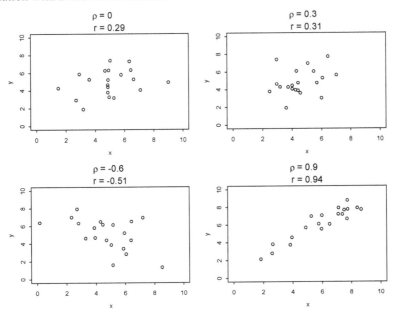

Figure 9.2   Four examples of correlation between two random variables X and Y based on samples of $n = 20$ paired observations

Estimating Covariance and Correlation

The covariance between two random variables X and Y can be estimated using either the unbiased or method of moments estimator. The unbiased estimator is given by:

$$\hat{\sigma}_{x,y} = \frac{1}{n-1} \sum_{i=1}^{n} \left(x_i - \mu_x\right)\left(y_i - \mu_y\right) \qquad (9.6)$$

and the method of moments estimator uses the multiplier $1/n$ instead of $1/(n-1)$. Based on the definition of correlation in Equation (9.5), it is easy to see that the estimate of correlation or *sample correlation*, usually denoted r, is simply the estimate of covariance divided by the product of the estimated standard deviations of X and Y. The estimator of correlation based on the unbiased estimators of covariance and variance is exactly the same as the estimator based on the method of moments estimators of covariance and variance:

$$\hat{\rho} = r = \frac{\hat{\sigma}_{x,y}}{\hat{\sigma}_x \hat{\sigma}_y}$$

$$\qquad (9.7)$$

$$= \frac{\sum_{i=1}^{n} (x_i - \bar{x})\left(y_i - \bar{y}\right)}{\sqrt{\sum_{i=1}^{n} (x_i - \bar{x})^2 \sum_{i=1}^{n} (y_i - \bar{y})^2}}$$

(Draper and Smith, 1998, p. 41; Zar, 1999, p. 378). Figure 9.2 displays the estimated correlations for each of the four examples.

Example 9.1: Correlation between Ozone Concentrations in Two Locations

In Example 7.3 of Chapter 7 we introduced the `ozone` data set, which contains daily maximum ozone concentrations (ppb) at ground level recorded between May 1 and September 30, 1974 at sites in Yonkers, New York and Stamford, Connecticut. Figure 9.3 displays a scatterplot of the concentrations in Stamford vs. the concentrations in Yonkers. The estimated correlation between these measurements is $r = 0.87$.

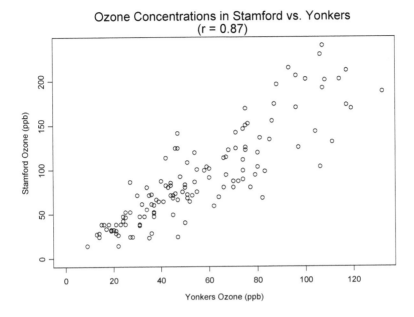

Figure 9.3   Maximum daily ozone concentrations in Stamford, CT vs. those in
Yonkers, NY

Menu

To compute the correlation between the ozone concentrations using the
S-PLUS pull-down menu, follow these steps.

1.   Find the data set **ozone** in the Object Explorer.
2.   On the S-PLUS menu bar, make the following menu choices:
     Statistics>Data Summaries>Correlations. This will bring up the
     Correlations and Covariances dialog box.
3.   For Data Set select **ozone**, for Variables select **<All>**, Under Statis-
     tic, for Type select **Correlations**. Click **OK** or **Apply**.

Command

To compute the correlation between the ozone concentrations using the
S-PLUS Command or Script Window, type this command.

```
cor(ozone)
```

Five Caveats about Correlation

Although correlation is a useful measure and is intuitively appealing, its simplicity can mask more complex relationships. The following five caveats about correlation are based on cautions in Chambers et al. (1983, pp. 76–79) and Helsel and Hirsch (1992, pp. 209–210):

- Correlation is a measure of the *linear* association between two random variables. A small absolute value of correlation does not imply two random variables are unrelated; two random variables may be strongly related in a curvilinear way and have a low level of correlation (see the upper left-hand plot in Figure 9.4).
- Sample correlation is affected by "outliers." In the upper right-hand plot of Figure 9.4, there is no true underlying relationship between the x- and y-values for the 19 values in the lower left-hand part of the scatterplot, but the one outlier forces a sample correlation of $r = 0.7$ for all 20 values.
- Sample correlation cannot account for complex relationships. For example, in the lower left-hand plot of Figure 9.4 the sample correlation is negative ($r = -0.08$), while the sample correlation for each individual line is $r = 1$!
- Sample correlation depends on the range of the variables. The relationship between two random variables may stay the same over a wide range of values for X and Y, but the estimated correlation will be smaller if the X values are confined to a limited range vs. looking at X values on a broader range. The lower right-hand plot of Figure 9.4 displays a subset of the ozone concentrations shown in Figure 9.3. In Figure 9.4 we have restricted the Yonkers ozone concentrations to lie between 60 and 80 ppb. Here the estimated correlation is $r = 0.41$, but for the full data set it is $r = 0.87$.
- Correlation does not imply cause. In some instances, two variables are correlated because one variable directly affects the other. For example, precipitation causes runoff and affects stream flow (Helsel and Hirsch, 1992). In other cases, two variables are correlated because they are both related to a third variable. For example, if you take a sample of cities and plot the number of crimes committed annually vs. the number of places of worship you may very well see a positive correlation. Of course both of these variables depend on the population of the city. Finally, two variables may exhibit correlation just by chance, with no immediate explanation as to why they do. As Helsel and Hirsch (1992, p. 210) state, "Evidence for causation must come from outside the statistical analysis – from the knowledge of the process involved."

These caveats make it clear that you should never report just an estimate of correlation without looking at the data first!

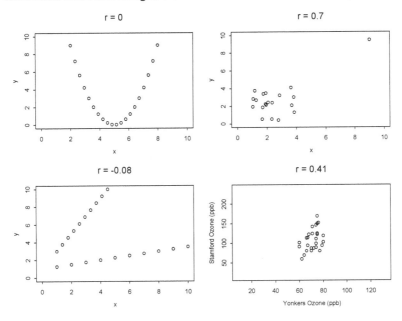

Figure 9.4   Four examples of the danger of reporting sample correlation without plotting the data

Confidence Intervals and Tests for Correlation

Zar (1999, Chapter 19) discusses how to create confidence intervals for the population correlation coefficient. To test the null hypothesis that the population correlation ρ is equal to 0, you can use the statistic:

$$t = \frac{r}{\sqrt{\dfrac{1 - r^2}{n - 2}}} \qquad (9.8)$$

which follows a Student's t-distribution with $n-2$ degrees of freedom under the null hypothesis (Sheskin, 2000, p. 766; Zar, 1999, p. 381). This test assumes the X and Y variables follow normal distributions, or that for each X value the Y values come from a normal distribution, with possibly different means for each value of X but the same variance over all X values. Deviations from the normality or homogeneity of variance assumption can affect the Type I error of this test (Zar, 1999, p. 380). In such cases, you probably want to consider using data transformations (see the subsection Using Transformations later in this chapter).

Example 9.2: Testing for Significant Correlation in the Ozone Data

In this example we will test the null hypothesis that the correlation between the Yonkers and Stamford ozone data is 0 vs. the one-sided alternative that the correlation is greater than 0. The computed t-statistic is 20.5 based on 130 degrees of freedom, yielding a p-value of essentially 0. Note that the assumption of equal variability among the Stamford concentrations for each fixed value of Yonkers concentrations is suspect; the Stamford observations show more scatter for larger values of Yonkers ozone concentrations (see Figure 9.3).

Menu

In the current version of S-Plus 2000 (Release 2), you cannot test the significance of the correlation using the S-Plus pull-down menu except by performing a simple linear regression analysis (see the section Simple Linear Regression below).

Command

To test the significance of the correlation between the Yonkers and Stamford ozone concentrations using the S-Plus Command or Script Window, type this command.

```
cor.test(ozone$yonkers, ozone$stamford,
    alternative="greater")
```

Other Measures of Monotonic Correlation

The estimator of correlation shown in Equation (9.7) is called ***Pearson's product-moment correlation coefficient***. Two other commonly used estimators of correlation are the Spearman rank correlation coefficient or ***Spearman's rho*** (Conover, 1980, p. 252; Hollander and Wolfe, 1999, p. 394; Sheskin, 2000, p. 863; Zar, 1999, p. 395), and ***Kendall's tau*** (Conover, 1980, p. 256; Hollander and Wolfe, 1999, p. 363; Sheskin, 2000, p. 881). Unlike the Pearson product-moment correlation coefficient, which only measures linear association, Spearman's rho and Kendall's tau are more general measures of any kind of monotonic relationship between X and Y. Also, both of these latter measures are based on ranks and are therefore not as sensitive to outliers.

Spearman's Rho

Spearman's rank correlation coefficient is simply the Pearson product-moment correlation coefficient based on the ranks of the observations instead of the observations themselves (where ranking is done separately within the X's and within the Y's):

$$\hat{\rho}_S = r_S = \frac{\sum_{i=1}^{n} (R_{x_i} - \bar{R}_x)\left(R_{y_i} - \bar{R}_y\right)}{\sqrt{\sum_{i=1}^{n} (R_{x_i} - \bar{R}_x)^2 \sum_{i=1}^{n} (R_{y_i} - \bar{R}_y)^2}} \tag{9.9}$$

In Equation (9.9), R_{x_i} denotes the rank of x_i and R_{y_i} denotes the rank of y_i. Conover (1980, p. 456), Hollander and Wolfe (1999, p. 732), Sheskin (2000, p. 962) and Zar (1999, pp. App116–App118) give tables of critical values of r_s that can be used to test the null hypothesis of independence between X and Y. As an alternative, you can employ the t-statistic of Equation (9.8) using r_s, which provides a reasonably good approximation as long as the sample size is greater than 10 (Sheskin, 2000, p. 868). Some references suggest using the test statistic:

$$z = \sqrt{n-1}\, r_S \tag{9.10}$$

which is approximately distributed as a standard normal random variable under the null hypothesis (Hollander and Wolfe, 1999, p. 395), but this approximation requires a sample size anywhere from 25 to 100 (Sheskin, 2000, pp. 868, 879).

Kendall's Tau

Kendall (1938, 1975) proposed a test for the hypothesis that the X and Y random variables are independent based on estimating the following quantity:

$$\tau = \left\{2\,\text{Pr}\left[(X_1 - X_2)(Y_1 - Y_2)\right] > 0\right\} - 1 \tag{9.11}$$

The quantity in Equation (9.11) is called Kendall's tau, although this term is more often applied to the estimate of τ (see Equation (9.12) below). If X and

Y are independent, then $\tau = 0$. Furthermore, for most distributions of interest, if $\tau = 0$ then the random variables X and Y are independent. (It can be shown that there exist some distributions for which $\tau = 0$ and the random variables X and Y are not independent (Hollander and Wolfe, 1999, p. 364)).

Note that Kendall's tau is similar to a correlation coefficient in that $-1 \leq \tau \leq 1$. If X and Y always vary in the same direction, that is if $X_1 < X_2$ always implies $Y_1 < Y_2$, then $\tau = 1$. If X and Y always vary in the opposite direction, that is if $X_1 < X_2$ always implies $Y_1 > Y_2$, then $\tau = -1$. If $\tau > 0$ this indicates X and Y are positively associated, and if $\tau < 0$ this indicates X and Y are negatively associated.

Kendall's tau can be estimated by:

$$\hat{\tau} = \frac{2S}{n(n-1)} \tag{9.12}$$

where

$$S = \sum_{i=1}^{n-1} \sum_{j=i+1}^{n} sign\left[\left(x_j - x_i\right)\left(y_j - y_i\right)\right] \tag{9.13}$$

$$sign\left(t\right) = \begin{cases} 1, & t > 0 \\ 0, & t = 0 \\ -1, & t < 0 \end{cases} \tag{9.14}$$

(Conover, 1980, pp. 256–260; Gibbons, 1994, Chapter 9; Gilbert, 1987, Chapter 16; Helsel and Hirsch, 1992, pp. 212–216; Hollander and Wolfe, 1999, Chapter 8). Note that the quantity S defined in Equation (9.13) is equal to the number of concordant pairs minus the number of discordant pairs. Hollander and Wolfe (1999, p. 364) use the notation K instead of S, and Conover (1980, p. 257) uses the notation T.

The null hypothesis of independence between the two random variables X and Y ($H_0 : \tau = 0$) can be tested using the S statistic defined above. Tables of critical values of S for small samples are given in Hollander and Wolfe (1999, pp. 724–731), Conover (1980, pp. 458–459), Gilbert (1987, p. 272), Helsel and Hirsch (1992, p. 469), and Gibbons (1994, p. 180). The large sample approximation to the distribution of S under the null hypothesis is given by:

$$z = \frac{S - E(S)}{Var(S)} \qquad (9.15)$$

where

$$E(S) = 0 \qquad (9.16)$$

$$Var(S) = \frac{n(n-1)(2n+5)}{18} \qquad (9.17)$$

Under the null hypothesis, the quantity z defined in Equation (9.15) is approximately distributed as a standard normal random variable. Both Kendall (1975) and Mann (1945) show that the normal approximation is excellent even for samples as small as $n = 10$, provided that the following continuity correction is used:

$$z = \begin{cases} \dfrac{S-1}{Var(S)} & , \ S > 0 \\ 0 & , \ S = 0 \\ \dfrac{S+1}{Var(S)} & , \ S < 0 \end{cases} \qquad (9.18)$$

In the case of tied observations in either the X's or Y's, the formula for the variance of S given in Equation (9.17) must be modified as shown in Equation (9.19) below. In this equation, g denotes the number of tied groups in the X observations, t_i is the size of the ith tied group in the X observations, h is the number of tied groups in the Y observations, and u_j is the size of the jth tied group in the Y observations. In the case of no ties in either the X or Y observations, Equation (9.19) reduces to Equation (9.17).

$$Var\ (S)\ =\ \frac{n\ (n\ -\ 1)\ (2n\ +\ 5)}{18}\ -\ \frac{\sum_{i=1}^{g} t_i\ (t_i\ -\ 1)\ (2t_i\ +\ 5)}{18}$$

$$-\ \frac{\sum_{j=1}^{h} u_j\ (u_j\ -\ 1)\ (2u_j\ +\ 5)}{18}\ +$$

$$\frac{\left[\sum_{i=1}^{g} t_i\ (t_i\ -\ 1)\ (t_i\ -\ 2)\right]\left[\sum_{j=1}^{h} u_j\ (u_j\ -\ 1)\ (u_j\ -\ 2)\right]}{9n\ (n\ -\ 1)\ (n\ -\ 2)}$$

$$+\ \frac{\left[\sum_{i=1}^{g} t_i\ (t_i\ -\ 1)\right]\left[\sum_{j=1}^{h} u_j\ (u_j\ -\ 1)\right]}{2n\ (n\ -\ 1)}$$

(9.19)

Example 9.3: Testing for Independence in the Ozone Data

Using the Stamford and Yonkers ozone data, Spearman's rho is equal to 0.88, yielding a t-statistic of 21.2 with 130 degrees of freedom. Kendall's tau is equal to 0.70, yielding a z-statistic of $z = 11.95$. Both of the associated one-sided p-values are essentially 0.

Menu

In the current version of S-PLUS 2000 (Release 2), you cannot test the significance of Spearman's rho or Kendall's tau using the S-PLUS pull-down menu.

Command

To test the significance of Spearman's rho and Kendall's tau for the Yonkers and Stamford ozone concentrations using the S-PLUS Command or Script Window, type these commands.

```
cor.test(ozone$yonkers, ozone$stamford,
   alternative="greater", method="spearman")
cor.test(ozone$yonkers, ozone$stamford,
   alternative="greater", method="kendall")
```

Note: For the test of Spearman's rho, ENVIRONMENTALSTATS for S-PLUS uses the t-statistic of Equation (9.8), while S-PLUS uses the z-statistic of Equation (9.10).

SIMPLE LINEAR REGRESSION

In Chapter 3 we looked at the relationship between ozone concentration and daily maximum temperature taken in New York based on 111 daily observations taken between May and September 1973. Figure 3.25 displays the scatterplot of ozone vs. temperature. If we are interested in modeling the relationship between ozone and temperature, one of the simplest models we can use is a simple straight line fit. Figure 9.5 displays the scatterplot along with a fitted line. In this section we will discuss how you calculate this fitted line and decide whether you need to somehow make your model better.

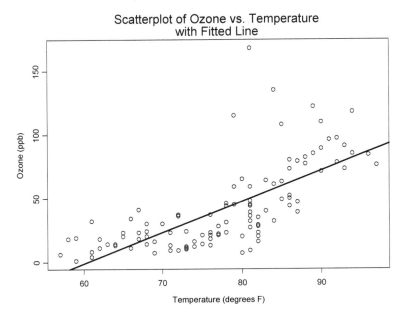

Figure 9.5   Scatterplot of ozone vs. temperature with fitted line

Mathematical Model

Simple linear regression assumes we have two variables X and Y, and that the relationship between X and Y can be expressed as follows:

$$Y = \beta_0 + \beta_1 X + \varepsilon \qquad (9.20)$$

where β_0 denotes the intercept of the line, β_1 denotes the slope of the line, and ε denotes a random variable with a mean of 0 and a standard deviation of σ_ε. That is, for a fixed value of X, say X^*, the average or expected value of Y is given by $\beta_0 + \beta_1 X^*$. Mathematically, this is written as:

$$\mu_{Y|X^*} = E\left(Y \mid X = X^*\right) = \beta_0 + \beta_1 X^* \qquad (9.21)$$

where the vertical line | is interpreted to mean "given." Also, for this fixed value of X, the Y values are bouncing around the mean with a standard deviation of σ_ε. Figure 9.6 illustrates these assumptions based on sampling 20 Y observations for each fixed value of X.

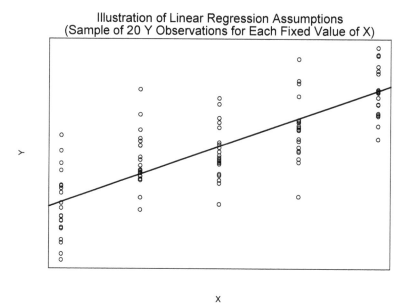

Figure 9.6   Illustration of the assumptions of linear regression

In the model shown in Equation (9.20), the variable Y is often called the **response variable** and the variable X is often called the **predictor variable**. The names **dependent** and **independent** variable are also often used to denote Y and X, but we avoid this terminology here to avoid confusion with the meaning of the term independent as applied to independent random variables.

Fitting the Line

The most common way to estimate the intercept and slope parameters of Equation (9.20) is to use a technique called least squares (Draper and Smith, 1998, Chapter 1; Zar, 1999, Chapter 17). This technique finds the fitted line that minimizes the sum of the squared vertical distances between the observed values of Y and the "predicted" values of Y based on the fitted line. Figure 9.7 illustrates this idea.

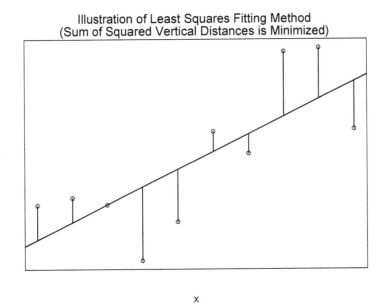

Illustration of Least Squares Fitting Method
(Sum of Squared Vertical Distances is Minimized)

Figure 9.7   Illustration of the idea behind the least squares method of fitting a straight line

Mathematically, the technique of least squares finds the estimated values of β_0 and β_1 that minimize the following quantity:

$$RSS = \sum_{i=1}^{n} (y_i - \hat{y}_i)^2$$

$$= \sum_{i=1}^{n} \left[y_i - \left(\hat{\beta}_0 + \hat{\beta}_1 x_i \right) \right]^2$$

(9.22)

where RSS stands for "Residual Sum of Squares." A *residual* is simply the observed value of Y minus its predicted value; so the ith residual is defined as:

$$r_i = y_i - \hat{y}_i \tag{9.23}$$

Thus, the technique of least squares minimizes the sum of the squared residuals. It can be shown that the formulas for the estimated slope and intercept are given by:

$$\hat{\beta}_1 = \frac{\sum\limits_{i=1}^{n} \left(x_i - \bar{x}\right)\left(y_i - \bar{y}\right)}{\sum\limits_{i=1}^{n} (x_i - \bar{x})^2} \tag{9.24}$$

$$\hat{\beta}_0 = \bar{y} - \hat{\beta}_1\bar{x} \tag{9.25}$$

(Draper and Smith, 1998, pp. 24–25; Zar, 1999, pp. 328, 330). Based on Equations (9.20) and (9.25), we can write:

$$\hat{Y} = \hat{\beta}_0 + \hat{\beta}_1 X = \bar{y} + \hat{\beta}_1 \left(X - \bar{x}\right) \tag{9.26}$$

which shows that the fitted line contains the point $(\bar{x}, \bar{y})$. Also, if the postulated model includes just an intercept and assumes a slope of 0 (i.e., $\beta_1 = 0$ so X is assumed to have no influence on Y), then the estimated intercept and fitted value of Y is simply the mean of the Y observations.

This method of fitting a straight line via least squares is commonly known as *linear regression*. Sir Francis Galton introduced this term in 1885 when he was describing the results of analyses he had performed looking at the relationship between the seed size of parents and offspring in plants, and the relationship between the heights of parents and offspring in humans (Draper and Smith, 1998, p. 45). Galton found that the seed size of offspring did *not* tend to resemble their parents' seed size, but instead "regressed towards mediocrity": If parents had larger than normal seeds, the seeds of the offspring tended to be smaller than those of their parents; if parents has smaller than normal seeds, the seeds of the offspring tended to be larger than those of their parents. So if you fit a straight line to a plot of offspring seed size vs. parent seed size, the slope of the line would be less than 1. Galton found a similar phenomenon in the heights of humans. The term linear regression now refers to the general practice of fitting linear models via least squares.

Example 9.4: Fitting a Straight Line to the Ozone Data

The formula for the fitted straight line shown in Figure 9.5 is given by:

$$\hat{O} = -147.6 + 2.4\,T \qquad\qquad \textbf{(9.27)}$$

That is, the intercept is −147.6 and the slope is 2.4. So the fitted straight line relationship between ozone and temperature says that for every 1 degree increase in temperature, ozone increases by 2.4 ppb.

Menu

To estimate the regression parameters of the straight line fitted to the ozone data using the S-PLUS pull-down menu, follow these steps.

1. Find **environmental** in the Object Explorer.
2. On the S-PLUS menu bar, make the following menu choices: **Statistics>Regression>Linear**. This will bring up the Linear Regression dialog box.
3. For Data Set select **environmental**, for Dependent select **ozone**, and for Independent select **temperature**. In the Save As box type **ozone.fit**. Click **OK** or **Apply**.

To reproduce Figure 9.5 using the pull-down menu or toolbars, follow these steps.

1. On the S-PLUS menu bar, make the following menu choices: **Graph>2D Plot**. This will bring up the Insert Graph dialog box. Under the Axes Type column **Linear** should be highlighted. Under the Plot Type column select **Fit-Linear Least Squares** and click **OK**. (Alternatively, left-click on the **2D Plots** button, then left-click on the **Linear Fit** button.)
2. The Curve Fitting Plot dialog box should appear. For Data Set select **environmental**, for x Columns select **temperature**, for y Columns select **ozone**, then click **OK** or **Apply**.

Command

To estimate the regression parameters of the straight line fitted to the ozone data using the S-PLUS Command or Script Window, type these commands.

```
ozone.fit <- lm(ozone ~ temperature,
   data=environmental)
ozone.fit
```

To reproduce Figure 9.5, type these commands.

```
x <- environmental$temperature
y <- environmental$ozone
plot(x, y, xlab="Temperature (degrees F)",
   ylab="Ozone (ppb)")
abline(ozone.fit)
title(main=paste("Scatterplot of Ozone vs.
   Temperature", "with Fitted Line", sep="\n"))
```

Hypothesis Test for the Significance of the Regression

In Equation (9.20), if we set the slope to 0 (i.e., $\beta_1 = 0$) then there is no assumed relationship between Y and X; knowing the value of X does not help us predict what the value of Y will be. In this case, the best predictor of Y is $\mu_y = E(Y)$, the average value of Y, and indeed if we minimize the residual sum of squares assuming $\beta_1 = 0$ we find that the estimated intercept is the sample mean. To test whether the variable X is really helping us to predict the value of Y, we need to test the null hypothesis H_0: $\beta_1 = 0$. To test this hypothesis, we have to make some kind of assumption about the distribution of the random error term ε in Equation (9.20). The most common assumption is that the errors, the deviations of the Y's from their expected values (how the Y's bounce around the line at a fixed value of X), come from a normal distribution.

The variance of the estimator of slope is given by:

$$Var\left(\hat{\beta}_1\right) = \sigma^2_{\hat{\beta}_1} = \frac{\sigma^2_\varepsilon}{\sqrt{\sum_{i=1}^{n}(x_i - \bar{x})^2}} \tag{9.28}$$

To test the significance of the slope, we can use the t-statistic

$$t = \frac{\hat{\beta}_1}{\hat{\sigma}_{\hat{\beta}_1}} = \frac{\hat{\beta}_1}{s / \sqrt{\sum_{i=1}^{n} (x_i - \bar{x})^2}} \qquad (9.29)$$

where

$$s = \hat{\sigma}_\varepsilon = \sqrt{\frac{RSS}{n-2}} \qquad (9.30)$$

This t-statistic follows a Student's t-distribution with $n-2$ degrees of freedom under the null hypothesis (Draper and Smith, 1998, p. 36; Zar, 1999, pp. 336–337). The results of testing the significance of a regression model are often presented in terms of an analysis of variance table. See Draper and Smith (1998) and Zar (1999) for details.

Example 9.5: Testing the Significance of the Straight Line Fit to the Ozone Data

In the previous example we showed that the estimated slope of the line fitted to the ozone data shown in Figure 9.5 is equal to 2.4. The estimated standard error of the slope is 0.24, so the t-statistic for this fit is 2.4/0.24 = 10 based on 111–2 = 109 degrees of freedom. This yields a two-sided p-value of essentially 0, so we would reject the null hypothesis that the slope of the regression line is 0.

Menu

To test the significance of the straight line fit using the S-PLUS pull-down menu, follow the same steps as in Example 9.4. The report window contains the estimated slope and intercept, along with their standard errors, associated t-statistics, and p-values.

Command

To test the significance of the straight line fit using the S-PLUS Command or Script Windows, follow the same steps as in Example 9.4 to produce the object ozone.fit, then type the following command.

```
summary(ozone.fit)
```

Relationship between Linear Regression and Correlation

It can be shown that

$$\hat{\rho} = r = \frac{\sqrt{\sum_{i=1}^{n} (x_i - \bar{x})^2}}{\sqrt{\sum_{i=1}^{n} (y_i - \bar{y})^2}} \hat{\beta}_1 \qquad (9.31)$$

where r is the sample correlation coefficient defined in Equation (9.7) (Draper and Smith, 1998, p. 42; Zar, 1999, p. 381). Also, the t-statistic defined in Equation (9.8) to test the significance of the correlation is identical to the t-statistic in Equation (9.29) to test the significance of the regression.

Confidence Intervals for the Slope, Intercept, Mean Value, and Line

Equation (5.87) shows a general formula for a two-sided confidence interval for some parameter θ based on the t-distribution. In general, this formula is approximate and relies on the Central Limit Theorem for its validity. In the case of the estimated slope for a line, this formula is exact because the distribution of the estimated slope is normal when the error terms are assumed to be normal. The two-sided $(1-\alpha)100\%$ confidence interval for the slope is given by:

$$\left[\hat{\beta}_1 - t_{n-2, 1-\alpha/2} \, \hat{\sigma}_{\hat{\beta}_1} \, , \, \hat{\beta}_1 + t_{n-2, 1-\alpha/2} \, \hat{\sigma}_{\hat{\beta}_1} \right] \qquad (9.32)$$

where $\hat{\sigma}_{\hat{\beta}_1}$ is defined in Equation (9.29). You can also construct confidence intervals for the intercept; see Draper and Smith (1998, p. 38).

Often, one of the major goals of regression analysis is to predict a future value of the response variable given a known value of the predictor variable. Equation (9.26) shows the equation for the predicted mean value of Y given a fixed value of X. To construct a confidence interval for this predicted mean value using Equation (5.87), we need to know the variance of the predicted mean value:

$$Var\left(\hat{\mu}_Y \mid X = X^*\right) = \sigma^2_{\hat{\mu}_Y \mid X^*}$$

(9.33)

$$= \sigma^2_\varepsilon \left[\frac{1}{n} + \frac{\left(X^* - \bar{x}\right)}{\sum\limits_{i=1}^{n} (x_i - \bar{x})^2} \right]$$

(Draper and Smith, 1998, p. 80) and the formula for the estimator of this variance:

$$\hat{\sigma}^2_{\hat{\mu}_Y \mid X^*} = s^2 \left[\frac{1}{n} + \frac{\left(X^* - \bar{x}\right)}{\sum\limits_{i=1}^{n} (x_i - \bar{x})^2} \right]$$

(9.34)

where s is defined in Equation (9.30). Thus, the two-sided $(1-\alpha)100\%$ confidence interval for the mean value of Y given $X = X^*$ is given by:

$$\left[\hat{Y} - t_{n-2,1-\alpha/2} \, \hat{\sigma}_{\hat{\mu}_Y \mid X^*} \quad , \quad \hat{Y} + t_{n-2,1-\alpha/2} \, \hat{\sigma}_{\hat{\mu}_Y \mid X^*} \right]$$

(9.35)

Just as we can plot the fitted line, we can connect all of the lower confidence limits over the range of X and all of the upper confidence limits over the range of X to produce lower and upper confidence limits for the line. Technically, however, the formula in Equation (9.35) is a confidence interval for the mean of Y for *one* fixed value of X. To create simultaneous confidence curves for the line over the range of X, the critical t-value $t_{n-2,1-\alpha/2}$ in Equation (9.35) has to be replaced with $[2F_{2,n-2,1-\alpha}]^{1/2}$ (Draper and Smith, 1998, p. 83; Miller, 1981a, p. 111):

$$\left[\hat{Y} \pm \sqrt{2F_{2,n-2,1-\alpha}} \, \hat{\sigma}_{\hat{\mu}_Y \mid X^*} \right]$$

(9.36)

Example 9.6: Confidence Intervals for the Ozone Data

Figure 9.8 displays a plot of ozone vs. temperature, along with the fitted line and 95% upper and lower simultaneous confidence bands. For the specific temperature of 70°, the predicted value of ozone is 23 ppb, and the 95% confidence intervals for the mean value of ozone is [17, 29] ppb. For a temperature of 90°, the predicted value is 72 ppb and the 95% confidence interval is [65, 79].

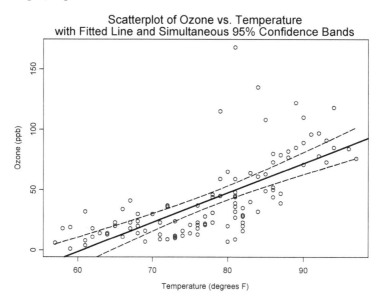

Figure 9.8   Ozone vs. temperature with fitted line and 95% simultaneous confidence bands

Menu

To compute the predicted values and the 95% confidence intervals for the mean value of ozone given temperatures of 70° and 90° using the S-PLUS pull-down menu, we first need to create a new data frame with a single column called **temperature** and two rows: 70 in the first row and 90 in the second row. We will call this data frame **New.Temperature**. To create this data frame, follow these steps.

1. On the S-PLUS tool bar, click on the **New** button, select **Data Set** and click **OK**.
2. In row 1 of column 1 type **70**, and in row 2 of column 1 type **90**. Right click on the top of column 1, choose **Properties**, and in the Name box type **temperature**. Close the data frame and do not save the changes to a file.

3. In the Object Explorer, rename the data frame **New.Temperature**.

To compute the predicted values and 95% confidence intervals, follow these steps.

4. On the S-PLUS menu bar, make the following menu choices: **Statistics>Regression>Linear**. This will bring up the Linear Regression dialog box.

5. For Data Set select **environmental**, for Dependent select **ozone**, and for Independent select **temperature**.

6. Click on the **Predict** tab. For New Data select **New.Temperature**, **check** the Predictions box, and **check** the Confidence Intervals box. Click **OK** or **Apply**.

To create the scatterplot with the fitted line and confidence bands shown in Figure 9.8, follow the same steps shown in Example 9.4 to create Figure 9.5, but modify Step 2 as follows:

2. The Curve Fitting Plot dialog box should appear. For Data Set select **environmental**, for x Columns select **temperature**, for y Columns select **ozone**. Click on the **By Conf Bound** tab. For level type **0.95**, and for Style select the 4th line down, then click **OK** or **Apply**.

Note: The confidence bands created via the S-PLUS menu are not simultaneous confidence bands based on Equation (9.36), but rather individual confidence bands based on Equation (9.35).

Command

To compute predicted values and the 95% confidence intervals for the mean value of ozone given temperatures of 70° and 90° using the S-PLUS Command or Script Windows, follow the same steps as in Example 9.4. to produce the object `ozone.fit`, then type the following commands.

```
predict.list <- predict(ozone.fit,
    newdata=data.frame(temperature=c(70,90)), se.fit=T)
pointwise(predict.list, coverage=0.95)
```

To create the scatterplot with the fitted line and simultaneous confidence bands shown in Figure 9.8, type the same commands shown in Example 9.4 to create Figure 9.5, but omit the call to the `title` function and add the following commands.

```
new.x <- seq(min(x), max(x), length=100)
predict.ozone <- predict(ozone.fit,
    newdata=data.frame(temperature=new.x), se.fit=T)
```

```
ci.ozone <- pointwise(predict.ozone, coverage=0.95,
    simultaneous=T)
lines(new.x, ci.ozone$lower, lty=4)
lines(new.x, ci.ozone$upper, lty=4)
title(main="Scatterplot of Ozone vs. Temperature\nwith
    Fitted Line and Simultaneous 95% Confidence Bands")
```

Note: In ENVIRONMENTALSTATS for S-PLUS the function `pointwise` has been modified to take the argument `simultaneous` so you can specify whether you want non-simultaneous (Equation (9.35)) or simultaneous (Equation (9.36)) confidence limits.

Prediction and Tolerance Intervals

In Chapter 6 we discussed how to construct a prediction interval for the next value of an observation (as opposed to constructing a confidence interval for the mean of the population). We can do the same thing under the linear model of Equation (9.20) using Equation (5.87). The formula for the variance of an estimated single future observation is given by:

$$Var\left(\hat{Y} \mid X = X^*\right) = \sigma^2_{\hat{Y}|X^*}$$

$$\text{(9.37)}$$

$$= \sigma^2_\varepsilon \left[1 + \frac{1}{n} + \frac{\left(X^* - \overline{x}\right)}{\sum\limits_{i=1}^{n} (x_i - \overline{x})^2} \right]$$

and the formula for the estimator of this variance is given by:

$$\hat{\sigma}^2_{\hat{Y}|X^*} = s^2 \left[1 + \frac{1}{n} + \frac{\left(X^* - \overline{x}\right)}{\sum\limits_{i=1}^{n} (x_i - \overline{x})^2} \right] \quad \text{(9.38)}$$

where s is defined in Equation (9.30). Thus, the two-sided $(1-\alpha)100\%$ prediction interval for a single future observation given $X = X^*$ is given by:

$$\left[\hat{Y} - t_{n-2,1-\alpha/2} \; \hat{\sigma}_{\hat{Y}|X^*} \;\;,\;\; \hat{Y} + t_{n-2,1-\alpha/2} \; \hat{\sigma}_{\hat{Y}|X^*} \right] \quad (9.39)$$

(Miller, 1981a, p. 115). We can connect all of the lower prediction limits over the range of X and all of the upper prediction limits over the range of X to produce lower and upper prediction limits for future observations; this is what is usually done in practice. Technically, however, the formula in Equation (9.39) is a prediction interval for a single future observation for *one* fixed value of X.

Miller (1981a, p. 115) gives a formula for simultaneous prediction intervals for k future observations. If we are interested in creating an interval that will encompass all possible future observations over the range of X with some specified probability however, we need to create simultaneous *tolerance* intervals. A formula for such an interval was developed by Lieberman and Miller (1963) and is given in Miller (1981a, p. 124):

$$\left[\hat{Y} \pm \sqrt{2F_{2,n-2,1-\alpha/2}} \; \hat{\sigma}_{\hat{\mu}_Y|X^*} + s \, z_{1-\alpha/2} \sqrt{\dfrac{n-2}{\chi^2_{n-2,1-\alpha/2}}} \right] \quad (9.40)$$

Example 9.7: Prediction Intervals for the Ozone Data

Figure 9.9 displays a plot of ozone vs. temperature, along with the fitted line, 95% upper and lower confidence bands, and 95% upper and lower (non-simultaneous) prediction bands. For the specific temperature of 70°, the predicted value of ozone is 23 ppb, and the 95% prediction interval is [−25, 71] ppb. For a temperature of 90°, the predicted value is 72 ppb and the 95% prediction interval is [24, 120]. These intervals are much wider compared to the confidence intervals for the average value of ozone computed in Example 9.6. Also, the prediction interval for a temperature of 70° includes negative values, which are impossible to observe. This is a reflection of the amount of variability in the observations, and the poor fit of our model (see the next section Regression Diagnostics).

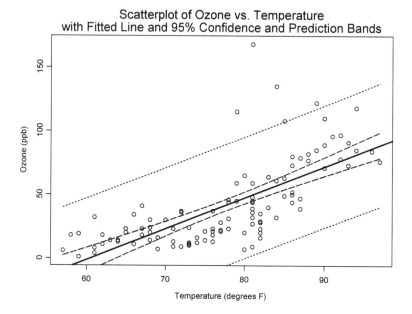

Scatterplot of Ozone vs. Temperature
with Fitted Line and 95% Confidence and Prediction Bands

Figure 9.9   Ozone vs. temperature with fitted line, 95% confidence bands, and 95% prediction bands

Menu

In the current version of S-PLUS 2000 Release 2, you cannot compute prediction limits for a linear model from the menu.

Command

To compute predicted values and the 95% prediction intervals for ozone given temperatures of 70° and 90° using the S-PLUS Command or Script Windows, follow the same steps as in Example 9.4. to produce the object ozone.fit, then type the following commands.

```
predict.list <- predict(ozone.fit,
   newdata=data.frame(temperature=c(70,90)), se.fit=T)
pointwise(predict.list, coverage=0.95, individual=T)
```

To create the scatterplot with the fitted line, confidence bands, and prediction bands shown in Figure 9.9, type the same commands shown in Example 9.4 to create Figure 9.5, but omit the call to the title function and add the following commands.

```
new.x <- seq(min(x), max(x), length=100)
```

```
predict.ozone <- predict(ozone.fit,
   newdata=data.frame(temperature=new.x), se.fit=T)
ci.ozone <- pointwise(predict.ozone, coverage=0.95)
lines(new.x, ci.ozone$lower, lty=4)
lines(new.x, ci.ozone$upper, lty=4)
pi.ozone <- pointwise(predict.ozone, coverage=0.95,
   individual=T)
lines(new.x, pi.ozone$lower, lty=8)
lines(new.x, pi.ozone$upper, lty=8)
title(main="Scatterplot of Ozone vs. Temperature\nwith
   Fitted Line and 95% Confidence and Prediction
   Bands")
```

Note: In ENVIRONMENTALSTATS for S-PLUS the function `pointwise` has been modified to take the argument `individual` so you can specify whether you want a confidence interval for the mean (`individual=F`) or a prediction interval for an individual observation (`individual=T`).

A Caution about Extrapolation

In the previous sections we discussed how to fit a straight line, test for the significance of the slope and intercept, and create confidence and prediction intervals. These discussions all assumed that the model shown in Equation (9.20) is correct and applied in the range of values we observed for the predictor variable. Although it is tempting to use regression models to predict future values of the response variable for values of the predictor variable outside its observed range, there is no way to statistically assess the validity of these predicted values, although sometimes common sense or knowledge of the physical process will indicate the predicted values are nonsense. For example, in Example 9.4 we showed that the fitted line relating ozone to temperature had an intercept of -147.6. That is, when the temperature is $0°$ F, the model predicts an observed value of ozone of -147.6 ppb, an impossible value. The observed values of temperature are between $57°$ and $97°$ F, and that is the only range of temperature values we can use to predict ozone with the linear model and still be able to validly indicate our uncertainty in these predictions.

REGRESSION DIAGNOSTICS

Our discussions of estimating the slope and intercept, confidence intervals for the mean value, and prediction intervals for individual observations have all been based on the following assumptions (Helsel and Hirsh, 1992, p. 225):

1. The linear relationship postulated between the response variable Y and the predictor variable X shown in Equation (9.20) is correct.
2. The set of n observed (X, Y) pairs used to fit the model are representative of the population.
3. The error terms in the model represented by ε in Equation (9.20) come from a distribution with an average value of 0 and a constant standard deviation of σ_ε. Furthermore, the distribution of the error terms does not depend on the value of the predictor variable X.
4. The error terms are independent of one another.
5. The error terms come from a normal distribution.

These assumptions are listed in more or less decreasing order of importance. To fit the model (i.e., estimate the slope and intercept) and produce a predicted value for a given value of X, you only need assumptions 1 and 2. To compute confidence or prediction intervals, or perform hypothesis tests, you need all five assumptions.

Several statistical and graphical procedures have been developed to help you check the assumptions of linear regression, and these tools are classified under the topic of ***regression diagnostics*** (e.g., Draper and Smith, 1998, Chapters 2 and 8). In this section we will discuss diagnostic plots, summary statistics, and the use of transformations.

Diagnostic Plots

Figure 9.10 displays six kinds of regression diagnostic plots based on the linear model relating ozone to temperature. We will discuss the meaning and interpretation of each of these plots below.

*   **Residuals vs. Fitted Values** (row 1, column 1). If the linear model of Equation (9.20) is appropriate, then the residuals should be bouncing around 0 and there should not be any kind of marked pattern to the plot, since the variance of the errors is supposed to be constant. For the ozone linear fit this plot displays both curvilinearity and several "outliers." The curvilinearity indicates that a straight line model is too simple. The outliers indicate that the assumption of a constant variance is suspect.
*   **Square-Root of Absolute Value of Residuals vs. Fitted Values** (row 1, column 2). This plot is similar to the first, but may be more helpful in detecting deviations from the assumption of a constant variance in the errors. If the error variance is constant, there should not be a marked pattern; if the error variance increases or decreases with the fitted values of Y, then you will see a pattern that looks a bit like a megaphone. For the ozone linear fit, there appears to be more variability for larger values of ozone.

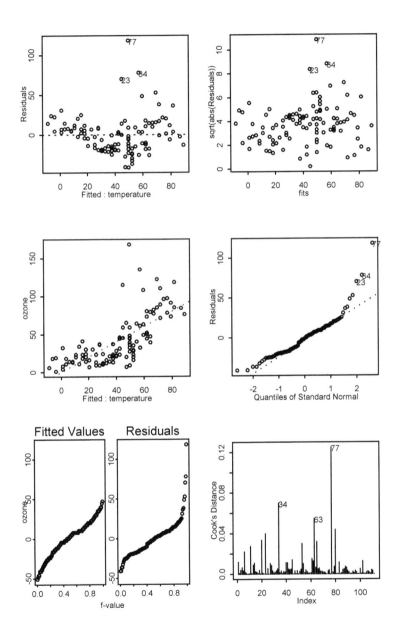

Figure 9.10   Six kinds of regression diagnostic plots for the linear model relating ozone to temperature

- **Observed vs. Fitted Values** (row 2, column 1). Ideally, this plot should show points bouncing very closely about the 0-1 line. For the ozone linear fit, both the curvilinearity and "outliers" are apparent.
- **Normal Q-Q Plot of Residuals** (row 2, column 2). This plot is used to determine the validity of the assumption that the errors come from a normal distribution. For the ozone linear fit, several larger observations fall off of the line, indicating that the normality assumption is suspect.
- **Residual-Fit Spread (r-f) Plot** (row 3, column 1). This plot is really two plots. The first plot is the quantile (empirical cdf) plot of the fitted values (centered by their mean). The second plot is the quantile plot of the residuals. (Note that these quantile plots have the x- and y-axes reversed compared to the way quantile plots are presented in Chapter 3.) The point of creating these two plots is to compare the spread between the two plots. If your fitted model is doing a good job, the spread of the fitted values should be much greater than the spread of the residuals. For the ozone linear fit, the spread is about the same in both plots, indicating our model is not too good.
- **Cook's Distances** (row 3, column 2). Cook's distance is a measure of how the fitted value or slope coefficient changes if you leave one of the observations out of the fit (Draper and Smith, 1998, pp. 211–212). Observations that do not greatly affect the fitted model have small Cook's distances, while observations that greatly affect the fitted model have large Cook's distances. For the ozone linear fit, observation 77 (temperature = 81°, ozone = 168 ppb) has a relatively large Cook's distance. Observations with large Cook's distances are often called ***high-leverage points***. These are not necessarily "bad" observations, they simply exert a large influence on the fitted model. Observations with large Cook's distances are usually associated with values of the predictor variable that are "far away" from the average value of the predictor variable.

Example 9.8: Creating Diagnostic Plots for the Ozone Simple Regression Model

This example explains how to create the diagnostic plots shown in Figure 9.10.

Menu

To create the regression diagnostic plots for the linear model based on the ozone data using the S-PLUS pull-down menu, follow these steps.

1. On the S-PLUS menu bar, make the following menu choices: **Statistics>Regression>Linear**. This will bring up the Linear Regression dialog box.
2. For Data Set select **environmental**, for Dependent select **ozone**, and for Independent select **temperature**.
3. Click on the **Plot** tab. Under Plots, **check** all but the last box. Under Options, **uncheck** the Include smooth box. Click **OK** or **Apply**.

Command

To create the regression diagnostic plots for the linear model based on the ozone data using the S-PLUS Command or Script Window, type this command.

```
plot(ozone.fit)
```

The R^2 Summary Statistic and Hypothesis Tests

A summary statistic that is often quoted to summarize the results of a regression is called R^2 (R-squared) or the *coefficient of determination* and is given by:

$$R^2 = 1 - \frac{\sum_{i=1}^{n} (y_i - \hat{y}_i)^2}{\sum_{i=1}^{n} (y_i - \bar{y})^2} = 1 - \frac{RSS}{SST} \qquad (9.41)$$

where SST denotes the Sum of Squares Total. The Sum of Squares Total is simply the Residual Sum of Squares for the linear model that has just an intercept (i.e., the slope is equal to 0; $\beta_1 = 0$). If the linear model with a slope and intercept is better than the model with just an intercept, then RSS (the Residual Sum of Squares) should be small compared to SST, and therefore R^2 should be close to 1. (Note: $RSS \leq SST$.) For the ozone linear fit, $R^2 = 0.49$.

For the case of a simple linear model with one predictor variable, it can be shown that the square of the sample correlation coefficient given in Equation (9.7) is the same as R^2 (Draper and Smith, 1998, pp. 42–43). Just as we have seen that the sample correlation coefficient is inadequate to fully describe the relationship between two variables, the summary statistic R^2 is inadequate to describe how well a linear model works.

Using Transformations

There are two approaches to improving a model fit: transform the Y and/or X variable, and/or use more predictor variables. We will discuss the first approach in this section and the second approach in the section Multiple Regression later in this chapter.

We are already familiar with the idea of using a transformation on the data in order to improve how well the data fit our pre-conceived model. For example, in Chapters 3 we discussed using Box-Cox transformations to decide what kind of transformation to use to make the observations look like they come from a normal distribution. Often in environmental data analysis we assume the observations come from a lognormal distribution and automatically take logarithms of the data.

If regression diagnostic plots indicate that a straight line fit is not adequate, but that the variance of the errors appears to be fairly constant, you may only need to transform the predictor variable X or perhaps use a quadratic or cubic model in X. On the other hand, if the diagnostic plots indicate that the constant variance and/or normality assumptions are suspect, you probably need to consider transforming the response variable Y. Data transformations for linear regression models are discussed in Draper and Smith (1998, Chapter 13) and Helsel and Hirsch (1992, pp. 228–229).

Example 9.9: Transforming the Ozone Data

In Chapter 3 we discussed how to compute "optimal" Box-Cox transformations based on the probability plot correlation coefficient (PPCC), the Shapiro-Wilk goodness-of-fit statistic, or the log-likelihood function. For linear models, you can compute "optimal" transformations of the response variable based on looking at the residuals from the fit. Figure 9.11 shows a plot of the PPCC for the residuals from the ozone linear model vs. various values of the transform power λ, where the ozone concentrations were transformed prior to the fit. Based on this plot, it looks like using a log-transformation ($\lambda = 0$) should improve the fit. The power that produces the largest PPCC is about 0.2. As we explained in Chapter 3, the S-PLUS built-in data set `air` contains exactly the same variables as `environmental`, except that the cube-root of ozone ($\lambda = 1/3$) is used instead of the raw values of ozone. Figure 9.12 shows the regression diagnostic plots based on modeling the cube-root of ozone as a linear function of temperature. Although the normal Q-Q plot for the residuals looks much better, the r-f plots indicate that the model can still be improved. Figure 3.27 shows that a loess smooth indicates a curvilinear instead of a linear relation. We will talk about how to fit a quadratic or higher order model later in this chapter in the section Polynomial Regression with One Predictor Variable.

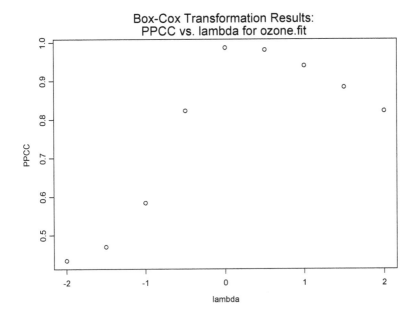

Figure 9.11   Probability plot correlation coefficient (PPCC) vs. Box-Cox transform power (λ) based on the residuals from the linear fit of the transformed ozone as a function of temperature

Menu

To create the plot of the PPCC vs. the transform power λ for the ozone linear fit using the ENVIRONMENTALSTATS for S-PLUS pull-down menu, follow these steps.

1. Find the object **ozone.fit** in the Object Explorer.
2. On the S-PLUS menu bar, make the following menu choices: **EnvironmentalStats>EDA>Box-Cox Transformations**. This will bring up the Box-Cox Transformations dialog box.
3. For Data to Use select **Linear Models**, in the Linear Model box select **ozone.fit**, then click **OK** or **Apply**.

To create the regression diagnostic plots for the linear model specifying the cube-root of ozone as a function of temperature, follow the same steps as in Example 9.8 but use the data set **air** instead of **environmental**. Also, in the Save As box type **cr.ozone.fit**.

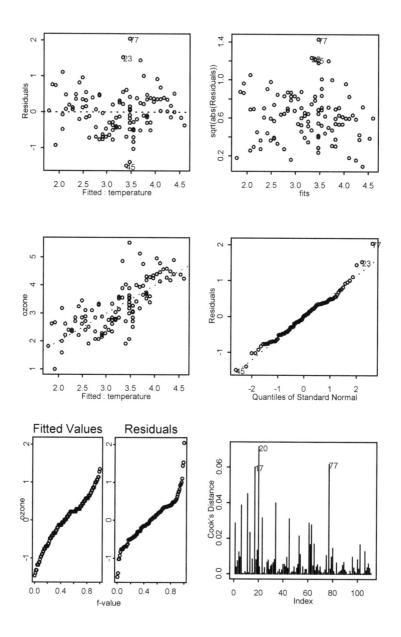

Figure 9.12   Diagnostic plots for the linear model relating the cube-root of ozone to temperature

Command

To create the plot of the PPCC vs. the transform power λ for the ozone linear fit using the ENVIRONMENTALSTATS for S-PLUS Command or Script Window, type these commands.

```
boxcox.list <- boxcox(ozone.fit)

plot(boxcox.list, plot.type="Objective vs. lambda")
```

To create the linear model with the cube-root of ozone as a function of temperature and then create the regression diagnostic plots, type these commands.

```
cr.ozone.fit <- lm(ozone ~ temperature, data=air)

plot(cr.ozone.fit)
```

Problems with Predicting on the Original Scale

As we stated in Chapter 3, a big problem with dealing with transformed observations is that it is usually not so easy to translate results back to the original scale unless we are only concerned with quantiles or prediction limits. Also, in the case of linear models, using transformations changes the nature of the underlying model. For example, if we decide to use the log-transformed observations, then our mathematical model is:

$$W = \ln(Y) = \beta_0 + \beta_1 X + \varepsilon \qquad (9.42)$$

which implies that the model on the original scale is:

$$Y = \exp(\beta_0 + \beta_1 X + \varepsilon) = e^{\beta_0} e^{\beta_1 X} e^{\varepsilon} \qquad (9.43)$$

so the errors are multiplicative on the original scale, not additive.

If you specify an "exponential model" like Equation (9.43) and ask for the predicted value of Y at a specific value of $X = X^*$, most statistics programs (including S-PLUS) will use Equation (9.42) to fit the model and then give you

$$\hat{Y}_{X^*} = \exp\left(\hat{W} \mid X = X^*\right) \qquad (9.44)$$

This predicted value is an estimate of the ***median*** value of Y at $X = X^*$, not the mean value of Y at $X = X^*$. Similarly, confidence intervals produced will be confidence intervals for the median of Y, not the mean. Cohen et al. (1989) discuss various methods for estimating the mean of Y for an exponential model. They recommend using the minimum variance unbiased estimator, which is an extension of the estimator shown in Equation (5.54).

CALIBRATION, INVERSE REGRESSION, AND DETECTION LIMITS

Determining the concentration of a chemical in a soil, water, or air sample is a very complex process that involves many steps. Also, we showed in Figure 1.2 that several steps are involved just to get the physical sample to the lab in the first place! Table 9.1 shows a list of some qualifiers commonly used by analytical chemists to report results. These qualifiers are related to the sources of variability shown in Figure 1.2 up to and including the categories *Sample Receipt and Storage at Laboratory*, *Sample Work Up*, and *Sample Analysis*. Within the *Sample Analysis* category, one of the sources of variability is labeled *Calibration Error*. See Keith (1996) for more information and details on sources of variability from field sampling, transport, and laboratory analysis.

Qualifier	Meaning
J	Result is of limited use due to discrepancies in holding times, blank analyses, duplicate analyses, spike analyses, or laboratory contamination problems
L	(Formerly known as "trace"); result is of limited use because it is between the instrument/method detection limit and the contract detection/quantitation limit
N	Result probably acceptable but is just outside the calibration range or the recovery is just outside the specification range
R	Results are unusable due to discrepancies in analytical technique/protocol, improper calibration, outside calibration range, outside specified recovery windows, or blunder
U	Reading was below instrument detection limit (inorganics), or method detection limit (organics)

Table 9.1   *Commonly used analytical laboratory qualifiers*

Almost always the process of determining concentration involves using some kind of machine that produces a signal, and this signal is related to the concentration of the chemical in the physical sample. The process of relating the machine signal to the concentration of the chemical is called ***calibration***. Once calibration has been performed, estimated concentrations in physical samples with unknown concentrations are computed using ***inverse regression***. The uncertainty in the process used to estimate the concentration may be quantified with *decision*, *detection*, and *quantitation limits*. We will discuss each of these topics in this section.

Linear Calibration and Inverse Regression

A simple and frequently used calibration model is a straight line:

$$S = \beta_0 + \beta_1 C + \varepsilon \tag{9.45}$$

where S denotes the signal of the machine, C denotes the true concentration in the physical sample, β_0 denotes the intercept, β_1 denotes the slope, and ε denotes the error term, which is assumed to follow a normal distribution with mean 0 and standard deviation σ_ε. Note that the average value of the signal for a blank ($C = 0$) is the intercept β_0. Figure 9.13 displays an example of a calibration line.

Example of a Calibration Line

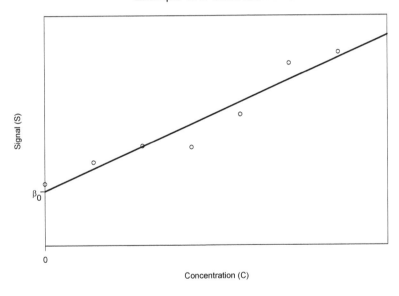

Figure 9.13   Idealized calibration line where the intercept and slope are known

In a typical setup, a small number of samples (e.g., $n = 6$) with known concentrations are measured and the signal is recorded. A sample with no chemical in it, called a **blank**, is also measured. (You have to be careful to define exactly what you mean by a "blank." A blank could mean a container from the lab that has nothing in it but is prepared in a similar fashion to containers with actual samples in them. Or it could mean a field blank: the container was taken out to the field and subjected to the same process that all other containers were subjected to, except a physical sample of soil or water was not placed in the container.) Usually, *replicate* measures at the same known concentrations are taken. (The term "replicate" must be well defined to distinguish between for example the same physical samples that are measured more than once vs. two different physical samples of the same known concentration.) Once the calibration line is fit, samples with unknown concentrations are measured and their signals are recorded. In order to produce estimated concentrations, you have to use inverse regression to map the signals to the estimated concentrations. Figure 9.14 illustrates the idea of inverse regression.

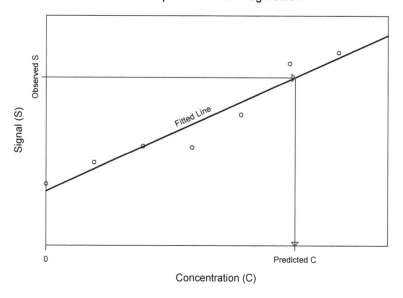

Figure 9.14   Idea of inverse regression

If a straight line model is used for the calibration, then Equation (9.45) can be rewritten to produce the following formula for the estimated concentration:

$$\hat{C} = \left(S - \hat{\beta}_0\right)\big/\hat{\beta}_1 \qquad (9.46)$$

(Draper and Smith, 1998, p. 83). There is uncertainty in the estimated concentration $\hat{C}$ because:

1. We do not know the true values of the slope β_1 and intercept β_0.
2. Even if we did know the true values of the slope and intercept, there is uncertainty relating the observed signal S to the true concentration C because of the error terms in the model in Equation (9.45).

We can quantify the uncertainty in the estimated concentration by combining inverse regression with prediction limits for the signal S. Using Equation (9.39) or (9.40), the lower and upper prediction limits C_L and C_U for the concentration given an observed signal S can be found (Draper and Smith, 1998, pp. 83–86; Hubaux and Vos, 1967). Figure 9.15 illustrates the process based on two-sided 99% non-simultaneous prediction limits.

99% Confidence Limits for Concentration

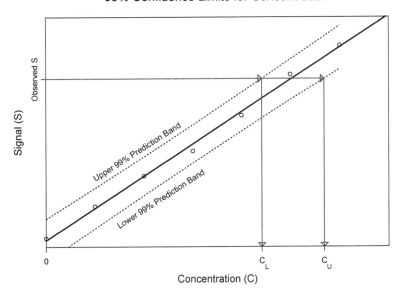

Figure 9.15   Constructing confidence limits for the true concentration

In practice, only the point estimate $\hat{C}$ is reported (along with a possible qualifier as shown in Table 9.1) without confidence bounds for C. This is most unfortunate because it gives the impression that there is no error associated with the reported concentration. Indeed, both the International Organization for Standardization (ISO) and the International Union of Pure and Applied Chemistry (IUPAC) recommend always reporting both the estimated concentration and the uncertainty associated with this estimate (Currie, 1997).

Example 9.10: Fitting a Calibration Line to Cadmium Data

Table 9.2 lists calibration data for cadmium at mass 111 that appeared in Gibbons et al. (1997b) and were provided to them by the U.S. EPA. In ENVIRONMENTALSTATS for S-PLUS these data are stored in the data frame `epa.97.cadmium.111.df`. Figure 9.16 displays a plot of these data along with the fitted calibration line and 99% non-simultaneous prediction limits. An observed signal of 60 results in an estimated value of cadmium of 59.97 ng/L and a confidence interval of [53.83, 66.15].

Concentration (C)	Signal (S)						
0	0.88	1.57	0.70	0.80	0.54	1.83	1.34
10	10.17	11.13	11.66	10.80	11.11	11.95	11.14
20	19.97	20.28	23.20	22.12	18.01	24.83	21.10
50	54.78	49.00	51.92	49.00	54.75	50.25	50.03
100	97.06	94.60	102.54	101.09	99.20	93.71	100.43

Table 9.2   Method 1638 ICPMS data from U.S. EPA for cadmium at mass 111 (ng/L) (as shown in Gibbons et al., 1997b, p. 3730)

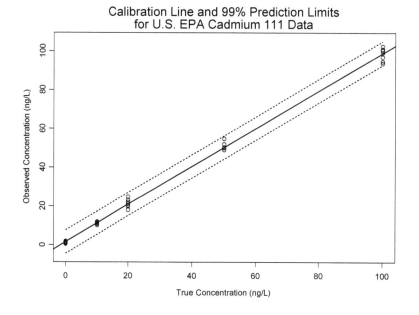

Figure 9.16   Cadmium 111 data with fitted calibration line and 99% non-simultaneous prediction limits

Menu

In the current version of S-PLUS 2000 Release 2, you cannot compute prediction limits for a linear model from the menu. Also, in the current version of ENVIRONMENTALSTATS for S-PLUS you cannot perform inverse prediction from the menu.

Command

To produce the plot shown in Figure 9.16 using the ENVIRONMENTALSTATS for S-PLUS Command or Script Window, type these commands.

```
Cadmium <- epa.97.cadmium.111.df$Cadmium

Spike <- epa.97.cadmium.111.df$Spike

calibrate.list <- calibrate(Cadmium ~ Spike,
    data=epa.97.cadmium.111.df, max.order=1)

newdata <- data.frame(Spike = seq(min(Spike),
    max(Spike), len=100))

pred.list <- predict(calibrate.list, newdata=newdata,
    se.fit=T)

pointwise.list <- pointwise(pred.list, coverage=0.99,
    individual=T)

plot(Spike, Cadmium, ylim=c(min(pointwise.list$lower),
    max(pointwise.list$upper)),
    xlab="True Concentration (ng/L)",
    ylab="Observed Concentration (ng/L)")

abline(calibrate.list, lwd=2)

lines(newdata$Spike, pointwise.list$lower, lty=8,
    lwd=2)

lines(newdata$Spike, pointwise.list$upper, lty=8,
    lwd=2)

title("Calibration Line and 99% Prediction Limits\nfor
    US EPA Cadmium 111 Data")
```

To estimate cadmium concentration based on a signal of 60 ng/L and compute confidence limits for the concentration based on 99% non-simultaneous prediction limits, type this command.

```
inverse.predict.calibrate(calibrate.list, obs.y=60,
    intervals=T, coverage=0.99, individual=T)
```

Caveats about Calibration Lines and the Measurement Process

There are several issues to consider when developing a calibration line and computing confidence bounds for the concentration.

- Figure 9.15 assumes that the linear model is appropriate. Sometimes a curvilinear model (e.g., a quadratic or higher order model) is more appropriate.
- Also, the model assumes a constant variance across the range of concentrations. This assumption is usually not met. Figure 9.17 displays diagnostic plots for the calibration model that was fit in Example 9.10 using the cadmium 111 data. The top two plots and Figure 9.16 clearly indicate that the assumption of a constant variance is not met.
- The confidence bounds for the concentration (C) shown in Figure 9.15 are based on non-simultaneous prediction intervals for the signal (S). You could instead base the confidence bounds for the concentration on simultaneous prediction intervals or tolerance intervals.
- Sometimes the instrument producing the signal is hard-wired to report a value of 0 for a signal that is negative or less than the decision limit (see the section Decision, Detection, and Quantification Limits below). This practice destroys important and useful information about variability at blank and near-zero concentrations and biases the fit of the calibration line.
- Although a physical sample may have a known concentration or amount of chemical, the process used to prepare the sample for measurement may involve losing some of the chemical. This is measured by the *percent recovery* associated with the preparation process. The percent recovery may change with concentration, which will bias the fit of the calibration line.
- The physical samples with known concentrations used to create the calibration line may not contain the full matrix of material that is associated with the physical samples from the field (this is almost always the case with blanks), and therefore the results of the calibration may not really be applicable to the physical samples with unknown concentrations that you want to measure.

The complexity of measuring chemical concentrations in a physical sample cannot be overemphasized. In the next section we describe some more complicated statistical models for fitting a calibration line.

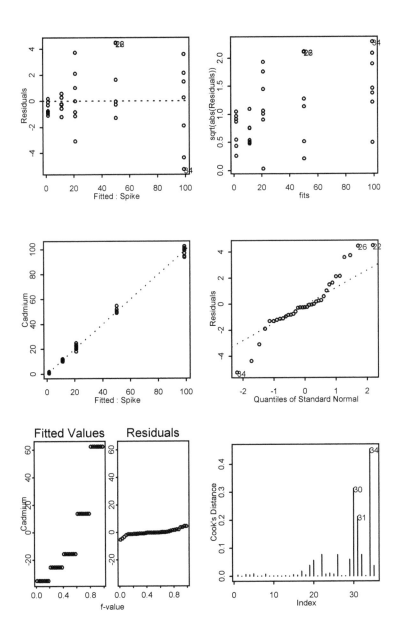

Figure 9.17   Diagnostic plots for the linear calibration line for cadmium 111

Linear Calibration Models with Nonconstant Variance

One commonly used model for linear calibration assumes that the standard deviation of the error terms is proportional to the amount of chemical present or to the average value of the signal (i.e., the coefficient of variation is constant). This model appears to work well to describe calibration lines computed based on using physical samples with non-negligible known concentrations. The problem with this model is that it implies the standard deviation decreases to 0 as the average value of the signal decreases to 0, which of course does not happen in practice (Rocke and Lorenzato, 1995). Instead, the standard deviation usually appears to be constant for blank and near-zero concentrations (Rocke and Lorenzato, 1995). Rocke and Lorenzato (1995) therefore suggested combining these two models into one:

$$S = \beta_0 + \beta_1 C e^{\eta} + \varepsilon \qquad (9.47)$$

The only difference between the model described in Equation (9.47) and the one described in Equation (9.45) is that the term e^{η} has been added. The variable η is assumed to follow a normal distribution with a mean of 0 and a standard deviation of σ_{η}. In this model, the standard deviation is relatively constant at near-zero concentrations (approximately equal to σ_{ε}), and proportional to the average value of the signal at higher concentrations. Rocke and Lorenzato (1995) explain how to use maximum likelihood estimation to estimate the parameters of their model and create prediction limits for the signal S. These prediction limits can then be used with inverse regression to compute a confidence interval for the true concentration C given an observed signal S. Programs to fit the model of Equation (9.47) are available on the World Wide Web at http://handel.cipic.ucdavis.edu/~dmrocke.

Gibbons et al. (1997a) discuss an approximation to the Rocke and Lorenzato's model and explain a weighted least-squares algorithm to fit it. Zorn et al. (1997) discuss a more general model in which the standard deviation is allowed to vary with the concentration (or average signal) but no constraints are placed on how it varies. They show how to use weighted least-squares to fit the model.

Decision, Detection, and Quantification Limits

Perhaps no other topic in environmental statistics has generated as much confusion or controversy as the topic of detection limits. After decades of disparate terminology, ISO and IUPAC provided harmonized guidance on the topic in 1995 (Currie, 1997). Intuitively, the idea of a detection limit is simple to grasp: the *detection limit* is "the smallest amount or concentration of a particular substance that can be reliably detected in a given type of sample or medium by a specific measurement process" (Currie, 1997, p. 152). Unfortunately, because of the exceedingly complex nature of measuring chemical concentrations, this simple idea is difficult to apply in practice.

Detection and quantification capabilities are fundamental performance characteristics of the **Chemical Measurement Process (CMP)** (Currie, 1996, 1997). In this subsection we will discuss some currently accepted definitions of the terms decision, detection, and quantification limits. For more details, the reader should consult the following references: Hubaux and Vos (1970), Glaser et al. (1981), Clayton et al. (1987), Massart et al. (1988), Porter et al. (1988), Lambert et al. (1991), Osborne (1991), Singh (1993), Clark and Whitfield (1994), Currie (1995, 1996, 1997), Gibbons (1995), Rocke and Lorenzato (1995), Davis (1997), Gibbons et al. (1997a,b; 1998), Kahn and White (1997), Kimbrough (1997), Zorn et al. (1997), Spiegelman (1997), and Kahn et al. (1998).

The idea of a decision limit and detection limit is directly related to calibration and can be framed in terms of a hypothesis test, as shown in Table 9.3. The null hypothesis is that the chemical is not present in the physical sample (i.e., $H_0: C = 0$).

	Reality	
Your Decision	H_0 True ($C = 0$)	H_0 False ($C > 0$)
Reject H_0 (Declare Chemical Present)	Mistake: Type I Error (Probability = α)	Correct Decision (Probability = $1-\beta$)
Do Not Reject H_0 (Declare Chemical Absent)	Correct Decision	Mistake: Type II Error (Probability = β)

Table 9.3   Hypothesis testing framework for decision and detection limits

Ideally, you would like to minimize both the Type I and Type II error rates. Just as we use critical values to compare against the test statistic for a hypothesis test, we need to use a critical signal level S_D called the **decision limit** to decide whether the chemical is present or absent. If the signal is less than or equal to S_D we will declare the chemical is absent, and if the signal is greater than S_D we will declare the chemical is present.

First, suppose no chemical is present (i.e., the null hypothesis is true). If we want to guard against the mistake of declaring that the chemical is present when in fact it is absent (Type I error), then we should choose S_D so that the probability of this happening is some small value α. Thus, the value of S_D depends on what we want to use for α (the Type I error rate), and the true (but unknown) value of σ_ε (the standard deviation of the errors assuming a constant standard deviation) (Massart et al., 1988, p. 111). Figure 9.18 displays the distribution of the signal S when the true concentration is 0 and shows that the decision limit is the $(1-\alpha)100^{th}$ percentile of the distribution. Note that the decision limit is on the scale of and in units of the signal S.

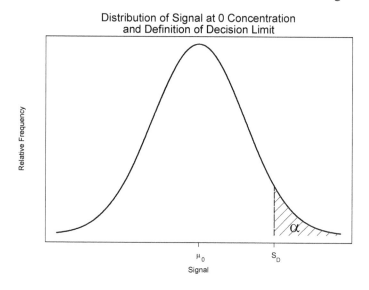

Figure 9.18   Distribution of the signal (S) at zero concentration ($C = 0$) and definition of the decision limit S_D

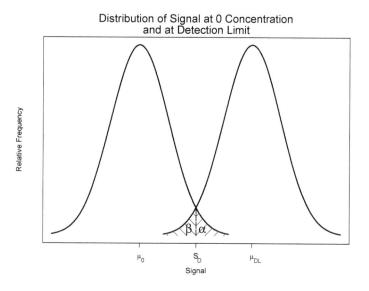

Figure 9.19   Distribution of the signal (S) at zero concentration ($C = 0$) and at the detection limit ($C = C_{DL}$)

Now suppose that in fact the chemical is present in some concentration C (i.e., the null hypothesis is false). If we want to guard against the mistake of declaring that the chemical is absent when in fact it is present (Type II error), then we need to determine a minimal concentration C_{DL} called the **detection limit (DL)** that we know will yield a signal less than the decision limit S_D only a small fraction of the time (β). Figure 9.19 displays the distribution of the signal S when the true concentration C is 0 and when the true concentration is equal to the detection limit C_{DL}. Note that the detection limit is on the scale of and in units of the concentration C.

In practice we do not know the true value of the standard deviation of the errors (σ_ε), so we cannot compute the true decision limit. Also, we do not know the true values of the intercept and slope of the calibration line (β_0 and β_1), so we cannot compute the true detection limit. Instead, we usually set $\alpha = \beta$ and estimate the decision and detection limits by computing prediction limits for the calibration line and using inverse regression. Figure 9.20 displays this method of estimating the decision and detection limits. Comparing this figure to Figure 9.15, we see that the estimated detection limit $\hat{C}_{DL}$ corresponds to the upper confidence bound on concentration given that the signal is equal to the estimated decision limit $\hat{S}_D$. Currie (1997) discusses other ways to define the detection limit, and Glaser et al. (1981) define a quantity called the **method detection limit**.

Estimating the Decision and Detection Limits

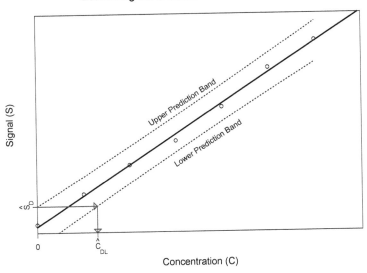

Figure 9.20 Graphical illustration of estimating the decision limit S_D and detection limit C_{DL}

The *quantification limit* is defined as the concentration C at which the coefficient of variation (also called relative standard deviation or RSD) for the distribution of the signal S is some small value, usually taken to be 10% (Currie, 1968, 1997). In practice the quantification limit is difficult to estimate because we have to estimate both the mean and the standard deviation of the signal S for any particular concentration, and, as we have said, usually the standard deviation varies with concentration. Variations of the quantification limit include the quantitation limit (Keith, 1991, p. 109), minimum level (USEPA, 1993), and alternative minimum level (Gibbons et al., 1997).

Example 9.11: Finding the Decision and Detection Limit Based on the Cadmium 111 Data

Using the cadmium 111 data of Example 9.10, the decision limit is 7.68 (in units of the signal) and the detection limit is 12.36 ng/L based on using $\alpha = \beta = 0.01$ and a model that assumes a constant variance. Of course we already pointed out that the assumption of constant variance is incorrect and a model like Rocke and Lorenzato's should be used instead.

Menu

In the current version of S-PLUS 2000 Release 2 and ENVIRONMENTALSTATS for S-PLUS you cannot compute decision and detection limits from the menu.

Command

To compute the decision and detection limits for the cadmium 111 data using the ENVIRONMENTALSTATS for S-PLUS Command or Script Window, follow the steps in Example 9.10 to create the object `calibrate.list`, then type these commands.

```
pred.list <- predict(calibrate.list,
    newdata=data.frame(Spike=0), se.fit=T)
decision.limit <- pointwise(pred.list, coverage=0.99,
    individual=T)$upper
detection.limit <-
    detection.limit.calibrate(calibrate.list)
```

Caveats about Decision, Detection, and Quantification Limits

We have illustrated decision and detection limits based on a single set of observations. Of course, if the process of taking measurements of known concentrations is repeated, the observed signals will be different, even if it is the same chemist using the same machine on the same day using the same standards. Because of all of the sources of variability that go into producing a single signal, the definitions of decision, detection, and quantification limits must include the general circumstances under which the measurements are taken (e.g., only for one chemist in one lab on one day vs. several chemists in several labs over several days). Kimbrough (1999) points out that required performance characteristics for the Chemical Measurement Process should be based on the Data Quality Objectives of the particular project, and cites the U.S. EPA Office of Drinking Water's Minimum Reporting Level as an example. Ideally, the data user should always have a clear understanding of what the chemist means by a value reported as "below detection limit."

MULTIPLE REGRESSION

In Example 9.4 we fit a model that used temperature to predict the value of ozone. The diagnostic plots of Figure 9.10 revealed that this model did not fit the usual assumptions of regression (e.g., errors that are normally distributed with a constant variance). In Example 9.9 we tried modifying this model by using the cube-root of ozone instead of ozone. The diagnostic plots in Figure 9.12 showed that while this model did a better job of satisfying the assumptions on the errors, we still were not doing very well at predicting ozone. This is not surprising because Figure 3.27 shows that a loess smooth indicates a curvilinear instead of a linear relation, and also the physical process that produces ozone at ground level involves several factors that cannot be measured by temperature alone. Multiple regression is the extension of simple regression from the one-predictor-variable model shown in Equation (9.20) to two or more predictor variables. In this section we will discuss both polynomial regression and general multiple regression.

Polynomial Regression with One Predictor Variable

The general model for polynomial regression with one predictor variable is given by:

$$Y = \beta_0 + \beta_1 X + \beta_2 X^2 + \cdots + \beta_p X^p + \varepsilon \qquad (9.48)$$

where p is some integer greater than or equal to 2. This model may be used as an alternative to or in conjunction with attempting to transform Y and/or X to induce linearity.

When fitting a model of this form, you have to decide on the order of the model (i.e., the value of p). One way to do this is based on your knowledge of the process. Another way to do this is informally by fitting a model that is linear, then quadratic, then cubic, etc., and inspecting regression diagnostic plots for each fit. A more formal method is to use the **partial F-test**. The partial F-test tests the null hypothesis that adding (or deleting) a particular term (or terms) does not substantially improve the fit of the model. For example, if our current model is a polynomial of order r, and we want to test whether adding a term of order $r+1$ will help, the null and alternative hypotheses are formally written as:

$$H_0 \; : \; \beta_{r+1} \; = \; 0 \; | \; \beta_0 \; \beta_1 \; \beta_2 \; ... \; \beta_r \qquad \text{(9.49)}$$

$$H_a \; : \; \beta_{r+1} \; \neq \; 0 \; | \; \beta_0 \; \beta_1 \; \beta_2 \; ... \; \beta_r \qquad \text{(9.50)}$$

That is, the null hypothesis is that given that we are using a polynomial of order r as our model, the coefficient for the term of order $r+1$ is equal to 0 vs. the alternative that is it not equal to 0.

The general form of the partial F-statistic is:

$$F \; = \; \frac{\left(RSS_0 \; - \; RSS_a\right)/\left(v_0 \; - \; v_a\right)}{RSS_a/v_a} \qquad \text{(9.51)}$$

where RSS_0 and RSS_a denote the residual sums of squares (see Equation (9.22)) under the null and alternative hypotheses, respectively, and v_0 and v_a denote the degrees of freedom under the null and alternative hypotheses. If the model under the alternative hypothesis is much better than the one under the null hypothesis, then the residual sums of squares under the alternative model should be substantially smaller than the residual sums of squares under the null model, and so the F-statistic should be large. On the other hand, if the alternative model is no better than the null model, then the residual sums of squares should be fairly similar and the F-statistic should be small.

Under the null hypothesis, the partial F-statistic follows an F-distribution with (v_0-v_a) and v_a degrees of freedom. For the particular null and alternative hypotheses shown in Equations (9.49) and (9.50), we have:

$$v_0 \; = \; n \; - \; (r \; + \; 1) \qquad \text{(9.52)}$$

$$v_a \; = \; n \; - \; (r \; + \; 2) \qquad \text{(9.53)}$$

An even more general formal method of deciding on a model involves using *stepwise regression* (Draper and Smith, 1998, pp. 335–339; Zar, 1999, p. 433) in which terms are added sequentially, but each time a new term is added partial F-tests are performed to determine whether other terms may be discarded.

Example 9.12: Fitting a Polynomial Regression Model to the Ozone Data

In Example 9.4 we fit a simple linear regression model relating ozone to temperature. Diagnostic plots in Figure 9.10 revealed that the assumption of a straight line is not adequate and neither is the assumption of a constant variance. In Example 9.9 we transformed the response variable and modeled the cube-root of ozone as a linear function of temperature. Diagnostic plots for this fit shown in Figure 9.12 revealed that although the assumption of constant variance is probably adequate, we can still probably do a better job of predicting the cube-root of ozone. This is not surprising since we saw in Figure 3.27 that a loess smooth indicates a curvilinear instead of a linear relation. In this example we will try modeling the cube-root of ozone as a quadratic and then a cubic function of temperature.

The quadratic model relating the cube-root of ozone to temperature is given by:

$$\hat{O}^{1/3} = 5.56 - 0.14\,T + 0.0013\,T^2 \qquad \textbf{(9.54)}$$

and the cubic model is given by:

$$\hat{O}^{1/3} = 25.4 - 0.93\,T + 0.012\,T^2 - 0.00045\,T^3 \quad \textbf{(9.55)}$$

The partial F-test comparing the simple linear model with the quadratic model yields an F-statistic of 6.3 on 1 and 108 degrees of freedom with an associated p-value of 0.01, indicating that the quadratic term is significant. On the other hand, the partial F-test comparing the quadratic model with the cubic model yields an F-statistic of 0.76 on 1 and 107 degrees of freedom with an associated p-value of 0.39, indicating that the cubic term does not substantially help to predict the cube-root of ozone.

Figure 9.21 displays the diagnostic plots for the quadratic fit. Although adding the quadratic term decreases the hint of curvilinearity in the residual plots, the r-f plot indicates that we still are not doing a very good job of predicting ozone.

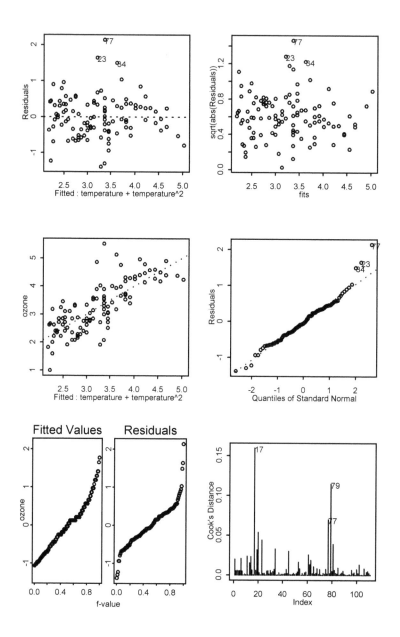

Figure 9.21   Diagnostic plots for the quadratic model relating the cube-root of ozone to temperature

Menu

To fit the quadratic model and produce diagnostic plots, follow these steps.

1. Find **air** in the Object Explorer.
2. On the S-PLUS menu bar, make the following menu choices: **Statistics>Regression>Linear**. This will bring up the Linear Regression dialog box.
3. For Data Set select **air**, for Save As type **cr.ozone.fit2**, then click on the **Create Formula** button.
4. For Choose Variables select **ozone** then click on the **Response** button so that `ozone~1` shows up in the Formula box. Next, for Choose Variables select **temperature** then click on the **Main Effect: (+)** button, and then click on the **Quadratic: (x^2)** button. The Formula box should now read `ozone~temperature+temperature^2`. Click on **OK**.
5. Click on the **Plot** tab. Under Plots, **check** all but the last box. Under Options, **uncheck** the Include smooth box. Click **OK** or **Apply**.

To fit the cubic model and produce diagnostic plots, follow the same steps as above, but in Step 3 in the Save As box type **cr.ozone.fit3**, and in Step 4 click on the **Cubic: (x^3)** button after you click on the **Quadratic: (x^2)** button.

To compare the simple linear model with the quadratic model using the partial F-test, follow these steps.

1. On the S-PLUS menu bar, make the following menu choices: **Statistics>Compare Models**. This will bring up the Compare Models dialog box.
2. For Model Objects, select **cr.ozone.fit** and **cr.ozone.fit2** (use CTRL-Click to select **cr.ozone.fit2** after you've selected **cr.ozone.fit**). For Test Statistic check the **F** box, then click **OK** or **Apply**.

To compare the quadratic model with the cubic model, follow the same steps as above, but in Step 2 select **cr.ozone.fit2** and **cr.ozone.fit3**.

Command

To fit the quadratic and cubic models and produce diagnostic plots, type these commands.

```
cr.ozone.fit2 <- lm(ozone ~ temperature +
    temperature^2, data=air)
plot(cr.ozone.fit2)
```

```
cr.ozone.fit3 <- lm(ozone ~ temperature +
   temperature^2 + temperature^3, data=air)
plot(cr.ozone.fit3)
```

To compare the simple linear model with the quadratic model, type this command.

```
anova(cr.ozone.fit, cr.ozone.fit2)
```

To compare the quadratic model with the cubic model, type this command.

```
anova(cr.ozone.fit2, cr.ozone.fit3)
```

Several Predictor Variables

The general model for multiple regression with several predictor variables is given by:

$$Y = \beta_0 + \beta_1 X_1 + \beta_2 X_2 + \cdots + \beta_p X_p + \varepsilon \qquad (9.56)$$

where X_1, X_2, ..., X_p denote the p predictor variables. Note that the model for polynomial regression in Equation (9.48) is a special case of multiple regression where $X_i = X^i$. Multiple regression is used to model or predict a response variable using two or more predictor variables that are believed to be related to the response variable. The choice of predictor variables to use is usually dictated by the purpose of the model. If the purpose of the model is to explain a physical phenomenon (e.g., how stream flow is affected by precipitation, geography, etc.), then usually the predictor variables are chosen based on your knowledge of the subject matter. On the other hand, if the purpose of the model is to simply come up with a good way to predict future values of the response variable, then several possible predictor variables may be available and chosen as candidates for the model based on exploratory data analysis. The topic of multiple regression is expansive and has a long history. A good reference is Draper and Smith (1998).

Example 9.13: Using Stepwise Regression to Model the Ozone Data

As we stated earlier, the physical process that produces ozone at ground level involves several factors that cannot be measured by temperature alone. Besides temperature, the `air` data frame in S-PLUS includes information on radiation (langleys) and wind speed (mph). Figure 9.22 displays the scatterplot matrix for these variables (note that this figure differs from Figure 3.34 because in the current figure we are using the cube-root of ozone, not ozone itself).

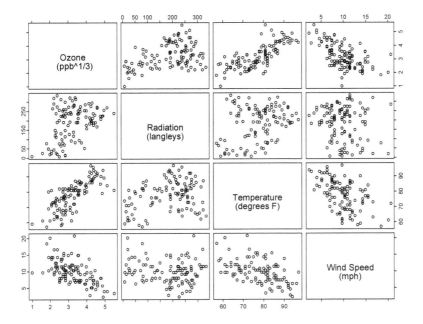

Figure 9.22   Scatterplot matrix for the variables in the `air` data frame

In Chapter 3 we created cloud, contour, surface, and multi-panel conditioning plots to look at ozone, temperature, and wind speed simultaneously. In this example, we will simply perform a stepwise regression to determine a model to use to predict the cube-root of ozone. We will use radiation, temperature, and wind as candidate predictor variables (i.e., R, T, and W), along with their quadratic terms (i.e., R^2, T^2, and W^2) and their possible interactions (i.e., RT, RW, TW, and RTW). We can start with the simplest model that includes only the intercept and use forward stepwise regression, or we can start with the full model that includes all of the candidate predictor variables and use backward stepwise regression. For this example, both methods happen to yield the same final model:

$$
\hat{O}^{1/3} = 6 - 0.00014\,R - 0.079\,T - 0.26\,W - 0.000013\,R^2 \tag{9.57}
$$
$$
+ \, 0.00069\,T^2 + 0.0084\,W^2 + 0.000089\,RT
$$

Figure 9.23 displays the diagnostic plot associated with this model. The residual plots look much nicer than for the previous models we considered in Example 9.9 and Example 9.12. Cleveland (1994) discusses other models for these data based on loess.

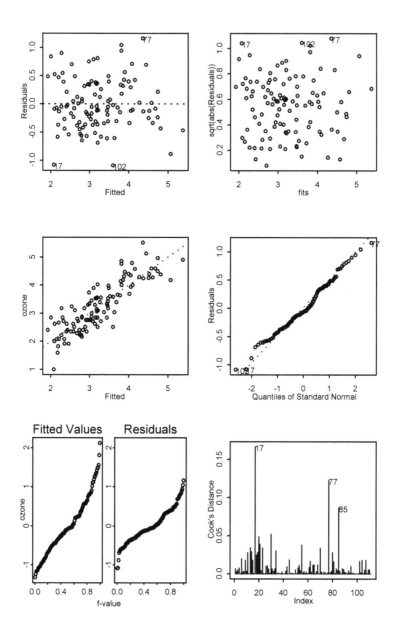

Figure 9.23   Diagnostic plots for the multiple regression model relating the cube-root of ozone to radiation, temperature, and wind speed

Note: Unlike most statistics programs which by default use the partial F-test to do stepwise regression, S-PLUS uses something called the Akaike Information Criterion (AIC). The built-in version of stepwise regression in S-PLUS, however, does not use the standard definition of the AIC and also can yield models with squared or higher order terms of a predictor variable in the model without the predictor variable itself in the model. You should therefore use the stepwise regression procedure available in the MASS library (Venables and Ripley, 1999) that comes with S-PLUS. To attach this library, choose **File>Load Library** from the S-PLUS menu and select **MASS**.

Menu

To perform the stepwise regression and create the diagnostic plots using the S-PLUS pull-down menu, follow these steps (make sure the MASS library has been loaded). First we have to create the model with all of the candidate predictor variables in it.

1. Find **air** in the Object Explorer.
2. On the S-PLUS menu bar, make the following menu choices: **Statistics>Regression>Linear**. This will bring up the Linear Regression dialog box.
3. For Data Set select **air**, for Save As type **cr.ozone.fit.full**, then click on the **Create Formula** button.
4. For Choose Variables select **ozone** then click on the **Response** button so that ozone~1 shows up in the Formula box. Next, for Choose Variables select **radiation**, **temperature** and **wind** (use Shift-Click), then click on the **Main + Interact.: (*)** button, and then click on the **Quadratic: (x^2)** button, then click on **OK**. Click on **OK** or **Apply**.

Now we can perform the stepwise regression.

5. On the S-PLUS menu bar, make the following menu choices: **MASS>StepAIC**. This will bring up the StepWise Model Fitting dialog box.
6. For Initial Object Name type **cr.ozone.fit.full**. For Save As type **cr.ozone.fit.stepwise**. For Stepping Direction choose backward. Click OK or Apply.

To produce the diagnostic plots for the final model, you have to manually create it again using **Statistics>Regression>Linear**. Once you have specified the formula for the model, click on the **Plot** tab, and **check** the boxes for the plots you want. Under Options, you may want to **check** or **uncheck** the Include smooth box. Click **OK** or **Apply**.

Command

To perform the stepwise regression and create the diagnostic plots using the S-Plus Command or Script Window, type this command (make sure the MASS library has been loaded).

```
cr.ozone.fit.full <- lm(ozone ~ radiation +
   temperature + wind + radiation^2 + temperature^2 +
   wind^2 + radiation*temperature + radiation*wind +
   temperature*wind + radiation*temperature*wind,
   data=air)
cr.ozone.fit.stepwise <- stepAIC(cr.ozone.fit.full,
   direction="backward")
plot(cr.ozone.fit.stepwise)
```

DOSE-RESPONSE MODELS: REGRESSION FOR BINARY OUTCOMES

In Chapter 13 we will talk about human health risk assessment, which involves estimating the amount of exposure to a chemical and estimating the relationship between exposure to the chemical and possibly increased chances of developing cancer or some other disease. One of the key components of risk assessment is something called the ***dose-response curve*** which describes the relationship between exposure to a known amount of chemical and the observed outcome in a group of test animals (e.g., mice, rats, etc.). Often the outcome is the number or proportion of animals that developed the disease at each dose level. Table 9.4 and Figure 9.24 display the results of an experiment in which different groups of rats were exposed to various concentration levels of ethylene thiourea (ETU), a decomposition product of a certain class of fungicides that can be found in treated foods (Graham et al., 1975; Rodricks, 1992, p. 133). In this experiment, the outcome of concern was the number of rats that developed thyroid tumors.

Dose (ppm/day)	Number That Developed Tumors	Number in Dose Group	Proportion That Developed Tumors
0	2	72	0.03
5	2	75	0.03
25	1	73	0.01
125	2	73	0.03
250	16	69	0.23
500	62	70	0.89

Table 9.4   Dose-response data for ETU experiment (Graham et al., 1975)

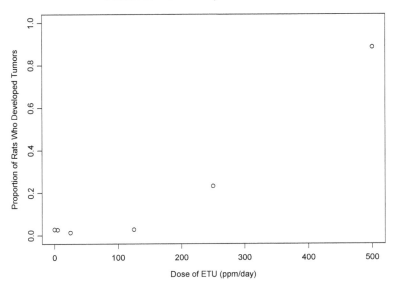

Observed Dose-Response for ETU Data

Figure 9.24   Dose-response data for ETU experiment (Graham et al., 1975)

Models for Binary Outcomes

The simple linear regression model of Equation (9.20) assumes that the response variable Y follows a normal distribution for a specified value of the predictor variable X. Obviously, when the outcome is binary (did the animal develop a tumor or not?) the distribution of Y follows a binomial distribution, not a normal distribution. In this case, the expected value of Y for a given value of X, $\mu_{Y|X}$, is the probability of observing the outcome (e.g., developing a tumor) for that value of X and must lie between 0 and 1. Looking at Equation (9.21) we see that using the conventional linear regression model does not guarantee this.

A commonly used general model for binary outcomes is based on the fact that the value of a cumulative distribution function (cdf) must always lie between 0 and 1:

$$
\begin{aligned}
\mu_{Y|X^*} &= E\left(Y \mid X = X^*\right) \\
&= P_{X^*} = F\left(\beta_0 + \beta_1 X^*\right)
\end{aligned}
\tag{9.58}
$$

where F denotes the cdf of some probability distribution (Piegorsch and Bailer, 1997, p. 313). Two commonly used cdfs are the standard normal cdf:

$$F(x) = \Phi(x) = \int_{-\infty}^{x} \frac{1}{\sqrt{2\pi}} e^{-t^2/2} dt$$

(9.59)

$$-\infty < x < \infty$$

and the standard logistic cdf:

$$F(x) = \frac{1}{1 + e^{-x}} \quad, \quad -\infty < x < \infty \qquad (9.60)$$

The above two models are usually called the probit and logit (or logistic) models.

Looking at Equations (9.58) to (9.60), you can see that both of these models imply that there is some non-zero probability of observing the response even at a dose of $X = 0$. An alternative model that guarantees a probability of 0 at a dose of 0 is based on the Weibull cdf:

$$p_{X^*} = 1 - \exp\left[-\left(\frac{X^*}{\beta_0} \right)^{\beta_1} \right]$$

(9.61)

The three models for dose-response curves discussed above are called *tolerance distribution models* because they are based on the idea that each organism possesses an intrinsic tolerance level to the toxic stimulus, and when this level is exceeded the organism responds (Piegorsch and Bailer, 1997, p. 313). Other models include mechanistic models and time-to-response models (Hallenbeck, 1993, p. 65).

Example 9.14: Fitting a Logistic Model to the ETU Data

In ENVIRONMENTALSTATS for S-PLUS, the data shown in Table 9.4 are stored in the data frame `graham.et.al.75.etu.df`. Figure 9.25 displays the observed data along with the fitted model, and Figure 9.26 displays the diagnostic plots from this fit. See the S-PLUS documentation for an explanation of how to interpret the diagnostic plots for logistic regression.

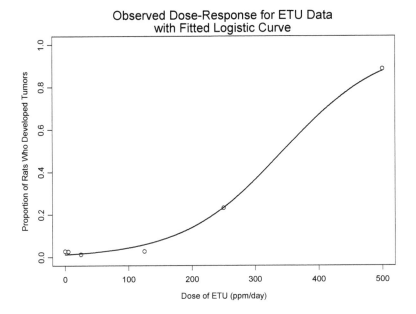

Figure 9.25   Observed dose-response ETU data with fitted logistic curve

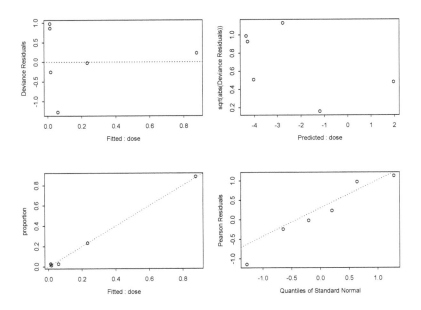

Figure 9.26   Diagnostic plots from the logistic fit

Menu

To fit the logistic model, produce summary statistics for the fit, and create diagnostic plots using the S-PLUS pull-down menu, follow these steps.

1. Find **graham.et.al.75.etu.df** in the Object Explorer.
2. On the S-PLUS menu bar, make the following menu choices: **Statistics>Regression>Logistic**. This will bring up the Logistic Regression dialog box.
3. For Data Set select **graham.et.al.75.etu.df**, for Weights select **n**, for Dependent select **proportion**, for Independent select **dose**, and for Save As type **etu.fit**.
4. Click on the **Plot** tab. Under Plots **check** all of the boxes. Then click **OK** or **Apply**.

Command

To fit the logistic model, produce summary statistics for the fit, and create diagnostic plots using the S-PLUS Command or Script Window, type these commands.

```
etu.fit <- glm(proportion ~ dose,
   data=graham.et.al.75.etu.df, family=binomial,
   weights=n)

summary(etu.fit)

plot(etu.fit)
```

OTHER TOPICS IN REGRESSION

Logistic and probit regression models are special examples of ***generalized linear models*** (GLMs). Generalized linear models were developed to handle regression in cases where the response variable is not assumed to follow a normal distribution, but rather a particular other kind of distribution such as binomial, Poisson, or gamma (McCullagh and Nelder, 1989). The general form of a GLM is:

$$g\left[E\left(Y \mid \underline{X}\right)\right] = \beta_0 + \beta_1 X_1 + \cdots + \beta_p X_p$$

$$(9.62)$$

$$Var\left(Y \mid \underline{X}\right) = \phi\, V\left[E\left(Y \mid \underline{X}\right)\right]$$

where g denotes the ***link function***, ϕ denotes the ***dispersion parameter***, and V denotes the ***variance function***. In the case where g is the identity function and the function V is identically 1 at all values, the GLM reduces to the standard multiple regression model of Equation (9.56) (in this case the dispersion parameter $\phi = \sigma^2$). For the logistic model, looking at Equations (9.58) and (9.60) you can see that the link function is:

$$g\left(p\right) \; = \; F^{-1}\left(p\right) \; = \; \log\left(\frac{p}{1-p}\right) \qquad \textbf{(9.63)}$$

and looking at Equation (4.40) you can see that the dispersion parameter is 1 and the variance function is:

$$V\left(p\right) \; = \; p\left(1-p\right) \qquad \textbf{(9.64)}$$

Generalized additive models (GAMs) are an extension of generalized linear models in that they allow the predictor variables to be nonparametric smooths such as loess smooths or splines. See Hastie and Tibshirani (1990), Venables and Ripley (1999), and the S-PLUS documentation for more information on GLMs and GAMs.

Other kinds of models we have not discussed in this chapter include nonlinear models, tree models, and robust models. See Venables and Ripley (1999) and the S-PLUS documentation for more information.

SUMMARY

- *Correlation* is a measure of the association between two variables. Positive or negative correlation between two variables does not imply these two variables are directly related by some causal mechanism (i.e., "correlation does not imply cause").
- The *Pearson product-moment correlation* is a measure of linear association and is affected by outliers. *Spearman's rho* and *Kendall's tau* are more general measures of any kind of monotonic relationship; they are based on ranks and are not as sensitive to outliers.
- A *simple linear regression model* assumes the response variable is a linear function of the predictor variable plus some error. The error is usually assumed to follow a normal distribution with a mean of 0 and a constant standard deviation.
- A linear regression model is usually fit using the *method of least squares*. It is then possible to create confidence intervals for the slope, intercept, mean value, and the line, as well as create prediction and tolerance intervals.

- Diagnostic plots help you determine deviations from the model assumptions.
- Transformations of the response and/or predictor variable(s) are sometimes useful to help satisfy the model assumptions.
- Determining the concentration of a chemical in a physical sample involves using a *calibration line* and *inverse regression*.
- The *decision limit* and *detection limit* are related to the Type I and Type II errors of a hypothesis test in which the null hypothesis is that the chemical is not present and the alternative hypothesis is that the chemical is present.
- *Multiple regression* involves two or more predictor variables.
- Models for a binary response include the *logistic* and *probit* model. These models are often used to create *dose-response curves*.

EXERCISES

9.1. Using the data in `environmental`, plot the relationship between ozone and temperature, then compute the correlation. Repeat these steps for the data in `air`. Why are the results different even though the names of the variables are the same?

9.2. Repeat the last exercise, but compute Spearman's rho instead of the Pearson product-moment correlation. Do the results based on the data in `environmental` differ from the results based on the data in `air`? Why or why not?

9.3. Using the data in `air`, perform a significance test on the correlation between ozone and temperature based on the Pearson product-moment estimator. Repeat using Spearman's rho, then Kendall's tau.

9.4. Plot the duration times vs. the waiting times for the data from Old Faithful contained in the data frame `geyser`. Perform a significance test on the correlation between waiting time and duration based on Spearman's rho. Does the sign (positive vs. negative) of the correlation seem to make sense? Why or why not?

9.5. Look at the gasoline consumption data in the data frame `fuel.frame` and look at the help file for these data.

    a. Plot Mileage vs. Weight. Fit a regression (possibly polynomial) model to these data. Look at the diagnostic plots.

    b. Fit a simple linear regression to these data, then determine an "optimal" Box-Cox transformation to make to Mileage to satisfy the assumption of normally distributed errors.

    c.   Plot Fuel vs. Weight. Fit a regression (possible polynomial) model to these data. Look at the diagnostic plots.

    d.   Discuss the pros and cons of the two different models you created in parts **a** and **c**.

9.6. Create a matrix scatterplot for the data in `fuel.frame`. Determine the "best" model for predicting Mileage (miles/gallon) as a function of one or more of the other predictor variables (excluding Fuel).

9.7. Repeat the last exercise but use Fuel as the response variable and exclude Weight as a predictor variable.

10 CENSORED DATA

Dealing with "Below Detection Limit" Data

In the last chapter we discussed calibration and detection limits. Often in environmental data analysis values are reported simply as being "below detection limit" along with the stated detection limit (e.g., USEPA, 1992c, p. 25; Porter et al., 1988). A sample of data contains ***censored observations*** if some of the observations are reported only as being below or above some censoring level. Although this results in some loss of information, we can still use data that contain nondetects for graphical and statistical analyses. Statistical methods for dealing with censored data have a long history in the field of survival analysis and life testing (Bain and Engelhardt, 1991; Kalbfleisch and Prentice, 1980; Lee, 1980; Miller, 1981b; Nelson, 1982; Parmar and Machin, 1995). In this chapter, we will discuss how to create graphs, estimate distribution parameters, perform goodness-of-fit tests, compare distributions, and fit linear regression models using censored data.

CLASSIFICATION OF CENSORED DATA

There are four major ways to classify censored data: truncated vs. censored, left vs. right vs. double, single vs. multiple (progressive), and censored Type I vs. censored Type II (Cohen, 1991, pp. 3–5). Most environmental data sets with nondetect values are either Type I left singly censored or Type I left multiply censored.

Truncated vs. Censored

A sample is ***left truncated*** (also called truncated on the left) if only values above a known truncation point T are reported or used. The number of observations falling below T and their actual values are unknown or ignored. A sample is ***right truncated*** (also called truncated on the right) if only values below a known truncation point T are reported or used.

A sample of N observations is ***left singly censored*** (also called singly censored on the left) if c observations are known only to fall below a known censoring level T, while the remaining n ($n = N-c$) uncensored observations falling above T are fully measured and reported. A sample of N observations is ***right singly censored*** (also called singly censored on the right) if c observations are known only to fall above a known censoring level T, while the remaining n uncensored observations falling below T are fully measured and reported.

Left vs. Right vs. Double

The definitions of left truncated, right truncated, left censored, and right censored are given above. A sample is **doubly truncated** if it is truncated both on the left and the right. A sample is **doubly censored** if it is censored both on the left and the right.

Single vs. Multiple (Progressive)

A sample is **singly truncated** (e.g., singly left truncated) if there is only one truncation point T. A sample is **singly censored** (e.g., singly left censored) if there is only one censoring level T.

A sample is **multiply truncated** or progressively truncated (e.g., multiply left truncated) if there are several truncation points T_1, T_2, ..., T_p, where $T_1 < T_2 < ... < T_p$. A sample is **multiply censored** or progressively censored (e.g., multiply left censored) if there are several censoring levels T_1, T_2, ..., T_p, where $T_1 < T_2 < ... < T_p$.

Type I vs. Type II Censoring

A censored sample has been subjected to **Type I censoring** if the censoring level(s) is(are) known in advance, so that given a fixed sample size N, the number of censored observations c (and hence the number of uncensored observations n) is a random outcome. Type I censored samples are sometimes called time-censored samples (Nelson, 1982, p. 248). A censored sample has been subjected to **Type II censoring** if the sample size N and number of censored observations c (and hence the number of uncensored observations n) are fixed in advance, so that the censoring level(s) are random outcomes. Type II censored samples are sometimes called failure-censored samples (Nelson, 1982, p. 248).

Examples of Different Kinds of Censored Data Sets

We have already come across a few data sets with censored observations. The following examples include some of these data sets as well as others from life testing and survival analysis.

Example 10.1: Type I Left Singly Censored Data

The benzene data presented in Table 5.2 represent type I left singly censored data with a single censoring level of 2 ppb. There are $N = 36$ observations, with $c = 33$ censored observations and $n = 3$ uncensored observations. The arsenic data presented in Table 6.4 have a single censoring level of 5 ppb, with $N = 20$, $c = 9$, and $n = 11$. Table 10.1 presents artificial TcCB concentrations based on the Reference area data shown in Table 3.2. For this data set, the concentrations of TcCB less than 0.5 ppb have been recoded as "<0.5," so there is a single censoring level of 0.5 ppb, with $N = 47$, $c = 19$, and $n = 28$. In ENVIRONMENTALSTATS for S-PLUS these data are stored in Modified.TcCB.df.

TcCB Concentrations (ppb)							
<0.5	<0.5	<0.5	<0.5	<0.5	<0.5	<0.5	<0.5
<0.5	<0.5	<0.5	<0.5	<0.5	<0.5	<0.5	<0.5
<0.5	<0.5	<0.5	0.5	0.5	0.51	0.52	0.54
0.56	0.56	0.57	0.57	0.6	0.62	0.63	0.67
0.69	0.72	0.74	0.76	0.79	0.81	0.82	0.84
0.89	1.11	1.13	1.14	1.14	1.2	1.33	

Table 10.1   Modified Reference area TcCB concentrations (see Table 3.2)

Example 10.2: Type I Left Multiply Censored Data

Table 10.2 displays copper concentrations (μg/L) in shallow groundwater samples from two different geological zones in the San Joaquin Valley, California (Millard and Deverel, 1988). The alluvial fan data include four different detection limits and the basin trough data include five different detection limits. Table 10.3 displays zinc concentrations (μg/L) from the same study for which there are two different detection limits. Table 10.4 displays 56 silver concentrations (μg/L) from an interlab comparison that include 34 values below one of 12 detection limits (Helsel and Cohn, 1988).

Zone	Copper (μg/L)										
Alluvial Fan	<1	<1	<1	<1	1	1	1	1	1	2	2
	2	2	2	2	2	2	2	2	2	2	2
	2	2	2	2	2	2	2	2	3	3	3
	3	3	3	4	4	4	<5	<5	<5	<5	<5
	<5	<5	<5	5	5	5	7	7	7	8	9
	<10	<10	<10	10	11	12	16	<20	<20	20	NA
	NA	NA									
Basin Trough	<1	<1	1	1	1	1	1	1	1	<2	<2
	2	2	2	2	3	3	3	3	3	3	3
	3	4	4	4	4	4	<5	<5	<5	<5	<5
	5	6	6	8	9	9	<10	<10	<10	<10	12
	14	<15	15	17	23	NA					

Table 10.2   Copper concentrations in shallow groundwater in two geological zones (Millard and Deverel, 1988)

Zone	Zinc (µg/L)									
Alluvial Fan	<3	5	7	8	9	<10	<10	<10	<10	<10 <10
	<10	<10	<10	<10	<10	<10	<10	<10	<10	10	10
	10	10	10	10	10	10	10	10	10	10	10
	10	10	10	10	10	10	10	11	11	12	17
	18	19	20	20	20	20	20	20	20	20	20
	20	20	20	20	20	23	29	30	33	40	50
	620	NA									
Basin Trough	<3	3	3	4	4	5	5	6	8	<10	<10
	<10	10	10	10	10	10	11	12	12	13	14
	15	17	17	20	20	20	20	20	20	20	20
	20	20	20	25	30	30	30	30	40	40	40
	50	50	60	60	70	90					

Table 10.3   Zinc concentrations in shallow groundwater in two geological zones (Millard and Deverel, 1988)

Silver Concentrations (µg/L)								
<0.1	<0.1	0.1	0.1	<0.2	<0.2	<0.2	<0.2	0.2
<0.3	<0.5	0.7	0.8	<1	<1	<1	<1	<1
<1	<1	<1	<1	<1	1	1	1	1.2
1.4	1.5	<2	2	2	2	2	<2.5	2.7
3.2	4.4	<5	<5	<5	<5	5	<6	<10
<10	<10	<10	<10	10	<20	<20	<20	<25
90	560							

Table 10.4   Silver concentrations from an interlab comparison (Helsel and Cohn, 1988)

Example 10.3: Type I Left Singly Truncated

If, in Example 10.1 above, the benzene concentrations reported as "<2" were eliminated from the sample completely, leaving only the three uncensored observations and no indication of how many observations were excluded, this would constitute a type I left singly truncated sample.

Example 10.4: Type I Right Singly Censored

This kind of sample frequently occurs in life testing experiments. For example, a set of N electronic components may be put "on-line" all at the same time. During the experiment, the time until failure is noted for each component. After a given time period T (e.g., 10 days) the experiment is terminated, even though not all of the components have failed by this time. For the components that have not failed, it is known only that their time until failure is greater than T.

Example 10.5: Type II Right Singly Censored

If we use the same set up as in Example 10.4, but terminate the experiment after n of the N components have failed, this produces a type II right singly censored sample.

Example 10.6: Type I Right Multiply Censored

This kind of sample frequently occurs in survival studies. A number of people are enrolled over the course of the entry period (e.g., 2 years) in a study of a new drug therapy whose purpose is to prolong the life of terminally ill patients. Individuals are assigned to the treatment or control group, and are periodically seen until the study ends, until they die, or until they stop coming for check ups (drop out of the study). At the end of the study, information is available on survival times: when or if individuals died during the study. Since not everyone entered the study at the same time, the survivors are censored at different censoring levels, and so are the people who dropped out of the study.

GRAPHICIAL ASSESSMENT OF CENSORED DATA

In Chapter 3 we discussed several ways of creating graphs for a single variable, including histograms, quantile (empirical cdf) plots, and probability (Q-Q) plots. When you have censored data, creating a histogram is not necessarily straightforward (especially with multiply censored data), but you can create quantile plots and probability plots, as well as determine "optimal" Box-Cox transformations.

Quantile (Empirical CDF) Plots for Censored Data

In Chapter 3 we explained that a quantile plot (also called an empirical cumulative distribution function plot or empirical cdf plot) plots the ordered data (the empirical quantiles) on the x-axis vs. the estimated cumulative probabilities (or plotting positions) on the y-axis. Various formulas for the plotting positions were shown in Equations (3.20) to (3.26) and Table 3.7. When you have censored data, the formulas for the plotting positions must be modified. For right-censored data, various formulas for the plotting positions are given by Kaplan and Meier (1958), Nelson (1972), and Michael and Schucany (1986). For left-censored data, formulas for the plotting positions are given by Michael and Schucany (1986) and Hirsch and Stedinger (1987).

When you have Type I left-censored data with only one censoring level, and all of the uncensored observations are larger than the censoring level, the computation of the plotting positions is straightforward because it is easy to order the uncensored observations. Table 10.5 illustrates this situation. When you have one or more uncensored observations with values less than one or more of the censoring levels, then the computation of the plotting positions becomes a bit trickier. Table 10.6 illustrates this situation. The help file for ppoints.censored in ENVIRONMENTALSTATS for S-PLUS gives a detailed explanation of the formulas for the plotting positions for censored data.

Observation	Blom Plotting Position	Observation	Michael-Schucany Plotting Position
1	0.12	<4	0.5
2	0.31	<4	0.5
5	0.5	5	0.5
10	0.69	10	0.69
15	0.88	15	0.88

Table 10.5   Plotting positions for uncensored and singly censored data

Observation	Blom Plotting Position	Observation	Michael-Schucany Plotting Position
1	0.12	<4	0.64
2	0.31	<4	0.64
5	0.5	5	0.64
10	0.69	<14	0.88
15	0.88	15	0.88

Table 10.6   Plotting positions for uncensored and multiply censored data

Example 10.7: Empirical CDF Plots of the Silver Data

Figure 10.1 displays the empirical cdf plot for the silver data shown in Table 10.4. This plot indicates the data are extremely skewed to the right. This is not surprising since looking at the data in Table 10.4 we see that all of the observations are less than 25 µg/L except for two that are 90 and 560 µg/L. Figure 10.2 displays the quantile plot based on the log-transformed observations. In both of these plots, the upside-down triangles indicate the censoring levels of observations that have been censored.

Menu

To create the quantile plot for the silver data using the ENVIRONMENTALSTATS for S-PLUS pull-down menu, follow these steps.

1. Find **helsel.cohn.88.silver.df** in the Object Explorer.

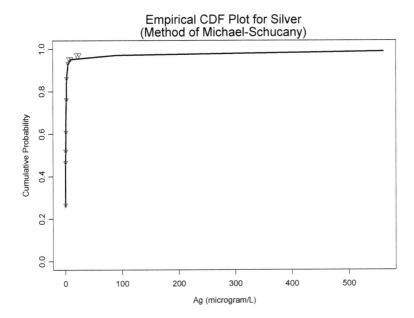

Figure 10.1   Empirical cdf plot of the silver data of Table 10.4

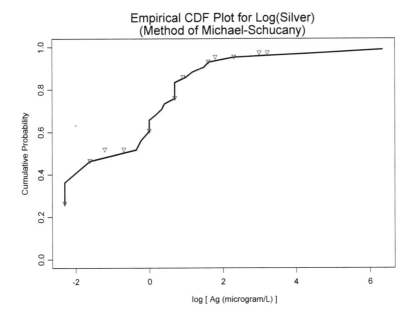

Figure 10.2   Empirical cdf plot of the log-transformed silver data of Table 10.4

2.   On the S-Plus menu bar, make the following menu choices: **EnvironmentalStats>Censored Data>EDA>CDF Plot>Empirical CDF**. This will bring up the Plot Empirical CDF for Censored Data dialog box.
3.   The Data Set box should display **helsel.cohn.88.silver.df**. For Variable select **Ag** and for Censor Ind. select **Censored**. Click on the **Options** tab, and **check** the Include Censored Values box, then click **OK** or **Apply**.

To create the quantile plot for the log-transformed silver data, repeat the above steps but in Step 3 for Variable select **log.Ag**.

Command

To create the quantile plots for the silver data using the ENVIRONMENTALSTATS for S-Plus Command or Script Window, type these commands.

```
attach(helsel.cohn.88.silver.df)
ecdfplot.censored(Ag, Censored,
   xlab="Ag (microgram/L)", main="Empirical CDF Plot
   for Silver\n(Method of Michael-Schucany)",
   include.cen=T)
ecdfplot.censored(log.Ag, Censored,
   xlab="log [Ag (microgram/L)]", main="Empirical CDF
   Plot for Silver\n(Method of Michael-Schucany)",
   include.cen=T)
detach()
```

Example 10.8: Comparing the Empirical CDF of Silver to a Lognormal CDF

Figure 10.3 compares the empirical cdf of the silver data with a lognormal cdf, where the parameters for the lognormal distribution are estimated from the data (see the section Estimating Distribution Parameters later in this chapter). Figure 10.4 compares the log-transformed silver data with a normal distribution. Both plots show the lognormal distribution appears to provide an adequate fit to these data.

Menu

To create the plot in Figure 10.3 using the ENVIRONMENTALSTATS for S-Plus pull-down menu, follow these steps.

1.   Find **helsel.cohn.88.silver.df** in the Object Explorer.

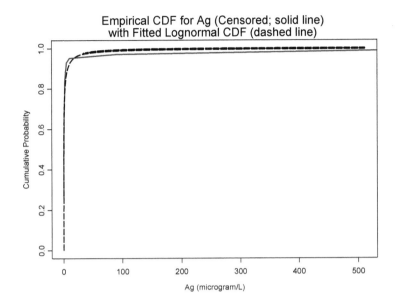

Figure 10.3 Empirical cdf of the silver data with a fitted lognormal distribution

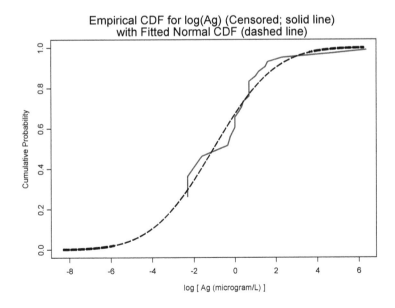

Figure 10.4 Empirical cdf of the log-transformed silver data with a fitted normal distribution

2. On the S-PLUS menu bar, make the following menu choices: **EnvironmentalStats>Censored Data>EDA>CDF Plot>Compare Two CDFs**. This will bring up the Compare Two CDFs for Censored Data dialog box.
3. The Data Set box should display **helsel.cohn.88.silver.df**. For Variable 1 select **Ag** and for Censor Ind. 1 select **Censored**. Under the Distribution Information Group, make sure that the **Estimate Parameters** box is checked. For Distribution select **Lognormal**.
4. Click **OK** or **Apply**.

To create the plot in Figure 10.4 follow the same steps as above, but in Step 2 for x Variable select **log.Ag** and for Distribution select **Normal**.

Command

To create the plot in Figure 10.3 and Figure 10.4 using the ENVIRONMENTALSTATS for S-PLUS Command or Script Window, type these commands.

```
attach(helsel.cohn.88.silver.df)
cdf.compare.censored(Ag, Censored,
    distribution="lnorm", xlab="Ag (microgram/L)")
cdf.compare.censored(log(Ag), Censored,
    distribution="norm", xlab="log [Ag (microgram/L)]")
detach()
```

Example 10.9: Comparing the Empirical CDF of Copper between Two Geological Zones

Figure 10.5 compares the empirical cdf of copper concentrations from the alluvial fan zone with those from the basin trough zone using the data shown in Table 10.2. This plot shows that the two distributions are fairly similar in shape and location.

Menu

To create the plot in Figure 10.5 using the ENVIRONMENTALSTATS for S-PLUS pull-down menu, follow these steps.

1. Find **millard.deverel.88.df** in the Object Explorer.
2. On the S-PLUS menu bar, make the following menu choices: **EnvironmentalStats>Censored Data>EDA>CDF Plot>Compare Two CDFs**. This will bring up the Compare Two CDFs for Censored Data dialog box.
3. For Compare Data select **Other Data**. The Data Set box should display **millard.deveral.88.silver.df**. For Variable 1 choose **Cu**, for

Censor Ind. 1 choose **Cu.censored**, for Variable 2 choose **Zone**, **check** the Variable 2 is a Grouping Variable box, then click **OK** or **Apply**.

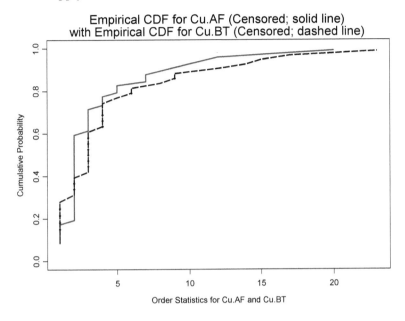

Figure 10.5   Empirical cdf's of copper concentrations in the alluvial fan and basin trough zones

Command

To create the plot in Figure 10.5 using the ENVIRONMENTALSTATS for S-PLUS Command or Script Window, type these commands.

```
attach(millard.deverel.88.df)
Cu.AF <- Cu[Zone=="Alluvial.Fan"]
Cu.AF.cen <- Cu.censored[Zone=="Alluvial.Fan"]
Cu.BT <- Cu[Zone=="Basin.Trough"]
Cu.BT.cen <- Cu.censored[Zone=="Basin.Trough"]
cdf.compare.censored(Cu.AF, Cu.AF.cen, "left", Cu.BT,
   Cu.BT.cen, "left")
detach()
```

Q-Q Plots for Censored Data

In Chapter 3 we explained that a probability or quantile-quantile (Q-Q) plot plots the ordered data (the empirical quantiles) on the y-axis vs. the corresponding quantiles from the assumed theoretical probability distribution on the x-axis, where the quantiles from the assumed distribution are computed based on the plotting positions. As for empirical cdf plots, when you have censored data, the formulas for the plotting positions must be modified.

Example 10.10: Comparing Arsenic Data to a Normal Distribution

Figure 10.6 shows the normal Q-Q plot for the background well arsenic data shown in Table 6.4, where the data from all of the background wells have been combined. This plot indicates that the normal distribution appears to provide an adequate fit to these data.

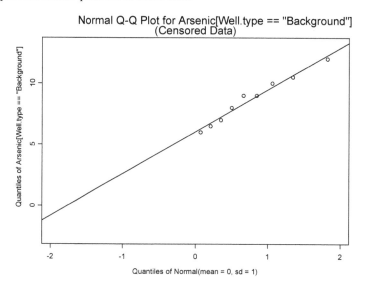

Figure 10.6   Normal Q-Q plot for the singly censored background well arsenic data of Table 6.4

Menu

To create the Q-Q plot for the background well arsenic data using the ENVIRONMENTALSTATS for S-PLUS pull-down menu, follow these steps.

1. Find **epa.92c.arsenic2.df** in the Object Explorer.
2. On the S-PLUS menu bar, make the following menu choices: **EnvironmentalStats>Censored Data>EDA>Q-Q Plot**. This will bring up the Q-Q Plot for Censored Data dialog box.

3. The Data Set box should display **epa.92c.arsenic2.df**. For Variable 1 choose **Arsenic**, for Censor Ind. 1 choose **Censored**, in the Subset Rows with box type **Well.type=="Background"**, and for Distribution select **Normal**.
4. Click on the **Plotting** tab. Click on the **Add a Line** box to select this option. Click **OK** or **Apply**.

Command

To create the Q-Q plot for the background well arsenic data using the ENVIRONMENTALSTATS for S-PLUS Command or Script Window, type these commands.

```
attach(epa.92c.arsenic2.df)
qqplot.censored(Arsenic[Well.type=="Background"],
   Censored[Well.type=="Background"], add.line=T)
detach()
```

Example 10.11: Comparing the Singly Censored TcCB Data to a Lognormal Distribution

Figure 10.7 shows the normal Q-Q plot for the log-transformed TcCB data shown in Table 10.1. This plot indicates that the lognormal distribution appears to provide an adequate fit to these data. This is not surprising since we already saw in Figure 3.19 that a lognormal distribution provides a good fit to the original data.

Menu

To create the Q-Q plot for the modified TcCB data using the ENVIRONMENTALSTATS for S-PLUS pull-down menu, follow these steps.

1. Find **Modified.TcCB.df** in the Object Explorer.
2. On the S-PLUS menu bar, make the following menu choices: **EnvironmentalStats>Censored Data>EDA>Q-Q Plot**. This will bring up the Q-Q Plot for Censored Data dialog box.
3. The Data Set box should display **Modified.TcCB.df**. For Variable 1 choose **TcCB**, for Censor Ind. 1 select **Censored**, and for Distribution select **Lognormal**.
4. Click on the **Plotting** tab. Click on the **Add a Line** box to select this option. Click **OK** or **Apply**.

Command

To create the Q-Q plot for the modified TcCB data using the ENVIRONMENTALSTATS for S-PLUS Command or Script Window, type these commands.

```
attach(Modified.TcCB.df)
qqplot.censored(TcCB, Censored, distribution="lnorm",
    add.line=T)
detach()
```

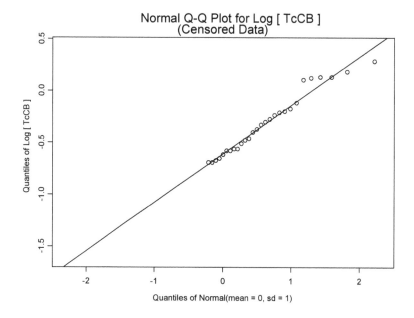

Figure 10.7   Normal Q-Q plot for the log-transformed singly left-censored TcCB data
of Table 10.1

Example 10.12: Comparing the Silver Data to a Lognormal Distribution

Figure 10.8 shows the normal Q-Q plot for the log-transformed silver data shown in Table 10.4. As in the case of the empirical cdf plot shown in Figure 10.3, this plot indicates that the lognormal distribution appears to provide an adequate fit to these data, although there is some suspect curvature.

Menu

To create the Q-Q plot for the silver data using the ENVIRONMENTALSTATS for S-PLUS pull-down menu, follow these steps.

1. Find **helsel.cohn.88.silver.df** in the Object Explorer.
2. On the S-PLUS menu bar, make the following menu choices: **EnvironmentalStats>Censored Data>EDA>Q-Q Plot**. This will bring up the Q-Q Plot for Censored Data dialog box.

3. The Data Set box should display **helsel.cohn.88.silver.df**. For Variable choose **Ag**, for Censor Ind. select **Censored**, and for Distribution select **Lognormal**.
4. Click on the **Plotting** tab. Click on the **Add a Line** box to select this option. Click **OK** or **Apply**.

Command

To create the Q-Q plot for the silver data using the ENVIRONMENTALSTATS for S-PLUS Command or Script Window, type these commands.

```
attach(helsel.cohn.88.silver.df)
qqplot.censored(Ag, Censored, distribution="lnorm",
   add.line=T)
detach()
```

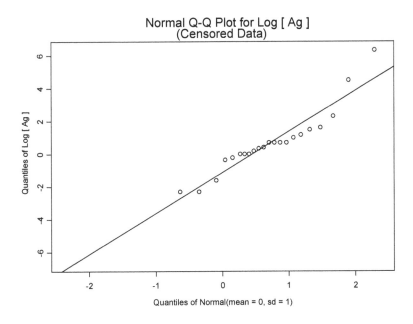

Figure 10.8   Normal Q-Q plot for the log-transformed multiply censored silver data of Table 10.4

Box-Cox Transformations for Censored Data

In Chapter 3 we discussed using Box-Cox transformations as a way to satisfy normality assumptions for standard statistical tests, and also sometimes to satisfy the linear assumption and/or the constant variance in the errors assumption for a standard linear regression model (see Equations (3.30) and (3.31)). We also discussed three possible criteria to use to decide on the power of the transformation: the probability plot correlation coefficient (PPCC), the Shapiro-Wilk goodness-of-fit test statistic, and the log-likelihood function. This idea can be extended to the case of singly and multiply censored data (e.g., Shumway et al., 1989). See the help files for `boxcox.singly.censored` and `boxcox.multiply.censored` in ENVIRONMENTALSTATS for S-PLUS for details.

Example 10.13: Determining the "Optimal" Transformation for the Modified TcCB Data

Figure 10.9 displays a plot of the probability plot correlation coefficient vs. various values of the transform power λ for the singly censored TcCB data shown in Table 10.1. For these data, the PPCC reaches its maximum between about $\lambda = 0$ (log transformation) and $\lambda = 0.5$ (square-root transformation). We saw a similar pattern for the original Reference area TcCB data in Figure 3.24, although in that figure the maximum appeared at about $\lambda = 0$.

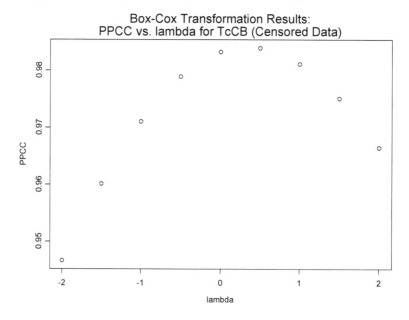

Figure 10.9   Probability plot correlation coefficient vs. Box-Cox transform power (λ) for the singly censored TcCB data of Table 10.1

Menu

To create the plot of the PPCC vs. the transformation power for the modified TcCB data using the ENVIRONMENTALSTATS for S-PLUS pull-down menu, follow these steps.

1. Find **Modified.TcCB.df** in the Object Explorer.
2. On the S-PLUS menu bar, make the following menu choices: **EnvironmentalStats>Censored Data>EDA>Box-Cox Transformations**. This will bring up the Box-Cox Transformations for Censored Data dialog box.
3. For Data to Use make sure the **Pre-Defined Data** button is selected. The Data Set box should display **Modified.TcCB.df**. For Variable choose **TcCB**, and for Censor Ind. select **Censored**, then click **OK** or **Apply**.

Command

To create the plot of the PPCC vs. the transformation power for the modified TcCB data using the ENVIRONMENTALSTATS for S-PLUS Command or Script Window, type these commands.

```
attach(Modified.TcCB.df)
boxcox.list <- boxcox.singly.censored(TcCB, Censored)
plot(boxcox.list, plot.type="Objective vs. lambda")
detach()
```

ESTIMATING DISTRIBUTION PARAMETERS

In Chapter 5 we discussed various methods of estimating distribution parameters, including the maximum likelihood estimator (MLE), method of moments estimator (MME), and minimum variance unbiased estimator (MVUE). It is fairly straightforward to extend maximum likelihood estimation to the case of censored data (e.g., Cohen, 1991; Schneider, 1986). More recently, researchers in the environmental field have proposed alternative methods of computing estimates and confidence intervals in addition to the classical ones such as maximum likelihood estimation.

Several authors have compared the performance of various estimators of the mean and standard deviation (based on bias and mean-squared error), including El-Shaarawi (1989), El-Shaarawi and Esterby (1992), Gilliom and Helsel (1986), Gleit (1985), Haas and Scheff (1990), and Newman et al. (1989). In practice, however, it is better to present a decision maker with a confidence interval for the mean or a joint confidence region for the mean and standard deviation, rather than rely on a single point-estimate of the mean. Since confidence intervals and regions depend on the properties of the estimators for *both* the mean and standard deviation, the results of studies that simply evaluated the performance of the mean and standard deviation separately cannot be readily extrapolated to predict the performance of various methods of constructing confidence intervals and regions. Furthermore, for several of the methods that have been proposed to estimate the mean based on Type I left-censored data, standard errors of the estimates are not available, making it problematic to construct confidence intervals (El-Shaarawi and Dolan, 1989).

Very few studies have been done to evaluate the performance of methods for constructing confidence intervals for the mean or joint confidence regions for the mean and standard deviation based on censored data. Schmee et al. (1985) studied Type II censoring for a normal distribution and noted that the bias and variances of the maximum likelihood estimators are of the order $1/N$, and that the bias is negligible for $N = 100$ and as much as 90% censoring. (If the proportion of censored observations is less than 90%, the bias becomes negligible for smaller sample sizes.) For small samples with moderate to high censoring, however, the bias of the MLEs causes confidence intervals based on them to be too narrow (so the associated hypothesis tests have inflated Type I errors). Schmee et al. (1985) provide tables for exact confidence intervals for sample sizes up to $N = 100$ that were created based on Monte Carlo simulation. Schmee et al. (1985) state that these tables should work well for Type I censored data as well.

Shumway et al. (1989) evaluated the coverage of 90% confidence intervals for the mean that were constructed from Type I left singly censored data. They used a Box-Cox transformation to induce normality, computed the MLEs based on the normal distribution, then computed the mean in the original scale. They considered three methods of constructing confidence intervals: the delta method, the bootstrap, and the bias-corrected bootstrap. Shumway et al. (1989) used three parent distributions in their study: Normal(3, 1), the square of this distribution, and the exponentiation of this distribution (i.e., a lognormal distribution). Based on sample sizes of 10 and 50 with a censoring level at the 10th or 20th percentile, they found that the delta method performed quite well and was superior to the bootstrap method.

In this section, we will discuss various methods for estimating distribution parameters for the normal and lognormal distribution and constructing confidence intervals for the mean based on left singly and multiply censored data.

Notation

Before discussing various methods for estimating distribution parameters, we need to introduce some notation. We will start with the left singly censored case, and then discuss the multiply censored case.

Notation for Left Singly Censored Data

Let $x_1, x_2, ..., x_N$ denote N observations from some distribution. Of these N observations, assume n $(0 < n < N)$ of them are known and c of them are censored below (left-censored) at some fixed censoring level T. Then

$$N = n + c \tag{10.1}$$

Let $x_{(1)}, x_{(2)}, ..., x_{(N)}$ denote the "ordered" observations, where now "observation" means either the actual observation (for uncensored observations) or the censoring level (for censored observations). If a censored observation has the same value as an uncensored one, the censored observation should be placed first. Thus,

$$x_{(1)} = x_{(2)} = \cdots = x_{(c)} = T \tag{10.2}$$

Notation for Left Multiply Censored Data

Again, let $x_1, x_2, ..., x_N$ denote N observations from some distribution. Of these N observations, assume n $(0 < n < N)$ of them are known and c of them are censored below (left-censored) at fixed censoring levels

$$T_1 < T_2 < \cdots < T_k \tag{10.3}$$

where $k \geq 2$. Let c_j denote the number of observations censored below censoring level T_j $(j = 1, 2, ..., k)$ so that

$$\sum_{j=1}^{k} c_j = c \tag{10.4}$$

As before, let $x_{(1)}, x_{(2)}, ..., x_{(N)}$ denote the "ordered" observations, where now "observation" means either the actual observation (for uncensored observations) or the censoring level (for censored observations). If a censored observation has the same value as an uncensored one, the censored observation should be placed first. Finally, let Ω (omega) denote the set of n subscripts in the "ordered" sample that correspond to uncensored observations.

Normal Distribution

In this section we will discuss various methods for estimating the mean and standard deviation of a normal distribution and constructing a confidence interval for the mean.

Estimation for Singly Censored Normal Data

Table 10.7 displays a list of estimation methods for Type I singly left-censored data assuming a normal distribution, along with references for these methods. All of these methods are explained in detail in the help file for enorm.singly.censored in ENVIRONMENTALSTATS for S-PLUS.

Method	References
MLE	Cohen (1959, 1991)
Bias-Corrected MLE	Saw (1961b) Schneider (1986) Haas and Scheff (1990) Bain and Engelhardt (1991)
Q-Q Regression (Probability Plot)	Nelson (1982) Gilbert (1987) Hass and Scheff (1990) Travis and Land (1990) Helsel and Hirsch (1992)
Q-Q Regression with Censoring Level	El-Shaarawi (1989)
Impute with Q-Q Regression (NR)	Hashimoto and Trussell (1983) Gilliom and Helsel (1986) El-Shaarawi (1989)
Impute with Q-Q Regression with Censoring Level (Modified NR)	El-Shaarawi (1989)
Impute with MLE (Modified MLE)	El-Shaarawi (1989)
Iterative Impute (Fill-In With Expected Values)	Gleit (1985)
Half Censoring Level	Gleit (1985) Hass and Scheff (1990) El-Shaarawi and Esterby (1992)

Table 10.7 Estimation methods for Type I left singly censored data assuming a normal distribution

MLE

For left singly censored data, the general maximum likelihood equation for any distribution with parameter vector θ is given by:

$$L\left(\theta \mid \underline{x}\right) = \binom{N}{c}\left[F\left(T\right)\right]^{c} \prod_{i=c+1}^{n} f\left[x_{(i)}\right] \qquad (10.5)$$

where f and F denote the pdf and cdf of the population, respectively. That is, the likelihood is product of (1) the probability of c observations out of N being less than T and (2) the product of the values of the pdf evaluated at the uncensored observations. In the case of a normal distribution, we have:

$$\theta = \left(\mu, \sigma\right) \qquad (10.6)$$

$$f\left(t\right) = \phi\left(\frac{t - \mu}{\sigma}\right) \qquad (10.7)$$

$$F\left(t\right) = \Phi\left(\frac{t - \mu}{\sigma}\right) \qquad (10.8)$$

where ϕ and Φ denote the pdf and cdf of the standard normal distribution, respectively. Cohen (1959, 1991) shows that the MLEs of μ and σ are the solutions to two simultaneous equations.

Bias-Corrected MLE

For censored data, the maximum likelihood estimates of μ and σ are biased. The bias tends to 0 as the sample size increases, but it can be considerable for small sample sizes, especially in the case of a large percentage of censored observations (Saw, 1961b). Schmee et al. (1985) note that the bias and variances of the MLEs are of the order $1/N$ (see for example, Bain and Engelhardt, 1991), and that for 90% censoring the bias is negligible if N is at least 100. (For less intense censoring, even fewer observations are needed.)

The exact bias of each estimator is extremely difficult to compute. Saw (1961b) however derived the first-order term (i.e., the term of order $1/N$) in the bias of each estimator and proposed bias-corrected MLEs. His bias-corrected estimators were derived for the case of Type II singly censored data. Schneider (1986, p. 110) and Haas and Scheff (1990) state that this bias correction should reduce the bias of the estimators in the case of Type I censoring as well.

Based on the tables of bias-correction terms given in Saw (1961b), Schneider (1986, pp. 107–110) performed a least-squares fit to produce computational formulas for the bias-corrected MLEs for both right- and left-censored data. For left-censored data, the formulas are as follows:

$$\hat{\mu}_{bcmle} = \hat{\mu}_{mle} + \frac{\hat{\sigma}_{mle}}{N+1} B_{\mu} \qquad (10.9)$$

$$\hat{\sigma}_{bcmle} = \hat{\sigma}_{mle} - \frac{\hat{\sigma}_{mle}}{N+1} B_{\sigma} \qquad (10.10)$$

where

$$B_{\mu} = -\exp\left(2.692 - 5.493 \frac{n}{N+1}\right) \qquad (10.11)$$

$$B_{\sigma} = -\left(0.312 + 0.859 \frac{n}{N+1}\right)^{-2} \qquad (10.12)$$

Note that the formulas and example given in Gibbons (1994, pp. 191–193) are incorrect.

Quantile-Quantile Regression

This method is usually called the Probability Plot method in the literature (Nelson, 1982, Chapter 3; Gilbert, 1987, pp. 134–136; Helsel and Hirsch, 1992, p. 361). In the case of no censoring, it is well known (e.g., Nelson, 1982, p. 113; Cleveland, 1993, p. 31) that for the standard normal Q-Q plot the intercept and slope of the fitted least-squares line estimate the mean and standard deviation, respectively. For example, the intercept and slope of the fitted line in Figure 3.19 estimate the mean and standard deviation of the log-transformed Reference area TcCB data.

For complete (no censoring) data, the quantile-quantile regression estimates of μ and σ are found by computing the least-squares estimates in the following linear model:

$$x_{(i)} = \mu + \sigma \, \Phi^{-1}\left(p_i\right) + \varepsilon_i \qquad\qquad \text{(10.13)}$$

where p_i denotes the plotting position associated with the ith largest value and is defined by

$$p_i = \frac{i - a}{N + 2a + 1} \qquad\qquad \text{(10.14)}$$

a is a constant such that $0 \le a \le 1$, and Φ denotes the cumulative distribution function (cdf) of the standard normal distribution. Usually, Blom plotting positions are used (i.e., $a = 0.375$).

This method can be adapted to the case of left singly censored data as follows. Plot the n uncensored observations against the n largest normal quantiles, where the normal quantiles are computed based on a sample size of N, as in Figure 10.6. Then fit the least-squares line to this plot, and estimate the mean and standard deviation from the intercept and slope, respectively. That is, use Equations (10.13) and (10.14), but use $i = (c+1)$, $(c+2)$, ..., N.

This method is discussed by Haas and Scheff (1990). In the context of lognormal data, Travis and Land (1990) suggest exponentiating the predicted 50th percentile from this fit to estimate the geometric mean (i.e., the median of the lognormal distribution).

Quantile-Quantile Regression with Censoring Level

This is a modification of the Quantile-Quantile Regression method and was proposed by El-Shaarawi (1989) in the context of lognormal data. El-Shaarawi's idea is to include the censoring level and an associated plotting position, along with the uncensored observations and their associated plotting positions, in order to include information about the value of the censoring level T. For left singly censored data, the modification involves adding the point $\left[\Phi^{-1}\left(p_c\right), \; T \right]$ to the plot before fitting the least-squares line.

Imputation with Quantile-Quantile Regression

This method involves using the Quantile-Quantile Regression method to fit a regression line (and thus initially estimate the mean and standard deviation), and then imputing the values of the c censored observations by predicting them from the regression equation. The final estimates of the mean and standard deviation are then computed using the usual formulas for the sample mean and standard deviation (see Equations (5.7) and (5.50)) based on the observed and imputed values. For left-censored data, the imputed values of the censored observations are computed as:

$$\hat{x}_{(i)} = \hat{\mu}_{qqreg} + \hat{\sigma}_{qqreg}\ \Phi^{-1}\left(p_i\right) \qquad (10.15)$$

for $i = 1, 2, ..., c$.

This is the NR method of Gilliom and Helsel (1986, p. 137). In the context of lognormal data, this method is discussed by Hashimoto and Trussell (1983), Gilliom and Helsel (1986), and El-Shaarawi (1989), and is referred to as the LR or Log-Probability Method.

Imputation with Quantile-Quantile Regression with Censoring Level

This is exactly the same method as Imputation with Quantile-Quantile Regression except that the Quantile-Quantile Regression with Censoring Level method is used to fit the regression line. In the context of lognormal data, this method is discussed by El-Shaarawi (1989), which he denotes as the Modified LR Method.

Imputation Using Maximum Likelihood

This is exactly the same method as Imputation with Quantile-Quantile Regression except that the maximum likelihood method is used to compute the initial estimates of the mean and standard deviation. In the context of lognormal data, this method is discussed by El-Shaarawi (1989), which he denotes as the Modified Maximum Likelihood Method.

Iterative Imputation with Quantile-Quantile Regression

This method is similar to the Imputation with Quantile-Quantile Regression method but iterates until the estimates of the mean and standard deviation converge. The algorithm is:

1. Compute the initial estimates of μ and σ using the Imputation with Quantile-Quantile Regression method. (Actually, any suitable estimates will do.)
2. Using the current values of μ and σ and Equation (10.15), compute new imputed values of the censored observations.
3. Use the new imputed values along with the uncensored observations to compute new estimates of μ and σ based on the usual formulas for the sample mean and standard deviation.
4. Repeat Steps 2 and 3 until the estimates converge.

This method is discussed by Gleit (1985), which he denotes as the Fill-In with Expected Values method.

Setting Censored Observations to Half the Censoring Level

This method involves simply replacing all the censored observations with half the detection limit, and then computing the mean and standard deviation with the usual formulas. This method is discussed by Gleit (1985), Haas and Scheff (1990), and El-Shaarawi and Esterby (1992). This method is not recommended. In particular, El-Shaarawi and Esterby (1992) show that these estimators are biased and inconsistent (i.e., the bias remains even as the sample size increases).

Confidence Intervals for the Mean for Singly Censored Normal Data

Table 10.8 displays a list of methods for constructing confidence intervals for the mean for Type I left singly censored data assuming a normal distribution, along with references for these methods. All of these methods are explained in detail in the help file for `enorm.singly.censored` in ENVIRONMENTALSTATS for S-PLUS.

Method	Reference
Normal Approximation	Cohen (1991)
Normal Approximation Using Covariance	Schneider (1986)
Bootstrap	Efron and Tibshirani (1993)

Table 10.8   Methods for constructing confidence intervals for the mean for Type I left singly censored data assuming a normal distribution

Normal Approximation

Equations (5.86) and (5.87) give general formulas for a confidence interval for a population parameter, assuming the distribution of the estimator of the parameter is approximately normal. We can use either of these formulas to construct an approximate confidence interval for the mean when we have left singly censored data by simply plugging in the estimated mean and standard error of the mean. If we want to use the formula in Equation (5.87), however, we have to decide on the value of the degrees of freedom, which we can think of as the assumed sample size (say m) minus 1.

For the maximum likelihood estimators, the standard deviation of the MLE of μ is estimated based on the observed or expected inverse of the Fisher Information matrix. For left-censored data, Cohen (1959; 1991, pp. 25–26) gives formulas for the asymptotic variances and covariance of the MLEs of μ and σ.

For the bias-corrected MLE, we can use the bias-corrected MLE of μ as the pivot point for the confidence interval and we can estimate the standard error of this estimate by using the estimated standard error of the MLE of μ. The true standard error of the bias-corrected MLE of μ is necessarily larger than the standard error of the MLE of μ (although the differences in the variances goes to 0 as the sample size gets large). Hence in theory this method of constructing a confidence interval leads to intervals that are too narrow for small sample sizes, but these interval should be better centered about the true value of μ.

For other estimators, the standard deviation of the estimated mean may be assumed to be approximated by

$$\hat{\sigma}_{\hat{\mu}} = \frac{\hat{\sigma}}{\sqrt{m}} \tag{10.16}$$

where, as already noted, m denotes the assumed sample size (e.g., the number of uncensored observations). This is simply an ad-hoc method of constructing confidence intervals and is not based on any published theoretical results.

Normal Approximation Using Covariance

This method was proposed by Schneider (1986, pp. 191–193) for the case of Type II censoring, but is applicable to any situation where the estimated mean and standard deviation are consistent estimators and are correlated. In particular, the MLEs of μ and σ are correlated under Type I censoring as well.

Bootstrap

As explained in Chapter 5, you can use the bootstrap to construct confidence intervals for any population parameter. In the context of Type I left singly censored data, this method of constructing confidence intervals was studied by Shumway et al. (1989).

Example 10.14: A Monte Carlo Simulation Study of Confidence Interval Coverage for Singly Censored Normal Data

Table 10.9 displays the results of a Monte Carlo simulation study of the coverage of two-sided approximate 95% confidence intervals for the mean based on using various estimators of the mean and using the normal approximation method of constructing the confidence interval. For each Monte Carlo trial, a set of observations from a N(3, 1) distribution were generated, and all observations less than the censoring level were censored. If there were less than three uncensored observations, this set of observations was not used. If none of the observations was censored, the usual method of constructing a confidence interval for the mean was used (Equation (5.78)). For each combination of estimation method, sample size, and censoring level, 1,000 trials were performed.

Estimation Method	Sample Size	Censoring Level (Percentile)	Root Mean Squared Error (%)	Coverage (%)
MLE	10	20	34	95 (z)
		50	35	95 (z)
	20	20	24	96 (z)
		50	30	97 (z)
Bias-Corrected MLE	10	20	37	96 (z)
		50	47	96 (z)
	20	20	24	97 (z)
		50	35	98 (z)
Q-Q Regression	10	20	34	96 (t)
		50	45	93 (t)
	20	20	24	96 (t)
		50	32	95 (t)
Q-Q Regression with Censoring Level	10	20	32	97 (t)
		50	36	95 (t)
	20	20	24	97 (t)
		50	30	95 (t)
Impute with Q-Q Regression	10	20	34	96 (t)
		50	45	92 (t)
	20	20	24	96 (t)
		50	32	94 (t)
Impute with Q-Q Regression with Censoring Level	10	20	33	96 (t)
		50	37	95 (t)
	20	20	24	96 (t)
		50	30	95 (t)
Impute with MLE	10	20	33	96 (t)
		50	35	96 (t)
	20	20	24	96 (t)
		50	30	96 (t)
Iterative Impute	10	20	32	95 (t)
		50	40	87 (t)
	20	20	23	95 (t)
		50	29	90 (t)
Half Censoring Level	10	20	37	95 (z)
		50	46	99 (z)
	20	20	27	96 (z)
		50	43	96 (z)

Table 10.9   Results of Monte Carlo simulation study based on Type I left singly censored data from a N(3, 1) parent distribution

The results in Table 10.9 show that the various estimators behave fairly similarly and fairly well in terms of root mean squared error and confidence interval coverage, except for the Iterative Imputation and Half Censoring Level methods, which do not perform as well. Note that some of the confidence intervals were constructed using the z-statistic (Equation (5.86)) and some were constructed using the t-statistic (Equation (5.87)).

Example 10.15: Estimating the Mean and Standard Deviation of Background Well Arsenic Concentrations

Table 10.10 displays various estimates of and confidence intervals for the mean of the background well arsenic data shown in Table 6.4. For these data, there are $N = 18$ observations, with $c = 9$ of them censored at 5 ppb. The estimates of the mean range from 5.1 to 6.5 ppb, and the estimates of the standard deviation range from 2.8 to 4.4 ppb. Both of the confidence intervals based on Schneider's (1986) method include 0, an impossible value!

Estimation Method	Estimated Mean, Sd	CI Method	CI
MLE	5.3, 4.0	Normal Approximation (z)	[2.3, 8.3]
		Normal Approximation with Covariance	[-0.7, 8.2]
Bias-Corrected MLE	5.1, 4.4	Normal Approximation (z)	[1.6, 8.6]
		Normal Approximation with Covariance	[-2.7, 8.3]
Q-Q Regression	6.0, 3.4	Normal Approximation (t)	[3.4, 8.7]
Q-Q Regression with Censoring Level	5.8, 3.6	"	[3.1, 8.6]
Impute with Q-Q Regression	6.0, 3.3	"	[3.5, 8.6]
Impute with Q-Q Regression with Censoring Level	5.9, 3.5	"	[3.2, 8.5]
Impute with MLE	5.5, 3.9	"	[2.5, 8.5]
Iterative Impute	6.5, 2.8	"	[4.3, 8.7]
Half Censoring Level	5.6, 3.5	"	[2.9, 8.2]

Table 10.10   Estimates of the mean and standard deviation, and confidence intervals for the mean for the background well arsenic data of Table 6.4

Menu

To compute estimates of the mean and standard deviation and construct a confidence interval for the mean for the background well arsenic data using the ENVIRONMENTALSTATS for S-PLUS pull-down menu, follow these steps.

1. Find **epa.92c.arsenic2.df** in the Object Explorer.
2. On the S-PLUS menu bar, make the following menu choices: **EnvironmentalStats>Censored   Data>Estimation>Parameters**. This will bring up the Estimate Distribution Parameters for Censored Data dialog box.
3. For Data to Use select **Pre-Defined Data**. The Data Set box should display **epa.92c.arsenic2.df**. For Variable select **Arsenic**, for Censor Ind. Select **Censored**, in the Subset Rows with box type **Well.type=="Background"**, for Distribution select **Normal**, and for Estimation Method select **mle**.
4. Under the Confidence Interval section **check** the Confidence Interval box, for CI Type select **two-sided**, for CI Method select **normal.approx**, for Confidence Level (%) type **95**, for Pivot Stat. select **z**, then click **OK** or **Apply**.

The above steps compute MLEs and base the confidence interval on the normal approximation using the z-statistic. To use a different estimation method change the value of Estimation Method in Step 3, to use a different method to construct the confidence interval change the value of CI Method in Step 4, and to use the t-statistic instead of the z-statistic for Pivot Stat. select **t** in Step 4.

Command

To compute estimates of the mean and standard deviation and construct a confidence interval for the mean for the background well arsenic data using the ENVIRONMENTALSTATS for S-PLUS Command or Script Window, type these commands.

```
attach(epa.92c.arsenic2.df)
enorm.singly.censored(
   Arsenic[Well.type=="Background"],
   Censored[Well.type=="Background"], method="mle",
   ci=T, ci.method="normal.approx",
   pivot.statistic="z")
```

The above command computes MLEs and bases the confidence interval on the normal approximation using the z-statistic. To use a different estimation method change the value of the argument `method`; to use a different method to construct the confidence interval change the value of the argument ci.method, and to use the t-statistic instead of the z-statistic set the argument pivot.statistic="t".

Estimation for Multiply Censored Normal Data

Table 10.11 displays a list of estimation methods for Type I multiply left-censored data assuming a normal distribution, along with references for these methods. All of these methods are explained in detail in the help file for enorm.multiply.censored in ENVIRONMENTALSTATS for S-PLUS.

Method	References
MLE	Cohen (1963, 1991)
Q-Q Regression (Probability Plot)	Extension from singly censored case
Impute with Q-Q Regression	Helsel and Cohn (1988)
Half Censoring Level	Helsel and Cohn (1988)

Table 10.11   Estimation methods for Type I left multiply censored data assuming a normal distribution

MLE

For left multiply censored data, the general maximum likelihood equation for any distribution with parameter vector θ is given by:

$$L\left(\theta \mid \underline{x}\right) = \begin{pmatrix} N \\ c_1 \; c_2 \; \cdots \; c_k \; n \end{pmatrix}$$

$$\text{(10.17)}$$

$$\prod_{j=1}^{k} \left[F\left(T_j\right)\right]^{c_j} \prod_{i \in \Omega}^{n} f\left[x_{(i)}\right]$$

where

$$\begin{pmatrix} N \\ c_1 \; c_2 \; \cdots \; c_k \; n \end{pmatrix} = \frac{N!}{c_1! \; c_2! \cdots c_k! \; n!} \qquad \text{(10.18)}$$

denotes the multinomial coefficient, and f and F denote the population pdf and cdf, as in Equation (10.5). In the case of a normal distribution, θ, f, and F are defined in Equations (10.6) to (10.8). Cohen (1963, 1991) shows that the MLEs of μ and σ are the solutions to two simultaneous equations.

Quantile-Quantile Regression

This method is an extension to multiply censored data of the Quantile-Quantile Regression method for singly censored data. The estimates of μ and σ are found by computing the least-squares estimates for the model shown in Equation (10.13) using the uncensored observations and their associated plotting positions (i.e., $i \in \Omega$). The formulas that can be used for the plotting position are an extension to the one for singly censored data shown in Equation (10.14). (See the ENVIRONMENTALSTATS for S-PLUS help file for `ppoints.censored` for details on the formulas.)

Imputation with Quantile-Quantile Regression

This method is an extension to multiply censored data of the Imputation with Quantile-Quantile Regression method for singly censored data. The imputed values of the c censored observations are computed using Equation (10.15), where $i \notin \Omega$. This method was developed in the context of lognormal data by Helsel and Cohn (1988) using the formulas for plotting positions given in Hirsch and Stedinger (1987) and Weibull plotting positions.

Confidence Intervals for the Mean for Multiply Censored Normal Data

Table 10.12 displays a list of methods for constructing confidence intervals for the mean for Type I left multiply censored data assuming a normal distribution, along with references for these methods. All of these methods are explained in detail in the help file for `enorm.multiply.censored` in ENVIRONMENTALSTATS for S-PLUS.

Method	Reference
Normal Approximation	Cohen (1991)
Bootstrap	Efron and Tibshirani (1993)

Table 10.12   Methods for constructing confidence intervals for the mean for Type I left multiply censored data assuming a normal distribution

Example 10.16: Estimating the Mean and Standard Deviation of Log-Transformed Silver Concentrations

We saw in Example 10.8 and Example 10.12 that the silver data of Table 10.4 appear to be adequately described by a lognormal distribution. Table 10.13 displays estimates of and confidence intervals for the mean of the log-transformed silver data based on using the normal approximation with the z-statistic. In this case all of the methods yield similar estimates and confidence intervals.

Estimation Method	Estimated Mean, Sd	CI
MLE	-1.0, 2.4	[-2.0, -0.1]
Q-Q Regression	-1.2, 2.5	[-2.2, -0.1]
Impute with Q-Q Regression	-1.1, 2.3	[-2.1, -0.2]

Table 10.13   Estimates of the mean and standard deviation, and confidence intervals for the mean for the log-transformed silver data of Table 10.4

Menu

To compute estimates of the mean and standard deviation and construct a confidence interval for the mean for the log-transformed silver data using the ENVIRONMENTALSTATS for S-PLUS pull-down menu, follow these steps.

1. Find **helsel.cohn.88.silver.df** in the Object Explorer.
2. On the S-PLUS menu bar, make the following menu choices: **EnvironmentalStats>Censored   Data>Estimation>Parameters**. This will bring up the Estimate Distribution Parameters for Censored Data dialog box.
3. For Data to Use select **Pre-Defined Data**. The Data Set box should display **helsel.cohn.88.silver.df**. For Variable select **log.Ag**, for Censor Ind. Select **Censored**, for Distribution select **Normal**, and for Estimation Method select **mle**.
4. Under the Confidence Interval section **check** the Confidence Interval box, for CI Type select **two-sided**, for CI Method select **normal.approx**, for Confidence Level (%) type **95**, for Pivot Stat. select **z**, then click **OK** or **Apply**.

The above steps compute MLEs and base the confidence interval on the normal approximation using the z-statistic. To use a different estimation method change the value of Estimation Method in Step 3, to use a different method to construct the confidence interval change the value of CI Method in Step 4, and to use the t-statistic instead of the z-statistic for Pivot Stat. select **t** in Step 4.

Command

To compute estimates of the mean and standard deviation and construct a confidence interval for the mean for the log-transformed silver data using the ENVIRONMENTALSTATS for S-PLUS Command or Script Window, type these commands.

```
attach(helsel.cohn.88.silver.df)
```

```
enorm.multiply.censored(log(Ag), Censored,
   method="mle", ci=T, ci.method="normal.approx",
   pivot.statistic="z")
```

The above command computes MLEs and bases the confidence interval on the normal approximation using the z-statistic. To use a different estimation method change the value of the argument method; to use a different method to construct the confidence interval change the value of the argument ci.method, and to use the t-statistic instead of the z-statistic set the argument pivot.statistic="t".

Lognormal Distribution

In Chapter 4 we explained two different ways to parameterize the lognormal distribution: one based on the parameters associated with the log-transformed random variable and one based on the parameters associated with the original random variable (see Equations (4.32) to (4.35)). If you are interested in characterizing the log-transformed random variable, you can simply take the logarithms of the observations and the censoring levels, treat them like they come from a normal distribution, and use the methods discussed in the previous section (this is what we did in Example 10.16 for the silver data). Often, however, we want an estimate of the mean and a confidence interval on the original scale. In this section we will discuss various methods for estimating the mean and coefficient of variation of a lognormal distribution and constructing confidence intervals for the mean.

Estimation for Singly Censored Lognormal Data

Table 10.14 displays a list of estimation methods for Type I left singly censored data assuming a lognormal distribution, along with references for these methods. All of these methods are explained in detail in the help file for elnorm.alt.singly.censored in ENVIRONMENTALSTATS for S-PLUS.

MLE

For the case of complete (no censored) data, the MLEs of the mean (θ), coefficient of variation (τ), and standard deviation (η) of a lognormal distribution are shown in Equations (5.44) to (5.46). These estimators are functions of the MLEs of the mean (μ) and standard deviation (σ) for the log-transformed random variable. In the case of censored data, you use the same equations to compute the MLEs of θ, τ, and η, but of course the MLEs of μ and σ are now based on the equations relevant to left singly censored data.

Method	References
MLE	Extension to Cohen (1959, 1991)
Quasi MVUE	Gilliom and Helsel (1986)
	Newman et al. (1989)
Bias-Corrected MLE	El-Shaarawi (1989)
Impute with Q-Q Regression (LR)	Hashimoto and Trussell (1983)
	Gilliom and Helsel (1986)
	El-Shaarawi (1989)
Impute with Q-Q Regression with Censoring Level (Modified LR)	El-Shaarawi (1989)
Impute with MLE (Modified MLE)	El-Shaarawi (1989)
Half Censoring Level	Gleit (1985)
	Hass and Scheff (1990)
	El-Shaarawi and Esterby (1992)

Table 10.14   Estimation methods for Type I left singly censored data assuming a log-normal distribution

Quasi Minimum Variance Unbiased Estimator (QMVUE)

The maximum likelihood estimators of the mean (θ) and variance (η^2) are biased. Even for complete (uncensored) samples these estimators are biased, as we noted in Chapter 5. The bias tends to 0 as the sample size increases, but it can be considerable for small sample sizes. (Cohn et al., 1989 demonstrate the bias for complete data sets.)

For the case of complete samples, the minimum variance unbiased estimators (MVUEs) of θ and η^2 were derived by Finney (1941) and are discussed in Gilbert (1987, pp. 164–167) and Cohn et al. (1989). The formulas for these estimators are shown in Equations (5.54) and (5.55) in Chapter 5. These estimators are based on the MVUEs of the mean (μ) and variance (σ^2) for the log-transformed random variable. For Type I left singly censored samples, the quasi minimum variance unbiased estimators (QMVUEs) of θ and η^2 are computed using the same equations, but the MLEs for μ and σ^2 based on censored data are used.

This is apparently the LM method of Gilliom and Helsel (1986, p. 137) (it is not clear from their description on page 137 whether their LM method is the straight MLE method or the QMVUE method). This method was also used by Newman et al. (1989, p. 915, equations 10 and 11).

Bias-Corrected MLE

This method was derived by El-Shaarawi (1989) and can be applied to complete or censored data sets. For complete samples, the exact relative bias of the MLE of the mean θ is given as:

$$B_{mle} = \frac{E\left(\hat{\theta}_{mle}\right)}{\theta}$$

$$= \exp\left[\frac{-\left(n-1\right)\sigma^2}{2n}\right]\left(1 - \frac{\sigma^2}{n}\right)^{-(n-1)/2}$$

(10.19)

For the case of complete or censored data, El-Shaarawi (1989) proposed the following "bias-corrected" maximum likelihood estimator:

$$\hat{\theta}_{bcmle} = \frac{\hat{\theta}_{mle}}{\hat{B}_{mle}} \tag{10.20}$$

where

$$\hat{B}_{mle} = \exp\left[\frac{1}{2}\left(\hat{V}_{11} + 2\hat{\sigma}_{mle}\hat{V}_{12} + \hat{\sigma}^2_{mle}\hat{V}_{22}\right)\right] \tag{10.21}$$

and V denotes the asymptotic variance-covariance of the MLEs of μ and σ.

Imputation with Quantile-Quantile Regression

This is exactly the same method as was discussed for the normal distribution earlier, except the observations and censoring levels are log-transformed prior to fitting the Q-Q regression model (Equation (10.15)), and once the log-transformed censored observations have been imputed they are transformed back (by exponentiating) prior to computing the sample mean and standard deviation based on the usual formulas. This method is discussed by Hashimoto and Trussell (1983), Gilliom and Helsel (1986), and El-Shaarawi (1989), and is referred to as the LR (Log-Regression) or Log-Probability Method.

Imputation with Quantile-Quantile Regression Including the Censoring Level

This is a modification of the previous method that simply includes the (log-transformed) censoring level in the fit to the Q-Q regression model. El-Shaarawi (1989) denotes this method as the Modified LR (MLR) Method.

Imputation with Maximum Likelihood

This is exactly the same method as Imputation with Quantile-Quantile Regression, except imputed values are computed based on the maximum likelihood estimates of μ and σ instead of the estimates based on the quantile-quantile regression. This method is discussed by El-Shaarawi (1989), which he denotes as the Modified Maximum Likelihood (MML) Method.

Confidence Intervals for the Mean for Singly Censored Lognormal Data

Table 10.15 displays a list of methods for constructing confidence intervals for the mean for Type I left singly censored data assuming a lognormal distribution, along with references for these methods. All of these methods are explained in detail in the ENVIRONMENTALSTATS for S-PLUS help file for `elnorm.alt.singly.censored`.

Method	Reference
Normal Approximation Based on Moment Estimators	
Normal Approximation Based on Delta Method	Shumway et al. (1989)
Normal Approximation Based on Cox's Method	El-Shaarawi (1989)
Bootstrap	Efron and Tibshirani (1993)

Table 10.15 Methods for constructing confidence intervals for the mean for Type I left singly censored data assuming a lognormal distribution

Normal Approximation Based on Moment Estimators

Equations (5.86) and (5.87) give general formulas for a confidence interval for a population parameter, assuming the distribution of the estimator of the parameter is approximately normal. We can use either of these formulas to construct an approximate confidence interval for the mean when we have left singly censored data by simply plugging in the estimated mean and standard error of the mean. If we want to use the formula in Equation (5.87), however, we have to decide on the value of the degrees of freedom, which we can think of as the assumed sample size (say m) minus 1.

In order to use Equation (5.68) or (5.87) we need to plug in the standard error of the mean. The standard deviation of the estimated mean may be assumed to be approximated by

$$\hat{\sigma}_{\hat{\theta}} = \frac{\hat{\eta}}{\sqrt{m}} \tag{10.22}$$

where, as already noted, m denotes the assumed sample size. This is simply an ad-hoc method of constructing confidence intervals and is not based on any published theoretical results.

Normal Approximation Based on the Delta Method

For the maximum likelihood estimator of θ, we can use the delta method to compute the standard error of the mean (Shumway et al., 1989). This method should also work approximately for the QMVUE and the Bias-Corrected MLE. The delta method yields

$$\hat{\sigma}_{\hat{\theta}} = \sqrt{\underline{a}' \hat{V} \underline{a}} \tag{10.23}$$

where V denotes the variance-covariance matrix of the MLEs of μ and σ, and $\underline{a}$ is a vector with the following elements:

$$\underline{a} = \begin{bmatrix} \exp\left(\hat{\mu} + \dfrac{\hat{\sigma}^2}{2}\right) \\ \\ \hat{\sigma}\exp\left(\hat{\mu} + \dfrac{\hat{\sigma}^2}{2}\right) \end{bmatrix} \tag{10.24}$$

Normal Approximation Based on Cox's Method

This method was proposed by El-Shaarawi (1989) and is an extension of the method derived by Cox and presented in Land (1972) for the case of complete data. The standard error of the mean is estimated by:

$$\hat{\sigma}_{\hat{\theta}} = \sqrt{\hat{V}_{11} + 2\hat{\sigma}\hat{V}_{12} + \hat{\sigma}^2\hat{V}_{22}} \tag{10.25}$$

Example 10.17: A Monte Carlo Simulation Study of Confidence Interval Coverage for Singly Censored Lognormal Data

Table 10.16 displays the results of a Monte Carlo simulation study of the coverage of one-sided upper approximate 95% confidence intervals for the mean based on using various estimators of the mean and methods of constructing the confidence interval.

Estimation Method	Sample Size	Censoring Level (Percentile)	CI Method	Coverage (%)
MLE	10	20	Delta	85
		50		91
	20	20		87
		50		93
	10	20	Cox	92
		50		96
	20	20		93
		50		97
QMVUE	10	20	Delta	83
		50		90
	20	20		84
		50		92
	10	20	Cox	90
		50		95
	20	20		92
		50		96
Bias-Corrected MLE	10	20	Delta	80
		50		84
	20	20		84
		50		87
	10	20	Cox	90
		50		94
	20	20		91
		50		95
Impute with Q-Q Regression	10	20	Normal Approx	78
		50		89
	20	20		80
		50		90
Impute with Q-Q Regression with Censoring Level	10	20	Normal Approx	78
		50		89
	20	20		80
		50		90

Table 10.16   Results of Monte Carlo simulation study based on Type I left singly censored data from a lognormal parent distribution with mean $\theta = 10$ and CV $\tau = 2$

For each Monte Carlo trial, a set of observations from a lognormal distribution with a mean of $\theta = 10$ and a coefficient of variation of $\tau = 2$ were generated, and all observations less than the censoring level were censored. If there were less than three uncensored observations, this set of observations was not used. Table 10.17 and Table 10.18 display what methods were used in the case when there were no censored observations. For each combination of estimation method, sample size, and censoring level, 1,000 trials were performed. All confidence intervals were constructed using the t-statistic (Equation (5.87)).

Censored Method	Complete Method
MLE	MLE
Quasi MVUE	MVUE
Bias-Corrected MLE	MVUE
Impute with Q-Q Regression	MMUE
Impute with Q-Q Regression with Censoring Level	MMUE

Table 10.17 Estimation methods used in the Monte Carlo simulation study when there were no censored observations

Censored Method	Complete Method
Delta	Parkin et al.
Cox	Cox
Normal Approximation	Normal Approximation

Table 10.18 Confidence interval methods used in the Monte Carlo simulation study when there were no censored observations

Estimation Method	Sample Size	Censoring Level (Percentile)	Bias	RMSE
MLE	10	20	1	8
		50	9	206
	20	20	1	5
		50	1	7
Quasi MVUE	10	20	0	5
		50	0	7
	20	20	0	4
		50	0	5
Bias-Corrected MLE	10	20	-2	5
		50	-3	5
	20	20	0	4
		50	1	4

Table 10.19 Bias and root-mean squared error of various estimators of the mean for a lognormal distribution based on the Monte Carlo study

Based on the results in Table 10.16, it looks like using either the MLE, QMVUE, or bias-corrected MLE to estimate the mean and using Cox's method to construct the confidence interval will provide reasonably good coverage. Looking at Table 10.19, however, we see that based upon the bias and root-mean squared error, the MLE should be avoided for small sample sizes with moderate to heavy censoring.

You should note that this Monte Carlo simulation always used an underlying lognormal distribution, so we have not investigated robustness to departures from this assumption. Helsel and Cohn (1988) performed a Monte Carlo simulation study in which they used lognormal, contaminated lognormal, gamma, and delta (zero-modified lognormal) parent distributions, created multiply censored data, and looked at bias and RMSE. They recommend using the Impute with Q-Q Regression estimator if you are not sure that the underlying population is lognormal.

Example 10.18: Confidence Interval for the Mean of the Reference Area TcCB Concentrations Based on Singly Censored Data

In Example 5.18 we computed two-sided 95% confidence intervals for the mean TcCB concentration (ppb) in the Reference area assuming the data come from a lognormal distribution (see Table 5.4). There are $n = 47$ uncensored observations in this data set. The modified TcCB data shown in Table 10.1 include $c = 19$ censored observations. Table 10.20 displays two-sided 95% confidence intervals for the mean based on these left singly censored data. Except for the confidence interval based on the Half Censoring Level method, all of these confidence intervals are very similar and similar to the ones based on the complete data as well.

Estimation Method	Confidence Interval Method	Confidence Interval
MLE	Cox	[0.51, 0.73]
Quasi MVUE	"	[0.50, 0.73]
Bias-Corrected MLE	"	[0.50, 0.73]
Impute with Q-Q Regression	Normal Approximation	[0.49, 0.71]
Impute with Q-Q Regression with Censoring Level	"	[0.49, 0.71]
Impute with MLE	"	[0.50, 0.71]
Half Censoring Level	"	[0.43, 0.68]

Table 10.20   Two-sided 95% confidence intervals for the mean TcCB concentration in the Reference area based on left singly censored data

Menu

To compute estimates of the mean and coefficient of variation and construct a confidence interval for the mean for the modified TcCB data using the ENVIRONMENTALSTATS for S-PLUS pull-down menu, follow these steps.

1. Find **Modified.TcCB.df** in the Object Explorer.
2. On the S-PLUS menu bar, make the following menu choices: **EnvironmentalStats>Censored   Data>Estimation>Parameters**. This will bring up the Estimate Distribution Parameters for Censored Data dialog box.
3. For Data to Use select **Pre-Defined Data**. The Data Set box should display **Modified.TcCB.df**. For Variable select **TcCB**, for Censor Ind. Select **Censored**, for Distribution select **Lognormal**, and for Estimation Method select **mle**.
4. Under the Confidence Interval section **check** the Confidence Interval box, for CI Type select **two-sided**, for CI Method select **cox**, for Confidence Level (%) type **95**, for Pivot Stat. select **t**, then click **OK** or **Apply**.

The above steps compute MLEs and base the confidence interval on Cox's method using the t-statistic. To use a different estimation method change the value of Estimation Method in Step 3, to use a different method to construct the confidence interval change the value of CI Method in Step 4, and to use the z-statistic instead of the t-statistic for Pivot Stat. select **z** in Step 4.

Command

To compute estimates of the mean and coefficient of variation and construct a confidence interval for the mean for the modified TcCB data using the ENVIRONMENTALSTATS for S-PLUS Command or Script Window, type these commands.

```
attach(Modified.TcCB.df)
elnorm.alt.singly.censored(TcCB, Censored,
    method="mle", ci=T, ci.method="cox",
    pivot.statistic="t")
```

The above command computes MLEs and bases the confidence interval on Cox's method using the t-statistic. To use a different estimation method change the value of the argument `method`; to use a different method to construct the confidence interval change the value of the argument `ci.method`, and to use the z-statistic instead of the t-statistic set the argument `pivot.statistic="z"`.

Estimation and Confidence Intervals for the Mean for Multiply Censored Lognormal Data

All of the methods of estimation for singly censored lognormal data shown in Table 10.14 and all of the methods of constructing confidence intervals shown in Table 10.15 can be extended to the case of multiply censored data. These methods are explained in detail in the help file for `elnorm.alt.multiply.censored` in ENVIRONMENTALSTATS for S-PLUS.

Example 10.19: Estimating the Mean Silver Concentration

In Example 10.16 we estimated and computed confidence intervals for the mean of the log-transformed silver concentration population. In this example we will do the same thing for the untransformed population. Table 10.21 displays estimates of and confidence intervals for the mean of the silver concentrations. We see that in this case different methods yield wildly different estimates of the mean and coefficient of variation. This is probably due to the large amount of censoring (61%) and because looking at Table 10.4 we see that all of the observations are less than 25 except for the two "outliers" of 90 and 560 (essentially one and two orders of magnitude greater than all of the other observations). The various methods of estimation will behave differently for different kinds of data with large "outliers."

Estimation Method	Estimated Mean, CV	95% CI
MLE	6, 16	[0.2 , 204]
Quasi MVUE	5, 8	[0.1 , 173]
Bias-Corrected MLE	1, 16	[0.01, 46]
Impute with Q-Q Regression	13, 6	[0 , 46]
Half Censoring Level	14, 5	[0 , 47]

Table 10.21   Estimates of the mean and coefficient of variation, and 95% confidence intervals for the mean using the silver data of Table 10.4

Menu

To compute estimates of the mean and coefficient of variation and construct a confidence interval for the mean for the silver concentrations using the ENVIRONMENTALSTATS for S-PLUS pull-down menu, follow these steps.

1. Find **helsel.cohn.88.silver.df** in the Object Explorer.
2. On the S-PLUS menu bar, make the following menu choices: **EnvironmentalStats>Censored   Data>Estimation>Parameters**. This will bring up the Estimate Distribution Parameters for Censored Data dialog box.

3. For Data to Use select **Pre-Defined Data**. The Data Set box should display **helsel.cohn.88.silver.df**. For Variable select **Ag**, for Censor Ind. Select **Censored**, for Distribution select **Logormal**, and for Estimation Method select **mle**.

4. Under the Confidence Interval section **check** the Confidence Interval box, for CI Type select **two-sided**, for CI Method select **cox**, for Confidence Level (%) type **95**, for Pivot Stat. select **t**, then click **OK** or **Apply**.

The above steps compute MLEs and base the confidence interval on Cox's method using the t-statistic. To use a different estimation method change the value of Estimation Method in Step 3, to use a different method to construct the confidence interval change the value of CI Method in Step 4, and to use the z-statistic instead of the t-statistic for Pivot Stat. select **z** in Step 4.

Command

To compute estimates of the mean and coefficient of variation and construct a confidence interval for the mean for the silver concentrations using the ENVIRONMENTALSTATS for S-PLUS Command or Script Window, type these commands.

```
attach(helsel.cohn.88.silver.df)
elnorm.alt.multiply.censored(Ag, Censored,
    method="mle", ci=T, ci.method="cox",
    pivot.statistic="t")
```

The above command computes MLEs and bases the confidence interval on Cox's method using the t-statistic. To use a different estimation method change the value of the argument `method`; to use a different method to construct the confidence interval change the value of the argument `ci.method`, and to use the z-statistic instead of the t-statistic set the argument `pivot.statistic="z"`.

Other Distributions

Methods for estimating distribution parameters for other distributions in the presence of censored data are discussed in Cohen (1991). In ENVIRONMENTALSTATS for S-PLUS you can use the pull-down menu or command line functions to estimate the rate parameter of the Poisson distribution in the presence of censored data. See the help files for the functions `epois.singly.censored` and `epois.multiply.censored` for details.

ESTIMATING DISTRIBUTION QUANTILES

In this section we will discuss how to estimate and construct confidence intervals for distribution quantiles (percentiles) based on censored data.

Parametric Methods for Estimating Quantiles

In Chapter 5 we discussed methods for estimating and constructing confidence intervals for population quantiles or percentiles. For a normal distribution, the estimated quantile and confidence interval are functions of the estimated mean and standard deviation (Equations (5.138) to (5.141)). For a lognormal distribution, they are functions of the estimated mean and standard deviation based on the log-transformed observations (Equations (5.142) to (5.146)).

In the presence of censored observations, we can construct estimates and confidence intervals for population percentiles using the same formulas as for complete data, but to estimate the mean and standard deviation we have to use the formulas that are appropriate for censored data. It is not clear how well the methods for estimating and constructing confidence intervals for population percentiles behave in the presence of censored data. This is an area that requires further research.

Example 10.20: Estimating and Constructing an Upper Confidence Limit for the 95th Percentile of the Background Well Arsenic Concentrations

In Example 10.10 we saw that the combined background well arsenic data shown in Table 6.4 appear to be adequately modeled by a normal distribution. In Example 10.15 we saw that the estimated mean and standard deviation are about 5 and 4 ppb, respectively, based on the MLEs. Using these estimates, the estimated 95th percentile is 11.8 ppb, and a one-sided upper 99% confidence limit for this percentile is 16.9 ppb.

Menu

To compute an estimate of and construct an upper 99% confidence limit for the 95th percentile of the background well arsenic concentrations using the ENVIRONMENTALSTATS for S-PLUS pull-down menu, follow the same steps as in Example 10.15 to estimate the mean and standard deviation, but in Step 4, **uncheck** the Confidence Interval box and in the Save As box type **estimation.list**. Then follow these next steps.

1. On the S-PLUS menu bar, make the following menu choices: **EnvironmentalStats>Estimation>Quantiles**. This will bring up the Estimate Distribution Quantiles dialog box.
2. For Data to Use select **Expression**. In the Expression box type **estimation.list**. In the Quantile(s) box type **0.95**.

3. Under the Confidence Interval group **check** the Confidence Interval box. For CI Type select **upper**. For CI Method select **exact**. For Confidence Level (%) type **99**. Click **OK** or **Apply**.

Command

To compute an estimate of and construct an upper 99% confidence limit for the 95th percentile of the background well arsenic concentrations using the ENVIRONMENTALSTATS for S-PLUS Command or Script Window type these commands.

```
attach(epa.92c.arsenic2.df)

estimation.list <- enorm.singly.censored(
   Arsenic[Well.type=="Background"],
   Censored[Well.type=="Background"], method="mle",
   ci=F)

eqnorm(estimation.list, p=0.95, ci=T, ci.type="upper",
   conf.level=0.99)

detach()
```

Nonparametric Methods for Estimating Quantiles

We saw in Chapter 5 that nonparametric estimates and confidence intervals for population percentiles are simply functions of the ordered observations (Equations (5.151) to (5.160)). Thus, for left censored data, you can still estimate quantiles and create one-sided upper confidence intervals as long as there are enough uncensored observations that can be ordered in a logical way (see Example 5.27).

PREDICTION AND TOLERANCE INTERVALS

In this section we will discuss how to construct prediction and tolerance intervals based on censored data.

Parametric Methods for Prediction and Tolerance Intervals

In Chapter 6 we discussed methods for constructing prediction and tolerance intervals. For a normal distribution, the prediction and tolerance intervals are functions of the estimated means and standard deviations (Equations (6.9) to (6.11)). For a lognormal distribution, they are functions of the estimated mean and standard deviation based on the log-transformed observations.

In the presence of censored observations, we can construct prediction and tolerance intervals using the same formulas as for complete data, but to estimate the mean and standard deviation we have to use the formulas that are appropriate for censored data. It is not clear how well the methods for constructing prediction and tolerance intervals behave in the presence of censored data. This is an area that requires further research.

Example 10.21: Using an Upper Tolerance Limit vs. an Upper Prediction Limit to Determine Contamination at a Cleanup Site Based on Singly Censored TcCB Data

In Example 6.7 we compared a one-sided upper tolerance limit with a one-sided upper prediction limit based on using the Reference area TcCB data and assuming a lognormal distribution. The tolerance limit was a 95% β-content tolerance limit with associated confidence level of 95%. The prediction limit was for the next $k = 77$ observations with associated confidence level of 95%. Table 10.22 compares these limits based on the complete data set vs. those based on the singly censored data set shown in Table 10.1. In this case there is very little difference between the limits based on the complete data vs. those based on the censored data.

Limit	Complete Data	Censored Data
Tolerance	1.42	1.38
Prediction	2.68	2.51

Table 10.22   Comparison of one-sided upper tolerance and prediction limits based on complete and singly censored TcCB data

Menu

To compute the one-sided upper 95% β-content tolerance limit with associated confidence level 95% using the ENVIRONMENTALSTATS for S-PLUS pull-down menu, follow these steps.

1. Find **Modified.TcCB.df** in the Object Explorer.
2. On the S-PLUS menu bar, make the following menu choices: **EnvironmentalStats>Censored   Data>Estimation>Parameters**. This will bring up the Estimate Distribution Parameters for Censored Data dialog box.
3. For Data to Use select **Pre-Defined Data**. The Data Set box should display **Modified.TcCB.df**. For Variable select **TcCB**, for Censor Ind. Select **Censored**, for Distribution select **Lognormal**, and for Estimation Method select **mle**.
4. Under the Confidence Interval section **uncheck** the Confidence Interval box, in the Save As box type **estimation.list**, then click **OK**.
5. On the S-PLUS menu bar, make the following menu choices: **EnvironmentalStats>Estimation>Tolerance Intervals**. This will bring up the Tolerance Interval dialog box.

6. For Data to Use select **Expression**. In the Expression box type **estimation.list**.

7. Click on the **Interval** tab. Under the Distribution group, for Type select **Parametric**, and for Distribution select **Lognormal**. Under the Tolerance Interval group, for Coverage Type select **Content**, for TI Type select **upper**, for Coverage (%) type **95**, and for Confidence Level (%) type **95**. Click **OK** or **Apply**.

To compute the one-sided upper prediction limit, follow these steps.

1. On the S-PLUS menu bar, make the following menu choices: **EnvironmentalStats>Estimation>Prediction Intervals**. This will bring up the Prediction Interval dialog box.

2. For Data to Use select **Expression**. In the Expression box type **estimation.list**.

3. Click on the **Interval** tab. Under the Distribution/Sample Size group, for Type select **Parametric** and for Distribution select **Lognormal**. Under the Prediction Interval group, **uncheck** the Simultaneous box, for # Future Obs type **77**, for PI Type select **upper**, for PI Method select **exact**, and for Confidence Level (%) type **95**. Click **OK** or **Apply**.

Command

To compute the one-sided upper 95% β-content tolerance limit with associated confidence level 95% using the ENVIRONMENTALSTATS for S-PLUS Command or Script Window type these commands.

```
attach(Modified.TcCB.df)
estimation.list <- elnorm.singly.censored(TcCB,
    Censored, method="mle", ci=F)
tol.int.lnorm(estimation.list, coverage=0.95,
    cov.type="content", ti.type="upper",
    conf.level=0.95)
```

To compute the 95% one-sided upper prediction limit for the next $k=77$ observations, type these commands.

```
pred.int.lnorm(estimation.list, k=77, method="exact",
    pi.type="upper", conf.level=0.95)
detach()
```

Nonparametric Prediction and Tolerance Intervals

We saw in Chapter 6 that nonparametric prediction and tolerance intervals are simply functions of the ordered observations (Equations (6.42) to

(6.48)). Thus, for left censored data, you can still estimate quantiles and create one-sided upper confidence intervals as long as there are enough uncensored observations that can be ordered in a logical way (see Example 6.3 and Example 6.10).

HYPOTHESIS TESTS

In this section we will discuss how to perform hypothesis tests in the presence of censored data.

Goodness-of-Fit Tests

In Chapter 5 we showed that the Shapiro-Francia goodness-of-fit statistic can be written as the square of the sample correlation coefficient between the ordered observations and the expected values of the normal order statistics (Equation (7.14)). In practice we use an approximation to the true expected values of the normal order statistics based on Blom plotting positions, so that the Shapiro-Francia goodness-of-fit statistic is the square of the probability plot correlation coefficient (Equations (7.15) to (7.17)), and is therefore equivalent to the goodness-of-fit test based on the probability plot correlation coefficient (the PPCC test). Similarly, the Shapiro-Wilk goodness-of-fit statistic can be written as the square of the sample correlation coefficient between the ordered observations and the expected values of the normal order statistics weighted by their variance-covariance matrix (Equation (7.10)).

Using these formulations, Royston (1993) extended both the Shapiro-Francia (PPCC) and Shapiro-Wilk goodness-of-fit tests to the case of singly censored data by computing these statistics based on the uncensored observations, similar to the way we explained how to construct a Q-Q plot for singly censored data earlier in this chapter in the section Q-Q Plots for Censored Data. Royston (1993) provides a method of computing p-values for these statistics based on tables given in Verrill and Johnson (1988). (For details, see the help files for `sw.singly.censored.gof` and `sf.singly.censored.gof` in ENVIRONMENTALSTATS for S-PLUS.) Although Verrill and Johnson (1988) produced their tables based on Type II censoring, Royston's (1993) approximation to the p-value of these tests should be fairly accurate for Type I censored data as well.

The PPCC test is also easily extendible to the case of multiply censored data, but it is not known how well Royston's (1993) method of computing p-values works in this case. See the ENVIRONMENTALSTATS for S-PLUS help file for `ppcc.norm.multiply.censored.gof` for details.

We noted in Chapter 7 that goodness-of-fit tests are of limited value for small sample sizes because there is usually not enough information to distinguish between different kinds of distributions. This also holds true even for moderate sample sizes if you have censored data, as the next example shows.

Example 10.22: Testing the Normality of the Singly Censored Reference Area TcCB Data

In Example 7.1 we showed that goodness-of-fit tests on the Reference area TcCB data indicated that the normal distribution was not appropriate, but that a lognormal distribution appeared to adequately model the data. In this example we will perform the same tests but use the modified TcCB data of Table 10.1. The Shapiro-Wilk test for normality yields a p-value of 0.34 and the test for lognormality yields a p-value of 0.28. Figure 10.10 and Figure 10.11 show companion plots for the tests for normality and lognormality, respectively. Compare these figures to Figure 7.1 and Figure 7.2. We see that unlike the case with complete data in Example 7.1, here censoring 40% of the observations leaves us unable to distinguish between a normal and lognormal distribution.

Menu

To perform the Shapiro-Wilk test for normality using the ENVIRONMENTALSTATS for S-PLUS pull-down menu, follow these steps.

1. Find **Modified.TcCB.df** in the Object Explorer.
2. On the S-PLUS menu bar, make the following menu choices: **EnvironmentalStats>Censored Data>Hypothesis Tests>GOF Tests>Shapiro-Wilk**. This will bring up the Shapiro-Wilk GOF Test for Censored Data dialog box.
3. For Data to Use, select **Pre-Defined Data**. For Data Set make sure **Modified.TcCB.df** is selected, for Variable select **TcCB**, and in Censor Ind. select **Censored**. In the Distribution box select **Normal**.
4. Click on the **Plotting** tab. Under Plotting Information, in the Significant Digits box type **3**, then click **OK** or **Apply**.

To perform the Shapiro-Wilk test for lognormality, repeat the above steps, but in Step 3 select **Lognormal** in the Distribution box.

Command

To perform the Shapiro-Wilk tests using the ENVIRONMENTALSTATS for S-PLUS Command or Script Window, type these commands.

```
attach(Modified.TcCB.df)
sw.list.norm <- sw.singly.censored.gof(TcCB, Censored)
sw.list.norm
sw.list.lnorm <- sw.singly.censored.gof(TcCB,
   Censored, dist="lnorm")
sw.list.lnorm
```

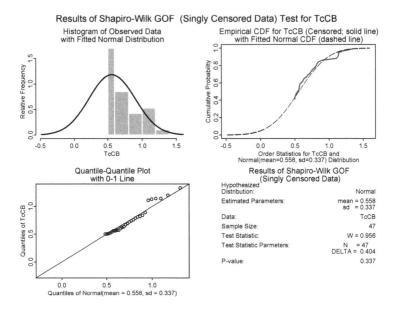

Figure 10.10 Companion plots for the Shapiro-Wilk test for normality for the singly censored Reference area TcCB data

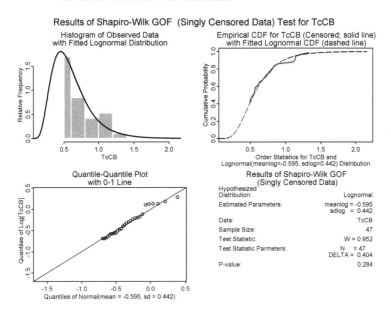

Figure 10.11 Companion plots for the Shapiro-Wilk test for lognormality for the singly censored Reference area TcCB data

To plot the results of these tests as shown in Figures 10.10 and 10.11, type these commands.

```
plot.gof.censored.summary(sw.list.norm, digits=3)
plot.gof.censored.summary(sw.list.lnorm, digits=3)
detach()
```

Hypothesis Tests on Distribution Parameters and Quantiles

In Chapter 7 we noted the one-to-one relationship between confidence intervals for a population parameter and hypothesis tests for the parameter (see Table 7.4). You can therefore perform hypothesis tests on distribution parameters and quantiles based on censored data by constructing confidence intervals for the parameter or quantile using the methods discussed in this chapter.

Nonparametric Tests to Compare Two Groups

In Chapter 7 we discussed various hypothesis tests to compare locations (central tendency) between two groups, including the Wilcoxon rank sum test, linear rank tests, and the quantile test. In the presence of censored observations, you can still use the Wilcoxon rank sum test or quantile test as long as there are enough uncensored observations that can be ordered in a logical way. For example, if both samples are singly censored with the same censoring level and all uncensored observations are greater than the censoring level, then all censored observations receive the lowest ranks and are considered tied observations. Actually you can use the Wilcoxon rank sum test even with multiply censored data as we will now discuss.

We stated in Chapter 7 that the Wilcoxon rank sum test is a particular kind of linear rank test (see Equation (7.90)). Several authors have proposed extensions of the Wilcoxon rank sum test to the case of singly or multiply censored data, mainly in the context of survival analysis (e.g., Breslow, 1970; Cox, 1972; Gehan, 1965; Mantel, 1966; Peto and Peto, 1972; Prentice, 1978). Prentice (1978) showed how all of these proposed tests are extensions of a linear rank test to the case of censored observations. As for the case of complete data, different linear rank tests use different score functions, and some may be better than others at detecting a small shift in location, depending upon the true underlying distribution.

Prentice and Marek (1979), Latta (1981), and Millard and Deverel (1988) studied the behavior of several linear rank tests for censored data. See the help file for `two.sample.linear.rank.test.censored` in ENVIRONMENTALSTATS for S-PLUS for details.

Example 10.23: Comparing Copper Concentrations between Two Geological Zones

In Example 10.9 we compared the empirical cumulative distribution functions of copper concentrations in the alluvial fan and basin trough zones (see Table 10.2 and Figure 10.5). The plot indicates the distributions of concentrations are fairly similar. The two-sample linear rank test based on normal scores and a hypergeometric variance yields a p-value of 0.2, indicating no significant difference.

Menu

To perform the two-sample linear rank test to compare copper concentrations using the ENVIRONMENTALSTATS for S-PLUS pull-down menu, follow these steps.

1. Find **millard.deverel.88.df** in the Object Explorer.
2. On the S-PLUS menu bar, make the following menu choices: **EnvironmentalStats>Censored Data>Hypothesis Tests>Compare Two Samples**. This will bring up the Compare Two Samples for Censored Data dialog box.
3. For Data to Use select **Pre-Defined Data**. The Data Set box should display **millard.deveral.88.silver.df**. For Variable 1 choose **Cu**, for Censor Ind. 1 choose **Cu.censored**, for Variable 2 choose **Zone**, and **check** the Variable 2 is a Grouping Variable box. For test select **normal.scores.2** and for variance select **hypergeometric**, then click **OK** or **Apply**.

Command

To perform the two-sample linear rank test to compare copper concentrations using the ENVIRONMENTALSTATS for S-PLUS Command or Script Window, type these commands.

```
attach(millard.deverel.88.df)
Cu.AF <- Cu[Zone=="Alluvial.Fan"]
Cu.AF.cen <- Cu.censored[Zone=="Alluvial.Fan"]
Cu.BT <- Cu[Zone=="Basin.Trough"]
Cu.BT.cen <- Cu.censored[Zone=="Basin.Trough"]
two.sample.linear.rank.test.censored(Cu.AF, Cu.AF.cen,
   Cu.BT, Cu.BT.cen, test="normal.scores.2",
   var="hypergeometric")
detach()
```

Regression

Methods for fitting linear regression models in the presence of singly and multiply censored response observations have been developed in the fields of econometrics and life testing (e.g., Amemiya, 1985, Chapter 10; Meeker and Escobar, 1998). In econometrics, these are sometimes called Tobit models (Amemiya, 1985, p. 360). In S-PLUS you can fit such models using the function `censorReg`. Also, SLIDA (S-PLUS Life Data Analysis) is a collection of S-PLUS functions that is a companion to Meeker and Escobar (1998) and is available at http://www.public.iastate.edu/~stat533/slida.html.

A NOTE ABOUT ZERO-MODIFIED DISTRIBUTIONS

In Chapter 4 we discussed zero-modified distributions and noted that they have sometimes been used to model environmental data with non-detects. In this case, the nondetects are assumed to actually be 0 values; that is, nondetects are set to 0. Methods for estimating distribution parameters and quantiles and performing goodness-of-fit tests for these kinds of distributions are available in ENVIRONMENTALSTATS for S-PLUS, but we do not recommend using them to model environmental data that contain nondetects.

SUMMARY

- Censored data can be classified as truncated vs. censored, left vs. right vs. double, singly vs. multiply, and Type I vs. Type II. Most environmental data sets with nondetect values are Type I left singly or multiply censored.
- You can use quantile (empirical cdf) plots and Q-Q plots to compare censored data with a theoretical distribution or compare two samples.
- Censored data have been studied extensively in the field of life testing and survival analysis. Several methods exist for estimating distribution parameters, creating confidence intervals, and performing hypothesis tests based on censored data.

EXERCISES

10.1. Create a plot of the empirical cdf for the background well arsenic data of Table 6.4 and compare it with the cdf of a normal distribution.

10.2. Create a plot of the empirical cdf for the singly censored TcCB data of Table 10.1 and compare it with the cdf of a normal distribution.

10.3. Create a plot of the empirical cdf for the singly censored TcCB data of Table 10.1 and compare it with the cdf of a lognormal distribution.

10.4. Consider the multiply censored zinc concentrations shown in Table 10.3.

    a. Compare the empirical cumulative distributions functions for the alluvial fan and basin trough zones.

    b. Use a two-sample linear rank test to test the equivalence of these two distributions.

10.5. Create normal and lognormal Q-Q plots for the alluvial fan copper data of Table 10.2. Do the same thing for the basin trough data.

10.6. Determine appropriate Box-Cox transformations (if any) for the copper data of Table 10.2 for each geological zone.

10.7. Create normal and lognormal Q-Q plots for the alluvial fan zinc data of Table 10.3. Do the same thing for the basin trough data.

10.8. Determine appropriate Box-Cox transformations (if any) for the zinc data of Table 10.3 for each geological zone.

10.9. Consider the silver data of Table 10.4. Determine an appropriate Box-Cox transformation (if any) for these data.

10.10. Compare different estimators of the mean and CV for the singly censored TcCB data of Table 10.1 assuming these data follow a lognormal distribution.

10.11. Estimate the 90th percentile of the silver concentrations based on the data of Table 10.4. Create a 95% confidence interval for this percentile. Use different methods of estimation and compare your results.

10.12. Compute a 99% prediction interval for the next $k = 4$ arsenic concentrations based on the arsenic data of Table 6.4.

10.13. Perform goodness-of-fit tests on the alluvial fan and basin trough copper data of Table 10.2 and create companion plots.

10.14. Perform goodness-of-fit tests on the alluvial fan and basin trough zinc data of Table 10.3 and create companion plots.

10.15. Perform goodness-of-fit tests on the silver data of Table 10.4 and create companion plots.

11 TIME SERIES ANALYSIS

How Are Things Changing over Time?

Classical statistical methods for constructing confidence intervals and performing hypothesis tests are based on two critical assumptions: that the observations are independent of one another and that they come from the same population (probability distribution). Parametric methods add the third assumption that the observations come from a particular probability distribution, usually the normal distribution. Of these three assumptions, the most important one (in terms of maintaining the assumed confidence or significance level) is the first one, the assumption of independent observations (e.g., Millard et al., 1985).

Environmental data are often collected sequentially over time (e.g., air quality data, groundwater monitoring data, etc.) or in close proximity in space (e.g., soil cleanup data). When observations are collected sequentially over time or close together in space, the assumption of independent observations may not be valid. This chapter discusses statistical methods for dealing with time series data, and the next chapter discusses methods for dealing with data collected in space. In this chapter we will discuss how to plot time series data, define autocorrelation and autocorrelation plots, explain how to test for and model autocorrelation, and how to estimate and test for trend, assuming either independent or autocorrelated observations.

CREATING AND PLOTTING TIME SERIES DATA

A time series is simply a set of observations collected sequentially over time. The observations may have been collected on a regular basis (e.g., monthly, weekly, daily, hourly, etc.) or on an irregular basis. It is quite common for time series data to contain missing values, so that a "regular" time series becomes an "irregular" time series. There are several ways you can create time series objects in S-PLUS from both the menu and the command line (see the S-PLUS documentation for details).

A time series plot is simply a plot of the observations on the y-axis vs. the times (or sequence in which) they were collected on the x-axis. Figure 7.7 in Chapter 7 is a time series plot of daily maximum ozone concentrations in Yonkers, New York between May 1 and September 30, 1974. Note that this time series contains several missing values.

As another example, recall from Chapters 3 and 9 that S-PLUS contains two built-in data frames called `air` and `environmental` that contain observations on ozone (ppb for environmental, ppb$^{1/3}$ for air), solar radiation (langleys), temperature (degrees Fahrenheit), and wind speed (mph) for 111 days between May 1 and September 30, 1973. Since there are 153 days in this time period, there are obviously missing values in this time series as well, but the air and environmental data frames do not include the dates the data were collected. These data are discussed in Chambers et al. (1983) and presented in an appendix (pp. 347–349). They are also discussed in Cleveland (1993, 1994). Chambers et al. (1983) present the original time series for all 153 days. In ENVIRONMENTALSTATS for S-PLUS, the full data set is stored in the multivariate calendar time series environmental.orig.cts. Figure 11.1 shows a time series plot of the ozone data and Figure 11.2 shows a time series plot of the temperature data for all 153 days. Note that the ozone data contain missing values.

As a final example, Figure 11.3 displays average monthly CO_2 concentrations (ppm) recorded on Mauna Loa, Hawaii between January 1959 and December 1990. In S-PLUS, these data are stored in the time series object co2. The original data contained missing values, but for the data stored in co2 missing values were imputed by a linear interpolation scheme (see the S-PLUS help file for co2).

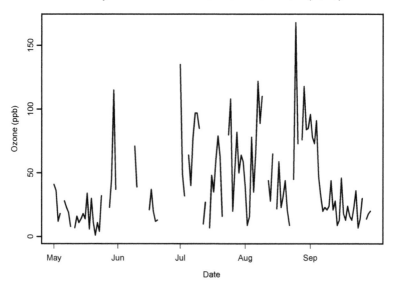

Daily Ozone Concentrations in New York (1973)

Figure 11.1   Daily ozone concentrations (ppb) between May 1 and September 30, 1973, in New York

Daily Maximum Temperature in New York (1973)

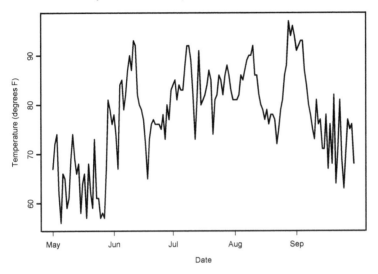

Figure 11.2   Daily maximum temperature (degrees Fahrenheit) between May 1 and September 30, 1973, in New York

Monthly Average CO_2 Concentrations at Mauna Loa

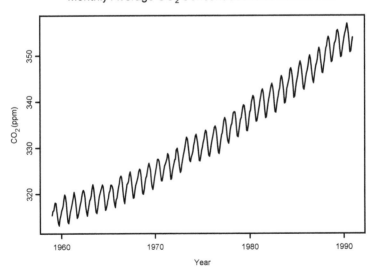

Figure 11.3   Average monthly CO_2 concentrations (ppm) on Mauna Loa, Hawaii between January 1959 and December 1990

Menu

To produce the ozone time series plot shown in Figure 11.1 using the S-PLUS pull-down menu or toolbar, follow these steps.

1. Find **environmental.orig.cts** in the Object Explorer.
2. On the S-PLUS menu bar, make the following menu choices: **Graph>2D Plot**. This will bring up the Insert Graph dialog box. Under the Axes Type column, **Linear** should be highlighted. Under the Plot Type column, select **Line Plot** and click **OK**. (Alternatively, left-click on the **2D Plots** button, then left-click on the **Line** button.)
3. The Line/Scatter Plot dialog box should appear. The data set box should display **environmental.orig.cts**. In the y Columns box, select **ozone**, then click **OK**.

To produce the temperature time series plot shown in Figure 11.2, repeat the above steps, but use the variable **temperature**. To produce the CO_2 time series plot shown in Figure 11.3, repeat the above steps, but use the data set **co2** and specify **co2** as the variable as well.

Command

To produce the ozone time series plot shown in Figure 11.1 using the S-PLUS Command or Script Window, type these commands.

```
ts.plot(environmental.orig.cts[, "ozone"],
   xlab="Date", ylab="Ozone (ppb)", xaxt="n",
   main="Daily Ozone Concentrations in New York
   (1973)", lwd=3)
date.ticks <- dates(paste(5:9, 1, 73, sep="/"))
axis(1, at=date.ticks,
   labels=as.character(months(date.ticks)))
```

To produce the time series plot shown in Figure 11.2, type these commands.

```
ts.plot(environmental.orig.cts[, "temperature"],
   xlab="Date", ylab="Temperature (degrees F)",
   xaxt="n", main="Daily Maximum Temperature in New
   York (1973)", lwd=3)
date.ticks <- dates(paste(5:9, 1, 73, sep="/"))
axis(1, at=date.ticks,
   labels=as.character(months(date.ticks)))
```

To produce the time series plot shown in Figure 11.3, type these commands.

```
tsplot(co2, xlab="Year", ylab="CO2 (ppm)",
    main="Monthly Average CO2 Concentrations at Mauna
    Loa", lwd=3)
```

Note that `tsplot` (as opposed to `ts.plot`) is *not* a typographical error. The object `co2` is an "old-style" time series object, so you should use `tsplot` to plot it.

AUTOCORRELATION

In Chapter 9 we defined the correlation between two random variables X and Y in Equation (9.5). We also defined the sample (estimated) correlation coefficient in Equation (9.7). We saw that correlation is a measure of how X and Y vary together about their respective means. For example, we saw in Example 9.1 that ozone concentrations in Yonkers and Stamford are positively correlated, so that when ozone is above its mean in Yonkers it is more likely to above its mean in Stamford as well. *Autocorrelation* (also called *serial correlation*) is simply a measure of how a random variable X varies over time relative to past observations of X. For example, we would not be surprised to find that if the ozone concentration at 1 P.M. is above the daily average ozone concentration, then is it very likely that the ozone concentration at 2 P.M. is also above the daily average as well.

Here we will assume we are taking observations on a random variable X at equally spaced times (e.g., once per hour, day, week, month, etc.), and we will let X_t denote the random variable at time t. Following Equation (9.5), the lag-1 autocorrelation for the time series is defined as:

$$\rho_1 = \frac{E\left[\left(X_t - \mu\right)\left(X_{t+1} - \mu\right)\right]}{\sqrt{Var\left(X_t\right) Var\left(X_{t+1}\right)}}$$

(11.1)

$$= \frac{Cov\left(X_t, X_{t+1}\right)}{Var\left(X_t\right)} = \frac{\gamma_1}{\gamma_0} = \frac{\gamma_1}{\sigma_X^2}$$

where μ and σ_X denote the mean and standard deviation of X, and γ_1 denotes the lag-1 autocovariance (Box and Jenkins, 1976, pp. 27–28). In general, the lag-k autocorrelation ($k = 0, 1, 2, 3, \ldots$) is defined as:

$$\rho_k = \frac{Cov\left(X_t, X_{t+k}\right)}{\sqrt{Var\left(X_t\right) Var\left(X_{t+k}\right)}} = \frac{\gamma_k}{\gamma_0} = \frac{\gamma_k}{\sigma_X^2} \quad \text{(11.2)}$$

In both of the above equations, we are assuming the time series is *stationary*, which means the average (mean) and standard deviation remain constant over time, and also that the value of the lag-k autocorrelation does not depend on what starting point we choose in the series.

As in the case of correlation between two separate variables, autocorrelation must lie between -1 and 1. Note that the lag-0 autocovariance is equal to the variance and the lag-0 autocorrelation is equal to 1. For values of k greater than or equal to 1, the autocorrelation lies strictly between -1 and 1:

$$-1 < \rho_k < 1 \ , \quad k = 1, 2, 3, \dots \tag{11.3}$$

If observations are independent of one another, then all lag-k covariances and autocorrelations are 0 for values of k greater than or equal to 1.

A Simple Model for Autocorrelation: The AR(1) Process

One of the simplest models for autocorrelation is the *first order autoregressive model*, also denoted as the *AR(1) model*:

$$X_t = \mu + \phi \left(X_{t-1} - \mu \right) + \varepsilon_t \tag{11.4}$$

where μ denotes the overall mean of the time series, ϕ (phi) denotes the lag-1 autocorrelation, and ε_t denotes an error term with mean 0 and constant standard deviation σ_ε. The error terms, also sometimes called the *random innovations*, are assumed to be independent of each other. If the lag-1 correlation ϕ is positive, then if an observation is above the mean value μ, the next observation is more likely to also be above the mean value μ compared to the case when the observations are independent. If the lag-1 correlation ϕ is negative, then if an observation is above the mean value μ, the next observation is more likely to be below the mean value μ compared to the case when the observations are independent. Note that if the lag-1 correlation ϕ is equal to 0 then the observations are independent of one another. Figure 11.4 displays an example of $n = 100$ observations from an AR(1) process with a lag-1 autocorrelation of $\phi = 0.8$, and Figure 11.5 displays an example with $\phi = -0.8$.

For an AR(1) process with a lag-1 autocorrelation of ϕ, the following property holds (Box and Jenkins, 1976, p. 57):

$$\rho_k = \phi^k \ , \quad k = 0, 1, 2, \dots \tag{11.5}$$

Example of an AR(1) Process with phi=0.8

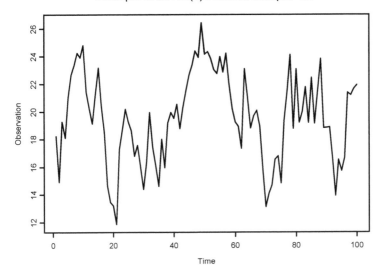

Figure 11.4   Example of an AR(1) process with a lag-1 autocorrelation of 0.8

Example of an AR(1) Process with phi=-0.8

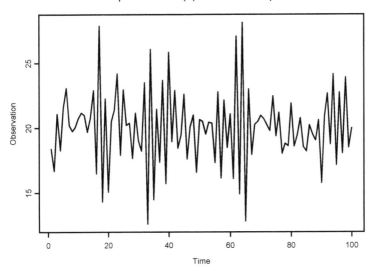

Figure 11.5   Example of an AR(1) process with a lag-1 autocorrelation of –0.8

Menu

In the current version of S-PLUS 2000 you cannot easily simulate time series models from the menu.

Command

To produce the simulated AR(1) process shown in Figure 11.4 using the S-PLUS Command or Script Window, type these commands.

```
set.seed(27)
series.ar.0.8 <-
   2 * arima.sim(n=100, model=list(ar=0.8)) + 20
```

To produce the simulated AR(1) process shown in Figure 11.5 type these commands.

```
set.seed(55)
series.ar.neg.0.8 <-
   2 * arima.sim(n=100, model=list(ar=-0.8)) + 20
```

The Autocorrelation Function (ACF)

A plot of the lag-k autocorrelation vs. k is called the ***autocorrelation function (ACF)***, and sometimes called the ***correlogram*** (Box and Jenkins, 1976, p. 30). Figure 11.6 displays the ACF for an AR(1) process with a lag-1 autocorrelation of 0.8, and Figure 11.7 displays the ACF for an AR(1) process with a lag-1 autocorrelation of –0.8 (see Equation (11.5)).

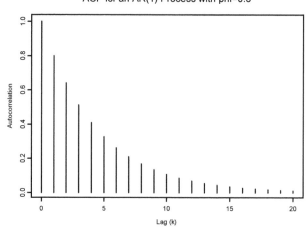

Figure 11.6   Autocorrelation function (ACF) for an AR(1) process with a lag-1 autocorrelation of 0.8

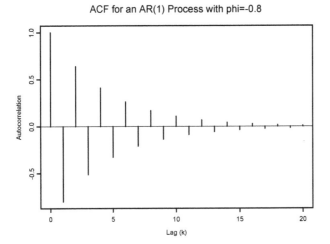

Figure 11.7   Autocorrelation function (ACF) for an AR(1) process with a lag-1 autocorrelation of –0.8

Estimating Autocorrelation

Although there are several methods for estimating autocorrelation, one of
the most commonly used estimators for the lag-k autocorrelation is the Yule-
Walker estimator given by:

$$\hat{\rho}_k = \frac{\hat{\gamma}_k}{\hat{\gamma}_0} = \frac{\dfrac{1}{n}\sum_{t=1}^{n-k}\left(x_t - \bar{x}\right)\left(x_{t+k} - \bar{x}\right)}{\dfrac{1}{n}\sum_{t=1}^{n}\left(x_t - \bar{x}\right)^2}$$

$$(11.6)$$

$$= \frac{\sum_{t=1}^{n-k}\left(x_t - \bar{x}\right)\left(x_{t+k} - \bar{x}\right)}{\sum_{t=1}^{n}\left(x_t - \bar{x}\right)^2}$$

(Box and Jenkins, 1976, p. 32). This method of estimation is analogous to
the method of estimating the correlation between two random variables X
and Y shown in Equation (9.7) in Chapter 9, except here we are estimating

the correlation between the series $\{X_{1+k}, X_{2+k}, \ldots, X_n\}$ and the lag-k version of this series $\{X_1, X_2, \ldots, X_{n-k},\}$.

Table 11.1 displays the true and estimated lag-k autocorrelations for the AR(1) time series shown in Figure 11.4 for the first five lags. Box and Jenkins (1976, p. 33) recommend having at least $n = 4k$ observations if you want to estimate the lag-k autocorrelation.

Lag	True Autocorrelation	Estimated Autocorrelation
1	0.8	0.75
2	0.64	0.60
3	0.51	0.43
4	0.41	0.27
5	0.33	0.18

Table 11.1   True and estimated autocorrelations for the time series shown in Figure 11.4

Sample Autocorrelation Function (SACF) and Lag Plots

Instead of looking at tables of estimated autocorrelations, it is easier to look at plots. A plot of the estimated lag-k autocorrelation vs. k is called the **sample autocorrelation function (SACF)**, or the **estimated correlogram**. Figure 11.8 displays the SACF for the AR(1) process of Figure 11.4 and Figure 11.9 displays the SACF for the AR(1) process of Figure 11.5. Compare these figures with the true ACF shown in Figure 11.6 and Figure 11.7.

SACF for Series of Figure 11.4

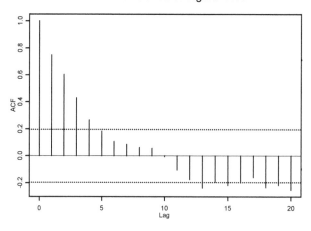

Figure 11.8   The sample autocorrelation function (SACF) for the AR(1) process of Figure 11.4

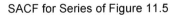

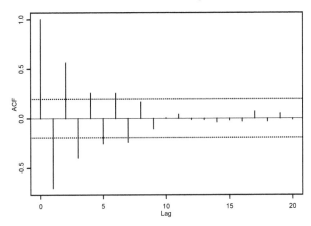

Figure 11.9   The sample autocorrelation function (SACF) for the AR(1) process of
Figure 11.5

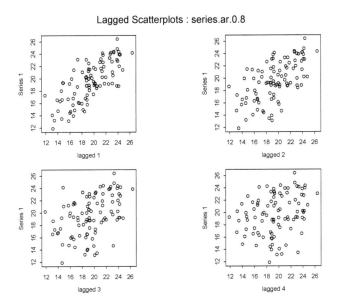

Figure 11.10   Lag plots for the AR(1) process of Figure 11.4

   As we discussed in Chapter 9, correlation is a measure of the linear association between two random variables, or in the case of time series analysis,
a time series and a lag of the time series. Besides looking at plots of the

SACF, you can also look at plots of the time series vs. the lag-k time series. Figure 11.10 displays lag plots for the first 4 lags of the AR(1) process of Figure 11.4.

Menu

To compute and plot the estimated autocorrelations for the time series of Figure 11.4 using the S-PLUS pull-down menu, follow these steps.

1. Find **series.ar.0.8** in the Object Explorer.
2. On the S-PLUS menu bar, make the following menu choices: **Statistics>Time Series>Autocorrelations**. This will bring up the Autocorrelations and Autocovariances dialog box.
3. The Data Set box should display **series.ar.0.8**. In the Variable box type **series.ar.0.8**, then click **OK** or **Apply**.

To produce the same thing for the time series of Figure 11.5, repeat the above steps but use the data set **series.ar.neg.0.8**.

To produce the lag plots of Figure 11.10, follow these steps.

1. On the S-PLUS menu bar, make the following menu choices: **Statistics>Time Series>Lag Plot**. This will bring up the Lag Plot dialog box.
2. The Data Set box should display **series.ar.0.8**. In the Variable box type **series.ar.0.8**, in the Lag box choose **4**, in the Rows box choose **2** and in the Columns box choose **2**, then click **OK** or **Apply**.

Command

To compute and plot the estimated autocorrelations for the time series of Figure 11.4 and Figure 11.5 using the S-PLUS Command or Script Window, type these commands.

```
acf.list.0.8 <- acf(series.ar.0.8, plot=F)
acf.plot(acf.list.0.8, lwd=3,
  main="SACF for Series of Figure 11.4")
acf.list.neg.0.8 <- acf(series.ar.neg.0.8, plot=F)
acf.plot(acf.list.neg.0.8, lwd = 3,
  main="SACF for Series of Figure 11.5")
```

The lists `acf.list.0.8` and `acf.list.neg.0.8` contain several components, including the estimated autocorrelations. You can also create the SACF by simply calling the `acf` function with `plot=T`.

To create the lag plots of Figure 11.10 , type this command.

```
lag.plot(series.ar.0.8, lags=4, layout=c(2,2))
```

Tests and Confidence Intervals for Autocorrelation

There are several ways to test for the presence of serial correlation, including simple tests for the presence of lag-1 or higher autocorrelation, and more involved methods that involve fitting a simple or complicated time series model and testing the significance of the parameters. In this section we will briefly discuss two simple hypothesis tests for the presence of autocorrelation and we will explain a simple method for creating a confidence interval for the lag-1 autocorrelation.

Test For Lag-k Autocorrelation Based on the Yule-Walker Estimate

The null hypothesis is that the lag-k autocorrelation is 0 for all values of k greater than 0 (i.e., the time series is purely random). For a specified value of k (e.g., $k = 1$), the three possible alternative hypotheses are the upper one-sided alternative

$$H_a : \rho_k > 0 \tag{11.7}$$

the lower one-sided alternative

$$H_a : \rho_k < 0 \tag{11.8}$$

and the two-sided alternative

$$H_a : \rho_k \neq 0 \tag{11.9}$$

Under the null hypothesis, the distribution of the Yule-Walker estimator of lag-k autocorrelation shown in Equation (11.6) is approximately normal with mean 0 and variance $1/n$, assuming the time series is stationary with normally distributed errors (Box and Jenkins, 1976, pp. 34–35). Thus, we can use the test statistic

$$z = \frac{\hat{\rho}_k - 0}{\sqrt{1/n}} = \sqrt{n}\, \hat{\rho}_k \tag{11.10}$$

which is distributed approximately as a standard normal random variable under the null hypothesis that the lag-k autocorrelation is 0. For the case of a two-sided alternative, we would reject the null hypothesis at level α if

$\hat{\rho}_k < -z_{1-\alpha/2}/\sqrt{n}$ or $\hat{\rho}_k > z_{1-\alpha/2}/\sqrt{n}$ where z_p denotes the pth quantile of the standard normal distribution. The horizontal dashed lines in Figures 11.8 and 11.9 are drawn at these values assuming a Type I error level of $\alpha = 5\%$. In both of these figures there are several estimated autocorrelations beyond these critical values, so we would conclude that autocorrelation is in fact present in both the these time series. You must be careful when interpreting this kind of plot, however, because of the multiple comparisons problem: in the case of no autocorrelation we would *not* expect *all 19* estimated autocorrelations of lag greater than 1 to fall within the dashed lines about 95% of the time. The results of a Monte Carlo simulation study confirm this. For 1000 trials, a random sample of 100 independent observations was generated for each trial and the lag-1 to lag-19 sample autocorrelations were computed. At least one of the lag-1 to lag-3 autocorrelations fell outside the dashed lines 13.4% of the time, and at least one of the lag-1 to lag-19 autocorrelations fell outside the dashed lines 50% of the time.

Test for Lag-1 Autocorrelation Based on the Rank von Neumann Ratio

The null distribution of the serial correlation coefficient may be badly affected by departures from normality in the underlying process (Cox, 1966; Bartels, 1977). It is therefore a good idea to consider using a nonparametric test for randomness if the normality of the underlying process is in doubt (Bartels, 1982).

Wald and Wolfowitz (1943) introduced the rank serial correlation coefficient, which for lag-1 autocorrelation is simply Equation (11.6) with $k = 1$ and the actual observations replaced with their ranks. von Neumann et al. (1941) introduced a test for randomness in the context of testing for trend in the mean of a process. Their statistic is given by:

$$VN = \frac{\sum_{t=1}^{n-1}(x_t - x_{t+1})^2}{\sum_{t=1}^{n}(x_t - \bar{x})^2}$$ (11.11)

which is the ratio of the sum of squared successive differences to the usual sum of squared deviations from the mean. This statistic is bounded between 0 and 4, and for a purely random process is symmetric about 2. Small values of this statistic indicate possible positive autocorrelation, and large values of this statistic indicate possible negative autocorrelation. Durbin and Watson (1950, 1951, 1971) proposed using this statistic in the context of checking the independence of residuals from a linear regression model and provided tables for the distribution of this statistic. This statistic is therefore often called the **Durbin-Watson statistic** (Draper and Smith, 1998, p. 181).

The rank version of the von Neumann ratio statistic simply replaces the observations with their ranks and is given by:

$$RVN = \frac{\sum_{t=1}^{n-1} \left(R_t - R_{t+1}\right)^2}{\sum_{t=1}^{n} (R_t - \bar{R})^2} \tag{11.12}$$

where R_t denotes the rank of the tth observation (Bartels, 1982). In the absence of ties, the denominator of this test statistic is equal to

$$\sum_{t=1}^{n} (R_t - \bar{R})^2 = \frac{n\left(n^2 - 1\right)}{12} \tag{11.13}$$

The range of the *RVN* test statistic is given by:

$$\left[\frac{12}{n\left(n+1\right)} , \ 4 - \frac{12}{n\left(n+1\right)} \right] \tag{11.14}$$

if n is even, with a negligible adjustment if n is odd (Bartels, 1982), so asymptotically the range is from 0 to 4, just as for the *VN* test statistic in Equation (11.11) above.

For any fixed sample size n, the exact distribution of the *RVN* statistic above can be computed by simply computing the value of *RVN* for all possible permutations of the serial order of the ranks. Based on this exact distribution, Bartels (1982) presents a table of critical values for the numerator of the RVN statistic for sample sizes between 4 and 10. Determining the exact distribution of *RVN* becomes impractical as the sample size increases. For values of n between 10 and 100, Bartels (1982) approximated the distribution of *RVN* by a beta distribution over the range 0 to 4, and he shows that asymptotically *RVN* has a normal distribution with mean 2 and variance $4/n$, but notes that a slightly better approximation is given by using a variance of $20/(5n + 7)$.

When ties are present in the observations and midranks are used for the tied observations, the distribution of the RVN statistic based on the assumption of no ties is not applicable. If the number of ties is small, however, they may not grossly affect the assumed p-value.

Confidence Interval for Lag-1 Autocorrelation

When lag-1 autocorrelation is present, the distribution of the estimated lag-1 autocorrelation is approximately normal with mean ρ_1, assuming normally distributed errors. The variance of the estimated lag-1 autocorrelation is approximately:

$$Var\left(\hat{\rho}_1\right) \approx \frac{1}{n}\left(1 - \rho_1^2\right) \tag{11.15}$$

(Box and Jenkins, 1976, p. 34). Thus, an approximate $100(1-\alpha)\%$ confidence interval for the lag-1 autocorrelation is given by:

$$\left[\hat{\rho}_1 - z_{1-\alpha/2}\sqrt{\frac{1 - \hat{\rho}_1^2}{n}} \; , \; \hat{\rho}_1 + z_{1-\alpha/2}\sqrt{\frac{1 - \hat{\rho}_1^2}{n}}\right] \tag{11.16}$$

Example 11.1: Testing for Serial Correlation in the New York Temperature Data

In this example we will test the null hypothesis of no serial correlation in the daily maximum temperature data shown in Figure 11.2 using the test based on the Yule-Walker estimate of lag-1 autocorrelation. Obviously there are monthly fluctuations in temperature, so if we perform the test on the raw data we would not be surprised if we find a statistically significant result. Here, we will perform the test both on the original data and what we will call the "monthly adjusted" data, which we will create by subtracting the monthly mean from each observation.

Figure 11.11 displays the ACF for the original temperature data, and Figure 11.12 displays the ACF for the adjusted data (compare these figures to Figure 11.8). Q-Q plots reveal that both the original temperature data and the adjusted data do not appear to deviate greatly from a normal distribution. The estimated lag-1 autocorrelation based on the original data is 0.81, yielding a p-value of 0 and a 95% confidence interval of [0.72, 0.90]. For the adjusted data, the estimated lag-1 autocorrelation is 0.60, yielding a p-value of 0 and a 95% confidence interval of [0.48, 0.73]. Thus, even for the adjusted data, there is evidence of serial correlation in the temperature data. This is not surprising; this says that if the temperature is above the monthly average on a particular day, it is more likely to be above (rather than below) the monthly average the next day as well.

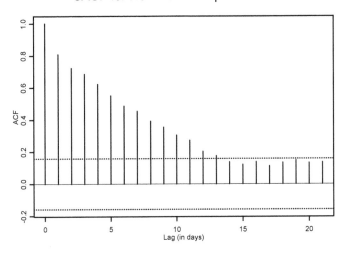

Figure 11.11   Sample autocorrelation function for the New York temperature data

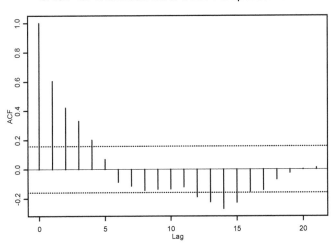

Figure 11.12   Sample autocorrelation function for the monthly adjusted New York temperature data

Menu

To perform the test of serial correlation for the temperature data using the ENVIRONMENTALSTATS for S-PLUS pull-down menu, follow these steps.

1. Find **environmental.orig.cts** in the Object Explorer.
2. On the S-PLUS menu bar, make the following menu choices: **EnvironmentalStats>Hypothesis Tests>Serial Correlation**. This will bring up the Serial Correlation Test dialog box.
3. For Data to Use select **Pre-Defined Data**. The Data Set box should display **environmental.orig.cts**. In the Variable box select **temperature**, for Test Method choose **AR(1) Y-W**, then click **OK** or **Apply**.

In the current version of S-PLUS 2000 it is not possible to create the monthly adjusted data from the menu, but steps for doing this from the command line are given below. To perform the test on the adjusted data, follow the same steps as for the raw data, but use the object **temperature.adjusted** instead of **environmental.orig.cts**.

Command

To perform the test of serial correlation for the temperature data using the ENVIRONMENTALSTATS for S-PLUS Command or Script Window, type these commands.

```
temperature <- environmental.orig.cts[,"temperature"]
serial.correlation.test(temperature, test="AR1.yw")
```

To perform the test on the adjusted data, type these commands.

```
months.temperature <- as.character(months(dates(
   as.character(time(temperature)))))
temperature.adjusted <- unlist(sapply(split(
   temperature, months.temperature), scale, scale=F))
serial.correlation.test(temperature.adjusted,
   test="AR1.yw")
```

Dealing with Missing Values

The Yule-Walker estimator of the lag-k autocorrelation shown in Equation (11.6) does not allow for missing values. There are two ways to deal with missing values when you are interested in assessing the presence of and accounting for serial correlation:

- Fill in the missing values with some kind of interpolation method. This was what was done for the Mauna Loa CO_2 data.

- Assume a particular autoregressive or moving average model (see below) and use maximum likelihood estimation to estimate the model parameters. Maximum likelihood estimation of time series parameters is discussed in standard textbooks on time series analysis (e.g., Box and Jenkins, 1976)

Example 11.2: Estimating the Lag-1 Autocorrelation for the Ozone Data

The ozone data shown in Figure 11.1 includes missing values, so we cannot use the Yule-Walker Equation (11.6) to estimate autocorrelations. In this example we will use maximum likelihood estimation to test the null hypothesis of independent observations vs. the alternative that the observations come from an AR(1) process (assuming a constant mean and variance). (Of course, we can do a better job of modeling these data by incorporating information on radiation, temperature, and wind speed; see Example 11.3). As we did for the temperature data, we will perform the test based on the original observations, and again for the "adjusted" data, which we will create by subtracting the monthly mean from each observation.

Q-Q plots reveal that the raw ozone data are skewed, while the log-transformed or cube-root-transformed observations appear to satisfy the assumption of normality. For the cube-root-transformed original data, the maximum likelihood estimate of lag-1 autocorrelation is 0.57, yielding a p-value of 0 and a 95% confidence interval of [0.42, 0.72]. For the adjusted data, the estimate lag-1 autocorrelation is 0.42, yielding a p-value of 0 and a 95% confidence interval of [0.25, 0.58]. As for the case of temperature, even the adjusted (transformed) ozone data display serial correlation.

Menu

To perform the test of serial correlation for the ozone data using the ENVIRONMENTALSTATS for S-PLUS pull-down menu, follow these steps.

1. Find **air.orig.cts** in the Object Explorer.
2. On the S-PLUS menu bar, make the following menu choices: **EnvironmentalStats>Hypothesis Tests>Serial Correlation**. This will bring up the Serial Correlation Test dialog box.
3. For Data to Use select **Pre-Defined Data**. The Data Set box should display **air.orig.cts**. In the Variable box select **ozone**, for Test Method choose **AR(1) MLE**, then click **OK** or **Apply**.

As for the temperature data, it is not possible to create the adjusted ozone data from the menu, steps for doing this at the command line are given below. To perform the test on the adjusted data, follow the same steps as for the raw data, but use the object **ozone.adjusted** instead of **air.orig.cts**.

Command

To perform the test of serial correlation for the ozone data using the ENVIRONMENTALSTATS for S-PLUS Command or Script Window, type these commands.

```
cr.ozone <- air.orig.cts[,"ozone"]
serial.correlation.test(cr.ozone, test="AR1.mle")
```

To perform the test on the adjusted data, type these commands.

```
months.cr.ozone <- as.character(months(dates(
    as.character(time(cr.ozone)))))
cr.ozone.adjusted <- unlist(sapply(split(
    cr.ozone, months.cr.ozone), scale, scale=F))
serial.correlation.test(cr.ozone.adjusted,
    test="AR1.mle")
```

Effect of Autocorrelation on Confidence Intervals and Test Statistics

What if we decide that our observations do indeed exhibit autocorrelation? So what? It turns out that ignoring autocorrelation can lead to incorrect confidence levels and Type I error rates (e.g., Millard et al., 1985). Here we will consider the effect of autocorrelation on the confidence interval for the mean and the Type I error level of the one-sample t-test.

We showed in Chapter 4 in Equation (4.27) that the variance of the sample mean is equal to the population variance divided by the sample size:

$$Var\left(\bar{x}\right) \;=\; \sigma_{\bar{x}}^{2} \;=\; \frac{\sigma_{\varepsilon}^{2}}{n} \qquad\qquad (11.17)$$

The above formula assumes the observations used to construct the sample mean are independent of one another. When observations come from a time series and are autocorrelated, the variance of the sample mean becomes:

$$Var\left(\bar{x}\right) \;=\; \frac{\sigma_{\bar{x}}^{2}}{n}\left[1 \;+\; \frac{2}{n}\sum_{k=1}^{n-1}\left(n-k\right)\rho_{k}\right] \qquad (11.18)$$

where σ_X^2 denotes the variance of an individual observation (Lettenmaier, 1976; Gilbert, 1987, p. 38). For an AR(1) process with a lag-1 autocorrelation of ϕ (Equation (11.4)), it can be shown that the variance of an individual observation X_t is:

$$Var\left(X\right) = \sigma_X^2 = \frac{\sigma_\varepsilon^2}{1 - \phi^2} \tag{11.19}$$

(Box and Jenkins, 1976, p. 58). Thus, by Equations (11.5), (11.18), and (11.19), for an AR(1) process the variance of the sample mean is given by:

$$Var\left(\overline{x}\right) = \frac{\sigma_\varepsilon^2}{n} \frac{1}{\left(1 - \phi\right)^2} \frac{1 - \phi^2 + \frac{2\phi}{n}\left(\phi^n - 1\right)}{\left(1 - \phi^2\right)} \tag{11.20}$$

which for large sample sizes is approximately equal to

$$Var\left(\overline{x}\right) \approx \frac{\sigma_\varepsilon^2}{n} \frac{1}{\left(1 - \phi\right)^2} \tag{11.21}$$

For both a confidence interval for the mean and the one-sample t-test on the mean, the variance of the sample mean is estimated by estimating the variance of an individual observation and dividing by the sample size:

$$\hat{\sigma}_{\overline{x}}^2 = \frac{\hat{\sigma}_X^2}{n} = \frac{s^2}{n} \tag{11.22}$$

and the average value of this estimator is the true variance for an individual observation divided by the sample size:

$$E\left(\hat{\sigma}_{\overline{x}}^2\right) = \frac{E\left(s^2\right)}{n} = \frac{\sigma_X^2}{n} \tag{11.23}$$

Looking at the ratio

$$\frac{Var\left(\overline{x}\right)}{E\left(\hat{\sigma}_{\overline{x}}^2\right)} = \frac{\sigma_{\overline{x}}^2}{\sigma_{X}^2/n} = \frac{\sigma_{\overline{x}}^2}{\dfrac{\sigma_{\varepsilon}^2}{\left(1-\phi^2\right)}\Big/n}$$

(11.24)

$$\approx \frac{\dfrac{\sigma_{\varepsilon}^2}{n}\dfrac{1}{\left(1-\phi\right)^2}}{\dfrac{\sigma_{\varepsilon}^2}{n}\dfrac{1}{\left(1-\phi^2\right)}} = \frac{\left(1-\phi^2\right)}{\left(1-\phi\right)^2}$$

we see that the estimator s^2/n of Equation (11.22) is a "good" estimator of the variance of the sample mean only when the lag-1 autocorrelation ϕ is 0.

Effect of Autocorrelation on Estimator of Variance of Sample Mean

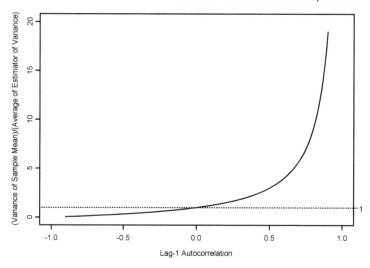

Figure 11.13 Approximate ratio of the true variance of the sample mean to the average value of its estimator as a function of the lag-1 autocorrelation

Figure 11.13 plots this ratio as a function of the lag-1 autocorrelation. When the lag-1 autocorrelation is positive ($\phi > 0$), s^2/n underestimates the true variance of the sample mean, so confidence intervals are too narrow and the one-sample t-statistic is too large, resulting in an inflated Type I error. When the lag-1 autocorrelation is negative, s^2/n overestimates the true variance of the sample mean, so confidence intervals are too wide and the one-sample t-statistic is too small, resulting in a Type I error rate less than the assumed rate. Both Lettenmaier (1976) and Gilbert (1987, p. 39) present sample size formulas to use in the presence of serial correlation.

DEALING WITH AUTOCORRELATION

So how do you deal with autocorrelation once you find it? There are two (possibly overlapping) approaches:

1. Eliminate autocorrelation by improving your model to account for more sources of variability.
2. Use a simple or complicated time series model to perform your statistical analyses.

Often, serial correlation can be reduced or eliminated by assuming seasonal effects, such as monthly or quarterly variations, by allowing for a trend (see the section Estimating and Testing for Trend later in this chapter), or by incorporating information from other covariates.

Eliminating Autocorrelation by Incorporating Information from Other Covariates

You can often eliminate serial correlation by incorporating information from other covariates. This can be as simple as allowing for seasonal fluctuations and/or incorporating an overall trend, or more complicated by using a regression model that incorporates several covariates. In the example below, we will fit a regression model to the ozone data we looked at earlier and then worry about whether there is serial correlation in the *residuals* from the fit.

Example 11.3: Eliminating Autocorrelation from the Ozone Data

In Example 11.2 we found evidence of serial correlation in the (transformed) ozone data, but we ignored the available information on radiation (R), temperature (T), and wind speed (W). Figure 9.22 in Chapter 9 displays the observed relationship between the cube-root of ozone and the other three variables. In Example 9.13 we found that an adequate model for predicting ozone (O) is given by:

$$O^{1/3} = \beta_0 + \beta_1 R + \beta_2 T + \beta_3 W + $$
$$\beta_4 R^2 + \beta_5 T^2 + \beta_6 W^2 + \beta_7 RT + \varepsilon \tag{11.25}$$

where β_i denotes the coefficient of the ith predictor variable and ε denotes the error in the model. Equation (9.57) shows the values of the estimated coefficients. If we perform a test of serial correlation on the residuals from this fitted model, we find that the maximum likelihood estimate of lag-1 autocorrelation is 0.12, yielding a p-value of 0.209 and a 95% confidence interval of [−0.06, 0.31]. Hence, once we take into account several sources that contribute to the variability of the ozone observations, the serial correlation vanishes.

Menu

To perform the test of serial correlation on the residuals of the linear model for ozone using the S-PLUS pull-down menu, we could use the model `cr.ozone.fit.stepwise` that we created in Example 9.13, but the data used to fit this model does not account for days with missing values. Therefore, we first have to construct the model based on the full time series data stored in `air.orig.cts`. Because the modeling capabilities in S-PLUS do not work with calendar time series, we have to copy the contents of `air.orig.cts` to a data frame, say `air.orig.df`. Currently, this cannot be done from the menu, but the Command section below shows how to do this from the command line. Once you have created `air.orig.df`, create the linear model by fol-lowing these steps.

1. Find **air.orig.df** in the Object Explorer.
2. On the S-PLUS menu bar, make the following menu choices: **Statistics>Regression>Linear**. This will bring up the Linear Regression dialog box.
3. For Data Set **air.orig.df** should be highlighted. In the Formula box type **ozone ~ radiation + temperature + wind + radiation^2 + temperature^2 + wind^2 + radiation*temperature**. In the Save As box type **lm.cr.ozone**, then click **OK** or **Apply**.

Once you have fit the model, follow these steps to test for serial correlation in the residuals.

4. On the S-PLUS menu bar, make the following menu choices: **EnvironmentalStats>Hypothesis Tests>Serial Correlation**. This will bring up the Serial Correlation Test dialog box.
5. For Data to Use select **Expression**. In the Expression box type **residuals(lm.cr.ozone)**, for Test Method choose **AR(1) MLE**, then click **OK** or **Apply**.

Command

To perform the test of serial correlation using the S-PLUS Command or Script Window, type these commands.

```
air.orig.df <- as.data.frame(air.orig.cts)

lm.cr.ozone <- lm(ozone ~ radiation + temperature +
    wind + radiation^2 + temperature^2 + wind^2 +
    radiation*temperature, data=air.orig.df,
    na.action=na.exclude)

serial.correlation.test(residuals(lm.cr.ozone),
    test="AR1.mle")
```

Accounting for Autocorrelation in the Model

The general model for multiple regression with one or more predictor variables is shown in Equation (9.56). This model assumes the error terms (the ε's) are independent. The p coefficients for this model are estimated using least squares, and the general formula in matrix notation is given by:

$$\hat{\beta} = (X'X)^{-1} X'Y \qquad (11.26)$$

where $\hat{\beta}$ denotes the $p \times 1$ vector of coefficients, Y denotes the $n \times 1$ vector of observations, X denotes the $n \times p$ matrix containing the values of the p predictor variables for each of the n observations, and X' denotes the transpose of X (Draper and Smith, 1998, pp. 135–136). When the error terms are correlated, you should use *generalized least squares* to estimate β:

$$\hat{\beta} = (X'V^{-1}X)^{-1} X'V^{-1}Y \qquad (11.27)$$

where V denotes the $p \times p$ variance-covariance matrix of the error terms (Draper and Smith, 1998, pp. 221–222). In practice the true values of the covariances are usually unknown and have to be estimated. Example 11.5 (later in this chapter) shows how to fit a parametric model for trend while accounting for correlated errors, and Example 11.7 shows how to do the same thing using a nonparametric model for trend.

MORE COMPLICATED MODELS: AUTOREGRESSIVE AND MOVING AVERAGE PROCESSES

So far in our discussion of testing for and estimating serial correlation, we have restricted ourselves to the simple first-order autoregressive or AR(1) model shown in Equation (11.4). A more complicated model is a pth *order autoregressive process*, also called an *AR(p) process*:

$$X_t = \mu + \phi_1 \left(X_{t-1} - \mu \right) + \phi_2 \left(X_{t-2} - \mu \right)$$

$$\text{(11.28)}$$

$$+ \cdots + \phi_p \left(X_{t-p} - \mu \right) + \varepsilon_t$$

where $p \geq 1$ (Box and Jenkins, 1976, p. 51). For $p > 1$, this model says that not only does the last observation help predict the value of the next observation, but also the 2nd to last, 3rd to last, and on up to the pth to last observation. Another model used in time series analysis is the qth *order moving average process*, also called an *MA(q) process*:

$$X_t = \mu + \varepsilon_t + \theta_1 \varepsilon_{t-1} + \theta_2 \varepsilon_{t-2}$$

$$\text{(11.29)}$$

$$+ \cdots + \theta_q \varepsilon_{t-q}$$

where $\varepsilon_t, \varepsilon_{t-1}, \ldots, \varepsilon_{t-q}$ are independent, identically distributed random variables with a constant mean of 0 and a constant variance (Box and Jenkins, 1976, p. 52). A *mixed autoregressive-moving average process of order (p,q)*, also called an *ARMA(p,q) process*, combines both of these models:

$$X_t = \mu + \phi_1 \left(X_{t-1} - \mu \right) + \phi_2 \left(X_{t-2} - \mu \right)$$

$$+ \cdots + \phi_p \left(X_{t-p} - \mu \right) + \varepsilon_t \qquad \text{(11.30)}$$

$$+ \theta_1 \varepsilon_{t-1} + \theta_2 \varepsilon_{t-2} + \cdots + \theta_q \varepsilon_{t-q}$$

(Box and Jenkins, 1976, pp. 52–53). Most textbooks on time series analysis discuss these models in detail, as well as methods you can use to assess what model appears to fit the data based on looking at the autocorrelation function (ACF) and partial autocorrelation function (PACF). See Box and Jenkins (1976) and the S-PLUS documentation for details.

ESTIMATING AND TESTING FOR TREND

Modeling trend is just a special case of regression in which one of the predictor variables is time: trend is the behavior of the response variable over time. In this section we will discuss tools you can use to estimate and test for trend in a time series.

Parametric Methods for Trend Analysis

The simplest parametric model for trend involves an intercept and a slope:

$$Y_t = \beta_0 + \beta_1 X_t + \varepsilon_t \qquad (11.31)$$

where β_0 denotes the intercept, $X_t = f(t)$ denotes some function of the time (e.g., the week, month, quarter, or year the tth observation was taken), β_1 denotes the change in the mean of Y per unit of time (i.e., the trend), and ε_t denotes the error at time t. As in Chapter 9, we assume the errors are independent and all come from a normal distribution with a constant mean of 0 and a standard deviation of σ_ε. For this model, the test for trend is simply the test for whether β_1 is significantly different from 0, and is based on the t-statistic shown in Equation (9.29) in Chapter 9. The confidence interval for the trend is the confidence interval for the slope, given in Equation (9.32). In some cases we may want to use a more sophisticated model for trend, for example by adding a quadratic term to Equation (11.31).

Accounting for Seasonality and Other Covariates

The simple regression model of Equation (11.31) can be modified to account for seasonality in the observations and/or information from other covariates. (It should be noted that in the time series literature, variations in the data that occur with a cyclical pattern are referred to as "seasonal" effects, whether or not the variation is truly quarterly (Fall, Winter, Spring, Summer) or on some other time scale (e.g., monthly).) The modification may be as simple as allowing for a quarterly or monthly effect, or a more complicated model that involves using the sine and/or cosine functions.

Example 11.4: Testing for Trend in Total Phosphorus at Station CB3.3e

Figure 11.14 displays monthly estimated total phosphorus mass (mg) within a water column at station CB3.3e for the 5-year time period October 1984 to September 1989 from a study on phosphorus concentration conducted in the Chesapeake Bay (Neerchal and Brunenmeister, 1993). In ENVIRONMENTALSTATS for S-PLUS these data are stored in the regular time series Total.P.rts and the data frame Total.P.df.

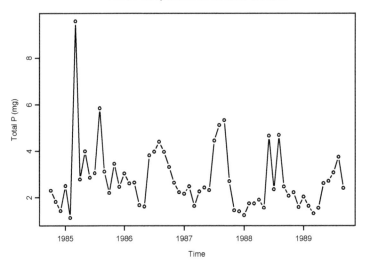

Figure 11.14 Monthly estimated total phosphorus mass within a water column at station CB3.3e

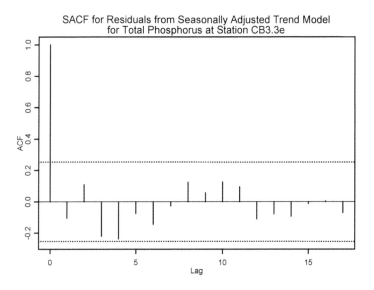

Figure 11.15 Sample autocorrelation function for the residuals from the fitted linear model allowing for trend and seasonality in the total phosphorus data at station CB3.3e

These data display seasonal variation, so we need to account for this while testing for trend. If we fit a model that allows for a different average value for each month as well as an annual linear trend, the estimated trend coefficient is –0.3 mg/year with a standard error of 0.1 and an associated p-value of 0.01. The 95% confidence interval for the trend is [–0.51, –0.07]. Diagnostic plots for the fit indicate that the observation for March 1985 is both an "outlier" and a "high-leverage" point. Figure 11.15 displays the sample autocorrelation function plot for the residuals from this fit. Neither the parametric nor nonparametric test indicates significant lag-1 autocorrelation in the residuals of the fitted model.

Menu

To fit the linear model that allows for separate monthly means and an annual trend using the S-PLUS pull-down menu, follow these steps.

1. Find **Total.P.df** in the Object Explorer.
2. On the S-PLUS menu bar, make the following menu choices: **Statistics>Regression>Linear**. This will bring up the Linear Regression dialog box.
3. For Data Set **Total.P.df** should be highlighted. In the Formula box type **CB3.3e ~ Month + Year**. In the Save As box type **Total.P.CB3.3e.fit**.
4. Click on the **Results** tab. Under Saved Results, **check** the Residuals box, and type **Total.P.CB3.3e.fit.residuals** in the Save In box. Click **OK** or **Apply**.

To create the sample autocorrelation plot for the residuals from this fit, follow these steps.

5. Find **Total.P.CB3.3e.fit.residuals** in the Object Explorer.
6. On the S-PLUS menu bar, make the following menu choices: **Statistics>Time Series>Autocorrelations**. This will bring up the Autocorrelations and Autocovariances dialog box.
7. The Data Set box should display **Total.P.CB3.3e.fit.residuals**. In the Variable box select **residuals**, then click **OK** or **Apply**.

To perform the parametric test for the presence of lag-1 autocorrelation in the residuals from this fit, follow these steps.

8. On the S-PLUS menu bar, make the following menu choices: **EnvironmentalStats>Hypothesis Tests>Serial Correlation**. This will bring up the Serial Correlation Test dialog box.
9. For Data to Use select **Pre-Defined Data**. The Data Set box should display **Total.P.CB3.3e.fit.residuals**. In the Variable box select **residuals**, for Test Method choose **AR(1) Y-W**, then click **OK** or **Apply**.

To perform the nonparametric test, follow the same steps as above, but for Test Method choose **Rank von Neumann**.

Command

To fit the linear model that allows for separate monthly means and an annual trend using the S-PLUS Command or Script Window, type these commands.

```
Total.P.CB3.3e.fit <- lm(CB3.3e ~ Month + Year,
    data=Total.P.df)
summary(Total.P.CB3.3e.fit)
```

To create the sample autocorrelation plot for the residuals from this fit, type this command.

```
acf(residuals(Total.P.CB3.3e.fit))
```

To perform the parametric and nonparametric tests for the presence of lag-1 autocorrelation in the residuals from this fit, type these commands.

```
serial.correlation.test(residuals(Total.P.CB3.3e.fit),
    test="AR1.yw")
serial.correlation.test(residuals(Total.P.CB3.3e.fit),
    test="rank.von.Neumann")
```

Accounting for Autocorrelation

If we fit a regression model for trend and find that the residuals exhibit serial correlation, we need to account for this fact. The example below gives an illustration.

Example 11.5: Testing for Trend in Total Phosphorus at Station CB3.1

Figure 11.16 displays monthly estimated total phosphorus mass (mg) within a water column at station CB3.1. If we fit a model that allows for a different average value for each month as well as an annual linear trend, the estimated trend coefficient is –0.63 mg/year with a standard error of 0.15 and an associated p-value of 0.0001. The 95% confidence interval for the trend is [–0.92, –0.33]. Diagnostic plots for the fit indicate that the variance of the errors appears to increase a bit with the size of the fitted values, so some transformation of the response variable (e.g., log) may be appropriate, but for the purpose of this example we will stick with the original data.

Figure 11.17 displays the sample autocorrelation function plot for the residuals from this fit. Neither the parametric nor nonparametric test indicates significant lag-1 autocorrelation in the residuals of the fitted model, but the

Total Phosphorus at Station CB3.1

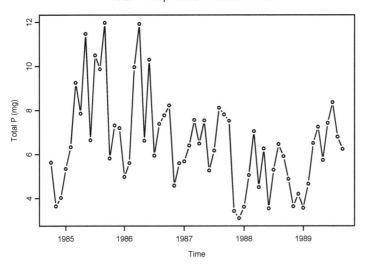

Figure 11.16 Monthly estimated total phosphorus mass within a water column at station CB3.1

SACF for Residuals from Seasonally Adjusted Trend Model for Total Phosphorus at Station CB3.1

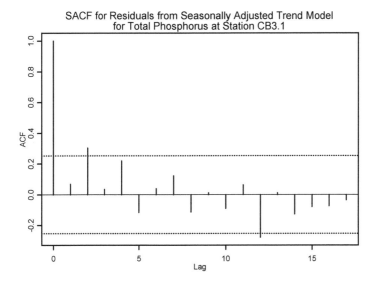

Figure 11.17 Sample autocorrelation function for the residuals from the fitted linear model allowing for trend and seasonality in the total phosphorus data at station CB3.1

SAFC indicates significant correlation at lag-2. This is may very well be a matter of the multiple comparisons problem we discussed earlier, but for the purposes of this example we will assume the autocorrelation is real and re-fit the model allowing for an AR(2) process in the model errors. In this case the estimated trend decreases very slightly to –0.56 mg/yr while the standard error increases to 0.23, and the associated p-value is 0.02. The 95% confidence interval for the trend becomes [–1.03, –0.1].

Menu

To fit the linear model that allows for separate monthly means and an annual trend using the S-PLUS pull-down menu, follow these steps.

1. Find **Total.P.df** in the Object Explorer.
2. On the S-PLUS menu bar, make the following menu choices: **Statistics>Regression>Linear**. This will bring up the Linear Regression dialog box.
3. For Data Set **Total.P.df** should be highlighted. In the Formula box type **CB3.1 ~ Month + Year**. In the Save As box type **Total.P.CB3.1.fit**.
4. Click on the **Results** tab. Under Saved Results, **check** the Residuals box, and type **Total.P.CB3.1.fit.residuals** in the Save In box. Click **OK** or **Apply**.

To create the sample autocorrelation plot for the residuals from this fit, follow these steps.

5. Find **Total.P.CB3.1.fit.residuals** in the Object Explorer.
6. On the S-PLUS menu bar, make the following menu choices: **Statistics>Time Series>Autocorrelations**. This will bring up the Autocorrelations and Autocovariances dialog box.
7. The Data Set box should display **Total.P.CB3.1.fit.residuals**. In the Variable box select **residuals**, then click **OK** or **Apply**.

To perform the parametric test for the presence of lag-1 autocorrelation in the residuals from this fit, follow these steps.

8. On the S-PLUS menu bar, make the following menu choices: **EnvironmentalStats>Hypothesis Tests>Serial Correlation**. This will bring up the Serial Correlation Test dialog box.
9. For Data to Use select **Pre-Defined Data**. The Data Set box should display **Total.P.CB3.1.fit.residuals**. In the Variable box select **residuals**, for Test Method choose **AR(1) Y-W**, then click **OK** or **Apply**.

To perform the nonparametric test, follow the same steps as above, but for Test Method choose **Rank von Neumann**.

In the current version of S-PLUS 2000, it is not trivial to use the menu to fit a new model that allows for an AR(2) process in the model errors. The menu selection **Statistics>Time Series>ARIMA Models** brings up the ARIMA Modeling dialog box, but there is no easy way to specify an intercept in the model unless you already have a column of 1's in the data frame (certainly not the usual case), and also the dialog does not properly handle predictor variables that are factors. Also the results from this menu selection do not include the standard errors associated with the estimated regression coefficients. You can, however, use the function gls at the command line, as explained below.

Command

To fit the linear model that allows for separate monthly means and an annual trend using the S-PLUS Command or Script Window, type these commands.

```
Total.P.CB3.1.fit <- lm(CB3.1 ~ Month + Year,
   data=Total.P.df)
summary(Total.P.CB3.1.fit)
```

To create the sample autocorrelation plot for the residuals from this fit, type this command.

```
acf(residuals(Total.P.CB3.1.fit))
```

To perform the parametric and nonparametric tests for the presence of lag-1 autocorrelation in the residuals from this fit, type these commands.

```
serial.correlation.test(residuals(Total.P.CB3.1.fit),
   test="AR1.yw")
serial.correlation.test(residuals(Total.P.CB3.1.fit),
   test="rank.von.Neumann")
```

To fit a new model that allows for an AR(2) process in the model errors, you can use the S+ function arima.mle:

```
Total.P.CB3.1.ar2.fit <- arima.mle(Total.P.df$CB3.1,
   model=list(order=c(2,0,0)),
   xreg=model.matrix(Total.P.CB3.1.fit))
```

Unfortunately, arima.mle does not return the standard errors associated with the estimated regression coefficients. You can, however, use the function gls from the NLME3 library that comes with S+2000 Release 2. To attach this library, choose **File>Load Library** from the S-PLUS menu and select **nlme3**. After loading the NLME3 library, type these commands:

```
Total.P.CB3.1.gls.fit <- gls(CB3.1 ~ Month + Year,
   data=Total.P.df, correlation=corARMA(p=2))
summary(Total.P.CB3.1.gls.fit)
```

Nonparametric Methods for Trend Analysis

If you are not willing to make the assumption that the errors in your model for trend follow a normal distribution, you can use nonparametric methods instead. The **Mann-Kendall test for trend** (Mann, 1945) is based on the following test statistic:

$$S = \sum_{i=1}^{n-1} \sum_{j=i+1}^{n} sign\left(Y_j - Y_i\right) \qquad \textbf{(11.32)}$$

where the `sign` function is defined in Equation (9.14). This test statistic is exactly the same as the statistic S defined in Equation (9.13) that is used to compute **Kendall's tau statistic** to test for the independence between two random variables X and Y (Kendall 1938, 1975), except here the X variable becomes time and is not needed in the equation because X_j is always bigger than X_i when $j > i$ so $(X_j - X_i)$ is always positive.

The observations Y_1, Y_2, ..., Y_n are assumed to be in order with respect to time, but the times at which they were observed do not have to be evenly spaced. Looking at Equation (11.32), you can see that every time an observation later in time is bigger than one taken earlier in time, a 1 is added to the sum. Also, every time an observation later in time is smaller than one taken earlier in time, a −1 is added to the sum. Hence, large values of S indicate a positive trend, and small (negative) values of S indicate a negative trend. Note that unlike the parametric case where we have to specify the form of the trend (e.g., linear, quadratic, exponential, etc.), the Mann-Kendall test for trend tests for any kind of monotonic trend.

Tables of critical values of S for small samples are given in Hollander and Wolfe (1999, pp. 724–731), Conover (1980, pp. 458–459), Gilbert (1987, p. 272), Helsel and Hirsch (1992, p. 469), and Gibbons (1994, p. 180). To compute p-values, you can also use the large sample normal approximation to the distribution of S. Under the null hypothesis of no trend, the statistic:

$$z = \frac{S}{\sqrt{Var\left(S\right)}} \qquad \textbf{(11.33)}$$

is approximately distributed as a standard normal random variable, where

$$Var(S) = \frac{n(n-1)(2n+5)}{18} \qquad (11.34)$$

In the case of tied observations, formula (11.34) is modified to:

$$Var(S) = \frac{n(n-1)(2n+5)}{18} -$$

$$\frac{\displaystyle\sum_{j=1}^{h} u_j(u_j-1)(2u_j+5)}{18} \qquad (11.35)$$

where u_j denotes the number of tied observations in the jth group of ties.

Both Kendall (1975) and Mann (1945) show that the normal approximation is excellent even for samples as small as $n = 10$, provided that the following continuity correction is used:

$$z = \begin{cases} \dfrac{S-1}{\sqrt{Var(S)}} & , \; S > 0 \\[4mm] 0 & , \; S = 0 \\[4mm] \dfrac{S+1}{\sqrt{Var(S)}} & , \; S < 0 \end{cases} \qquad (11.36)$$

Although the Mann-Kendall test for trend is useful because you do not need to worry about the normality assumption, it is useful to have some kind of estimate of trend. For the simple model of a linear trend shown in Equation (11.31), Theil (1950) proposed the following nonparametric estimator of the slope:

$$\hat{\beta}_1 = \underset{i<j}{Median} \left(\frac{Y_j - Y_i}{X_j - X_i} \right) \tag{11.37}$$

This estimator is the median of all N estimates of slope formed by taking all possible pairs of observations and estimating the slope based on just these two points. In the case of no ties in the X values, there are

$$N = \binom{n}{2} = \frac{n(n-1)}{2} \tag{11.38}$$

possible pairs of points.

Sen (1968) generalized this estimator to the case where there are possibly tied observations in the X values. In this case, Sen simply ignores the two-point estimated slopes where the X's are tied and computes the median based on the remaining N' two-point estimated slopes. That is, Sen's estimator is given by:

$$\hat{\beta}_1 = \underset{\substack{i<j \\ X_i \neq X_j}}{Median} \left(\frac{Y_j - Y_i}{X_j - X_i} \right) \tag{11.39}$$

(Hollander and Wolfe, 1999, pp. 421–423). It is interesting to note that if we modified the Theil-Sen estimator of slope to use the mean instead of the median, we would get the least squares estimate of slope.

Conover (1980, p. 267) suggests the following estimator for the intercept:

$$\hat{\beta}_0 = Y_{0.5} - \hat{\beta}_1 X_{0.5} \tag{11.40}$$

where $X_{0.5}$ and $Y_{0.5}$ denote the sample medians of the X's and Y's, respectively. With these estimators of slope and intercept, the estimated regression line passes through the point $(X_{0.5}, Y_{0.5})$.

Theil (1950) and Sen (1968) proposed methods to compute a confidence interval for the true slope (see Hollander and Wolfe, 1999, pp. 424–426). Gilbert (1987, p. 218) illustrates a simpler method than the one given by Sen (1968) that is based on a normal approximation. Gilbert's (1987) method is an extension of the one given in Hollander and Wolfe

(1999, p. 424) that allows for ties and/or multiple observations per time period. This method is valid for a sample size as small as $n = 10$ unless there
are several tied observations.

Let N' denote the number of defined two-point estimated slopes that are
used in Equation (11.39) above (if there are no tied X values then $N' = N$),
and let

$$\hat{\beta}_{1_{(1)}} \, , \; \hat{\beta}_{1_{(2)}} \, , \; \cdots \, , \; \hat{\beta}_{1_{(N')}} \tag{11.41}$$

denote the N' ordered slopes. For Gilbert's (1987) method, a $100(1-\alpha)\%$
two-sided confidence interval for the true slope is given by

$$\left[\, \hat{\beta}_{1_{(M_1)}} \, , \; \hat{\beta}_{1_{(M_2+1)}} \, \right] \tag{11.42}$$

where

$$M_1 \; = \; \frac{N' - C_\alpha}{2} \tag{11.43}$$

$$M_2 \; = \; \frac{N' + C_\alpha}{2} \tag{11.44}$$

$$C_\alpha \; = \; z_{1-\alpha/2} \, \sqrt{Var\left(S\right)} \tag{11.45}$$

Usually the quantities M_1 and M_2 will not be integers, in which case Gilbert
(1987, p. 219) suggests interpolating between adjacent values.

Accounting for Seasonality

Hirsch et al. (1982) introduced a modification of the Mann-Kendall test
for trend that allows for seasonality in observations collected over time.
They call this test the **seasonal Kendall test**. Their test is appropriate for
testing for trend within each season when the trend is in the same direction

across all seasons. Assuming there are p seasons, the seasonal Kendall statistic is simply the sum of the Kendall statistics for each season:

$$S' = \sum_{j=1}^{p} S_j \qquad (11.46)$$

To compute p-values, you can use the z-statistic of Equation (11.36), using S' instead of S, where

$$Var\left(S'\right) = \sum_{j=1}^{p} Var\left(S_j\right) \qquad (11.47)$$

To estimate the overall slope, Hirsch et al. (1982, p. 117) suggest using the median of the p Theil-Sen estimates of slopes computed for each season (see Equation (11.39)). Also, to estimate the overall intercept, you can use the median of the p estimates of intercept (see Equation (11.40)). Gilbert (1987, pp. 227–228) extends his method of computing a confidence interval for the slope to the case of seasonal observations by using the same formulas as in Equations (11.41) to (11.45) but using all two-point estimated slopes from all seasons, and using S' instead of S in Equation (11.45).

Van Belle and Hughes (1984) suggest using the following statistic to test for heterogeneity in trend prior to applying the seasonal Kendall test:

$$\chi^2_{het} = \sum_{j=1}^{p} z_j^2 - p\bar{z}^2 \qquad (11.48)$$

where z_j denotes the z-statistic for the jth season computed without the continuity correction (Equation (11.33)), and

$$\bar{z} = \frac{1}{p} \sum_{j=1}^{p} z_j \qquad (11.49)$$

Under the null hypothesis of no trend in any season, the statistic defined in Equation (11.48) is approximately distributed as a chi-square random variable with $p-1$ degrees of freedom. The continuity correction is not used to compute the z-statistics since using it results in an unacceptably conservative test (van Belle and Hughes, 1984, p. 132). Van Belle and Hughes (1984) imply that their heterogeneity statistic may be used to test the null hypothesis of the same amount of (possibly nonzero) trend in each season. For this case, however, the distribution of the test statistic in Equation (11.48) is unknown since it depends on the unknown true value of the trend. The heterogeneity chi-square statistic of Equation (11.48) may be assumed to be approximately distributed as chi-square with $p-1$ degrees of freedom under the null hypothesis of equal trend in each season, but further study is needed to determine how well this approximation works.

Example 11.6: Testing for Trend in Total Phosphorus at Station CB3.3e

In this example we will repeat our analyses of Example 11.4, but here we will use the nonparametric seasonal Kendall trend test to test for trend. The van Belle-Hughes test for heterogeneity in trend between seasons is not significant, with a p-value of 0.95. The estimated trend coefficient is –0.23 mg/year with an associated p-value of 0.006. The 95% confidence interval for the trend is [–0.36, –0.06]. Note that the estimated trend and lower confidence limit for the trend are smaller in magnitude (absolute value) because the nonparametric test is not affected by the March 1985 outlier.

Menu

To perform the nonparametric test for annual trend in the total phosphorus data and also allow for separate monthly effects using the ENVIRONMENTALSTATS for S-PLUS pull-down menu, follow these steps.

1. Find **Total.P.df** in the Object Explorer.
2. On the S-PLUS menu bar, make the following menu choices: **EnvironmentalStats>Hypothesis Tests>Trend Tests>Nonparametric**. This will bring up the Nonparametric Trend Test dialog box.
3. For Data to Use select **Pre-Defined Data**. The Data Set box should display **Total.P.df**. In the Variable box select **CB3.3e**, **check** the Seasonal box, for Season select **Month** and for Year select **Year**, then click **OK** or **Apply**.

Command

To test for annual trend in the total phosphorus data and also allow for separate monthly effects using the ENVIRONMENTALSTATS for S-PLUS Command or Script Window, type these commands.

```
attach(Total.P.df)
seasonal.kendall.trend.test(CB3.3e, Month, Year)
```

Accounting for Autocorrelation

Hirsch and Slack (1984) introduced a modification of the seasonal Kendall test that is robust against serial dependence (in terms of type I error) except when the observations have a very strong long-term persistence (very large autocorrelation) or when the sample sizes are small (e.g., less than 6 years of monthly data). This modification is based on a multivariate test introduced by Dietz and Killeen (1981).

In the case of serial dependence, the equation for $Var(S')$ (Equation (11.47)) must be modified as follows:

$$Var(S') = \sum_{j=1}^{p} Var(S_j) +$$

(11.50)

$$\sum_{g=1}^{p-1} \sum_{h=g+1}^{p} Cov(S_g, S_h)$$

where $Cov(S_g, S_h)$ denotes the covariance between the Kendall statistic for season g and the Kendall statistic for season h. In order to compute p-values for the seasonal Kendall test statistic when serial correlation is present, we need estimates of the covariances

$$\sigma_{gh} = Cov(S_g, S_h)$$ (11.51)

Let R denote the $n \times p$ matrix of ranks for the Y observations, where the Y's are ranked within season:

$$R = \begin{pmatrix} R_{11} & R_{12} & \cdots & R_{1p} \\ R_{21} & R_{22} & \cdots & R_{2p} \\ \vdots & & & \\ R_{n1} & R_{n2} & \cdots & R_{np} \end{pmatrix}$$ (11.52)

where

$$R_{ij} = \frac{n_j + 1 + \sum\limits_{k=1}^{n_j} sign\left(Y_{ij} - Y_{kj}\right)}{2} \qquad (11.53)$$

and n_j denotes the number of (X, Y) pairs without missing values for season j. By this definition, missing values are assigned the mid-rank of the non-missing values. Hirsch and Slack (1984) suggest using the following formula, given by Dietz and Killeen (1981), in the case where there are no missing values:

$$\hat{\sigma}_{gh} = \frac{K_{gh} + 4\sum\limits_{i=1}^{n} R_{ig}R_{ih} - n(n + 1)^2}{3} \qquad (11.54)$$

where

$$K_{gh} = \sum\limits_{i=1}^{n-1} \sum\limits_{j=i+1}^{n} sign\left[\left(Y_{jg} - Y_{ig}\right)\left(Y_{jh} - Y_{ih}\right)\right] \quad (11.55)$$

For the case of missing values, Hirsch and Slack (1984) derive the following modification of Equation (11.54):

$$\hat{\sigma}_{gh} = \frac{K_{gh} + 4\sum\limits_{i=1}^{n} R_{ig}R_{ih} - n\left(n_g + 1\right)\left(n_h + 1\right)}{3} \qquad (11.56)$$

Technically, the estimates in Equations (11.54) and (11.56) are not correct estimators of covariance, because the model Dietz and Killeen (1981) use assumes that observations within a year (within a row of R) may be correlated, but observations between years (between rows of R) are independent. Serial dependence induces correlation between all of the observations. In most cases, however, the serial dependence shows an exponential decay in correlation across time and so these estimates work fairly well.

Hirsch and Slack (1984) performed a Monte Carlo study to determine the empirical significance level and power of their modified test. For $p = 12$

seasons, they found their modified test gave correct significance levels for n ≥ 10 as long as the lag-one autocorrelation was 0.6 or less, while the original test that assumes independent observations yielded highly inflated significance levels.

The seasonal and over-all estimates of slope and intercept can be computed using the same methods as in the case of serial independence. Also, the method for computing the confidence interval for the slope is the same as in the case of serial independence. Note that the serial dependence is accounted for in the term $\mathtt{Var(S')}$ in Equation (11.45).

Example 11.7: Testing for Trend in Total Phosphorus at Station CB3.1

In this example we will repeat our analyses of Example 11.5, but here we will use the nonparametric seasonal Kendall trend test to test for trend. Table 11.2 compares the results of the parametric and nonparametric tests both when we do not assume autocorrelation is present and when we do assume it is present. For all four approaches, the estimated trend is similar, but for the methods that incorporate autocorrelation the confidence interval for the trend is wider, and for the nonparametric test that incorporates autocorrelation the trend is not significant at the 0.10 level.

Test Assumptions	Trend and P-Value	95% CI
Parametric, No Autocorrelation	-0.63 (0.0001)	[-0.92, -0.33]
Parametric, Autocorrelation	-0.56 (0.02)	[-1.03, -0.10]
Nonparametric, No Autocorrelation	-0.56 (0.0002)	[-0.87, -0.33]
Nonparametric, Autocorrelation	-0.56 (0.11)	[-1.57, 0.20]

Table 11.2   Comparison of results for trend tests on total phosphorus at station CB3.1

In Example 11.6 and Example 11.7 we are using data with "small" sample sizes because we have less than 10 years of data per season. On the other hand, for these particular data there is not much difference in the estimated trend between the parametric and nonparametric tests nor between the tests that do and do not assume autocorrelation.

Menu

To perform the nonparametric test for annual trend in the total phosphorus data, allow for separate monthly effects, and allow for autocorrelation using the ENVIRONMENTALSTATS for S-PLUS pull-down menu, follow these steps.

1. Find **Total.P.df** in the Object Explorer.
2. On the S-PLUS menu bar, make the following menu choices: **EnvironmentalStats>Hypothesis Tests>Trend Tests>Nonparametric**. This will bring up the Nonparametric Trend Test dialog box.
3. For Data to Use select **Pre-Defined Data**. The Data Set box should display **Total.P.df**. In the Variable box select **CB3.1**, **check** the Seasonal box, **uncheck** the Independent Obs box, and for Season select **Month**, then click **OK** or **Apply**.

Command

To perform the nonparametric test for annual trend in the total phosphorus data, allow for separate monthly effects, and allow for autocorrelation using the ENVIRONMENTALSTATS for S-PLUS Command or Script Window, type these commands.

```
attach(Total.P.df)
seasonal.kendall.trend.test(CB3.1, Month,
   independent.obs=F)
detach()
```

SUMMARY

- Classical statistical methods for constructing confidence intervals and performing hypothesis tests assume the observations are independent. Environmental data are often collected sequentially over time and therefore may not satisfy this assumption.
- A time series plot is simply a plot of the observations on the y-axis vs. the times (or sequence in which) they were collected on the x-axis.
- *Autocorrelation* (also called *serial correlation*) is simply a measure of how a random variable X varies over time relative to past observations of X.
- Two properties of a *stationary* time series are (1) the mean and standard deviation remain constant over time, and (2) the value of the lag-k autocorrelation does not depend on what starting point we choose in the series.
- One of the simplest models for autocorrelation is the *first order autoregressive model*, also denoted as the *AR(1) model*. More complicated models include p-order autoregressive models (AR(p) models), q-order moving average models (MA(q) models), and ARMA(p,q) models.

- Three tools to graphically assess the presence of autocorrelation are the sample autocorrelation function (SACF), the sample partial autocorrelation function (SPACF), and lag plots.
- There are several ways to test for the presence of autocorrelation, including simple tests for lag-1 autocorrelation, as well as fitting complicated ARMA models and testing for the significance of the model parameters.
- Ignoring autocorrelation leads to inflated type I errors and confidence intervals that are too narrow in the case of positive serial correlation.
- There are two ways to deal with autocorrelation: (1) eliminate it by improving your model to account for more sources of variability, and/or (2) use a simple or complicated time series model to perform your statistical analyses.
- Modeling trend is just a special case of regression in which one of the predictor variables is time: trend is the behavior of the response variable over time.
- To test for and estimate trend, you can use parametric regression models or nonparametric tests based on Kendall's tau statistic.

EXERCISES

11.1. Give an example of a time series from your experience. Would it be reasonable to assume this time series is stationary?

11.2. Give an example of a seasonal time series from your experience. If you seasonally adjusted the data would it be reasonable to assume the time series is stationary?

11.3. Generate a set of 30 independent observations from a normal distribution, then look at the sample autocorrelation function for these observations. Do this 100 times to build up an internal "gestalt" of what the SAFC looks like when the observations are independent.

11.4. Repeat Exercise 11.3 using different sample sizes.

11.5. Repeat Exercises 11.2 and 11.3 using data from other distributions (e.g., lognormal, gamma, Weibull, etc.).

11.6. Using the function `arima.sim`, generate observations from an autocorrelated and/or moving average process and look at the SAFC. Do this 100 times.

11.7. Do a simulation study on the Type I error of the one-sample t-test when the underlying observations come from an AR(1) process. Vary the value of the lag-1 autocorrelation.

11.8.   Repeat Exercise 11.7 but use the parametric test for trend that does not assume autocorrelation is present (i.e., the model based on Equation (11.31)).

11.9.   Repeat Exercise 11.8, but use the nonparametric test for trend that does not assume autocorrelation is present.

11.10.   Consider the model of Example 11.4. Using the data from this example, fit a model that allows for quarterly variations (instead of monthly variations), as well as an annual trend. Compare the model with quarterly variations with the model that allows for monthly variations using analysis of variance.

11.11.   Consider the Mauna Loa CO_2 data shown in Figure 11.3. Fit a model to these data that accounts for seasonality and trend. Do the residuals from your model exhibit autocorrelation?

12   SPATIAL STATISTICS

How Are Things Changing over Space?

In Chapter 11 we discussed methods for dealing with data collected over time, including testing for trend and how to accommodate serial correlation. Environmental studies often generate data collected over space as well as time. Sometimes we are interested not only in characterizing chemical concentrations in an area by simple summary statistics, but also by describing the spatial pattern of concentration. Just as we fit models to describe changes in concentration over time, we can also fit models to describe changes in concentration over space. In this chapter we will talk about statistical tools for dealing with spatial data.

OVERVIEW: TYPES OF SPATIAL DATA

Spatial data can be classified by how the location associated with each observation is defined: by a point (e.g., latitude and longitude of the location) or by an area or region (e.g., King County, Washington, USA) (Cressie, 1993; Kaluzny et al., 1998). Point spatial data can be further subclassified by whether it is geostatistical or a spatial point pattern. *Geostatistical data* have point locations associated with them and usually one or more variables are measured at each location. For example, data from an air quality monitoring network may include observations of ozone, particulate matter, temperature, humidity, radiation, wind speed, and time of day for each monitoring location. *Spatial point pattern data* have point locations associated with them and the locations themselves are the variable of interest. Often one of the main hypotheses of interest is whether the locations are random, clustered, or regular. For example, if suddenly several new cases of a type of cancer or infectious disease appear in a town or county, is this apparent "clustering" real or is it something that could reasonably happen by chance given the incidence rate of the disease for the country as a whole? *Lattice data* are observations associated with an area or region, such as the vegetation index of a pixel on a remote-sensing image (regular lattice data) or the cancer rate within each county of a state (irregular lattice data). Table 12.1 summarizes the various classifications of spatial data.

In this chapter we will use data on benthic characteristics collected at point locations to illustrate a few tools for spatial statistics that are available in S+SPATIALSTATS. We will also discuss how to use S-PLUS for ArcView GIS, a link between S-PLUS and the geographical information system (GIS) software package ArcView GIS. More tools for spatial analysis are described in the S+SPATIALSTATS User's Manual. General references for spatial statistics include Arlinghaus (1996), Cliff and Ord (1981), Cox et al. (1997), Cressie (1993), David (1977), Davidson (1994), Diggle (1983), Haining (1993), Isaaks and Srivastava (1989), Journel and Huijbregts (1978), Kitanidis (1997), Kaluzny et al. (1998), and Ripley (1981). A few references that deal specifically with spatial statistics and sampling chemicals in the environment include Burmaster and Thompson (1997), Ginevan and Splitstone (1997), and Qian (1997).

Location Type	Spatial Data Type	Examples
Point	Geostatistical	Air quality at monitoring stations TcCB concentrations on a grid Benthic index at irregularly-spaced sampling sites
	Spatial Point Pattern	Locations of bird nests Locations of cancer cases
Areal (Regional)	Lattice	Number of cancer cases for a county Vegetation index for a pixel on a remote-sensing image

Table 12.1   Classifications and examples of spatial data

THE BENTHIC DATA

The data frame `benthic.df` in ENVIRONMENTALSTATS for S-PLUS contains data from the Long Term Benthic Monitoring Program of the Chesapeake Bay. The data consist of measurements of benthic characteristics and a computed index of benthic health for several locations in the bay. Sampling methods and designs of the program are discussed in Ranasinghe et al. (1992). The United States Environmental Protection Agency (USEPA) established an initiative for the Chesapeake Bay in partnership with the states bordering the bay in 1984. The goal of the initiative is the restoration (abundance, health, and diversity) of living resources to the bay by reducing nutrient loadings, reducing toxic chemical impacts, and enhancing habitats. USEPA's Chesapeake Bay Program Office is responsible for implementing this initiative and has established an extensive monitoring program that includes traditional water chemistry sampling, as well as collecting data on living resources to measure progress towards meeting the restoration goals.

Sampling benthic invertebrate assemblages has been an integral part of the Chesapeake Bay monitoring program due to their ecological importance and their value as biological indicators. The condition of benthic assemblages is a measure of the ecological health of the bay, including the effects of multiple types of environmental stresses. Nevertheless, regional-scale assessment of ecological status and trends using benthic assemblages are limited by the fact that benthic assemblages are strongly influenced by naturally variable habitat elements, such as salinity, sediment type, and depth. Also, different state agencies and USEPA programs use different sampling methodologies, limiting the ability to integrate data into a unified assessment. To circumvent these limitations, USEPA has standardized benthic data from several different monitoring programs into a single database, and from that database developed a Restoration Goals Benthic Index that identifies whether benthic restoration goals are being met.

Table 12.2 lists the variables contained in the data frame `benthic.df`. The data represent observations collected at 585 separate point locations (sites). The sites are divided into 31 different strata, numbered 101 through 131, each strata consisting of geographically close sites of similar degradation conditions. The benthic index values range from 1 to 5 on a continuous scale, where high values correspond to healthier benthos. Salinity was measured in parts per thousand (ppt), and silt content is expressed as a percentage of clay in the soil with high numbers corresponding to muddy areas.

Variable	Description (Units)
Site.ID	Site ID
Stratum	Stratum Number (101–131)
Latitude	Latitude (degrees North)
Longitude	Longitude (negative values; degrees West)
Index	Benthic Index
Salinity	Salinity (ppt)
Silt	Silt Content (% Clay in Soil)

Table 12.2   Description of variables in the data frame `benthic.df`

Example 12.1: Visualizing the Benthic Data

In Chapter 3 we discussed several tools for visualizing data. Here, we will apply these tools to the benthic data. Figure 12.1 displays a histogram of the benthic index and Figure 12.2 displays the quantile plot. You can see that about 40% of the values are between 1 and 2, indicating poor benthic health. Figure 12.3 displays the 585 sampling locations and Figure 12.4 displays a bubble plot, where the size of the plotting symbol is proportional to the value of the benthic index. Because of the placement of the sampling locations, the bubble plot is difficult to interpret, but it looks like low values of the benthic index occur at the southern stations, and also between latitudes 38.5 to 39 at longitude 76.4. Figure 12.5 displays a contour plot of the benthic index and Figure 12.6 displays a surface plot. Here it is easier to see the

Histogram of Benthic Index

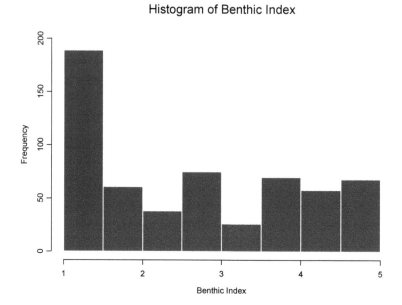

Figure 12.1   Histogram of the benthic index values

Empirical CDF of Benthic Index

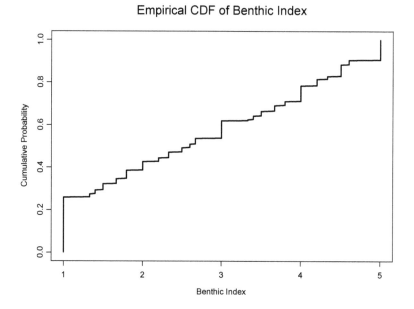

Figure 12.2   Quantile plot of the benthic index values

Sampling Station Locations

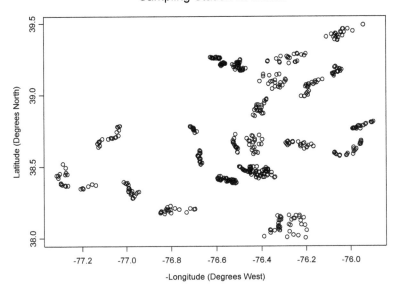

Figure 12.3   Sampling locations for the benthic data

Bubble Plot for Benthic Data

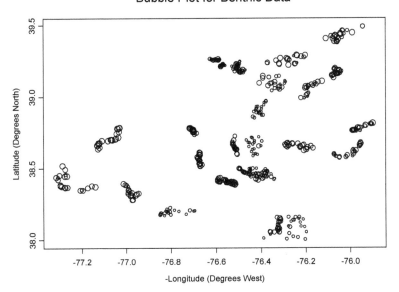

Figure 12.4   Bubble plot for the benthic index values

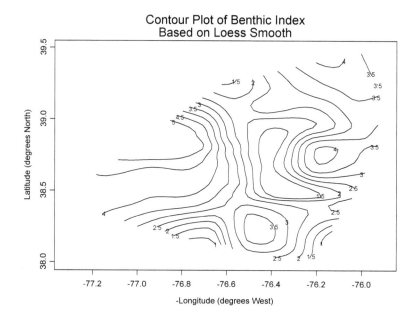

Figure 12.5   Contour plot of the benthic index values

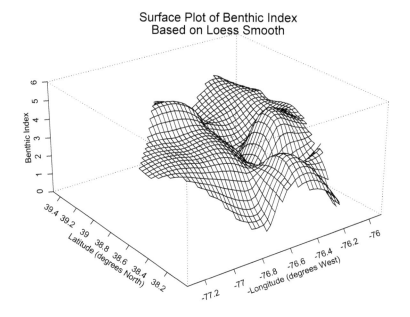

Figure 12.6   Surface plot of the benthic index values

areas of low index values. Both of these plots are based on a loess smooth in three dimensions in which quadratic surfaces are fit locally within neighborhoods of a point (we talked about loess smooths for scatterplots in Chapter 3).

Command

See Chapter 3 for information on how to produce histograms, quantile plots, scatterplots, and bubble plots. Here we will show you how to produce the contour and surface plots. Note that both plots show only values within the convex hull of station locations where the index was measured.

```
loess.fit <- loess(Index ~ Longitude * Latitude,
   data=benthic.df, normalize=F, span=0.25)

lat <- benthic.df$Latitude

lon <- benthic.df$Longitude

Latitude <- seq(min(lat), max(lat), length=50)

Longitude <- seq(min(lon), max(lon), length=50)

predict.list <- list(Longitude=Longitude,
   Latitude=Latitude)

predict.grid <- expand.grid(predict.list)

predict.fit <- predict(loess.fit, predict.grid)

index.chull <- chull(lon, lat)

index.poly <- list(x=lon[index.chull],
   y=lat[index.chull])

inside <- points.in.poly(predict.grid$Longitude,
   predict.grid$Latitude, index.poly)

predict.fit[!inside] <- NA

contour(Longitude, Latitude, predict.fit,
   levels=seq(1, 5, by=0.5), labex=0.75,
   xlab="-Longitude (degrees West)",
   ylab="Latitude (degrees North)")

title(main=paste("Contour Plot of Benthic Index",
   "Based on Loess Smooth", sep="\n"))

persp(Longitude, Latitude, predict.fit,
   xlab="-Longitude (degrees West)",
   ylab="Latitude (degrees North)",
   zlab="Benthic Index")

title(main=paste("Surface Plot of Benthic Index",
   "Based on Loess Smooth", sep="\n"))
```

MODELS FOR GEOSTATISTICAL DATA

In Chapter 11 we discussed models for time series data (data collected over time), including simple or multiple linear regression models. Sometimes we explicitly included time as a predictor variable in the model (e.g., the linear trend model of Equation (11.31)) and sometimes we included variables in the model that vary with time without explicitly including time as a predictor variable (e.g., Equation (11.25)). We can extend these same ideas to geostatistical data (data collected over space).

Notation

With time series models we often use t as a subscript or variable to denote the time at which an observation was taken. For example, Y_t or $Y(t)$ may denote the observed value at time t. For geostatistical data we will follow the convention of Cressie (1993) and Kaluzny et al. (1993) and use s to denote the location, where

$$s = (X, Y) \qquad (12.1)$$

and (X, Y) denotes the Cartesian coordinates of the location. The observation at location s is denoted by $Z(s)$. Sometimes we will use the notation:

$$Z_i = Z(s_i) = Z(X_i, Y_i) \qquad (12.2)$$

A general form for a model for geostatistical data is given by:

$$Z(W) = f(W) + \varepsilon \qquad (12.3)$$

where W denotes a vector of predictor variables such as the location variables X and Y and any other variables thought be important, f denotes some parametric or nonparametric function of the predictor variables, and ε denotes the error term. This model partitions spatial variation into large-scale variation, $f(W)$, and small-scale variation, ε. The small-scale variation reflects measurement error and local fluctuations.

Parametric Models for Geostatistical Data

Just as the simplest parametric model for a time trend involves fitting a straight line (Equation (11.31)), the simplest parametric model for a spatial trend involves fitting a plane:

$$Z_i = \beta_0 + \beta_1 X_i + \beta_2 Y_i + \varepsilon_i \qquad (12.4)$$

where Z denotes the response variable, (X, Y) denotes the location, and ε_i denotes the error for the ith observation, assumed to have a mean of 0 and a constant standard deviation σ_ε. A more sophisticated model fits a quadratic surface:

$$Z_i = \beta_0 + \beta_1 X_i + \beta_2 Y_i +$$

$$(12.5)$$

$$\beta_3 X_i^2 + \beta_4 Y_i^2 + \beta_5 X_i Y_i + \varepsilon_i$$

and in general, you can fit any kind of polynomial surface to describe changes in the response variable Z as a function of location. You can also add other predictor variables to the model, and in fact adding other predictor variables to the model will probably change your model for spatial trend and may even make the trend disappear altogether, obviating the need for the location variables (i.e., X and Y) as predictor variables in the model.

Nonparametric Models for Geostatistical Data

In Chapter 3 we explained the loess algorithm for fitting a smooth line to a scatterplot. The idea of loess can be extended to three or more dimensions. In three dimensions, planar surfaces (Equation (12.4)) or quadratic surfaces (Equation (12.5)) are fit locally within neighborhoods of a point. You can also include other variables besides (or instead of) location in the loess fit.

Another nonparametric model for geostatistical data is based on Delaunay triangulation (Ripley, 1981; Venables and Ripley, 1999, pp. 437–438). This is the algorithm used by the S-PLUS interp function and is called the "Bivariate" algorithm in the dialog boxes for contour, level, and surface plots.

Example 12.2: Quartic Surface vs. High-Order Loess Fit

The loess model we used to construct the contour and surface plots in Figures 12.5 and 12.6 uses a span argument of 0.25 and uses up about 20

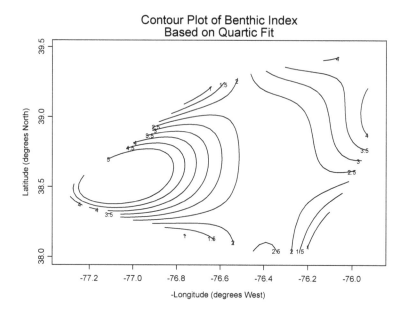

Figure 12.7   Contour plot for the quartic polynomial model for benthic index

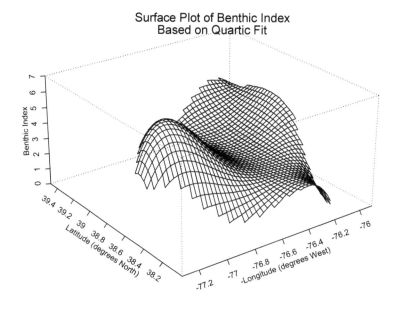

Figure 12.8   Surface plot for the quartic polynomial model for benthic index

"equivalent number of parameters," making it comparable to about a 5th order parametric polynomial fit. If we want to use a smoother surface, we can
either increase the value of the span argument for the loess fit, or fit a lower
order polynomial surface. In this example, we will fit a 4th degree (quartic)
polynomial surface. Figures 12.7 and 12.8 display the contour and surface
plots for this fit.

Command

To fit the quartic surface using the S-PLUS Command or Script Window,
type this command.

```
poly4.fit <- lm(Index ~ poly(Longitude, Latitude, 4),
    data=benthic.df)
```

(Note: You can also use the function surf.ls in the MASS library to fit
polynomial surfaces.) To create the contour and surface plots, follow the
same commands as those shown in Example 12.1, except use poly4.fit
in place of loess.fit and modify the titles of the plots. You will also
need to explicitly call predict.gam to create predict.fit:

```
predict.fit <- predict.gam(poly4.fit, predict.grid)
```

MODELING SPATIAL CORRELATION

In Chapter 11 we noted that sometimes you can eliminate autocorrelation
by incorporating information from other predictor variables (e.g., Example
11.3), but that you should always be on the look out for correlated errors in
time series data because of their effect on standard confidence intervals and
hypothesis tests. The same ideas apply to spatial data. Observations close
together in space will probably look more similar to one another than observations collected farther away from one another, so when you fit any kind of
trend surface to these data, the errors from this model may be correlated.
This section discusses tools you can use to model spatial correlation.

Covariogram, Correlogram, and Variogram

In Chapter 11 we defined the lag-k covariance and correlation in Equation (11.2). These measure the covariance and correlation between two observations that are k units apart in time for a stationary time series (i.e., the
time series has a constant mean and the covariance and correlation depend
only on the lag k, not where you start in the time series).

Extending this idea to spatial data, we might assume that the observations
represent a snapshot from a random process over space. Weak stationarity
implies that the surface has a constant mean (so for the general model in
Equation (12.3) assume for now that $f(W)$ is a constant) and that the covari-

ance between two observations depends only on the distance (and perhaps direction) between the locations of these observations (Kaluzny et al., 1998). The covariance is *isotropic* if it depends only on the distance between the locations and not on the direction:

$$Cov\left[Z\left(\mathbf{s}_1\right), Z\left(\mathbf{s}_2\right)\right] = C\left(h\right) \qquad (12.6)$$

where h denotes the distance (e.g., Euclidean distance) between location $\mathbf{s}_1$ and location $\mathbf{s}_2$ and C denotes some monotonic function bounded below by 0 (it must also satisfy certain other properties as well). The function C (or its estimate) is called the *covariogram* and is analogous to the autocovariance function for time series data. The function:

$$\rho\left(h\right) = \frac{C\left(h\right)}{C\left(0\right)} \qquad (12.7)$$

is called the *correlogram* and is analogous to the autocorrelation function (also called the correlogram) for time series data.

An estimator of the lag-k autocovariance for time series data is shown in Equation (11.6). For spatial data, the estimate of covariance between two locations that are h units apart is given by:

$$\hat{C}\left(h\right) = \frac{1}{\left|N\left(h\right)\right|} \sum_{N(h)} \left[Z\left(\mathbf{s}_i\right) - \bar{Z}\right]\left[Z\left(\mathbf{s}_j\right) - \bar{Z}\right] \qquad (12.8)$$

where $N(h)$ denotes the set of all pairs of locations that are h units apart, $|N(h)|$ denotes the number of pairs in this set, and $\bar{Z}$ denotes the average of all of the observations at all locations (Cressie, 1993, p. 70). When the data are unevenly spaced in two dimensions, this estimator is usually modified so that $N(h)$ denotes the set of all pairs of locations that are at least $h-\tau$ units apart and no more than $h+\tau$ units apart, where τ ($\tau > 0$) is some tolerance parameter. An estimator of correlation between two locations that are h units apart is given by:

$$\hat{\rho}\left(h\right) = \frac{\hat{C}\left(h\right)}{\hat{C}\left(0\right)} \qquad (12.9)$$

Rather than looking at covariograms and correlograms, people who deal with spatial data often use the *variogram*, which is defined as:

$$Var\left[Z\left(\boldsymbol{s}_1\right) - Z\left(\boldsymbol{s}_2\right)\right] =$$

$$Var\left[Z\left(\boldsymbol{s}_1\right)\right] + Var\left[Z\left(\boldsymbol{s}_2\right)\right] -$$

$$\text{(12.10)}$$

$$2\,Cov\left[Z\left(\boldsymbol{s}_1\right),\,Z\left(\boldsymbol{s}_2\right)\right] =$$

$$2\left[C\left(0\right) - C\left(h\right)\right] = 2\,\gamma\left(h\right)$$

The function

$$\gamma\left(h\right) = C\left(0\right) - C\left(h\right) \qquad \text{(12.11)}$$

is called the *semivariogram*, but some texts (including the S+SPATIALSTATS User's Manual) also refer to this as the variogram, which can cause a bit of confusion. The empirical semivariogram is computed as:

$$\hat{\gamma}\left(h\right) = \frac{1}{2\left|N\left(h\right)\right|} \sum_{N(h)} \left[Z\left(\boldsymbol{s}_i\right) - Z\left(\boldsymbol{s}_j\right)\right]^2 \qquad \text{(12.12)}$$

Figure 12.9 displays three examples of a theoretical covariogram, and Figure 12.10 displays the corresponding semivariograms. These plots illustrate three properties of covariograms and semivariograms:

- **Sill**. The sill is equal to the variance of the process, i.e., the covariance at distance $h = 0$.
- **Range**. The range is the distance at which observations are no longer correlated. The range may be finite or infinite.
- **Nugget Effect**. The nugget effect represents micro-scale variation and/or measurement error.

In these figures, the range is finite for examples 1 and 3 and infinite for example 2. There is a nugget effect in example 3, but none in examples 1 and 2.

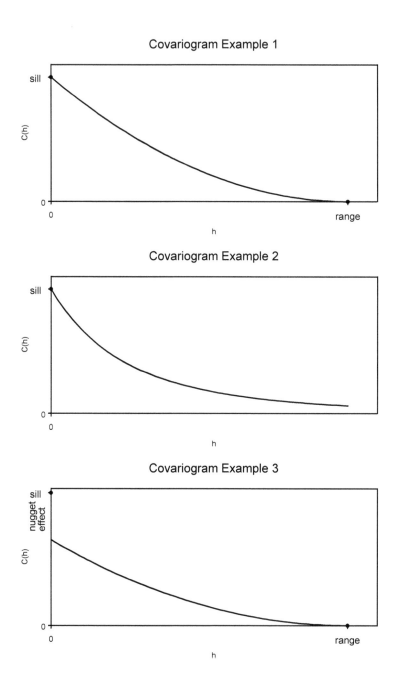

Figure 12.9   Three examples of theoretical covariograms

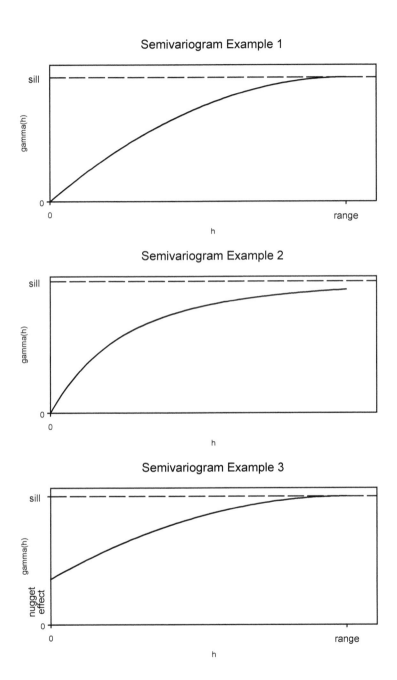

Figure 12.10   Three examples of theoretical semivariograms

Example 12.3: Plotting the Empirical Covariogram and Semivariogram for the Benthic Data

Figures 12.11 and 12.12 show the omnidirectional empirical covariogram and semivariogram for the benthic data. Looking at Figure 12.12, we might guess a nugget effect of about 0.75, the sill to be at about 2.25 and the range to be about 0.7. Both figures indicate the presence of spatial correlation, which is not surprising because these plots are based on the assumption of a constant average value over the area where the benthos was sampled, and we know from the loess fit shown in Figures 12.5 and 12.6 that this is not the case. Figures 12.13 and 12.14 show the omnidirectional empirical semivariograms for the residuals from the quartic and loess fits to the benthic data, respectively. Spatial correlation appears to be present in the residuals from the quartic fit, but once we remove the trend using the loess model, spatial correlation is no longer present.

Menu

To create the omnidirectional empirical covariogram for the benthic index using the S+SPATIALSTATS pull-down menu, follow these steps.

1. Find **benthic.df** in the Object Explorer.
2. On the S-PLUS menu bar, make the following menu choices: **Spatial>Empirical Variogram**. This will bring up the Empirical Variogram dialog box.
3. The Data Set box should display **benthic.df**. For Variable select **Index**, for Location 1 select **Longitude**, for Location 2 select **Latitude**, for Type select **covariogram**, then click **OK** or **Apply**.

To create the empirical semivariogram, follow the same steps as above, but in Step 3 for Type select **variogram** and for Save As type **vg.benthic**.

Command

To create the omnidirectional empirical covariogram and semivariogram for the benthic index using the S-PLUS Command or Script Window, type these commands.

```
cg.benthic <- covariogram(Index ~
   loc(Longitude, Latitude), data=benthic.df)
plot(cg.benthic, main="Empirical Covariogram for
   Benthic Index")
vg.benthic <- variogram(Index ~
   loc(Longitude, Latitude), data=benthic.df)
plot(vg.benthic, main="Empirical Semivariogram for
   Benthic Index")
```

Empirical Covariogram for Benthic Index

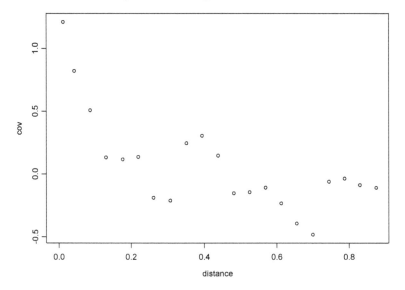

Figure 12.11 Empirical covariogram for the benthic index

Empirical Semivariogram for Benthic Index

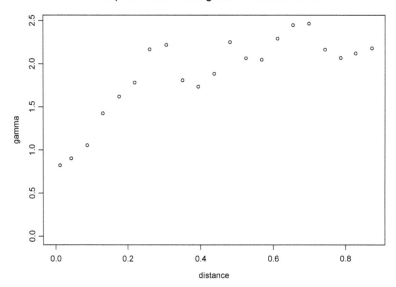

Figure 12.12 Empirical semivariogram for the benthic index

Empirical Semivariogram for Quartic Residuals

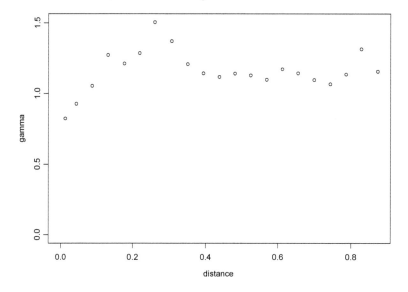

Figure 12.13   Empirical semivariogram for the benthic quartic fit residuals

Empirical Semivariogram for Loess Residuals

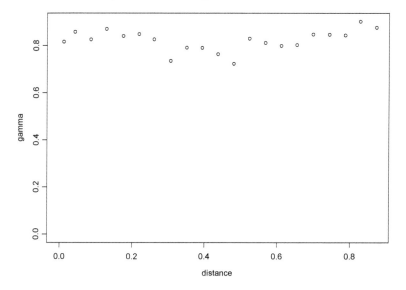

Figure 12.14   Empirical semivariogram for the benthic loess fit residuals

To create the omnidirectional empirical semivariogram for the residuals from the quartic and loess fits, type these commands.

```
vg.q.resid <- variogram(poly4.fit$residuals ~
   loc(Longitude, Latitude), data=benthic.df)
plot(vg.q.resid, main="Empirical Semivariogram for
   Quartic Residuals")
vg.l.resid <- variogram(loess.fit$residuals ~
   loc(Longitude, Latitude), data=benthic.df)
plot(vg.l.resid, main="Empirical Semivariogram for
   Loess Residuals")
```

Example 12.4: Using Directional Semivariograms to Explore the Assumptions of Stationarity and Isotropy

The empirical covariograms and semivariograms that we created in Example 12.3 assume the spatial process is stationary and that the spatial correlation is isotropic.   We can use directional semivariograms and semivariogram clouds to explore whether these assumptions appear to be valid for our data. Figure 12.15 displays the empirical semivariogram for the benthic index for six difference directions. Figures 12.16 and 12.17 show the same thing for the residuals from the quartic and loess fits.

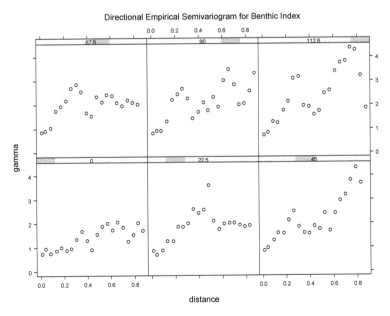

Figure 12.15 Directional empirical semivariograms for the benthic index

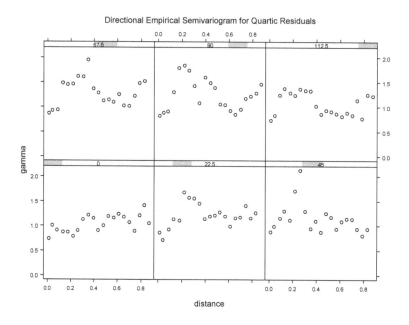

Figure 12.16   Directional empirical semivariograms for the quartic fit residuals

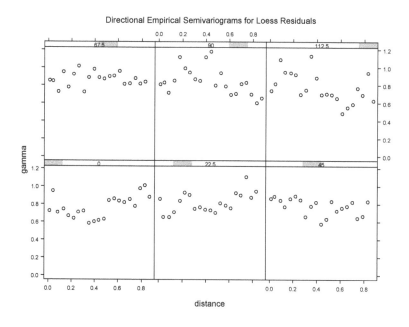

Figure 12.17   Directional empirical semivariograms for the loess fit residuals

Not surprisingly, the form of the semivariogram appears to be different for different directions based on the raw benthic index data (Figure 12.15). Differences in the semivariogram in different directions are caused by the presence of trend (which we already know is present) and/or anisotropy (different forms of spatial correlation in different directions). Figure 12.16 indicates that there may be some anisotropy in the residuals from the quartic fit.

Menu

To create the directional empirical semivariograms for the benthic index using the S+SPATIALSTATS pull-down menu, follow these steps.

1. Find **benthic.df** in the Object Explorer.
2. On the S-PLUS menu bar, make the following menu choices: **Spatial>Empirical Variogram**. This will bring up the Empirical Variogram dialog box.
3. The Data Set box should display **benthic.df**. For Variable select **Index**, for Location 1 select **Longitude**, for Location 2 select **Latitude**, for Azimuth type **c(0, 22.5, 45, 67.5, 90, 112.5)**, for Azimuth Tolerance type **11.25**, for Type select **variogram**, then click **OK** or **Apply**.

Command

To create the directional empirical semivariogram for the benthic index using the S-PLUS Command or Script Window, type these commands.

```
az <- c(0, 22.5, 45, 67.5, 90, 112.5)
d.vg.benthic <- variogram(Index ~
    loc(Longitude, Latitude), data=benthic.df,
    azimuth=az, tol.azimuth=11.25)
plot(d.vg.benthic, main="Directional Empirical
    Semivariograms for Benthic Index")
```

To create the directional empirical semivariograms for the residuals from the quartic and loess fits, type these commands.

```
d.vg.q.resid <- variogram(poly4.fit$residuals ~
    loc(Longitude, Latitude), data=benthic.df,
    azimuth=az, tol.azimuth=11.25)
plot(d.vg.q.resid, main="Directional Empirical
    Semivariograms for Quartic Residuals")
d.vg.l.resid <- variogram(loess.fit$residuals ~
    loc(Longitude, Latitude), data=benthic.df,
    azimuth=az, tol.azimuth=11.25)
```

```
plot(d.vg.l.resid, main="Directional Empirical
    Semivariograms for Loess Residuals")
```

Example 12.5: Semivariogram Clouds for the Benthic Data

A semivariogram cloud displays the distribution of the variance between points that are h units apart vs. the distance h. By default, S+SPATIALSTATS plots $[Z(s_1) - Z(s_2)]^2/2$ vs. h, where h is the distance between s_1 and s_2. For dense variogram clouds, you can use boxplots to display the distributions for intervals of distance. Semivariogram clouds are useful for identifying outliers, trend, and non-constant variance. If the spatial process has a constant mean and the errors have a constant variance, then the semivariogram cloud should look consistent across all distances. Figure 12.18 displays the semivariogram cloud for the benthic index. Figure 12.19 and Figure 12.20 show the same thing for the residuals from the quartic and loess fits, respectively.

We do not expect to see any gross outliers in these data because the benthic index is constrained to lie between 1 and 5. Figure 12.18 shows that points closer to one another tend to look more like one another than points further away, which we already know because we know there is spatial trend in these data. Like Figure 12.13, Figure 12.19 indicates that the quartic fit may still not have removed all of the local-scale trend in the data.

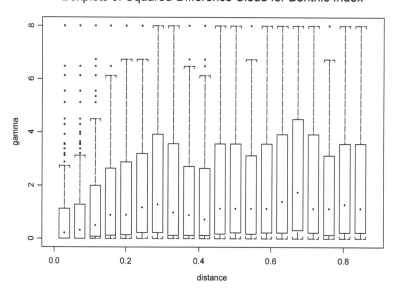

Boxplots of Squared-Difference Cloud for Benthic Index

Figure 12.18 Semivariogram cloud for the benthic index

Boxplots of Squared-Difference Cloud for Quartic Residuals

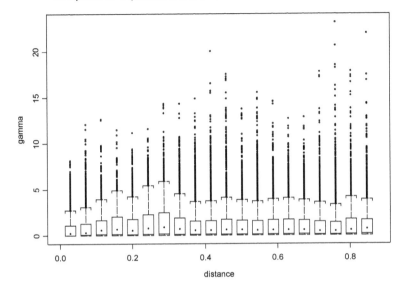

Figure 12.19   Semivariogram cloud for the quartic fit residuals

Boxplots of Squared-Difference Cloud for Loess Residuals

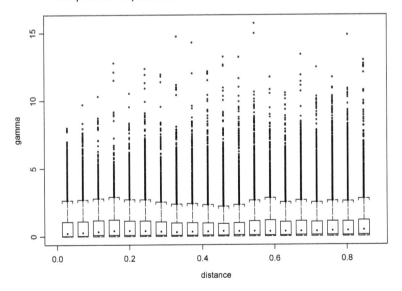

Figure 12.20   Semivariogram cloud for the loess fit residuals

Menu

To create the semivariogram cloud for the benthic index using the S+SPATIALSTATS pull-down menu, follow these steps.

1. Find **benthic.df** in the Object Explorer.
2. On the S-PLUS menu bar, make the following menu choices: **Spatial>Variogram Cloud**. This will bring up the Variogram Cloud dialog box.
3. The Data Set box should display **benthic.df**. For Variable select **Index**, for Location 1 select **Longitude**, for Location 2 select **Latitude**, for Plots check **Box Plot**, then click **OK** or **Apply**.

Command

To create the semivariogram cloud for the benthic index using the S-PLUS Command or Script Window, type these commands.

```
vg.cld.benthic <- variogram.cloud(Index ~
   loc(Longitude, Latitude), data=benthic.df)

boxplot(vg.cld.benthic, main="Boxplots of Squared-
   Difference Cloud for Benthic Index")
```

To create the semivariogram clouds for the residuals from the quartic and loess fits, type these commands.

```
vg.cld.q.resid <- variogram.cloud(
   poly4.fit$residuals ~ loc(Longitude, Latitude),
   data=benthic.df)

boxplot(vg.cld.q.resid, main="Boxplots of Squared-
   Difference Cloud for Quartic Residuals")

vg.cld.l.resid <- variogram.cloud(
   loess.fit$residuals ~ loc(Longitude, Latitude),
   data=benthic.df)

boxplot(vg.cld.l.resid, main="Boxplots of Squared-
   Difference Cloud for Loess Residuals")
```

Models for the Covariogram and Semivariogram

Several models have been postulated for isotropic spatial correlation. These models can be written either in terms of the covariogram or the semivariogram. Three common models are the exponential, Gaussian, and spherical (Kaluzny et al., 1998, p. 91; Venables and Ripley, 1999, p. 441). Figure 12.21 displays the covariograms and semivariograms for each of these three models.

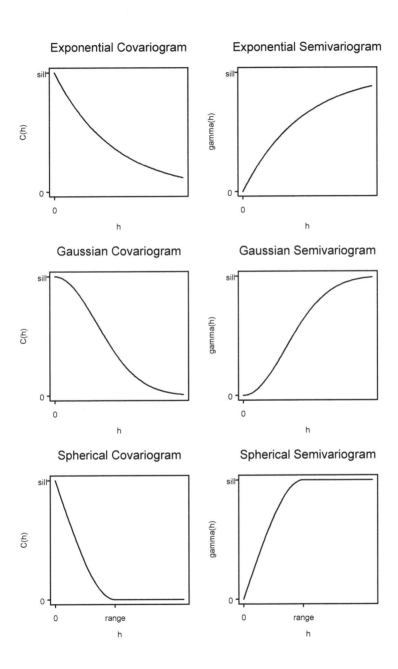

Figure 12.21   Covariograms and semivariograms for the exponential, Gaussian, and spherical models

The exponential covariogram is given by:

$$C(h) = \sigma^2 \exp\left(\frac{-h}{r}\right) \tag{12.13}$$

where σ^2 denotes the variance of the process (i.e., the sill; the covariance at distance $h = 0$), and r denotes a constant. The Gaussian covariogram is given by:

$$C(h) = \sigma^2 \exp\left[-\left(\frac{h}{r}\right)^2\right] \tag{12.14}$$

and the spherical covariogram is given by:

$$C(h) = \begin{cases} \sigma^2\left(1 - \dfrac{3h}{2r} + \dfrac{h^3}{2r^3}\right) & , \ h \le r \\ \\ 0 & , \ h > r \end{cases} \tag{12.15}$$

Note that the range for the exponential and Gaussian models is infinite, while for the spherical model it is equal to the constant r. A nugget effect can be added to each of these models. In this case, the covariogram is written as:

$$c^*(h) = \begin{cases} C(0) + g \, , \ h < \eta \\ \\ C(h) \quad\quad , \ h \ge \eta \end{cases} \tag{12.16}$$

where g denotes the nugget effect, and η denotes the maximum distance up to which the nugget effect occurs.

Example 12.6: Fitting Variograms to the Benthic Data

In S+SPATIALSTATS there are two ways to fit theoretical variogram models: by eye or by nonlinear least squares (Cressie, 1993, p. 97). Figure 12.22 shows the fitted variogram for the benthic index using the Gaussian model for covariance. Figure 12.23 shows the fitted variogram for the residuals from the quartic model using the spherical model for covariance. Note that the fit shown in Figure 12.22 is not really valid because we showed in Figure 12.15 that the spatial correlation for the raw benthic index data does not appear to be isotropic.

Menu

To fit the Gaussian variogram to the benthic index using the S+SPATIALSTATS pull-down menu, follow these steps.

1. Find **vg.benthic** in the Object Explorer. (You created this in Example 12.3).
2. On the S-PLUS menu bar, make the following menu choices: **Spatial>Model Variogram**. This will bring up the Model Variogram dialog box.
3. The Variogram Object box should display **vg.benthic**. For Function select **Gaussian**, **check** the Fit Parameters box, then click **OK** or **Apply**.

Command

To fit the Gaussian variogram to the benthic index data using the S+SPATIALSTATS Command or Script Window, type these commands.

```
vg.benthic <- variogram(Index ~
   loc(Longitude, Latitude), data=benthic.df)
vg.benthic.fit <- variogram.fit(vg.benthic,
   fun=gauss.vgram)
plot(vg.benthic)
plot(vg.benthic.fit, add=T)
title(main="Empirical and Fitted Gaussian
   Semivariogram\nfor Benthic Index")
```

To fit the spherical variogram to the residuals from the quartic fit, type these commands.

```
vg.q.resid <- variogram(poly4.fit$residuals ~
   loc(Longitude, Latitude), data=benthic.df)
```

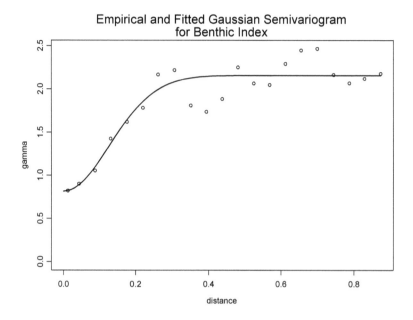

Figure 12.22   Fitted Gaussian variogram for the benthic index

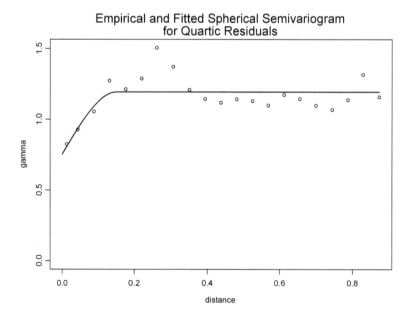

Figure 12.23   Fitted spherical variogram for the residuals from the quartic fit

```
vg.q.resid.fit <- variogram.fit(vg.q.resid,
    param=c(range=0.4, sill=1.5,
    nugget=0.75),fun=spher.vgram)
plot(vg.q.resid)
plot(vg.q.resid.fit, add=T)
title(main="Empirical and Fitted Spherical
    Semivariogram\nfor Quartic Residuals")
```

PREDICTION FOR GEOSTATISTICAL DATA

Two common methods for predicting values at particular locations that take spatial correlation into account are generalized least squares and kriging (Venables and Ripley, 1999, chapter 14). We will discuss each of these.

Generalized Least Squares

If the model for our geostatistical data is parametric (e.g., Equation (12.4) or (12.5)), we can use standard least squares to estimate the coefficients if we do not think the errors are correlated (see Equation (11.26)). Once we have the estimated coefficients, we can predict values at any location by simply plugging in the location coordinates into the fitted model. (Of course the caution we made in Chapter 9 against extrapolating outside the range of the predictor variable(s) applies to models for geostatistical data as well.) Figures 12.7 and 12.8 were constructed by using standard least squares to fit a quartic model to the benthic index data, predicting values on a grid, and then displaying only those values within the convex hull defined by the observed longitudes and latitudes.

If we think the errors are correlated, we need to use generalized least squares to estimate the coefficients (Equation (11.27)). As we discussed in Chapter 11, generalized least squares requires you to specify or estimate the covariance between each observation. You can use the function surf.gls in the MASS library to fit a surface using generalized least squares.

Kriging

For a parametric model that is fit using least squares or generalized least squares, the predicted value at any location is the fitted value $\hat{f}(w)$; the predicted value of the error term ε in Equation (12.3) is 0. An alternative method of predicting values, called *kriging*, was developed in the early 1960s by G. Matheron, who named this method after D.G. Krige, a South African mining engineer (Cressie, 1993, p. 106; Venables and Ripley, 1999, p. 439). *Universal kriging* is a method of predicting values at any location where the ε term is not assumed to be 0, but is instead predicted as well

(Venables and Riply, 1999, p. 493). ***Ordinary kriging***, usually just called kriging, assumes a constant trend surface, so it is usually performed on residuals from a trend surface model.

For any location s_0, the (ordinary) kriging prediction of the response variable is given by:

$$\hat{Z}\left(s_0\right) = \sum_{i=1}^{n} \lambda_i \, Z\left(s_i\right) \qquad (12.17)$$

where

$$\sum_{i=1}^{n} \lambda_i = 1 \qquad (12.18)$$

That is, the predicted value is simply a weighed average of all of the observed values. The weights $\lambda_1, \lambda_2, ..., \lambda_n$ depend on the spatial correlation and can be written in terms of either the variogram or covariogram function (Cressie, 1993, pp. 122–123).

In terms of the covariogram, the weights are given by:

$$\lambda' = \left(c + 1\, m\right)' \Sigma^{-1} \qquad (12.19)$$

where c is an $n \times 1$ vector with i^{th} value

$$c_i = C\left(s_0 - s_i\right) \qquad (12.20)$$

1 denotes an $n \times 1$ vector of 1's, m is a scalar defined by

$$m = \frac{1 - 1'\Sigma^{-1}c}{1'\Sigma^{-1}1} \qquad (12.21)$$

and Σ denotes the $n \times n$ variance-covariance matrix with ij^{th} element

$$\Sigma_{ij} = C\left(s_i - s_j\right) \tag{12.22}$$

and $s_i - s_j$ denotes the distance between locations s_i and s_j.
    In terms of the variogram, the weights are given by:

$$\lambda' = \left(\gamma - 1\,m^*\right)' \Gamma^{-1} \tag{12.23}$$

where γ is an $n \times 1$ vector with i^{th} value

$$\gamma_i = \gamma\left(s_0 - s_i\right) \tag{12.24}$$

m^* is a scalar defined by

$$m^* = -\frac{1 - 1'\Gamma^{-1}\gamma}{1'\Gamma^{-1}1} \tag{12.25}$$

and Γ denotes the $n \times n$ matrix with ij^{th} element

$$\Gamma_{ij} = \gamma\left(s_i - s_j\right) \tag{12.26}$$

The variance of the predicted response, termed the **kriging (or prediction) variance**, is given by

$$Var\left[\hat{Z}\left(s_0\right)\right] = \sum_{i=1}^{n} \lambda_i \gamma_i + m^* \tag{12.27}$$

$$= C\left(0\right) - \sum_{i=1}^{n} \lambda_i c_i + m$$

Fitting a model that incorporates spatial correlation using either generalized least squares or kriging can be an iterative process. You can fit a surface without assuming spatial correlation is present, then look for spatial correlation in the residuals. You can then fit a new surface that incorporates spatial correlation, and based on the residuals from this new fit you may or may not want to change the specification of the surface and the spatial correlation structure.

Example 12.7: Fitting a Quartic Surface to the Benthic Data Using Kriging

Figure 12.24 displays the contour plot for the benthic index based on using a quartic model and universal kriging, assuming the spherical covariance function we fit to the residuals in Example 12.6. Figure 12.25 displays the companion surface plot. Comparing these figures to Figures 12.5 to 12.8, we see that the kriging model displays a lot more local variation than the loess or quartic model. Figure 12.26 displays the contour plot of the standard errors of the predictions, and Figure 12.27 displays the corresponding surface plot.

Menu

In Version 1.5 of S+SpatialStats, you can only fit up to a 2nd order polynomial (i.e., quadratic) surface with universal kriging. Use the menu choices **Spatial>Universal Kriging**.

Command

To fit the universal kriging model with the quartic trend surface using the S+SPATIALSTATS Command or Script Window, type these commands.

```
krige.poly4.fit <- krige(Index ~
   loc(Longitude, Latitude) + Longitude + Latitude +
   Longitude^2 + Longitude*Latitude + Latitude^2 +
   Longitude^3 + Longitude^2*Latitude +
   Longitude*Latitude^2 + Latitude^3 + Longitude^4 +
   Longitude^3*Latitude + Longitude^2*Latitude^2 +
   Longitude*Latitude^3 + Latitude^4, data=benthic.df,
   covfun=spher.cov,
   range=vg.q.resid.fit$parameters["range"],
   sill=vg.q.resid.fit$parameters["sill"],
   nugget=vg.q.resid.fit$parameters["nugget"])
```

To compute the predicted values on a 50×50 grid, type these commands.

```
lat <- benthic.df$Latitude
lon <- benthic.df$Longitude
Latitude <- seq(min(lat), max(lat), length=50)
```

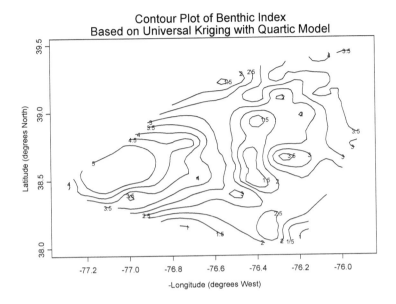

Figure 12.24   Contour plot of benthic index based on universal kriging using the quartic polynomial model for surface trend and a spherical covariance function for spatial correlation

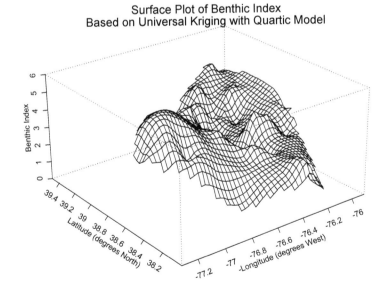

Figure 12.25   Corresponding surface plot

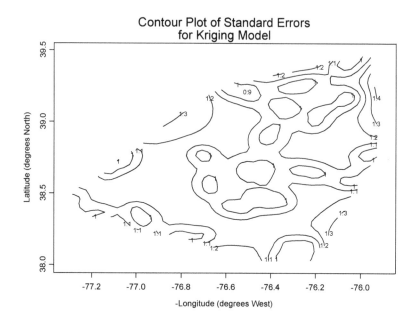

Figure 12.26   Contour plot of standard errors of predicted values

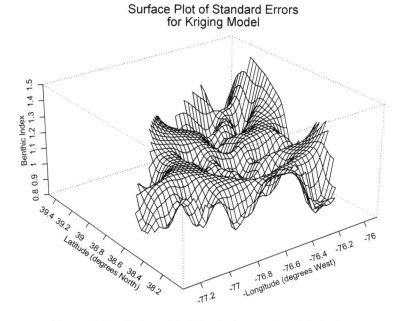

Figure 12.27   Surface plot of standard errors of predicted values

```
Longitude <- seq(min(lon), max(lon), length=50)

predict.list <- list(Longitude=Longitude,
   Latitude=Latitude)

predict.grid <- expand.grid(predict.list)

predict.fit <- predict(krige.poly4.fit,
   newdata=predict.grid, se.fit=T)
```

To create the contour and surface plots, displaying only those values within the convex hull defined by the observed longitudes and latitudes, type these commands.

```
index.chull <- chull(lon, lat)

index.poly <- list(x=lon[index.chull],
   y=lat[index.chull])

inside <- points.in.poly(predict.grid$Longitude,
   predict.grid$Latitude, index.poly)

predict.fit[!inside, c("fit", "se.fit")] <- NA

contour(Longitude, Latitude,
   matrix(predict.fit$fit, 50, 50),
   levels=seq(1, 5, by=0.5), labex=0.75,
   xlab="-Longitude (degrees West)",
   ylab="Latitude (degrees North)")

title(main=paste("Contour Plot of Benthic Index",
   "Based on Universal Kriging", sep="\n"))

persp(Longitude, Latitude,
   matrix(predict.fit$fit, 50, 50),
   xlab="-Longitude (degrees West)",
   ylab="Latitude (degrees North)",
   zlab="Benthic Index")

title(main=paste("Surface Plot of Benthic Index",
   "Based on Kriging", sep="\n"))
```

To create the contour and surface plots of the standard errors of the predicted values, call contour and persp as above, but use predit.fit$se.fit instead of predict.fit$fit and change the names of the titles. Also, in the call to contour, leave out the argument levels=seq(1, 5, by=0.5).

USING S-PLUS FOR ARCVIEW GIS

ArcView® GIS software from Environmental Systems Research Institute (ESRI) Inc. is a geographic information system (GIS) that navigates complex

GIS databases and links them with user-supplied data. It provides a platform to visualize, explore, query, analyze, and map the data, and provides tools to facilitate the preparation of presentations and reports. S-PLUS for ArcView GIS from MathSoft Inc. is a link between ArcView GIS and S-PLUS that lets you share data and functionality between these two software tools. You can access S-PLUS directly from ArcView to analyze and graph data stored in ArcView. You can also export data from ArcView into S-PLUS, perform analyses, and import the results back into ArcView.

The *S-PLUS for ArcView GIS User's Guide* discusses several examples of exporting data from ArcView to S-PLUS and performing data analysis. In this section we will give a brief explanation of ArcView GIS and the nomenclature associated with GIS objects, then we will show you how to import the benthic data from S-PLUS into ArcView and present the spatial locations in the context of a map.

ArcView GIS Objects

In ArcView geographic data (maps) are organized in *tables*. A map showing the locations of roads, buildings, parks, rivers, etc., can be coded as a table with each row defining a specific geographic *feature* and the columns defining the *attributes* of that feature. Some features such as buildings may be adequately represented by points, some such as roads and rivers may require lines, and others such as parks and lakes may require polygons.

Clearly, whether or not a park is represented by a polygon or a point will depend on the level of detail (scale) of the map. Similarly, the attributes listed for a feature will depend on the feature. The attributes of a park might include the location of the park and the size of the park. The attributes of a city might include the location, the size, and the population.

ArcView manages geographic information in sets of related features and their attributes called *themes*. A theme is simply a collection of all features of a particular type, for example all roads on that map or all building on that map. The union of all of the themes makes up the entire table (GIS data base). Themes do not have to be mutually exclusive sets of features. For example, the set of all schools (one theme) is a subset of the set of all buildings (another theme).

In ArcView, you can store all related maps, charts, and tables in a *project*. When an existing project or a new (blank) project is opened in ArcView, icons for Views, Tables, Charts, Layouts, and Scripts are displayed (Figure 12.28). These are the basic components of a project:

- **Views** display themes as interactive maps. Opening a view simultaneously displays the Table of Contents which shows what is available.

- **Tables** display the tabular data of maps: features are listed in the rows and attributes for a feature a listed in the columns.
- **Charts** are graphical representations of tabular data. You can get charts from either views or tables by more clicks.
- **Layouts** are presentation-ready displays of views, tables, charts, and images. They may be sent to the printer or integrated into presentation software.
- **Scripts** are programs (macros) written in ArcView's programming language called *Avenue*.

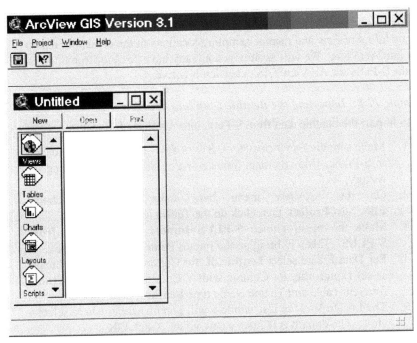

Figure 12.28   The top level ArcView GIS window

Linking S-PLUS and ArcView GIS

ArcView GIS augments its desktop capabilities by allowing other standalone software tools to be available within its environment. These are referred to as ***extensions***. S-PLUS for ArcView GIS is an ArcView extension. If you have successfully installed ArcView GIS, S-PLUS, and S-PLUS for ArcView GIS, then you can access S-PLUS from within ArcView.

When you first start ArcView, if S-PLUS for ArcView has not yet been activated, then the ArcView Window will have the title "ArcView GIS Version X.X," where X.X denotes the version of ArcView you are running. On the ArcView menu choose **File>Extensions**. This will bring up the Extensions dialog box. Scroll down until you find S-PLUS for ArcView GIS and **check** the box, then click **OK**. Once S-PLUS for ArcView has been activated, the ArcView Window title changes to S-PLUS for ArcView GIS, and an additional menu with the title S-PLUS is added. If you have S+SPATIALSTATS, you will also see a menu with the title Spatial Statistics. Now you are ready to use S-PLUS for ArcView GIS. Note: If you have more than one version of S-PLUS on your computer, you should start the version of S-PLUS you want to use with ArcView *before* you start ArcView.

Importing Data from S-PLUS into ArcView

In this section we will show you how to import the benthic data from S-PLUS into ArcView and display sampling locations in the context of a map built into ArcView. We will assume you already have ArcView running and that the S-PLUS for ArcView GIS extension is active.

Example 12.8: Importing the Benthic Data into ArcView

To import the benthic data from S-PLUS into ArcView, follow these steps.

1. Make sure the ENVIRONMENTALSTATS for S-PLUS module is loaded in S-PLUS. Copy the data frame `benthic.df` to the working directory.
2. On the ArcView menu bar, make the menu choice **File>New Project**, then click on the Tables icon.
3. Make the menu choice **S-PLUS>Import Point Theme from S-PLUS**. This will bring up the Import Point Theme dialog box.
4. For Data Frame select **benthic.df**, for Column with X Coordinates select **Longitude**, for Column with Y Coordinates select **Latitude**, for New Table and Theme Name type **Benthos**, then click **OK**.
5. The Locate View to Use window will pop up asking you to select a view. The default is [Create a new view]. Click **OK**.
6. There is now a table called Benthos in your still-untitled project. Click on the Views icon and you will see a view called View1. Make the menu choice **Project>Rename 'View1'** and rename the view **Sampling Locations**.
7. Make the menu choice **File>Save Project As**. This brings up the Save Project As dialog box. For File Name type **example.apr**, then click **OK**.

Click on the **Open** button and the Sampling Locations window will pop up. If you click on the Benthos box the sampling locations will be displayed, as shown in Figure 12.29. You can change the plotting characters by making the menu selection **Theme>Edit Legend**. In this case we changed the point size of the symbol to 4.

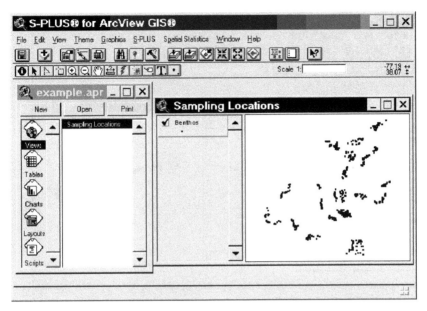

Figure 12.29   The view displaying the sampling locations of the benthic data

Example 12.9:   Plotting the Benthic Sampling Locations on a Map

To plot the benthic sampling locations on a map, follow these steps.

1.   Make the menu choice **View>Add Theme**. This will bring up the Add Theme dialog box.

2.   Browse to the directory **ESRI HOME\esri\esridata\usa**, where ESRI HOME denotes the directory where ESRI software is stored. Select the file **States.shp** (a shape file containing all of the state boundaries), and click **OK**. **Check** the States.shp box. The Sampling Locations view now has two themes: States.shp and Benthos.

If you want to store only the four states surrounding the Chesapeake Bay, follow these steps.

3.   **Uncheck** the Benthos theme box. Click on the **Zoom to Full Extent** button, so that the view looks like what is shown in Figure 12.30.

4.   Click on the **Zoom In** button, then click on the eastern part of the map to enlarge it. Click on the **Select Feature** button and use Shift-click to simultaneously highlight the states Delaware, Maryland, West Virginia, and Virginia.

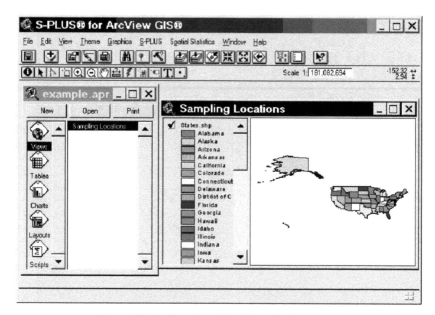

Figure 12.30   The view of the shape file States.shp

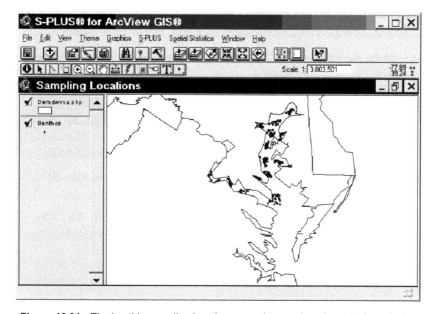

Figure 12.31   The benthic sampling locations superimposed on the state boundaries

5. Make the menu choice **Theme>Convert to Shapefile**. Call this file **DEMDWVVA.shp**. In response to "Add shapefile as theme to the view?" click on **Yes**. Make the menu choice **Edit>Delete Themes** and click on **Yes** to delete States.shp from View 1.
6. The Sampling Locations view now has two themes: Demdwvva.shp and Benthos. **Check** the Demdwvva.shp box to display the four states, and click on the name **Demdwvva.shp** to activate the option buttons. Click on the **Edit Legend** button, and color the map white to show only the boundaries of the states surrounding the Chesapeake Bay.
7. **Check** the Benthos box to display the sampling stations. The results should look like Figure 12.31.

Once you have the data from S-PLUS imported into ArcView, you can use the Charts icon to add graphics to the views.

SUMMARY

- Spatial data can be classified as *geostatistical* data, *spatial point pattern* data, or *lattice* data.
- A general form for a model for geostatistical data is the sum of large-scale variation plus small-scale variation (Equation (12.3)). The small-scale variation reflects measurement error and local fluctuations.
- Parametric models for geostatistical data use polynomials for the large-scale variation term and assume the small-scale variation follows a normal distribution. Usually the polynomial involves the variables that determine location in space (e.g., latitude and longitude).
- Nonparametric models for geostatistical data include loess and local fitting based on Delaunay triangulation.
- The *covariogram*, *correlogram*, and *semivariogram* (often called the *variogram*) are three tools you can use to investigate the presence of and model spatial correlation. There is a one-to-one relationship between the theoretical covariogram and semivariogram.
- The covariance is *isotropic* if it depends only on the distance between the locations and not on the direction.
- Three properties of covariograms and semivariograms are the *sill*, the *range*, and the *nugget effect*.
- Three common models for the covariogram and semivariogram are the exponential, Gaussian, and spherical.
- In S+SPATIALSTATS there are two ways to fit theoretical variogram models: by eye or by nonlinear least squares.

- Two common methods for predicting values at particular locations that take spatial correlation into account are generalized least squares and *kriging*.
- *Ordinary kriging* assumes the large scale variation term is constant. For ordinary kriging, the predicted value is simply a weighed average of all of the observed values, where the weights depend on the assumed or estimated covariance function.
- *Universal kriging* fits both the large-scale variation term and the small-scale variation term simultaneously.
- S-Plus for ArcView GIS is a link between ArcView GIS and S-Plus that lets you share data and functionality between these two software tools.

EXERCISES

12.1. Use S+Spatial Stats to plot exponential, Gaussian, and spherical variogram models. Vary the values of the sill, range, and nugget effect. Are there cases where two or all three of these models can look very similar to each other?

12.2. Use S+Spatial Stats to generate spatially correlated data, then plot the empirical variogram and compare it with the true underlying variogram.

12.3. Perform a Monte Carlo simulation where for each Monte Carlo trial you generate spatially correlated data on a grid, then estimate the mean and construct a 95% confidence interval for the mean based on assuming no spatial correlation is present. What is the observed coverage of the 95% confidence interval? Perform the experiment for various degrees of spatial correlation.

12.4. Plot the empirical and directional empirical variograms for the salinity and silt variables that are part of the benthic data. Compare these plots to the empirical variogram of the benthic index.

12.5. Look at contour and surface plots of the silt variable. Try fitting parametric and nonparametric models for large-scale variation.

12.6. Perform ordinary kriging for the silt variable, then perform ordinary kriging on the residuals from a model you fit in the previous exercise, or perform universal kriging.

12.7. Create a pairwise scatterplot matrix to look at the relationship between benthic index, salinity, and silt. Do you think either salinity or silt may help predict benthic index?

13 MONTE CARLO SIMULATION AND RISK ASSESSMENT

Attempting to Quantify Uncertainty in a Mathematical Model

In the first 12 chapters of this book we have discussed several statistical tools for looking at data, modeling probability distributions, estimating distribution parameters and quantiles, constructing prediction and tolerance intervals, comparing two or more groups, and modeling trend and spatial patterns. All of our examples have concentrated on assessing how much chemical is in the environment and comparing chemical concentrations to "background." But given that chemicals are in the environment, what happens when people or other living organisms are exposed to these chemicals?

Of course, "chemicals" in the environment are part of our everyday lives: they are in the food we eat, the water we drink, the air we breathe. Some are natural and others are synthetic, having been added either on purpose or as a by-product of a manufacturing process. There is no doubt that the chemical revolution of the past century has improved our lives immensely. But we have also learned that some chemicals that have improved some facet of our lives can also have devastating consequences upon our health and environment.

Several government agencies are charged with evaluating the potential health and ecological effects of environmental toxicants. Based on their assessments, these agencies set standards for acceptable concentration levels of these toxicants in air, water, soil, food, etc. The process of modeling exposure to a toxicant and predicting health or ecological effects is termed *risk assessment*.

Risk assessment is a process where science, politics, and psychology all intersect. Not surprisingly, it is also a field full of controversy. References discussing risk assessment and the concept of risk include Cothern (1996), Everitt (1999), Hallenbeck (1993), Laudan (1997), Lewis (1990), Lundgren and McMakin (1998), Molak (1997), Neely (1994), Rodricks (1992), Shaffner (1999), Suter (1993), and Walsh (1996). In this chapter, we will introduce basic mathematical models used in risk assessment and talk about Monte Carlo simulation and probabilistic risk assessment.

OVERVIEW

Human and ecological risk assessment involves characterizing the exposure to a toxicant for one or several populations, quantifying the relationship between exposure (dose) and health or ecological effects (response), determining the risk (probability) of a health or ecological effect given the observed level(s) of exposure, and characterizing the uncertainty associated with the estimated risk (Hallenbeck, 1993, p. 1). Risk assessment is an enormous field of research and practice. The merits and disadvantages of particular exposure models (including fate and transport models) and dose-response models are not discussed in this chapter. Several journals, textbooks, and web sites are dedicated to risk assessment. Three standard texts include Hallenbeck (1993), Suter (1993), and Rodricks (1992). Other references are listed in the ENVIRONMENTALSTATS for S-PLUS help file References: Risk Assessment.

In the past, estimates of risk were often based solely on setting values of the input variables (e.g., body weight, dose, etc.) to particular point estimates and producing a single point estimate of risk, with little, if any, quantification of the uncertainty associated with the estimated risk. More recently, several practitioners have advocated "probabilistic" risk assessment, in which the input variables are considered random variables, so the result of the risk assessment is a probability distribution for predicted risk or exposure.

Usually, the equation describing risk or exposure is so complicated that it is not feasible to determine the output distribution using analytical methods, so the distribution of risk or exposure is derived via Monte Carlo simulation. This chapter discusses the concepts of Monte Carlo simulation, sensitivity and uncertainty analysis, and risk assessment, and shows you how to use S-PLUS and ENVIRONMENTALSTATS for S-PLUS to perform probabilistic risk assessment.

Note: The ENVIRONMENTALSTATS for S-PLUS menus that deal with risk assessment were under construction at the time this book was being written. Therefore, many of the examples in this chapter show only how to use S-PLUS and ENVIRONMENTALSTATS for S-PLUS from the Command or Script Window. See the ENVIRONMENTALSTATS for S-PLUS help file or User's Manual for information on how to use the pull-down menus to duplicate the examples in this chapter.

MONTE CARLO SIMULATION

Monte Carlo simulation is a method of investigating the distribution of a random variable by simulating random numbers (Gentle, 1985). Usually, the random variable of interest, say Y, is some function of one or more other random variables:

$$Y = h\left(\underline{X}\right) = h\left(X_1, X_2, \ldots, X_k\right) \tag{13.1}$$

For example, Y may be an estimate of the median of a population with a Cauchy distribution, in which case the vector of random variables $\underline{X}$ represents k independent and identically distributed observations from some particular Cauchy distribution. As another example, Y may be the incremental lifetime cancer risk due to ingestion of soil contaminated with benzene (Thompson et al., 1992; Hamed and Bedient, 1997). In this case the random vector $\underline{X}$ may represent observations from several kinds of distributions that characterize exposure and dose-response, such as benzene concentration in the soil, soil ingestion rate, average body weight, the cancer potency factor for benzene, etc. These distributions may or may not be assumed to be independent of one another (Smith et al., 1992; Bukowski et al., 1995).

Sometimes the input variables X_1, X_2, ..., X_k are called ***input parameters***. This terminology can be confusing, however, since the input variables are often random variables and therefore have ***distribution parameters*** associated with their probability distributions (e.g., mean and standard deviation for a normally distributed input variable).

Sometimes the distribution of Y in Equation (13.1) can be derived analytically based on statistical theory (Springer, 1979; Slob, 1994). Often, however, the function h is complicated and/or the elements of the random vector $\underline{X}$ involve several kinds of probability distributions, making it difficult or impossible to derive the exact distribution of Y. In this case, Monte Carlo simulation can be used to approximate the distribution of Y. Monte Carlo simulation is often used in risk assessment, specifically in sensitivity and uncertainty analysis.

Monte Carlo simulation involves creating a large number of realizations of the random vector $\underline{X}$, say n, and computing Y for each of the n realizations of $\underline{X}$. The resulting distribution of Y, or some characteristic of this distribution (e.g., the mean), is then assumed to be "close" to the true distribution or distribution characteristic of Y. The adequacy of the approximation depends on a number of factors, including how well the mathematical relationship described in Equation (13.1) reflects the true relationship between Y and $\underline{X}$, how well the specified distribution of $\underline{X}$ reflects its true distribution (including any possible dependencies between the individual elements of $\underline{X}$), and how many Monte Carlo samples or trials (n) are created.

Usually, Monte Carlo simulation involves generating random numbers from some specified theoretical probability distribution, such as a normal, lognormal, beta, etc. When the simulation is done based on an empirical distribution, this is also called bootstrapping (Efron and Tibshirani, 1993), as we discussed in Chapter 5.

Vose (1996, p. 40) claims that the term "Monte Carlo" comes from the code name of an American project on the atom bomb during World War II, and not from the town of the same name in Monaco that is well known for its casinos. This, however, begs the question of the origin of the code name. Gentle (1985, p. 612) notes that Monte Carlo simulation was used extensively during World War II and immediately after during work on the development of the atomic bomb. He states that the name "Monte Carlo" dates from that time period and comes from the casino in Monaco with the same name. Hayes (1993, p. 114) states that the modern form of Monte Carlo simulation was invented at Los Alamos shortly after World War II by Stanislaw Ulam while he was working on a problem of the diffusion of neutrons, and that the term Monte Carlo method is "named in honor of the well-known generator of random integers (between 0 and 36) in the Mediterranean principality." References that address the issues of how to properly perform and report the results of a Monte Carlo simulation study include Hoaglin and Andrews (1975), Law and Kelton (1991), Burmaster and Anderson (1994), and Vose (1996).

Example 13.1: Simulating the Distribution of the Sum of Two Normal Random Variables

Suppose X_1 and X_2 are two independent standard normal random variables. Then the distribution of

$$Y = h\left(X_1, X_2\right) = X_1 + X_2 \qquad (13.2)$$

is normal with a mean of 0 and a variance of 2 (see Equations (4.21) to (4.22)). Suppose, however, that we do not know how to derive the distribution of Y. We can use Monte Carlo simulation to investigate the shape of the distribution of Y, as well as compute characteristics of the distribution (e.g., mean, median, standard deviation, quantiles, etc.)

Figure 13.1 displays the empirical and true distribution of Y, where the empirical distribution of Y was derived by using Monte Carlo simulation with 1,000 trials. That is, for each trial, two random numbers from a standard normal distribution were generated and added together. Figure 13.2 displays the empirical distribution based on 10,000 trials along with the true distribution. Table 13.1 displays some summary statistics for the two empirical distributions and compares them with the true population values. As we increase the number of Monte Carlo trials, the simulated distribution tends to get "closer" to the true distribution. This is called the ***Law of Large Numbers***, as discussed in Chapter 4.

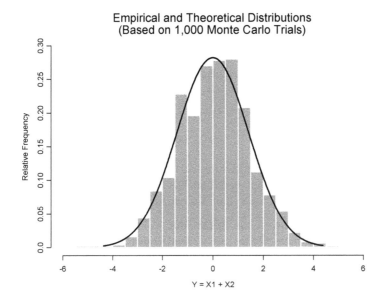

Figure 13.1   Empirical and theoretical distribution of the sum of two independent N(0, 1) random variables based on 1,000 Monte Carlo trials

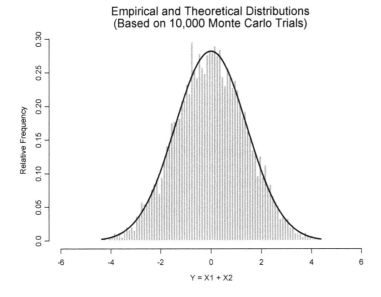

Figure 13.2   Empirical and theoretical distribution of the sum of two independent N(0, 1) random variables based on 10,000 Monte Carlo trials

Parameter	Empirical (1,000)	Empirical (10,000)	Population N(0, 2)
Mean	0.05	0.00	0
Standard Deviation	1.42	1.41	1.41
5th Percentile	-2.37	-2.33	2.33
95th Percentile	2.42	2.31	-2.33

Table 13.1   Comparison of empirical and population summary statistics

Menu

To generate the empirical distribution of the sum of two independent standard normal random variables based on 1,000 Monte Carlo trials using the ENVIRONMENTALSTATS for S-PLUS pull-down menu follow these steps.

1. On the S-PLUS menu bar, make the following menu choices: **EnvironmentalStats>Probability Distributions and Random Numbers>Random Numbers>Multivariate>Normal**. This will bring up the Multivariate Normal Random Numbers dialog box.
2. For Sample Size type **1000**, for Set Seed with type **20**, for Number of Vars type **2**, **uncheck** the Print Results box, and for Save As type **rmvnorm.mat**. Click **OK**.
3. Find **rmvnorm.mat** in the Object Explorer.
4. On the S-PLUS menu bar, make the following menu choices: **EnvironmentalStats>Monte Carlo Simulation**. This will bring up the Monte Carlo Simulation dialog box.
5. In the Data Set box select **rmvnorm.mat**, in the Expression box type **Var.1+Var.2**, and click **OK** or **Apply**.

To produce the empirical distribution based on 10,000 Monte Carlo trials, in Step 2 above for Sample Size type **10000**.

Command

To generate the empirical distribution of the sum of two independent standard normal random variables based on 1,000 Monte Carlo trials using the S-PLUS Command or Script Window, type these commands.

```
set.seed(20)
rmvnorm.mat <-rmvnorm(n=1000)
y <- apply(rmvnorm.mat, 1, sum)
```

or you can use the ENVIRONMENTALSTATS for S-PLUS functions simu-late.mv.matrix and eval.expr:

```
rmvnorm.mat <-simulate.mv.matrix(n=1000, seed=20)
y <- eval.expr(Var.1+Var.2, data=rmvnorm.mat)
```

To produce the empirical distribution based on 10,000 Monte Carlo trials, set n=10000 in the call to rmvnorm or simulate.mv.matrix.

GENERATING RANDOM NUMBERS

A random number is a realization of a random variable, say X. For many people, the term random number initially conjures up an image of somehow choosing an integer between a specified lower and upper bound (e.g., 1 and 10), where each number is equally likely to be chosen. In that case, the random variable X is a discrete uniform random variable with probability density (mass) function given by:

$$f(x) = \Pr(X = x) = \frac{1}{10} \; ; \; x = 1, \; 2, \dots, 10 \quad \textbf{(13.3)}$$

In general, a random number can be a realization of a random variable from any kind of probability distribution (e.g., uniform, normal, lognormal, gamma, empirical, etc.)

Generating Random Numbers from a Uniform Distribution

The S-PLUS function runif generates *pseudo-random numbers* from a (continuous) uniform distribution. Uniform random number generation in S-PLUS is adapted from Marsaglia et al. (1973), which couples a multiplicative-congruential generator with a feedback shift register. References that discuss generating pseudo-random numbers include Kennedy and Gentle (1980), Rubenstein (1981), Ripley (1987), Law and Kelton (1991), Hayes (1993), and Barry (1996). Venables and Ripley (1999, pp. 116–117) explain the S-PLUS pseudo-random number generator in detail.

Pseudo-random number generators start with an initial *seed*, and then appear to generate random numbers, although these numbers are actually generated by a deterministic mechanism. Each time a set of random numbers is generated, the value of the seed changes. If you start with the same seed, you will get the same sequence of pseudo-random numbers. You can use the S-PLUS function set.seed to set the seed of the random number generator. Also, S-PLUS and ENVIRONMENTALSTATS for S-PLUS dialog boxes for generating random numbers allow you to set the seed.

The *period* of a random number generator is the number of random numbers that can be generated before the sequence repeats itself. For most starting seeds, the period of the random number generator in S-PLUS is $2^{30} \times 4,292,868,097 \approx 4.6 \times 10^{18}$ (about 4.6 quintillion) (Venables and Ripley, 1999, p. 117).

Generating Random Numbers from an Arbitrary Distribution

The S-PLUS and ENVIRONMENTALSTATS for S-PLUS functions of the form `rabb` (where `abb` denotes the abbreviation of the distribution) generate random numbers from several theoretical probability distributions. For example, the function `rnorm` generates random numbers from a normal distribution. You can generate random numbers from the S-PLUS menu by selecting **Data>Random Numbers**, and you can generate random numbers from the ENVIRONMENTALSTATS for S-PLUS menu by selecting **Environmental-Stats>Probability Distributions and Random Numbers>Random Numbers**.

To generate random numbers from a specified probability distribution, most computer software programs use the inverse transformation method (Rubenstein, 1981, pp. 38–43; Law and Kelton, 1991, pp. 465–474; Vose, 1996, pp. 39–40). Suppose the random variable U has a U[0,1] distribution, that is, a uniform distribution over the interval [0,1]. Let F_X denote the cumulative distribution function (cdf) of the specified probability distribution. Then the random variable X defined by:

$$X = F_X^{-1}(U) \tag{13.4}$$

has the specified distribution, where the quantity F_X^{-1} denotes the inverse of the cdf function F_X. Thus, to generate a set of random numbers from any distribution, all you need is a set of random numbers from a U[0,1] distribution and a function that computes the inverse of the cdf function for the specified distribution. Figure 13.3 illustrates the inverse transformation method for a standard normal distribution, with $U = 0.8$. In this case, the random number generated is $\Phi^{-1}(0.8) = 0.8416212$.

Latin Hypercube Sampling

Latin Hypercube sampling, sometimes abbreviated LHS, is a method of sampling from a probability distribution (one random variable) or a joint probability distribution (several random variables) that ensures all portions of the probability distribution are represented in the sample. It was introduced in the published literature by McKay et al. (1979). Other references include Iman and Conover (1980, 1982), and Vose (1996, pp. 42–45). Latin Hypercube sampling is an extension of quota sampling, and when applied to the joint distribution of k random variables, can be viewed as a k-dimensional extension of Latin square sampling, thus the name (McKay et al., 1979, p. 240).

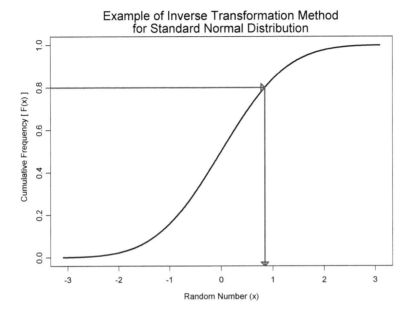

Example of Inverse Transformation Method
for Standard Normal Distribution

Random Number (x)

Figure 13.3   Example of the inverse transformation method of generating random
numbers from a specified distribution

Latin Hypercube sampling was introduced to overcome the following problem in Monte Carlo simulation based on simple random sampling (SRS). Suppose we want to generate random numbers from a specified distribution. If we use simple random sampling, there is a low probability of getting very many observations in an area of low probability of the distribution. For example, if we generate n observations from the distribution, the probability that none of these observations falls into the upper 98th percentile of the distribution is 0.98^n. So, for example, there is a 13% chance that out of 100 random numbers, none will fall at or above the 98th percentile. If we are interested in reproducing the shape of the distribution, we will need a very large number of observations to ensure that we can adequately characterize the tails of the distribution (Vose, 1996, p. 40).

Latin Hypercube sampling was developed in the context of using computer models that required enormous amounts of time to run and for which only a limited number of Monte Carlo simulations could be implemented. In cases where it is fairly easy to generate tens of thousands of Monte Carlo trials, Latin Hypercube sampling may not offer any real advantage.

Algorithm for Latin Hypercube Sampling

Latin Hypercube sampling works as follows for a single probability distribution. If we want to generate n random numbers from the distribution, the distribution is divided into n intervals of equal probability $1/n$. A random number is then generated from each of these intervals. For k independent probability distributions, LHS is applied to each distribution, and the resulting random numbers are matched at random to produce n random vectors of dimension k.

The three figures below illustrate Latin Hypercube sampling for a sample size of $n = 4$, assuming a standard normal distribution. Figure 13.4 shows the four equal-probable intervals for a standard normal distribution in terms of the probability density function, and Figure 13.5 shows the same thing in terms of the cumulative distribution function. Figure 13.6 shows how Latin Hypercube sampling is accomplished using the inverse transformation method for generating random numbers. In this case, the interval [0,1] is divided into the four intervals [0, 0.25], [0.25, 0.5], [0.5, 0.75], and [0.75, 1]. Next, a uniform random number is generated within each of these intervals. For this example, the four numbers generated are (to two decimal places) 0.04, 0.35, 0.70, and 0.79. Finally, the standard normal random numbers associated with the inverse cumulative distribution function of the four uniform random numbers are computed.

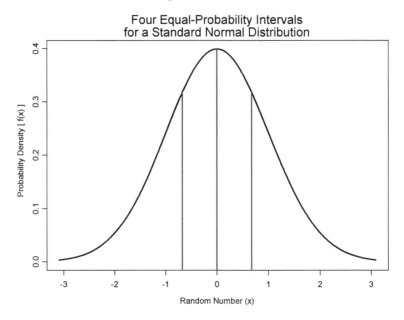

Figure 13.4   Four equal-probability intervals for a N(0, 1) distribution

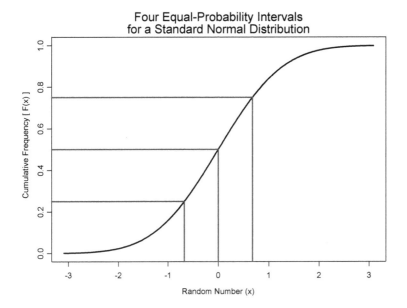

Figure 13.5   Four equal-probability intervals for a N(0, 1) distribution

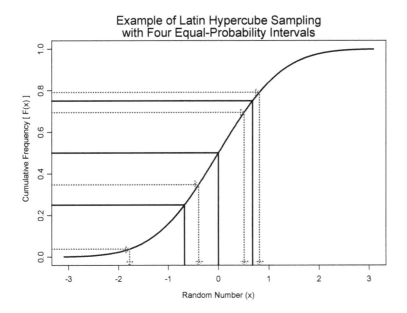

Figure 13.6   Visual explanation of generating four random numbers from a N(0, 1) distribution using Latin Hypercube sampling

Example 13.2: Simple Random Sampling vs. Latin Hypercube Sampling

Figure 13.7 displays a histogram of 50 observations based on a simple random sample from a standard normal distribution. Figure 13.8 displays the same thing based on a Latin Hypercube sample. You can see that the form of the histogram constructed with the Latin Hypercube sample more closely resembles the true underlying distribution.

Menu

To generate the simple random sample used to create the histogram in Figure 13.7 using the ENVIRONMENTALSTATS for S-PLUS pull-down menu, follow these steps.

1. On the S-PLUS menu bar, make the following menu choices: **EnvironmentalStats>Probability Distributions and Random Numbers>Random Numbers>Univariate**. This will bring up the Univariate Random Number Generation dialog box.
2. For Sample Size type **50**, for Set Seed with type **209**, for Sample Method select **Random**, **uncheck** the Print Results box, and for Save As type **x.srs**.

To generate the Latin Hypercube sample used to create the histogram in Figure 13.8 follow the same steps as above, but in Step 2 for Sample Method select **Latin Hypercube** and for Save As type **x.lhs**.

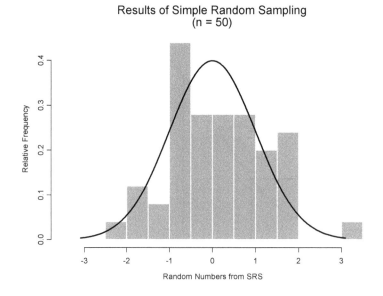

Figure 13.7   Results of simple random sampling from a N(0, 1) distribution

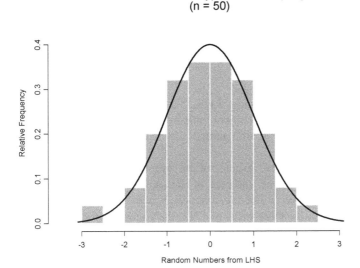

Figure 13.8   Results of Latin Hypercube sampling from a N(0, 1) distribution

Command

To generate the simple random sample and Latin Hypercube sample used to create the histograms in Figure 13.7 and Figure 13.8 using the ENVIRONMENTALSTATS for S-PLUS Command or Script Window, type these commands.

```
x.srs <- simulate.vector(50, seed=209)
x.lhs <- simulate.vector(50, sample.method="LHS",
   seed=209)
```

Properties of Latin Hypercube Sampling

Let Y denote the outcome variable for one trial of a Monte Carlo simulation, and suppose Y is a function of k independent random variables as shown in Equation (13.1) above. McKay et al. (1979) consider the class of estimators of the form

$$T = T\left(\underline{Y}\right) = T\left(Y_1, Y_2, \ldots, Y_n\right) = \frac{1}{n}\sum_{i=1}^{n} g\left(Y_i\right) \quad \textbf{(13.5)}$$

where g is an arbitrary function. This class of estimators includes the mean, the rth sample moment, and the empirical cumulative distribution function. Setting

$$\tau = E\left[T\left(\underline{Y}\right)\right] \tag{13.6}$$

McKay et al. (1979) show that under LHS, T is an unbiased estimator of τ, and also, if h in Equation (13.1) is monotonic in each of its arguments and g is monotonic, then the variance of T under LHS is less than or equal to the variance of T under SRS.

Stein (1987) shows that the variance of the sample mean of Y under LHS is asymptotically less than the variance of the sample mean under simple random sampling whether or not the function h is monotonic in its arguments. Unfortunately, for most cases of LHS, the formula for the true variance of the sample mean is difficult to derive, and thus a good estimate of true variance is not available. Using the usual formula of dividing the sample variance by the sample size will usually overestimate the true variance of the sample mean.

Iman and Conover (1980) and Stein (1987) suggest producing several independent Latin Hypercube samples, say N, and for each Latin Hypercube sample computing the sample mean based on the n observations within that sample. (They call this method *replicated Latin Hypercube sampling.*) You can then estimate the variance of the sample mean by computing the usual sample variance of these N sample means. Note that this method can be applied to any quantity of interest, such as the median, 95th percentile, etc.

UNCERTAINTY AND SENSITIVITY ANALYSIS

Uncertainty analysis and sensitivity analysis are terms used to describe various methods of characterizing the behavior of a complex mathematical/computer model. The model in Equation (13.1) above is different from most conventional statistical models, where the form of the model is:

$$Y = h\left(\underline{X}\right) + \varepsilon \tag{13.7}$$

(e.g., linear regression models, generalized linear models, nonlinear regression models, etc.). In Equation (13.7), the vector $\underline{X}$ is assumed to be set or observed at fixed values, and for fixed values of $\underline{X}$ the response variable Y deviates about its mean value according to the distribution of the error term ε. This kind of model is useful when we are interested in the specific relationship between Y and $\underline{X}$, and we want to predict the value of Y for a specified value of $\underline{X}$. Furthermore, this kind of model is fit using paired observations of Y and $\underline{X}$.

In Equation (13.1), Y is assumed to be observed without error, that is, the value of Y is deterministic for a set value of $\underline{X}$. The output variable Y, however, is a random variable because the input variables X_1, X_2, ..., X_k are assumed to be a combination of random variables and constant terms. This kind of model is useful when we are interested in describing the distribution of Y taken over all possible (read as "reasonable and realistic") combinations of the input variables. Furthermore, paired observations of Y and $\underline{X}$ are usually not available to validate this kind of model, hence, there is some amount of unquantifiable uncertainty associated with this kind of model.

Uncertainty analysis involves describing the variability or distribution of values of the output variable Y that is due to the collective variation in the input variables $\underline{X}$ (Iman and Helton, 1988, p. 72). This description usually involves graphical displays such as histograms, empirical density plots, and empirical cdf plots, as well as summary statistics such as the mean, median, standard deviation, coefficient of variation, 95th percentile, etc.

Sensitivity analysis involves determining how the distribution of Y changes with changes in the individual input variables. It is used to identify which input variables contribute the most to the variation or uncertainty in the output variable Y (Iman and Helton, 1988, p. 72). Sensitivity analysis is also used in a broader sense to determine how changing the distributions of the input variables and/or their assumed correlations or even changing the form of the model affects the output (Thompson et al., 1992; Smith et al., 1992; Cullen, 1994; Shlyakhter, 1994; Bukowski et al., 1995; Hamed and Bedient, 1997; USEPA, 1997).

Important vs. Sensitive Parameters

Two useful concepts associated with sensitivity analysis are important parameters (variables) and sensitive parameters (variables) (Crick et al., 1987 as cited in Hamby, 1994, p. 137). *Sensitive parameters* have a substantial influence on the resulting distribution of the output variable Y, that is, small changes in the value of a sensitive parameter result in substantial changes in Y. *Important parameters* have some amount of uncertainty and/or variability associated with them and this variability contributes substantially to the resulting variability in the output variable Y.

Figure 13.9 below illustrate these two concepts for the simple case of three input variables. The top plot is an example of an important variable. An important variable is always sensitive. The middle plot is an example of a variable that is not sensitive, and hence not important. The bottom plot is an example of a variable that is not important. This variable may not be sensitive (like the one in the middle plot), or it may be sensitive like the one in the top plot but it is not important because of its limited variability.

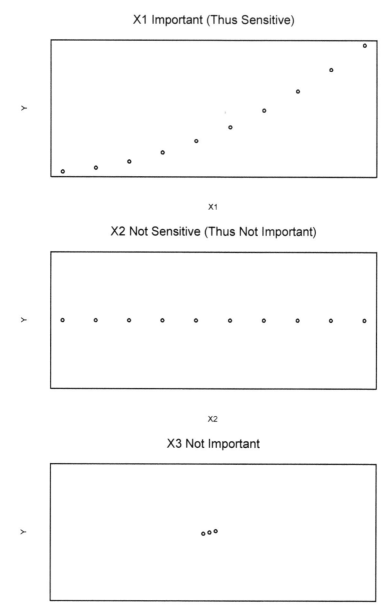

Figure 13.9   Three examples of the concept of important and sensitive parameters

Uncertainty vs. Variability

The terms uncertainty and variability have specific meanings in the risk assessment literature that do not necessarily match their meanings in the statistical literature or everyday language. The term *variability* refers to the inherent heterogeneity of a particular variable (parameter). For example, there is natural variation in body weight and height between individuals in a given population. The term *uncertainty* refers to a lack of knowledge about specific parameters, models, or factors (Morgan and Henrion, 1990; Hattis and Burmaster, 1994; Rowe, 1994; Bogen, 1995; USEPA, 1997, p. 10). For example, we may be uncertain about the true distribution of exposure to a toxic chemical in a population (parameter uncertainty due to lack of data, measurement errors, sampling errors, systematic errors, etc.), or we may be uncertain how well our model of incremental lifetime cancer risk reflects reality (model uncertainty due to simplification of the process, misspecification of the model structure, model misuse, use of inappropriate surrogate variables, etc.), or we may be uncertain about whether a chemical is even present at a site of concern (scenario uncertainty due to descriptive errors, aggregation errors, errors in professional judgment, incomplete analysis, etc.).

We can usually reduce uncertainty through further measurement or study. We cannot reduce variability, since it is inherent in the variable. Note that in the risk assessment literature, measurement error contributes to uncertainty; we can decrease uncertainty by decreasing measurement error. In the statistical literature, measurement error is one component of the variance of a random variable. Note that a parameter (input variable) may have little or no variability associated with it, yet still have uncertainty associated with it (e.g., the speed of light is constant, but we only know its value to a given number of decimal places).

This meaning of the term "uncertainty" should not be confused with the term "uncertainty analysis" described above. Uncertainty analysis characterizes the distribution of the output variable Y. The output variable Y varies due to the fact that the input variables are random variables. The distributions of the input random variables reflect both variability (inherent heterogeneity) and uncertainty (lack of knowledge).

Sensitivity Analysis Methods

Sensitivity analysis methods can be classified into three groups: one-at-a-time deviations from a baseline case, factorial design and response surface modeling, and Monte Carlo simulation (Hamby, 1994). Each of these kinds of sensitivity analysis is briefly discussed below. For more detailed information, see the section Sensitivity Analysis Methods in the help file Glossary: Uncertainty and Sensitivity Analysis in ENVIRONMENTALSTATS for S-PLUS. Several studies indicate that using Monte Carlo simulation in conjunction with certain sensitivity measures usually provides the best method of determining sensitivity of the parameters.

One-at-a-Time Deviations from a Baseline Case

These sensitivity analysis methods include differential analysis and measures of change in output to change in input. Differential analysis is simply approximating the variance of the output variable Y at a particular value of the input vector $\underline{X}$ (called the **baseline case**) by using a first-order Taylor series expansion (Kotz and Johnson, 1985, Volume 8, pp. 646–647; Downing et al., 1985; Seiler, 1987; Iman and Helton, 1988; Hamby, 1994, pp. 138–139). This approximating equation for the variance of Y is useful for quantifying the proportion of variability in Y that is accounted for by each input variable. Unfortunately, the approximation is usually good only in a small region close to the baseline case, and the relative contribution of each input variable to the variance of Y may differ dramatically for differently chosen baseline cases. Also, differential analysis requires the calculation of partial derivatives, which may or may not be a simple task, depending on the complexity of the input function h in Equation (13.1).

Measures of change in output to change in input include the ratio of percent change in Y to percent change in X_i (Hamby, 1994, pp. 139–140), the ratio of percent change in Y to change in X_i in units of the standard deviation of X_i (Hamby, 1994, p. 140; Finley and Paustenbach, 1994), the percent change in Y as X_i ranges from its minimum to maximum value (Hamby, 1994, p. 140), and spider plots, which are plots of Y vs. percent change in X_i, or Y vs. percentiles of X_i (Vose, 1996, pp. 282–283).

Factorial Design and Response Surface Modeling

The concepts of factorial designs and response surfaces come from the field of experimental design (Box et al., 1978). In the context of sensitivity analysis for computer models, n distinct values of the input vector $\underline{X}$ are chosen (usually reflecting the possible ranges and medians of each of the k input variables), and the model output Y is recorded for each of these input values. Then a multiple linear regression model (a **response surface**) is fit to these data. The fitted model is called the fitted response surface, and this response surface is used as a replacement for the computer model (Downing et al., 1985; Iman and Helton, 1988; Hamby, 1994, p. 140). The sensitivity analysis is based on the fitted response surface. The reason for using a response surface to replace the actual model output is that some computer models are, or used to be, very costly to run, whereas computing output based on a response surface is relatively inexpensive.

One way to rank the importance of the variables in the response surface is to simply compare the magnitudes of the estimated coefficients. The estimated coefficients, however, depend on the units of the predictor variables in the model, so most analysts use standardized regression coefficients (Iman and Helton, 1988; Hamby, 1994, p. 146). The standardized regression coefficients are simply the coefficients that are obtained from fitting the response surface model based on the "standardized" output variable and the "standardized" predictor (input) variables. That is, for each variable, each observation is replaced by subtracting the mean (for that variable) from the observation and dividing by the standard deviation (for that variable).

Weisberg (1985, p. 186) warns against using standardized coefficients to determine the importance of predictor variables because they depend on the range of the predictor variables. So, for example, the variable X_3 in Figure 13.9 above may be very important but it has a very limited range. In the context of sensitivity analysis for computer models, however, if the analyst is comfortable with assuming the range of input variables will not change appreciably in the future, then standardized coefficients are an acceptable way to rank the importance of variables.

Iman and Helton (1988) compared uncertainty and sensitivity analysis of several models based on differential analysis, factorial design with a response surface model, and Monte Carlo simulation using Latin Hypercube sampling. The main outcome they looked at for uncertainty analysis was estimating the cumulative distribution function. They found that the models were too mathematically complex to be adequately represented by a response surface. Also, the results of differential analysis gave widely varying results depending on the values chosen for the baseline case. The method based on Monte Carlo simulation gave the best results.

Monte Carlo Simulation

Monte Carlo simulation is used to produce a distribution of Y values based on generating a large number of values of the input vector $\underline{X}$ according to the joint distribution of $\underline{X}$. There are several possible ways to produce a distribution for Y, including varying all of the input parameters (variables) simultaneously, varying one parameter at a time while keeping the others fixed at baseline values, or varying the parameters in one group while keeping the parameters in the other groups at fixed baseline values. Sensitivity methods that can be used with Monte Carlo simulation results include the following:

- **Histograms, Empirical CDF Plots, Percentiles of Output**. A simple graphical way to assess the effect of different input variables or groups of input variables on the distribution of Y is to look at how the histogram and empirical cdf of Y change as you vary one parameter at a time or vary parameters by groups (Thompson et al., 1992). Various quantities such as the mean, median, 95th percentile, etc. can be displayed on these plots as well.

- **Scatterplots**. Another simple graphical way to assess the effect of different input variables on the distribution of Y and their relationship to each other is to look at pair-wise scatterplots.
- **Correlations and Partial Correlations**. A quantitative measure of the relationship between Y and an individual input variable X_i is the correlation between these two variables, computed based on varying all parameters simultaneously (Saltelli and Marivoet, 1990; Hamby, 1994, pp. 143–144). Recall from Chapter 9 that the Pearson product moment correlation measures the strength of *linear* association between Y and X_i, while the Spearman rank and Kendall's tau correlation measures the strength of any *monotonic* relationship between Y and X_i. Vose (1996, pp. 280–282) suggests using **tornado charts**, which are simply horizontal barcharts displaying the values of the correlations. Individual correlations are hard to interpret when some or most of the input variables are highly related to one another. One way to get around this problem is to look at partial correlation coefficients. See the section Sensitivity Analysis Methods in the ENVIRONMENTALSTATS for S-PLUS help file Glossary: Uncertainty and Sensitivity Analysis Methods for more information.
- **Change in Output to Change in Input**. Any of these types of measures that are described above under the subsection One-at-a-Time Deviations from a Baseline Case can be adapted to the results of a Monte Carlo simulation. Additional measures include relative deviation, in which you vary one parameter at a time and compute the coefficient of variation (CV) of Y for each case (Hamby, 1994, p. 142), and relative deviation ratio, in which you vary one parameter at a time and compute the ratio of the CV of Y to the CV of X_i (Hamby, 1994, pp. 142–143).
- **Response Surface**. This methodology that was described above in the subsection Factorial Design and Response Surface Modeling can be adapted to the results of a Monte Carlo simulation. In this case, use the model input and output to fit a regression equation (possibly stepwise) and then use standardized coefficients to rank the input variables (Iman and Helton, 1988; Saltelli and Marivoet, 1990).
- **Comparing Groupings within Input Distributions Based on Partitioning the Output Distribution**. One final method of sensitivity analysis that has been used with Monte Carlo simulation is to divide the distribution of the output variable Y into two or more groups, and then to compare the distributions of an input variable that has been split up based on these groupings

(Saltelli and Marivoet, 1990; Hamby, 1994, pp. 146–150; Vose, 1996, pp. 296–298). For example, you could divide the distribution of an input variable X_i into two groups based on whether the values yielded a value of Y below the median of Y or above the median of Y. You could then use any of the Trellis functions to compare the distributions of these two groups, compare these distributions with a goodness-of-fit test, or compare the means or medians of these distributions with the t-test or Wilcoxon rank sum test. A significant difference between the two distributions is an indication that the input variable is important in determining the distribution of Y.

Uncertainty Analysis Methods

A specific model such as Equation (13.1) with specific joint distributions of the input variables leads to a specific distribution of the output variable. The process of describing the distribution of the output variable is called *uncertainty analysis* (Iman and Helton, 1988, p. 72). This description usually involves graphical displays such as histograms, empirical distribution plots, and empirical cdf plots, as well as summary statistics such as the mean, median, standard deviation, coefficient of variation, 95th percentile, etc.

Sometimes the distribution of Y in Equation (13.1) can be derived analytically based on statistical theory (Springer, 1979; Slob, 1994). For example, if the function h describes a combination of products and ratios, and all of the input variables have a lognormal distribution, then the output variable Y has a lognormal distribution as well, since products and ratios of lognormal random variables have lognormal distributions. Many risk models, however, include several kinds of distributions for the input variables, and some risk models are not easily described in a closed algebraic form. In these cases, the exact distribution of Y can be difficult or almost impossible to derive analytically.

The rest of this section briefly describes some methods of uncertainty analysis based on Monte Carlo simulation. For more information on uncertainty analysis, see the section Uncertainty Analysis Methods in the help file Glossary: Uncertainty and Sensitivity Analysis in ENVIRONMENTALSTATS for S-PLUS.

Quantifying Uncertainty with Monte Carlo Simulation

When the distribution of Y cannot be derived analytically, it can usually be estimated via Monte Carlo simulation. Given this simulated distribution, you can construct histograms or empirical density plots and empirical cdf plots, as well as compute summary statistics. You can also compute confidence bounds for specific quantities, such as percentiles. These confidence bounds are based on the assumption that the observed values of Y are randomly selected based on simple random sampling. When Latin Hypercube sampling is used to generate input variables and hence the output variable Y, the statistical theory for confidence bounds based on simple random sampling is not truly applicable (Easterling, 1986; Iman and Helton, 1991, p. 593; Stein, 1987). Most of the time, confidence bounds that assume simple random sampling but are applied to the results of Latin Hypercube sampling will probably be too wide.

Quantifying Uncertainty by Repeating the Monte Carlo Simulation

One way around the above problem with Latin Hypercube sampling is to use replicated Latin Hypercube sampling, that is, repeat the Monte Carlo simulation numerous times, say N, so that you have a collection of N empirical distributions of Y, where each empirical distribution is based on n observations of the input vector $\underline{X}$. You can then use these N replicate distributions to assess the variability of the sample mean, median, 95th percentile, empirical cdf, etc. Obviously, this is a computer intensive process.

A simpler process is to repeat the simulation just twice and compare the values of certain distribution characteristics, such as the mean, median, 5th, 10th, 90th and 95th percentiles, and also to graphically compare the two empirical cdf plots. If the values between the two simulations are within a small percentage of each other, then you can be fairly confident about characterizing the distribution of the output variable. Iman and Helton (1991) did this for a very complex risk assessment for a nuclear power plant and found a remarkable agreement in the empirical cdf's. Thompson et al. (1992) did the same thing for a risk assessment for incremental lifetime cancer risk due to ingestion or dermal contact with soil contaminated with benzene.

You may also want to compare the results of the original simulation with a simulation that uses say twice as many Monte Carlo trials (e.g., Thompson et al., 1992). Barry (1996) warns that if the moments of the simulated distribution do not appear to stabilize with an increasing number of Monte Carlo trials, this can mean that they do not exist. For most risk models, however, the true distributions of any random variables involved in a denominator in Equation (13.1) are bounded above 0, so the moments will exist. If a random variable involved in the denominator has a mean or median that is close to 0, it is important to use a bounded or truncated distribution to assure the random variable stays sufficiently far away from 0.

Quantifying Uncertainty Based on Mixture Distributions

To account for the uncertainty in specifying the distribution of the input variables, some authors suggest using mixture distributions to describe the distributions of the input variables (e.g., Hoffman and Hammonds, 1994; Burmaster and Wilson, 1996). For each input variable, a distribution is specified for the parameter(s) of the input variable's distribution. Some authors call random variables with this kind of distribution second-order random variables (e.g., Burmaster and Wilson, 1996).

For example we may assume the first input variable comes from a lognormal distribution with a certain mean and coefficient of variation (CV). The mean is unknown to a certain degree, and so we may specify that the mean comes from a uniform distribution with a given set of upper and lower bounds. We can also specify a distribution for the CV. In this case, the Monte Carlo simulation can be broken down into two stages. In the first stage, a set of parameters is generated for each input distribution. In the second stage, n realizations of the input vector $\underline{X}$ are generated based on this one set of distribution parameters. This two-stage process is repeated N times, so that you end up with N different empirical distributions of the output variable Y, and each empirical distribution is based on n observations of the input vector $\underline{X}$.

For a fixed number of Monte Carlo trials nN, the optimal combination of n and N will depend on how the distribution of the output variable Y changes relative to variability in the input distribution parameter(s) vs. variability in the input variables themselves. For example, if the distribution of Y is very sensitive to changes in the input distribution parameters, then N should be large relative to n. On the other hand, if the distribution of Y is relatively insensitive to the values of the parameters of the input distributions, but varies substantially with the values of the input variables, then N may be small relative to n.

Caveat

An important point to remember is that no matter how complex the mathematical model in Equation (13.1) is, or how extensive the uncertainty and sensitivity analyses are, there is always the question of how well the mathematical model reflects reality. The only way to attempt to answer this question is with data collected directly on the input and output variables. Often, however, it is not possible to do this, which is why the model was constructed in the first place.

For example, in order to attempt to directly verify a model for incremental lifetime cancer risk for a particular exposed population within a particular geographical region, you have to collect data on lifetime exposure for each person and the actual proportion of people who developed that particular cancer within their lifetime, accounting for competing risks as well, and compare these data to similar data collected on a proper control population. A controlled experiment that involves exposing a random subset of a particular human population to a toxin and following the exposed and control group throughout their lifetimes for the purpose of a risk assessment is not possible to perform for several reasons, including ethical and practical ones. Rodricks (1992) is an excellent text that discusses the complexities of risk assessment based on animal bioassay and epidemiological studies.

RISK ASSESSMENT

This section discusses the concepts and practices involved in risk assessment, and gives examples of how to use ENVIRONMENTALSTATS for S-PLUS to perform probabilistic risk assessment. This information is also contained in the help file Glossary: Risk Assessment.

Definitions

It will be helpful to start by defining common terms and concepts used in risk assessment.

Risk

The common meaning of the term **risk** when used as a noun is "the chance of injury, damage, or loss." Thus, risk is a probability, since "chance" is another term for "probability."

Risk Assessment

Risk assessment is the practice of gathering and analyzing information in order to predict future risk. Risk assessment has been commonly used in the fields of insurance, engineering, and finance for quite some time. In the last couple of decades it has been increasingly applied to the problems of predicting human health and ecological effects from exposure to toxicants in the environment (e.g., Hallenbeck, 1993; Suter, 1993). In this chapter, the term risk assessment is applied in the context of human health and ecological risk assessment.

The basic model that is often used as the foundation for human health and ecological risk assessment is:

$$Risk = Dose \times \Pr\left(Effect \text{ per } Unit\ Dose\right) \qquad \textbf{(13.8)}$$

That is, the risk of injury (the **effect**) to an individual is equal to the amount of toxicant the individual absorbs (the **dose**) times the probability of the effect occurring for a single unit of the toxicant. If the effect is some form of cancer, the second term on the right-hand side of Equation (13.8) is often called the **cancer slope factor** (abbreviated CSF) or the **cancer potency factor** (abbreviated CPF).

The first term on the right-hand side of Equation (13.8), the dose, is estimated by identifying sources of the toxicant and quantifying their concentrations, identifying how these sources will expose an individual to the toxicant (via fate and transport models), quantifying the amount of exposure an individual will receive, and estimating how much toxicant the individual will absorb at various levels of exposure. Sometimes the dose represents the amount of toxicant absorbed over a lifetime, and sometimes it represents the amount absorbed over a shorter period of time.

The second term on the right-hand side of Equation (13.8), the probability of an effect (CSF or CPF), is estimated from a ***dose-response curve***, a model that relates the probability of the effect to the dose received. Dose-response curves are developed from controlled laboratory experiments on animals or other organisms, and/or from epidemiological studies of human populations (Hallenbeck, 1993; Piegorsch and Bailer, 1997). In Example 9.14 we fit a dose-response curve for the probability that a rat develops thyroid tumors as a function of dose of ethylene thiourea (ETU).

Risk assessment involves three major steps (Hallenbeck, 1993, p. 1; USEPA, 1995c):

- **Hazard Identification**. Describe the effects (if any) of the toxicant on laboratory animals, humans, and/or wildlife species, based on documented studies. Describe the quality and relevance of the data from these studies. Describe what is known about how the toxicant produces these effects. Describe the uncertainties and subjective choices or assumptions associated with determining the degree of hazard of the toxicant.

- **Dose-Response Assessment**. Describe what is known about the biological mechanism that causes the health or ecological effect. Describe what data, models, and extrapolations have been used to develop the dose-response curve for laboratory animals, humans, and/or wildlife species. Describe the routes and levels of exposure used in the studies to determine the dose-response curve, and compare them to the expected routes and levels of exposure in the population(s) of concern. Describe the uncertainties and subjective choices or assumptions associated with characterizing the dose-response relationship.

- **Exposure Assessment**. Identify the sources of environmental exposure to the population(s) of concern. Describe what is known about the principal paths, patterns, and magnitudes of exposure. Determine average and "high end" levels of exposure. Describe the characteristics of the population(s) that is(are) potentially exposed. Determine how many members of the population are likely to be exposed. Describe the uncertainties and subjective choices or assumptions associated with characterizing the exposure for the population of concern.

Once these steps have been completed, some form of Equation (13.8) is usually used to estimate the risk for a particular population of concern. Both sensitivity analysis and uncertainty analysis should be applied to the risk assessment model to quantify the uncertainty associated with the estimated risk.

Risk Characterization

USEPA (1995c) distinguishes between the process of risk assessment and risk characterization. *Risk characterization* is the summarizing step of risk assessment that integrates all of the information from the risk assessment, including uncertainty and sensitivity analyses and a discussion of uncertainty vs. variability, to form an overall conclusion about the risk. Risk assessment is the tool that a risk assessor uses to produce a risk characterization. A risk characterization is the product that is delivered to the risk assessor's client: the risk manager.

Risk Management

Risk management is the process of using information from risk characterizations (calculated risks), perceived risks, regulatory policies and statutes, and economic and social analyses in order to make and justify a decision (USEPA, 1995c). If the risk manager decides that the risk is not acceptable, he or she will order or recommend some sort of action to decrease the risk. If the risk manager decides that the risk poses minimal danger to the population of concern, he or she may recommend that no further action is needed at the present time.

Because the risk manager is the client of the risk assessor, the risk manager must be involved in the risk assessment process from the start, helping to determine the scope and endpoints of the risk assessment (USEPA, 1995c). Also, the risk manager must interact with the risk assessor throughout the risk assessment process, so that he or she may take responsibility for critical decisions. The risk manager, however, must be careful not to let non-scientific (e.g., political) issues influence the risk assessment. Non-scientific issues are dealt with at the risk management stage, not the risk assessment stage.

Risk Communication

USEPA (1995c) defines *risk communication* as exchanging information with the public. While the communication of risk from the risk assessor to the risk manager is accomplished through risk characterization, the communication of risk between the risk manager (or representatives of his or her agency) and the public is accomplished through risk communication. The risk characterization will probably include highly technical information, while risk communication should concentrate on communicating basic ideas of risk to the public.

Building a Risk Assessment Model

Building a risk assessment model involves all three steps outlined in the definition of risk assessment (hazard identification, dose-response assessment, and exposure assessment). Once these steps have been completed, some form of Equation (13.8) is usually used to estimate the risk for a particular population of concern.

The form of the two terms on the right-hand side of Equation (13.8) may be very complex. Estimation of dose involves identifying sources of exposure, postulating pathways of exposure from these sources, estimating exposure concentrations, and estimating the resulting dose for a given exposure. Estimation of dose-response involves using information from controlled laboratory experiments on animals and/or epidemiological studies. Given a set of dose-response data, there are several possible statistical models that can be used to fit these data, including tolerance distribution models, mechanistic models, linear-quadratic-exponential models, and time-to-response models (Hallenbeck, 1993, Chapter 4).

Probably the biggest controversy in risk assessment involves the extrapolation of dose-response data from high-dose to low-dose and from one species to another (e.g., between mice and humans). Rodricks (1992) discusses these problems in detail. A very recent example of this problem is the case of saccharin, which was shown in the late 1970s to produce bladder tumors in male rats that were fed extremely large concentrations of the chemical. These studies led the FDA to call for a ban on saccharin, but Congress placed a moratorium on the ban that was renewed periodically. A little over two decades later, the National Institute of Environmental Health Sciences states that new studies show "no clear association" between saccharin and human cancer (The Seattle Times, Tuesday, May 16, 2000) and has taken saccharin off of its list of cancer-causing chemicals.

Many risk assessment models have the general form

$$Risk = Dose \times \Pr\left(Effect \text{ per } Unit\ Dose\right)$$

$$(13.9)$$

$$= h\left(X_1, X_2, \ldots, X_k\right) \times CPF$$

That is, the risk is assumed to be proportional to dose, and the dose term is a function of several input variables. For example, USEPA (1991b, p. 2) uses the following general equation for intake (here intake is equivalent to exposure and dose):

$$Intake = \frac{C \times IR \times EF \times ED}{BW \times AT} \qquad \textbf{(13.10)}$$

where C is the chemical concentration, IR is the intake or contact rate, EF is the exposure frequency, ED is the exposure duration, BW is body weight, and AT is the averaging time (equal to exposure duration for non-carcinogens and 70 years for carcinogens).

Usually, many of the input variables and sometimes the cancer potency factor (CPF) in Equation (13.9) are themselves assumed to be random variables because they exhibit inherent heterogeneity (variability) within the population (e.g., body weight, fluid intake, etc.), and because there is a certain amount of uncertainty associated with their values. The choice of what distribution to use for each of the input variables is based on a combination of available data and expert judgment.

Example 13.3: Cancer Risk from PCE in Groundwater

McKone and Bogen (1991) present a risk assessment model for lifetime cancer risk from exposure to perchloroethylene (PCE) in groundwater in California. In order to estimate dose, they estimated the concentration of PCE in California water supplies, identified three pathways of exposure (ingestion of tap water, indoor inhalation from tap water, and dermal contact from bathing and showering), estimated concentrations and absorption rates for each pathway, and estimated actual dose based on exposure concentration for each pathway. To estimate the cancer potency factor (CPF), McKone and Bogen (1991) used eight rodent-bioassay data sets and a multistage dose-response extrapolation model. They assumed the input variables for dose and the CPF follow certain probability distributions, and they present two point estimates of risk (using the mean value of each input variable and the 95th percentile of each input variable) as well as a distribution of risk based on a Monte Carlo simulation.

Example 13.4: Incremental Lifetime Cancer Risk from Ingesting Soil Contaminated with Benzene

Thompson et al. (1992) present a simplified case study of a risk assessment for incremental lifetime cancer risk (ILCR) for children (ages 8 to 18) ingesting soil contaminated with benzene. They use the following model:

$$ILCR = Dose \times CPF$$

$$= \frac{Cs \cdot SIngR \cdot RBA \cdot DpW \cdot WpY \cdot YpL \cdot CF}{BW \cdot DinY \cdot YinL} \text{(13.11)}$$

$$\times CPF$$

Symbol	Variable Name	Units	Distribution	Point Estimate
Cs	Benzene Concentration	mg/kg	Lognormal (0.84, 0.77)	3.39
$SIngR$	Soil Ingestion Rate	mg/dy	Lognormal (3.44, 0.80)	50
RBA	Relative Bioavailability			1
DpW	Exposure Days per Week	dy/wk		1
WpY	Exposure Weeks per Year	wk/yr		20
YpL	Exposure Years per Life	yr/life		10
CF	Conversion Factor	kg/mg		1×10^{-6}
BW	Average Body Weight	kg	Normal (47, 8.3)	47
$DinY$	Days in a Year	dy/yr		364
$YinL$	Years in a Lifetime	yr/life		70
CPF	Cancer Potency Factor	(kg dy)/mg	Lognormal (-4.33, 0.67)	0.029

Table 13.2   Description of terms in Equation (13.11) (Thompson et al., 1992)

Table 13.2 explains what the input variables are and their assumed distributions. In this table, the normal distribution is parameterized by its mean and standard deviation, and the lognormal distributions are parameterized by the mean and standard deviation of the log-transformed random variable. The point estimates for the four input variables with associated probability distributions, benzene concentration (Cs), soil ingestion rate ($SIngR$), average body weight (BW), and cancer potency factor (CPF) are, respectively, the 95% upper confidence limit for the mean, the 72nd percentile, the mean, and the 88th percentile. Thompson et al. (1992) used the model in Equation (13.11) (along with a model for ILCR due to dermal contact with the soil) to demonstrate how to use Monte Carlo simulation to perform risk assessment.

Point Estimates of Risk

USEPA (1989c, pp. 6-4 to 6-5) states that

> Actions at Superfund sites should be based on an estimate of *the rea-sonable maximum exposure (RME)* expected to occur under both *current* and *future* land-use conditions. The reasonable maximum exposure is defined here as the highest exposure that is reasonably expected to occur at a site. RMEs are estimated for individual pathways. If a population is exposed via more than one pathway, the combination of exposure pathways also must represent an RME. ... The intent of the RME is to estimate a conservative exposure case (i.e., well above the average case) that is still within the range of possible exposures.

In the context of estimating the exposure term of a risk assessment model, the document further specifies that

> Each intake variable in the equation has a range of values. For Superfund exposure assessments, intake variable values for a given pathway should be selected so that the combination of all intake variables results in an estimate of the reasonable maximum exposure for that pathway. ... Under this approach, some intake variables may not be at their individual maximum values but when in combination with other variables will result in estimates of the RME.

(USEPA, 1989c, p. 6-19). The document then goes on to suggest using the 95% upper confidence limit of the mean for exposure concentration, the 95th or 90th percentile for contact rate, the 95th percentile for exposure frequency and duration, and the average body weight over the exposure period (USEPA, 1989c, pp. 6-19, 6-22). This guidance is intended to yield an estimate of risk that is both "protective and reasonable." USEPA (1991b, p. 2) reiterates these suggestions.

The above type risk estimate is a "conservative" ***point estimate of risk***. This type of estimate of risk has been frequently criticized for the following reasons:

- There is no way of assessing the degree of conservatism in this kind of estimate (Thompson et al., 1992).
- The input variables may be set to a specific combination that will rarely, if ever, occur in reality (Thompson et al., 1992).
- Traditional sensitivity analyses (one-at-a-time-deviations from the baseline case) are meaningless for this baseline case since many of the input variables are already at or close to their maximum values (Thompson et al., 1992).

- This estimate of risk is often larger by one or several orders of magnitude than the 95th percentile of the population risk (McKone and Bogen, 1991; Burmaster and Harris, 1993; Cullen, 1994).

Probabilistic Risk Assessment

Because of the shortcomings of point estimates of risk, most practitioners in the field of risk assessment now advocate presenting not only the results of points estimates but also the whole distribution of risk and the characteristics of this distribution (Eschenroeder and Faeder, 1988; McKone and Bogen, 1991; Thompson et al., 1992; Burmaster and Harris, 1993; Finley and Paustenbach, 1994; Smith, 1994; McKone, 1994). Presenting the distribution of the output variable (risk in this case) and characterizing this distribution is in fact the definition of uncertainty analysis (Iman and Helton, 1988). *Probabilistic risk assessment is simply uncertainty analysis applied to a risk assessment model that assumes some of the input variables in the model are random variables.*

Sometimes the distribution of risk can be derived analytically for simple models (Springer, 1979; Slob, 1994). Usually, however, the distribution of risk must be derived by Monte Carlo simulation. We will give a specific example later in this chapter.

Guidelines for Conducting and Reporting a Probabilistic Risk Assessment

Because of the complexity of probabilistic risk assessment, it is important to carefully conduct the risk assessment, document exactly how the risk assessment was performed, and effectively communicate the results. Hoaglin and Andrews (1975) suggest certain procedures to follow when reporting the results of a Monte Carlo simulation. More recent references specific to probabilistic risk assessment include Burmaster and Anderson (1994), Vose (1996, Chapter 12), and USEPA (1997). The following guidelines are based on these last three references.

Conducting a Probabilistic Risk Assessment

1. Identify the purpose and scope of the risk assessment, and define the endpoints.
2. Create a model or several models for the risk assessment based on the three steps outlined in the definition of risk assessment (hazard identification, dose-response assessment, and exposure assessment). Document the assumptions of each model.
3. Conduct preliminary sensitivity analyses to determine if structural differences (different models) have substantial effects on the output

distribution (in both the region of central tendency and the tails). For a given model, determine which input variables contribute the most to the variability of the output distribution. If warranted, determine the effects of changing the shape of the input distributions and/or accounting for correlations between the input variables (Smith et al., 1992; Bukowski et al., 1995; Hammed and Bedient, 1997).

4. Simplify the Monte Carlo simulation by setting unimportant input variables (determined from the previous step) to fixed values.

5. Develop distributions for important input variables based on available data and expert judgment. Document the sources of the data sets and what assumptions were used to derive the input distributions. Perform goodness-of-fit tests and graphical assessments of goodness-of-fit.

6. Quantify and distinguish between the variability and uncertainty in the input parameters. For important input parameters, it may be necessary to gather additional data to reduce uncertainty.

7. Perform the Monte Carlo simulations, always recording the random number seeds. Perform sensitivity analyses to determine which variables or group of variables contribute substantially to the variability of the output variable (e.g., exposure or risk). Distinguish between variability and uncertainty in the input parameters (one way to do this is to model the input distributions as mixture distributions; see Hoffman and Hammonds, 1994 and Burmaster and Wilson, 1996).

8. Verify the numerical stability of the Monte Carlo simulation results by performing at least two simulations (using different starting random seeds, of course) or dividing the simulation into non-overlapping sub-sequences and comparing results.

Presenting the Results of a Probabilistic Risk Assessment

1. Taylor the document to the client. Results presented to a risk manager will include many technical details that are usually omitted for presentations to the general public. Results presented to a risk manager should be detailed enough so that the risk assessment may be independently duplicated and verified.

2. Use a tiered presentation style. Put the question and results up front, put details in appendices.

3. Use both a histogram and empirical cumulative distribution function plot to present the distribution of risk on one page, using identical horizontal scales for each plot. On each plot, identify the mean, the 95th percentile, and the point estimate of risk based on established regulatory guidance. Accompany these plots with a table of sum-

mary statistics (e.g., minimum, 5th percentile, median, mean, 95th percentile, and maximum).

4. Present and explain the model. Include all equations, describe all terms and their units of measure, and, where appropriate, include schematic diagrams documenting the structure of the model. Outline model assumptions and shortcomings. Document references for the model.

5. Document how each input distribution was selected, including sources of any relevant data sets. Include results of goodness-of-fit tests and graphical assessments of goodness-of-fit. Also discuss the relative contributions of uncertainty and variability for each input distribution. For each input distribution, present a plot of the probability density function and cumulative distribution function on one page, using identical horizontal axes for each plot. Indicate the location of the mean on each plot. Also include a table of relevant statistics for the input distribution, including the minimum, 5th percentile, median, mean, 95th percentile, and maximum.

6. Discuss the results of sensitivity and uncertainty analyses, including the effects of changing the shape of input distributions and/or including correlations among the input variables, as well as analyses that separate uncertainty from variability.

7. Document the software that was used to perform the Monte Carlo simulation, and present the name and statistical quality (i.e., the period) of the random number generator used by the software.

Case Study: Risk Assessment for Benzene-Contaminated Soil

In this section we will give an abbreviated example of performing a risk assessment based on Thompson et al.'s (1992) model for incremental lifetime cancer risk (ILCR) for children ingesting soil contaminated with benzene. The model is described in Equation (13.11) and Table 13.2. A more detailed analysis is presented in the ENVIRONMENTALSTATS for S-PLUS User's Manual and in the ENVIRONMENTALSTATS for S-PLUS help file Glossary: Risk Assessment – Using ENVIRONMENTALSTATS for S-PLUS for Probabilistic Risk Assessment.

Point Estimates of Exposure and Risk

Using the point estimates of the input variables shown in Table 13.2, the point estimates of exposure and risk are about 3×10^{-8} mg/(kg·d) and 8×10^{-10}, respectively, so the point estimate for incremental lifetime cancer risk is about 8 in 10 billion.

Command

To compute the point estimates of exposure and risk using the S-PLUS Command or Script Window, type the following commands. First we will create a function called `exposure.fcn` to model exposure, and then we will create a function called `ilcr.fcn` to model incremental lifetime cancer risk. Finally, we will plug in the points estimates of the input variables.

```
exposure.fcn <- function(Cs, S.Ing.R, BW, RBA=1,
   DpW=1, WpY=20, YpL=10, CF=1e-006, DinY=364,
   YinL=70) {
     (Cs * S.Ing.R * RBA * DpW * WpY * YpL * CF) /
     (BW * DinY * YinL)
   }

ilcr.fcn <- function(Cs, S.Ing.R, BW, CPF, RBA=1,
   DpW=1, WpY=20, YpL=10, CF=1e-006, DinY=364,
   YinL=70) {
     exposure.fcn(Cs=Cs, S.Ing.R=S.Ing.R, BW=BW,
       RBA=RBA, DpW=DpW, WpY=WpY, YpL=YpL, CF=CF,
       DinY=DinY, YinL=YinL) * CPF
   }

exposure.pt.est <- exposure(Cs=3.39, S.Ing.R=50,
   BW=47)

ilcr.pt.est <- ilcr(Cs=3.39, S.Ing.R=50, BW=47,
   CPF=2.9e-2)
```

Characterizing the Distributions of the Input Variables

Figures 13.10 and 13.11 display the hypothesized probability density function and cumulative distribution function for the benzene concentration in the soil (*Cs*). In both of these figures we have added a vertical line to indicate the value of the mean. (Note that we must use Equation (4.32) to compute the mean of the distribution since it has been parameterized in terms of the mean and standard deviation of the log-transformed random variable.) The minimum, 5th percentile, median, 95th percentile, and maximum of the distribution are: 0, 0.7, 2.3, 8.2, and ∞. You can create pdf and cdf plots and compute population quantiles for the other three input variables that have associated distributions as well (i.e., for *SingR*, *BW*, and *CPF*).

Menu

See Chapter 4 for examples on how to use the ENVIRONMENTALSTATS for S-PLUS pull-down menu to plot probability density functions, plot cumulative distribution functions, and compute quantiles of a distribution.

Assumed PDF for Benzene Concentration

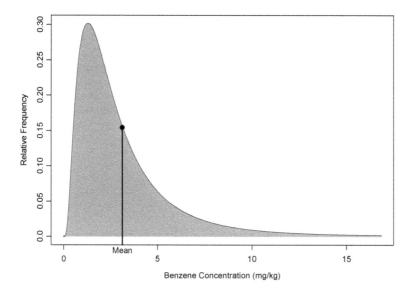

Figure 13.10   Assumed pdf for benzene concentration in soil

Assumed CDF for Benzene Concentration

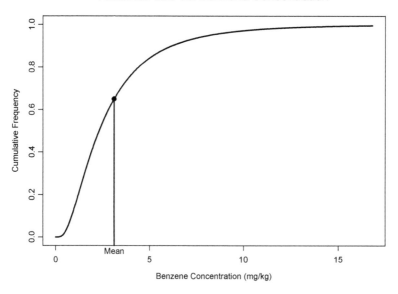

Figure 13.11   Assumed cdf for benzene concentration in soil

Command

See Chapter 4 for examples on how use the ENVIRONMENTALSTATS for
S-PLUS Command or Script Window to plot probability density functions,
plot cumulative distribution functions, and compute quantiles of a distribution. Here we will show you how to create the pdf and cdf plots and add the
line showing the value of the mean. To create the pdf plot, type these commands.

```
pdfplot(dist="lnorm", param.list=list(meanlog=0.84,
    sdlog=0.77), right.tail.cutoff=0.005, xlab="Benzene
    Concentration (mg/kg)", main="Assumed PDF for
    Benzene Concentration")
avg <- exp(0.84 + (0.77^2)/2)
f.avg <- dlnorm(avg, 0.84, 0.77)
segments(avg, par("usr")[3], avg, f.avg)
points(avg, f.avg, pch=16, cex=1.25*par("cex"))
mtext("Mean", side=1, at=avg)
```

To create the cdf plot, type these commands.

```
cdfplot(dist = "lnorm", param.list=list(meanlog=0.84,
    sdlog=0.77), right.tail.cutoff=0.005, xlab="Benzene
    Concentration (mg/kg)", main="Assumed CDF for
    Benzene Concentration")
F.avg <- plnorm(avg, 0.84, 0.77)
segments(avg, par("usr")[3], avg, F.avg)
points(avg, F.avg, pch=16, cex=1.25*par("cex"))
mtext("Mean", at=avg, side=1)
```

Simulating Risk

Figure 13.12 displays the empirical probability distribution function (i.e.,
the histogram) of incremental lifetime cancer risk based on performing a
Monte Carlo simulation using the model in Equation (13.11) and the specified probability distributions and point estimates shown in Table 13.2.
Figure 13.13 displays the empirical cumulative distribution function. In both
of these plots we have identified the mean and 95th percentile of the simulated distribution, along with the point estimate of risk.

Note that all four variables with assumed probability distributions (Cs,
$SingR$, BW, and CPF) are assumed to be independent of each other in this
model. In reality, soil ingestion rate is probably positively correlated with
body weight. The ENVIRONMENTALSTATS for S-PLUS function simu-
late.mv.matrix allows you to specify rank correlations between varia-

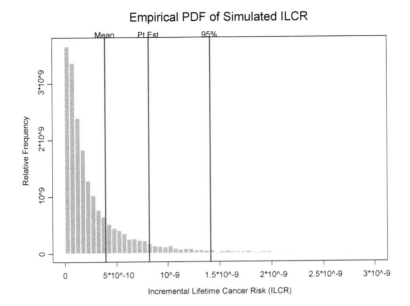

Figure 13.12 Simulated probability density function of ILCR

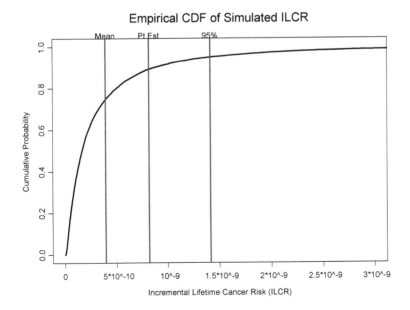

Figure 13.13 Simulated cumulative distribution function of ILCR

bles, but both Smith et al. (1992) and Bukowski et al. (1995) show that for many circumstances the shapes of the chosen input probability distributions have much more effect on the resultant distribution of risk than does including or excluding correlations between the input variables.

Command

To produce the simulated distribution of ILCR using the ENVIRONMENTALSTATS for S-PLUS Command or Script Window, type these commands.

```
input.matrix <- simulate.mv.matrix(10000,
    distributions = c(Cs       = "lnorm",
                      S.Ing.R = "lnorm",
                      BW       = "norm",
                      CPF =     "lnorm"),
    param.list = list(
      Cs       = list(meanlog =  0.84, sdlog = 0.77),
      S.Ing.R = list(meanlog =  3.44, sdlog = 0.80),
      BW       = list(mean    = 47,    sd    = 8.3),
      CPF      = list(meanlog = -4.33, sdlog = 0.67)),
    cor.mat = diag(4), seed = 47)
simulated.ilcr <- eval.expr(ilcr.fcn(Cs, S.Ing.R, BW,
    CPF), data = input.matrix)
```

To computed the mean and 95th percentile of the simulated distribution, type these commands.

```
ilcr.avg <- mean(simulated.ilcr)
ilcr.95.pct <- quantile(simulated.ilcr, 0.95)
```

To plot the histogram of the simulated distribution and add the mean, point estimate, and 95th percentile, type these commands.

```
hist(simulated.ilcr, nclass = 500, prob = T,
    xlim=c(0, 3e-9), xlab="Incremental Lifetime Cancer
    Risk (ILCR)", ylab="Relative Frequency",
    main="Empirical PDF of Simulated ILCR")
abline(v=ilcr.avg)
abline(v=ilcr.95.pct)
abline(v=ilcr.pt.est)
box()
mtext("Mean", at=ilcr.avg)
mtext("95%", at=ilcr.95.pct)
```

```
mtext("Pt Est", at=ilcr.pt.est)
```

To plot the empirical cdf of the simulated distribution and add the mean, point estimate, and 95th percentile, type these commands.

```
ecdfplot(simulated.ilcr, xlim = c(0, 3e-9),
    xlab="Incremental Lifetime Cancer Risk
    (ILCR)",main="Empirical CDF of Simulated ILCR")
abline(v=ilcr.avg)
abline(v=ilcr.95.pct)
abline(v=ilcr.pt.est)
mtext("Mean", at=ilcr.avg)
mtext("95%", at=ilcr.95.pct)
mtext("Pt Est", at=ilcr.pt.est)
```

Sensitivity Analysis

As discussed earlier in this chapter, there are several methods of performing sensitivity analysis for risk assessment. For this example here, we will simply look at the change in distribution of incremental lifetime cancer risk as we let one input variable vary and keep the other variables fixed at their point estimates (shown in Table 13.2). Figure 13.14 displays the original histogram for ILCR, along with the four others produced by varying only average body weight (BW), benzene concentration in soil (Cs), soil ingestion rate ($SIngR$), or cancer potency factor (CPF) one at a time. Similar plots appear in Thompson et al. (1992).

Based on these figures, you can see that average body weight (BW) has the smallest effect on the variability of ILCR compared to the other three input variables. Also, the bulk of the distribution based on varying any of the other three variables is below the point estimate of risk, with the cancer potency factor (CPF) having the greatest influence.

Command

To produce the simulated distribution of ILCR where only average body weight (BW) is varied using the S-PLUS Command or Script Window, type these commands.

```
hist(ilcr.fcn(Cs=3.39, S.Ing.R=50,
    BW=input.matrix[, "BW"], CPF=0.029), nclass=100,
    prob=T, xlim=c(0, 3e-9), xlab="Incremental Lifetime
    Cancer Risk (ILCR)", ylab="Relative
    Frequency",main="Simulated ILCR Varying BW")
abline(v = ilcr.pt.est)
```

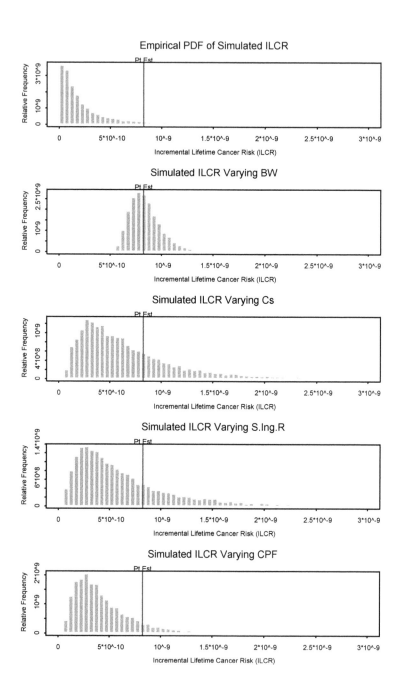

Figure 13.14 Sensitivity analysis for the model for ILCR

```
box()
mtext("Pt Est", at=ilcr.pt.est)
```

To produce the other histograms, type the same commands but subscript the appropriate column from `input.matrix`.

Uncertainty Analysis

As discussed earlier in this chapter, there are several methods of performing uncertainty analysis for risk assessment. We have already presented the simulated distribution of incremental lifetime cancer risk in Figures 13.12 and 13.13. The two-sided nonparametric 95% confidence interval for the median ILCR is $[1.6, 1.7] \times 10^{-10}$, and for the 95th percentile it is $[1.3, 1.5] \times 10^{-9}$. Figure 13.15 compares the empirical cumulative distribution of ILCR based on this simulation with a second simulation. As you can see, the two simulated distributions are virtually identical.

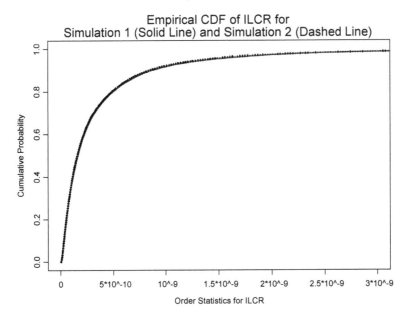

Figure 13.15 Comparison of the empirical cdf of ILCR based on two separate simulations with 10,000 Monte Carlo trials

We could also run another simulation with an increased number of Monte Carlo trials, and/or use mixture distributions for the input variables and compare the results with our original simulation. Both of these methods of uncertainty analysis are illustrated in the ENVIRONMENTALSTATS for S-PLUS help file and User's Manual.

SUMMARY

- Human and ecological risk assessment involves characterizing the exposure to a toxicant for one or several populations, quantifying the relationship between exposure (dose) and health or ecological effects (response), determining the risk (probability) of a health or ecological effect given the observed level(s) of exposure, and characterizing the uncertainty associated with the estimated risk.

- In the past, estimates of risk were often based solely on setting values of the input variables (e.g., body weight, dose, etc.) to particular point estimates and producing a single point estimate of risk, with little, if any, quantification of the uncertainty associated with the estimated risk. More recently, several practitioners have advocated "probabilistic" risk assessment, in which the input variables are considered random variables, so the result of the risk assessment is a probability distribution for predicted risk or exposure.

- Most risk assessment models follow the form of Equations (13.1) and (13.9), where the output variable (risk or exposure) is a function of several input variables, and some or all of the input variables are considered to be random variables.

- *Uncertainty analysis* involves describing the variability or distribution of values of the output variable that is due to the collective variation in the input variables. This description usually involves graphical displays such as histograms, empirical density plots, and empirical cdf plots, as well as summary statistics such as the mean, median, standard deviation, coefficient of variation, 95th percentile, etc.

- *Sensitivity analysis* involves determining how the distribution of the output variable changes with changes in the individual input variables. It is used to identify which input variables contribute the most to the variation or uncertainty in the output variable. Sensitivity analysis is also used in a broader sense to determine how changing the distributions of the input variables and/or their assumed correlations or even changing the form of the model affects the output.

- *Variability* refers to the inherent heterogeneity of a particular variable (parameter). For example, there is natural variation in body weight and height between individuals in a given population. *Uncertainty* refers to a lack of knowledge about specific parameters, models, or factors. We can usually reduce uncertainty through further measurement or study. We cannot reduce variability, since it is inherent in the variable.

- Sensitivity analysis methods include one-at-a-time deviations from a baseline case, factorial design and response surface modeling, and Monte Carlo simulation.

- Uncertainty analysis methods include describing the empirical distribution of risk, repeating the simulation using a different set of random numbers, and using mixture distributions.
- An important point to remember is that no matter how complex the mathematical model in Equation (13.1) or (13.9), or how extensive the uncertainty and sensitivity analyses, there is always the question of how well the mathematical model reflects reality. The only way to attempt to answer this question is with data collected directly on the input and output variables.

EXERCISES

13.1. Suppose five independent random variables have standard normal distributions. The sum of the squares of these random variables has a chi-square distribution with 5 degrees of freedom, but suppose you do not know this. Use Monte Carlo simulation to generate a simulated distribution. Plot the histogram and overlay the true theoretical distribution.

13.2. Simulate the distribution of the product $X_1 X_2$, where X_1 is normal with mean 10 and standard deviation 2, and X_2 is lognormal with mean 10 and CV 2.

13.3. Simulate the distribution of ILCR based on Thompson et al.'s (1992) model based on $n = 100$; 500; 1,000; and 5,000 Monte Carlo trials. Compare the results of the simulated distributions and sensitivity analyses.

13.4. Find a risk assessment model in the environmental literature and reproduce the results.

13.5. Find tables or references that list annual number of deaths by cause in the United States (or your country).

    a. Which category accounts for the most deaths?

    b. Has the death rate due to cancer (all forms) increased, decreased, or stayed relatively constant over the past 5 years? 10 years? 20 years? 50 years?

    c. Rank the risk of dying for the following modes of transportation: airplane, train, automobile, bus, boat, etc. Now find a table that gives you the actual risks. Are you surprised? Were your rankings close to the truth? If not, what do you think influenced your rankings?

13.6.   When a government agency or the media reports that exposure to a particular chemical increases the risk of developing cancer by some specific number (e.g., 1 in 1 million), how accurate do you think this number is? What questions would you ask the person who came up with this number?

13.7.   This exercise asks you to compare risks and think about the psychology of risk perception.

    a.   What is the estimated risk of developing lung cancer during your lifetime from smoking cigarettes (say one pack of cigarettes/day or some other specified amount)?

    b.   What is the estimated risk of dying from cancer or some other complication due to radiation exposure from a nuclear power plant (given that you live "close" to a nuclear power plant)?

    c.   What is the estimated risk of dying in an auto accident over the course of a lifetime?

    d.   Why do you think people are more concerned with dying from radiation leakage from a nuclear power plant than with dying in an auto accident or dying from the effects of smoking (given that you smoke), even though the risk of the first is much smaller than the risk of the latter two?

REFERENCES

ASTM. (1996). *PS 64-96: Provisional Standard Guide for Developing Appropriate Statistical Approaches for Ground-Water Detection Monitoring Programs.* American Society for Testing and Materials, West Conshohocken, PA.

Aitchison, J. (1955). On the Distribution of a Positive Random Variable Having a Discrete Probability Mass at the Origin. *Journal of the American Statistical Association* **50**, 901–908.

Aitchison, J., and J.A.C. Brown (1957). *The Lognormal Distribution (with special references to its uses in economics).* Cambridge University Press, London, 176 pp.

Akritas, M., T. Ruscitti, and G.P. Patil. (1994). Statistical Analysis of Censored Environmental Data. In Patil, G.P., and C.R. Rao, eds., *Handbook of Statistics, Vol. 12: Environmental Statistics.* North-Holland, Amsterdam, Chapter 7, 221–242.

Amemiya, T. (1985). *Advanced Econometrics.* Harvard University Press, Cambridge, MA, 521 pp.

Arlinghaus, S.L., ed. (1996). *Practical Handbook of Spatial Statistics.* CRC Press, Boca Raton, FL, 307 pp.

Bain, L.J., and M. Engelhardt. (1991). *Statistical Analysis of Reliability and Life-Testing Models.* Marcel Dekker, New York, 496 pp.

Balakrishnan, N., and A.C. Cohen. (1991). *Order Statistics and Inference: Estimation Methods.* Academic Press, San Diego, CA, 375 pp.

Barclay's California Code of Regulations. (1991). Title 22, §66264.97 [concerning hazardous waste facilities] and Title 23, §2550.7(e)(8) [concerning solid waste facilities]. Barclay's Law Publishers, San Francisco, CA.

Barnett, V., and A. O'Hagan. (1997). *Setting Environmental Standards: The Statistical Approach to Handling Uncertainty and Variation.* Chapman & Hall, London, 111 pp.

Barnett, V., and K.F. Turkman, eds. (1997). *Statistics for the Environment 3: Pollution Assessment and Control.* John Wiley & Sons, New York, 345 pp.

Barr, D.R., and T. Davidson. (1973). A Kolmogorov-Smirnov Test for Censored Samples. *Technometrics* **15**(4), 739–757.

Barry, T.M. (1996). Recommendations on the Testing and Use of Pseudo-Random Number Generators Used in Monte Carlo Analysis for Risk Assessment. *Risk Analysis* **16**, 93–105.

Bartels, R. (1982). The Rank Version of von Neumann's Ratio Test for Randomness. *Journal of the American Statistical Association* **77**(377), 40–46.

Berthouex, P.M., and I. Hau. (1991a). A Simple Rule for Judging Compliance Using Highly Censored Samples. *Research Journal of the Water Pollution Control Federation* **63**(6), 880–886.

Berthouex, P.M., and I. Hau. (1991b). Difficulties Related to Using Extreme Percentiles for Water Quality Regulations. *Research Journal of the Water Pollution Control Federation* **63**(6), 873–879.

Berthouex, P.M., and L.C. Brown. (1994). *Statistics for Environmental Engineers*. Lewis Publishers, Boca Raton, FL, 335 pp.

Birnbaum, Z. W., and F.H. Tingey. (1951). One-Sided Confidence Contours for Probability Distribution Functions. *Annals of Mathematical Statistics* **22**, 592–596.

Blom, G. (1958). *Statistical Estimates and Transformed Beta Variables*. John Wiley & Sons, New York.

Bogen, K.T. (1995). Methods to Approximate Joint Uncertainty and Variability in Risk. *Risk Analysis* **15**(3), 411–419.

Bose, A., and N. K. Neerchal. (1996). *Unbiased Estimation of Variance Using Ranked Set Sampling*. Technical Report # 96-11, Department of Statistics, Purdue University, West Lafayette, IN.

Boswell, M. T., and G. P. Patil. (1987). A Perspective of Composite Sampling. *Communications in Statistics—Theory and Methods* **16**, 3069–3093.

Box, G.E.P., and D.R. Cox. (1964). An Analysis of Transformations (with Discussion). *Journal of the Royal Statistical Society, Series B* **26**(2), 211–252.

Box, G.E.P., W.G. Hunter, and J.S. Hunter. (1978). *Statistics for Experimenters: An Introduction to Design, Data Analysis, and Model Building*. John Wiley & Sons, New York, 653 pp.

Box, G.E.P., and G.M. Jenkins. (1976). *Time Series Analysis: Forecasting and Control*. Prentice-Hall, Englewood Cliffs, NJ, 575 pp.

Bradu, D., and Y. Mundlak. (1970). Estimation in Lognormal Linear Models. *Journal of the American Statistical Association* **65**, 198–211.

Brainard, J., and D.E. Burmaster. (1992). Bivariate Distributions for Height and Weight of Men and Women in the United States. *Risk Analysis* **12**(2), 267–275.

Bukowski, J., L. Korn, and D. Wartenberg. (1995). Correlated Inputs in Quantitative Risk Assessment: The Effects of Distributional Shape. *Risk Analysis* **15**(2), 215–219.

Burmaster, D.E., and P.D. Anderson. (1994). Principles of Good Practice for the Use of Monte Carlo Techniques in Human Health and Ecological Risk Assessments. *Risk Analysis* **14**(4), 477–481.

Burmaster, D.E., and R.H. Harris. (1993). The Magnitude of Compounding Conservatisms in Superfund Risk Assessments. *Risk Analysis* **13**(2), 131–134.

Burmaster, D.E., and K.M. Thompson. (1997). Estimating Exposure Point Concentrations for Surface Soils for Use in Deterministic and Probabilistic Risk Assessments. *Human and Ecological Risk Assessment* **3**(3), 363–384.

Burmaster, D.E., and A.M. Wilson. (1996). An Introduction to Second-Order Random Variables in Human Health Risk Assessments. *Human and Ecological Risk Assessment* **2**(4), 892–919.

Byrnes, M.E. (1994). *Field Sampling Methods for Remedial Investigations*. Lewis Publishers, Boca Raton, FL, 254 pp.

Calitz, F. (1973). Maximum Likelihood Estimation of the Parameters of the Three-Parameter Lognormal Distribution—a Reconsideration. *Australian Journal of Statistics* **15**(3), 185–190.

Callahan, B.G., ed. (1998). Special Issue: Probabilistic Risk Assessment. *Human and Ecological Risk Assessment* **4**(2).

Callahan, B.G., D.E. Burmaster, R.L. Smith, D.D. Krewski, and B.A.F. DelloRusso, eds. (1996). Commemoration of the 50th Anniversary of Monte Carlo, *Human and Ecological Risk Assessment* **2**(4).

Castillo, E. (1988). *Extreme Value Theory in Engineering*. Academic Press, New York, 389 pp.

Castillo, E., and A. Hadi. (1994). Parameter and Quantile Estimation for the Generalized Extreme-Value Distribution. *Environmetrics* **5**, 417–432.

Caulcutt, R., and R. Boddy. (1983). *Statistics for Analytical Chemists*. Chapman & Hall, London, 253 pp.

Chambers, J.M., W.S. Cleveland, B. Kleiner, and P.A. Tukey. (1983). *Graphical Methods for Data Analysis*. Duxbury Press, Boston, MA, 395 pp.

Chen, J.J., D.W. Gaylor, and R.L. Kodell. (1990). Estimation of the Joint Risk from Multiple-Compound Exposure Based on Single-Compound Experiments. *Risk Analysis* **10**(2), 285–290.

Chen, L. (1995a). A Minimum Cost Estimator for the Mean of Positively Skewed Distributions with Applications to Estimation of Exposure to Contaminated Soils. *Environmetrics* **6**, 181–193.

Chen, L. (1995b). Testing the Mean of Skewed Distributions. *Journal of the American Statistical Association* **90**(430), 767–772.

Cheng, R.C.H., and N.A.K. Amin. (1982). Maximum Product-of-Spacings Estimation with Application to the Lognormal Distribution. *Journal of the Royal Statistical Society, Series B* **44**, 394–403.

Chew, V. (1968). Simultaneous Prediction Intervals. *Technometrics* **10**(2), 323–331.

Chou, Y.M., and D.B. Owen. (1986). One-sided Distribution-Free Simultaneous Prediction Limits for p Future Samples. *Journal of Quality Technology* **18**, 96–98.

Chowdhury, J.U., J.R. Stedinger, and L. H. Lu. (1991). Goodness-of-Fit Tests for Regional Generalized Extreme Value Flood Distributions. *Water Resources Research* **27**(7), 1765–1776.

Clark, M.J.R., and P.H. Whitfield. (1994). Conflicting Perspectives about Detection Limits and about the Censoring of Environmental Data. *Water Resources Bulletin* **30**(6), 1063–1079.

Clayton, C.A., J.W. Hines, and P.D. Elkins. (1987). Detection Limits with Specified Assurance Probabilities. *Analytical Chemistry* **59**, 2506–2514.

Cleveland, W.S. (1993). *Visualizing Data*. Hobart Press, Summit, NJ, 360 pp.

Cleveland, W.S. (1994). *The Elements of Graphing Data*. Revised Edition. Hobart Press, Summit, NJ, 297 pp.

Cochran, W.G. (1977). *Sampling Techniques*. Third Edition. John Wiley & Sons, New York, 428 pp.

Code of Federal Regulations. (1996). Definition and Procedure for the Determination of the Method Detection Limit—Revision 1.11. Title 40, Part 136, Appendix B, 7-1-96 Edition, pp. 265–267.

Cogliano, V.J. (1997). Plausible Upper Bounds: Are Their Sums Plausible? *Risk Analysis* **17**(1), 77–84.

Cohen, A.C. (1951). Estimating Parameters of Logarithmic-Normal Distributions by Maximum Likelihood. *Journal of the American Statistical Association* **46**, 206–212.

Cohen, A.C. (1959). Simplified Estimators for the Normal Distribution When Samples are Singly Censored or Truncated. *Technometrics* **1**(3), 217–237.

Cohen, A.C. (1963). Progressively Censored Samples in Life Testing. *Technometrics* **5**, 327–339.

Cohen, A.C. (1988). Three-Parameter Estimation. In Crow, E.L., and K. Shimizu, eds. *Lognormal Distributions: Theory and Applications.* Marcel Dekker, New York, Chapter 4.

Cohen, A.C. (1991). *Truncated and Censored Samples.* Marcel Dekker, New York, 312 pp.

Cohen, A.C., and B.J. Whitten. (1980). Estimation in the Three-Parameter Lognormal Distribution. *Journal of the American Statistical Association* **75**, 399–404.

Cohen, A.C., B.J. Whitten, and Y. Ding. (1985). Modified Moment Estimation for the Three-Parameter Lognormal Distribution. *Journal of Quality Technology* **17**, 92–99.

Cohen, M.A., and P.B. Ryan. (1989). Observations Less Than the Analytical Limit of Detection: A New Approach. *JAPCA* **39**(3), 328–329.

Cohn, T.A. (1988). *Adjusted Maximum Likelihood Estimation of the Moments of Lognormal Populations form Type I Censored Samples.* U.S. Geological Survey Open-File Report 88-350, 34 pp.

Cohn, T.A., L.L. DeLong, E.J. Gilroy, R.M. Hirsch, and D.K. Wells. (1989). Estimating Constituent Loads. *Water Resources Research* **25**(5), 937–942.

Cohn, T.A., E.J. Gilroy, and W.G. Baier. (1992). Estimating Fluvial Transport of Trace Constituents Using a Regression Model with Data Subject to Censoring. *Proceedings of the American Statistical Association, Section on Statistics and the Environment.*

Conover, W.J. (1980). *Practical Nonparametric Statistics.* Second Edition. John Wiley & Sons, New York, 493 pp.

Conover, W.J., M.E. Johnson, and M.M. Johnson. (1981). A Comparative Study of Tests for Homogeneity of Variances, with Applications to the Outer Continental Shelf Bidding Data. *Technometrics* **23**(4), 351–361.

Cothern, C.R. (1994). How Does Scientific Information in General and Statistical Information in Particular Input to the Environmental Regulatory Process? In Patil, G.P., and C.R. Rao, eds., *Handbook of Statistics, Vol. 12: Environmental Statistics.* North-Holland, Amsterdam, Chapter 25, 791–816.

Cothern, C.R. (1996). *Handbook for Environmental Risk Decision Making: Values, Perceptions, & Ethics.* Lewis Publishers, Boca Raton, FL, 408 pp.

Cothern, C.R., and N.P. Ross. (1994). *Environmental Statistics, Assessment, and Forecasting.* Lewis Publishers, Boca Raton, FL, 418 pp.

Cox, D.D., L.H. Cox, and K.B. Ensor. (1997). Spatial Sampling and the Environment: Some Issues and Directions. *Environmental and Ecological Statistics* **4**, 219–233.

Cox, D.R., and D.V. Hinkley. (1974). *Theoretical Statistics.* Chapman & Hall, New York, 511 pp.

Cox, L.H., and W.W. Piegorsch. (1996). Combining Environmental Information, I: Environmental Monitoring, Measurement and Assessment. *Environmetrics* **7**, 299–308.

Cressie, N.A.C. (1993). *Statistics for Spatial Data.* Revised Edition. John Wiley & Sons, New York, 900 pp.

Crow, E.L., and K. Shimizu, eds. (1988). *Lognormal Distributions: Theory and Applications.* Marcel Dekker, New York, 387 pp.

Csuros, M. (1997). *Environmental Sampling and Analysis, Lab Manual.* Lewis Publishers, Boca Raton, FL, 373 pp.

Cullen, A.C. (1994). Measures of Compounding Conservatism in Probabilistic Risk Assessment. *Risk Analysis* **14**(4), 389–393.

Cunnane, C. (1978). Unbiased Plotting Positions—A Review. *Journal of Hydrology* **37**(3/4), 205–222.

Currie, L.A. (1968). Limits for Qualitative Detection and Quantitative Determination: Application to Radiochemistry. *Annals of Chemistry* **40**, 586–593.

Currie, L.A. (1988). *Detection in Analytical Chemistry: Importance, Theory, and Practice.* American Chemical Society, Washington, D.C.

Currie, L.A. (1995). Nomenclature in Evaluation of Analytical Methods Including Detection and Quantification Capabilities. *Pure & Applied Chemistry* **67**(10), 1699–1723.

Currie, L.A. (1996). Foundations and Future of Detection and Quantification Limits. *Proceedings of the Section on Statistics and the Environment*, American Statistical Association, Alexandria, VA.

Currie, L.A. (1997). Detection: International Update, and Some Emerging Di-Lemmas Involving Calibration, the Blank, and Multiple Detection Decisions. *Chemometrics and Intelligent Laboratory Systems* **37**, 151–181.

D'Agostino, R.B. (1970). Transformation to Normality of the Null Distribution of g1. *Biometrika* **57**, 679–681.

D'Agostino, R.B. (1971). An Omnibus Test of Normality for Moderate and Large Size Samples. *Biometrika* **58**, 341–348.

D'Agostino, R.B. (1986a). Graphical Analysis. In D'Agostino, R.B., and M.A. Stephens, eds. *Goodness-of-Fit Techniques*. Marcel Dekker, New York, Chapter 2, pp. 7–62.

D'Agostino, R.B. (1986b). Tests for the Normal Distribution. In D'Agostino, R.B., and M.A. Stephens, eds. *Goodness-of-Fit Techniques*. Marcel Dekker, New York, Chapter 9, pp. 367–419.

D'Agostino, R.B., and E.S. Pearson (1973). Tests for Departures from Normality. Empirical Results for the Distributions of $\beta2$ and $\beta1$. *Biometrika* **60**(3), 613–622.

D'Agostino, R.B., and M.A. Stephens, eds. (1986). *Goodness-of-Fit Techniques*. Marcel Dekker, New York, 560 pp.

D'Agostino, R.B., and G.L. Tietjen (1973). Approaches to the Null Distribution of Öb1. *Biometrika* **60**(1), 169–173.

Dallal, G. E., and L. Wilkinson. (1986). An Analytic Approximation to the Distribution of Lilliefor's Test for Normality. *The American Statistician* **40**, 294–296.

Danziger, L., and S. Davis. (1964). Tables of Distribution-Free Tolerance Limits. *Annals of Mathematical Statistics* **35**(5), 1361–1365.

David, M. (1977). *Geostatistical Ore Reserve Estimation*. Elsevier, New York, 364 pp.

Davidson, J.R. (1994). *Elipgrid-PC: A PC Program for Calculating Hot Spot Probabilities*. ORNL/TM-12774. Oak Ridge National Laboratory, Oak Ridge, TN.

Davis, C.B. (1994). Environmental Regulatory Statistics. In Patil, G.P., and C.R. Rao, eds., *Handbook of Statistics, Vol. 12: Environmental Statistics*. North-Holland, Amsterdam, Chapter 26, 817–865.

Davis, C.B. (1997). Challenges in Regulatory Environmetrics. *Chemometrics and Intelligent Laboratory Systems* **37**, 43–53.

Davis, C.B. (1998a). *Ground-Water Statistics & Regulations: Principles, Progress and Problems*. Second Edition. Environmetrics & Statistics Limited, Henderson, NV.

Davis, C.B. (1998b). Personal Communication, September 3, 1998.

Davis, C.B. (1998c). *Power Comparisons for Control Chart and Prediction Limit Procedures*. EnviroStat Technical Report 98-1, Environmetrics & Statistics Limited, Henderson, NV.

Davis, C.B., and R.J. McNichols. (1987). One-sided Intervals for at Least p of m Observations from a Normal Population on Each of r Future Occasions. *Technometrics* **29**, 359–370.

Davis, C.B., and R.J. McNichols. (1994a). Ground Water Monitoring Statistics Update: Part I: Progress Since 1988. *Ground Water Monitoring and Remediation* **14**(4), 148–158.

Davis, C.B., and R.J. McNichols. (1994b). Ground Water Monitoring Statistics Update: Part II: Nonparametric Prediction Limits. *Ground Water Monitoring and Remediation* **14**(4), 159–175.

Davis, C.B., and R.J. McNichols. (1999). Simultaneous Nonparametric Prediction Limits (with Discusson). *Technometrics* **41**(2), 89–112.

Davis, C.S., and M.A. Stephens. (1978). Algorithm AS 128. Approximating the Covariance Matrix of Normal Order Statistics. *Applied Statistics* **27**, 206–212.

Diggle, P.J. (1983). *Statistical Analysis of Spatial Point Patterns*. Academic Press, London.

Downing, D.J., R.H. Gardner, and F.O. Hoffman. (1985). An Examination of Response-Surface Methodologies for Uncertainty Analysis in Assessment Models. *Technometrics* **27**(2), 151–163. See also Easterling, R.G. (1986). Discussion of Downing, Gardner, and Hoffman (1985). *Technometrics* **28**(1), 91–93.

Downton, F. (1966). Linear Estimates of Parameters in the Extreme Value Distribution. *Technometrics* **8**(1), 3–17.

Draper, N., and H. Smith. (1981). *Applied Regression Analysis*. Second Edition. John Wiley & Sons, New York, 709 pp.

Draper, N., and H. Smith. (1998). *Applied Regression Analysis*. Third Edition. John Wiley & Sons, New York, 706 pp.

Dunnett, C.W. (1955). A Multiple Comparisons Procedure for Comparing Several Treatments with a Control. *Journal of the American Statistical Association* **50**, 1096–1121.

Dunnett, C.W. (1964). New Tables for Multiple Comparisons with a Control. *Biometrics* **20**, 482–491.

Dunnett, C.W., and M. Sobel. (1955). Approximations to the Probability Integral and Certain Percentage Points of a Multivariate Analogue of Student's t-Distribution. *Biometrika* **42**, 258–260.

Edland, S.D., and G. van Belle. (1994). Decreased Sampling Costs and Improved Accuracy with Composite Sampling. In Cothern, C.R., and N.P. Ross, eds., *Environmental Statistics, Assessment, and Forecasting*, Lewis Publishers, Boca Raton, FL, Chapter 2, pp. 29–55.

Efron, B. (1979a). Bootstrap Methods: Another Look at the Jackknife. *Annals of Statistics* **7**, 1–26.

Efron, B. (1979b). Computers and the Theory of Statistics: Thinking the Unthinkable. *Society for Industrial and Applied Mathematics* **21**, 460–480.

Efron, B. (1981). Nonparametric Standard Errors and Confidence Intervals. *Canadian Journal of Statistics* **9**, 139–172.

Efron, B. (1982). *The Jackknife, the Bootstrap and Other Resampling Plans*. Society for Industrial and Applied Mathematics, Philadelphia, PA, 92 pp.

Efron, B., and R.J. Tibshirani. (1993). *An Introduction to the Bootstrap*. Chapman & Hall, New York, 436 pp.

Einax, J.W., H.W. Zwanziger, and S. Geib. (1997). *Chemometrics in Environmental Analysis*. VCH Publishers, John Wiley & Sons, New York, 384 pp.

Ellison, B.E. (1964). On Two-Sided Tolerance Intervals for a Normal Distribution. *Annals of Mathematical Statistics* **35**, 762–772.

El-Shaarawi, A.H. (1989). Inferences about the Mean from Censored Water Quality Data. *Water Resources Research* **25**(4) 685–690.

El-Shaarawi, A.H. (1993). Environmental Monitoring, Assessment, and Prediction of Change. *Environmetrics* **4**(4), 381–398.

El-Shaarawi, A.H., and D.M. Dolan. (1989). Maximum Likelihood Estimation of Water Quality Concentrations from Censored Data. *Canadian Journal of Fisheries and Aquatic Sciences* **46**, 1033–1039.

El-Shaarawi, A.H., and S.R. Esterby. (1992). Replacement of Censored Observations by a Constant: An Evaluation. *Water Research* **26**(6), 835–844.

El-Shaarawi, A.H., and A. Naderi. (1991). Statistical Inference from Multiply Censored Environmental Data. *Environmental Monitoring and Assessment* **17**, 339–347.

Eschenroeder, A.Q., and E.J. Faeder. (1988). A Monte Carlo Analysis of Health Risks from PCB-Contaminated Mineral Oil Transformer Fires. *Risk Analysis* **8**(2), 291–297.

Esterby, S.R. (1993). Trend Analysis Methods for Environmental Data. *Environmetrics* **4**(4), 459–481.

Evans, M., N. Hastings, and B. Peacock. (1993). *Statistical Distributions*. Second Edition. John Wiley & Sons, New York, 170 pp.

Everitt, B.S. (1998). *The Cambridge Dictionary of Statistics*. Cambridge University Press, New York, 360 pp.

Everitt, B.S. (1999). *Chance Rules: An Informal Guide to Probability, Risk, and Statistics*. Springer-Verlag, New York, 202 pp.

Fertig, K.W., and N.R. Mann. (1977). One-Sided Prediction Intervals for at Least p out of m Future Observations from a Normal Population. *Technometrics* **19**, 167–177.

Fill, H.D., and J.R. Stedinger. (1995). *L* Moment and Probability Plot Correlation Coefficient Goodness-of-Fit Tests for the Gumbel Distribution and Impact of Autocorrelation. *Water Resources Research* **31**(1), 225–229.

Filliben, J.J. (1975). The Probability Plot Correlation Coefficient Test for Normality. *Technometrics* **17**(1), 111–117.

Finley, B., and D. Paustenbach. (1994). The Benefits of Probabilistic Exposure Assessment: Three Case Studies Involving Contaminated Air, Water, and Soil. *Risk Analysis* **14**(1), 53–73.

Finley, B., D. Proctor, P. Scott, N. Harrington, D. Paustenback, and P. Price. (1994). Recommended Distributions for Exposure Factors Frequently Used in Health Risk Assessment. *Risk Analysis* **14**(4), 533–553.

Finney, D.J. (1941). On the Distribution of a Variate Whose Logarithm is Normally Distributed. *Supplement to the Journal of the Royal Statistical Society* **7**, 155–161.

Fisher, L.D., and G. van Belle. (1993). *Biostatistics: A Methodology for the Health Sciences*. John Wiley & Sons, New York, 991 pp.

Fleiss, J. L. (1981). *Statistical Methods for Rates and Proportions*. Second Edition. John Wiley & Sons, New York, 321 pp.

Fleming, T.R., and D.P. Harrington. (1981). A Class of Hypothesis Tests for One and Two Sample Censored Survival Data. *Communications in Statistics—Theory and Methods* **A10**(8), 763–794.

Fleming, T.R., and D.P. Harrington. (1991). *Counting Processes and Survival Analysis*. John Wiley & Sons, New York, 429 pp.

Fletcher, D.J., L. Kavalieris, and B.F.J. Manly, eds. (1998). *Statistics in Ecology and Environmental Monitoring 2: Decision Making and Risk Assessment in Biology*. Otago Conference Series No. 6, University of Otago Press, Dunedin, New Zealand, 220 pp.

Fletcher, D.J., and B.F.J. Manly, eds. (1994). *Statistics in Ecology and Environmental Monitoring*. Otago Conference Series No. 2. University of Otago Press, Dunedin, New Zealand, 269 pp.

Flora, J.D. (1991). Statistics and Environmental Regulations. *Environmetrics* **2**(2), 129–137.

Freudenburg, W.R., and J.A. Rursch. (1994). The Risks of "Putting the Numbers in Context": A Cautionary Tale. *Risk Analysis* **14**(6), 949–958.

Gaylor, D.W., and J.J. Chen. (1996). A Simple Upper Limit for the Sum of the Risks of the Components in a Mixture. *Risk Analysis* **16**(3), 395–398.

Gentle, J.E. (1985). Monte Carlo Methods. In Kotz, S., and N.L. Johnson, eds. *Encyclopedia of Statistics, Volume 5*. John Wiley & Sons, New York, 612–617.

Gibbons, R.D. (1987a). Statistical Prediction Intervals for the Evaluation of Ground-Water Quality. *Ground Water* **25**, 455–465.

Gibbons, R.D. (1987b). Statistical Models for the Analysis of Volatile Organic Compounds in Waste Disposal Sites. *Ground Water* **25**, 572–580.

Gibbons, R.D. (1990). A General Statistical Procedure for Ground-Water Detection Monitoring at Waste Disposal Facilities. *Ground Water* **28**, 235–243.

Gibbons, R.D. (1991a). Some Additional Nonparametric Prediction Limits for Ground-Water Detection Monitoring at Waste Disposal Facilities. *Ground Water* **29**, 729–736.

Gibbons, R.D. (1991b).   Statistical Tolerance Limits for Ground-Water Monitoring. *Ground Water* **29**, 563–570.

Gibbons, R.D. (1994).   *Statistical Methods for Groundwater Monitoring.* John Wiley & Sons, New York, 286 pp.

Gibbons, R.D. (1995).   Some Statistical and Conceptual Issues in the Detection of Low-Level Environmental Pollutants (with Discussion). *Environmetrics* **2**, 125–167.

Gibbons, R.D. (1996).   Some Conceptual and Statistical Issues in Analysis of Groundwater Monitoring Data. *Environmetrics* **7**, 185–199.

Gibbons, R.D., and J. Baker. (1991).   The Properties of Various Statistical Prediction Intervals for Ground-Water Detection Monitoring. *Journal of Environmental Science and Health* **A26**(4), 535–553.

Gibbons, R.D., D.E. Coleman, and R.F. Maddalone. (1997a).   An Alternative Minimum Level Definition for Analytical Quantification. *Environmental Science and Technology* **31**(7), 2071–2077.   Comments and Discussion in Volume 31(12), 3727–3731, and Volume 32(15), 2346–2353.

Gibbons, R.D., D.E. Coleman, and R.F. Maddalone. (1997b).   Response to Comment on "An Alternative Minimum Level Definition for Analytical Quantification." *Environmental Science and Technology* **31**(12), 3729–3731.

Gibbons, R.D., D.E. Coleman, and R.F. Maddalone. (1998).   Response to Comment on "An Alternative Minimum Level Definition for Analytical Quantification." *Environmental Science and Technology* **32**(15), 2349–2353.

Gibbons, R.D., N.E. Grams, F.H. Jarke, and K.P. Stoub. (1992).   Practical Quantitation Limits. *Chemometrics Intelligent Laboratory Systems* **12**, 225–235.

Gibbons, R.D., F.H. Jarke, and K.P. Stoub. (1991).   Detection Limits: For Linear Calibration Curves with Increasing Variance and Multiple Future Detection Decisions. In Tatsch, D.E., ed. *Waste Testing and Quality Assurance: Volume 3.*   American Society for Testing and Materials, Philadelphia, PA.

Gibrat, R. (1930).   Une loi des repartition economiques: l'effect proportionnel. *Bull. Statist. Gen. Fr.* **19**, 469ff.

Gilbert, R.O. (1987).   *Statistical Methods for Environmental Pollution Monitoring.* Van Nostrand Reinhold, New York, 320 pp.

Gilbert, R.O. (1994). An Overview of Statistical Issues Related to Environmental Cleanup. In Patil, G.P., and C.R. Rao, eds., *Handbook of Statistics, Vol. 12: Environmental Statistics*. North-Holland, Amsterdam, Chapter 27, 867–880.

Gilliom, R.J., and D.R. Helsel. (1986). Estimation of Distributional Parameters for Censored Trace Level Water Quality Data: 1. Estimation Techniques. *Water Resources Research* **22**, 135–146.

Gilliom, R.J., R.M. Hirsch, and E.J. Gilroy. (1984). Effect of Censoring Trace-Level Water-Quality Data on Trend-Detection Capability. *Environmental Science and Technology* **18**, 530–535.

Gilroy, E.J., R.M. Hirsch, and T.A. Cohn. (1990). Mean Square Error of Regression-Based Constituent Transport Estimates. *Water Resources Research* **26**(9), 2069–2077.

Ginevan, M.E., and D.E. Splitstone. (1997). Improving Remediation Decisions at Hazardous Waste Sites with Risk-Based Geostatistical Analysis. *Environmental Science and Technology* **31**(2), 92A–96A.

Glasser, J.A., D.L. Foerst, G.D. McKee, S.A. Quave, and W.L. Budde. (1981). Trace Analyses for Wastewaters. *Environmental Science and Technology* **15**, 1426–1435.

Gleit, A. (1985). Estimation for Small Normal Data Sets with Detection Limits. *Environmental Science and Technology* **19**, 1201–1206.

Graham, R.C. (1993). *Data Analysis for the Chemical Sciences: A Guide to Statistical Techniques*. VCH Publishers, New York, 536 pp. Distributed by John Wiley & Sons, New York.

Graham, S.L., K.J. Davis, W.H. Hansen, and C.H. Graham. (1975). Effects of Prolonged Ethylene Thiourea Ingestion on the Thyroid of the Rat. *Food and Cosmetics Toxicology* **13**(5), 493–499.

Greenwood, J.A., J.M. Landwehr, N.C. Matalas, and J.R. Wallis. (1979). Probability Weighted Moments: Definition and Relation to Parameters of Several Distributions Expressible in Inverse Form. *Water Resources Research* **15**(5), 1049–1054.

Griffiths, D.A. (1980). Interval Estimation for the Three-Parameter Lognormal Distribution via the Likelihood Function. *Applied Statistics* **29**, 58–68.

Gringorten, I.I. (1963). A Plotting Rule for Extreme Probability Paper. *Journal of Geophysical Research* **68**(3), 813–814.

Gupta, A.K. (1952). Estimation of the Mean and Standard Deviation of a Normal Population from a Censored Sample. *Biometrika* **39**, 260–273.

Guttman, I. (1970). *Statistical Tolerance Regions: Classical and Bayesian*. Hafner Publishing Co., Darien, CT, 150 pp.

Haas, C.N. (1997). Importance of Distributional Form in Characterizing Inputs to Monte Carlo Risk Assessments. *Risk Analysis* **17**(1), 107–113.

Haas, C.N., and P.A. Scheff. (1990). Estimation of Averages in Truncated Samples. *Environmental Science and Technology* **24**(6), 912–919.

Hahn, G., and W. Nelson. (1973). A Survey of Prediction Intervals and Their Applications. *Journal of Quality Technology* **5**, 178–188.

Hahn, G.J. (1969). Factors for Calculating Two-Sided Prediction Intervals for Samples from a Normal Distribution. *Journal of the American Statistical Association* **64**(327), 878–898.

Hahn, G.J. (1970a). Additional Factors for Calculating Prediction Intervals for Samples from a Normal Distribution. *Journal of the American Statistical Association* **65**(332), 1668–1676.

Hahn, G.J. (1970b). Statistical Intervals for a Normal Population, Part I: Tables, Examples and Applications. *Journal of Quality Technology* **2**(3), 115–125.

Hahn, G.J. (1970c). Statistical Intervals for a Normal Population, Part II: Formulas, Assumptions, Some Derivations. *Journal of Quality Technology* **2**(4), 195–206.

Hahn, G.J., and W.Q. Meeker. (1991). *Statistical Intervals: A Guide for Practitioners*. John Wiley & Sons, New York, 392 pp.

Haimes, Y.Y., T. Barry, and J.H. Lambert, eds. (1994). When and How Can You Specify a Probability Distribution When You Don't Know Much? *Risk Analysis* **14**(5), 661–706.

Hall, I.J., and R.R. Prairie. (1973). One-Sided Prediction Intervals to Contain at Least m out of k Future Observations. *Technometrics* **15**, 897–914.

Hall, I.J., R.R. Prairie, and C.K. Motlagh. (1975). Non-Parametric Prediction Intervals. *Journal of Quality Technology* **7**(3), 109–114.

Hallenbeck, W.H. (1993). *Quantitative Risk Assessment for Environmental and Occupational Health*. Second Edition. Lewis Publishers, Boca Raton, FL, 224 pp.

Hamby, D.M. (1994). A Review of Techniques for Parameter Sensitivity Analysis of Environmental Models. *Environmental Monitoring and Assessment* **32**, 135–154.

Hamed, M., and P.B. Bedient. (1997). On the Effect of Probability Distributions of Input Variables in Public Health Risk Assessment. *Risk Analysis* **17**(1), 97–105.

Harrington, D.P., and T.R. Fleming. (1982). A Class of Rank Test Procedures for Censored Survival Data. *Biometrika* **69**(3), 553–566.

Harter, H.L., and A.H. Moore. (1966). Local-Maximum-Likelihood Estimation of the Parameters of Three-Parameter Lognormal Populations from Complete and Censored Samples. *Journal of the American Statistical Association* **61**, 842–851.

Hashimoto, L.K., and R.R. Trussell. (1983). Evaluating Water Quality Data Near the Detection Limit. Paper presented at the Advanced Technology Conference, American Water Works Association, Las Vegas, Nevada, June 5–9, 1983.

Hastie, T.J., and R.J. Tibshirani. (1990). *Generalized Additive Models*. Chapman & Hall, New York.

Hattis, D. (1990). Three Candidate "Laws" of Uncertainty Analysis. *Risk Analysis* **10**(1), 11.

Hattis, D., and D.E. Burmaster. (1994). Assessment of Variability and Uncertainty Distributions for Practical Risk Analyses. *Risk Analysis* **14**(5), 713–730.

Hayes, B. (1993). The Wheel of Fortune. *American Scientist* **81**, 114–118.

Hazardous Materials Training and Research Institute (HMTRI). (1997). *Site Characterization: Sampling and Analysis*. Van Nostrand Reinhold, 316 pp. Distributed by John Wiley & Sons, New York.

Helsel, D.R. (1987). Advantages of Nonparametric Procedures for Analysis of Water Quality Data. *Hydrological Sciences Journal* **32**, 179–190.

Helsel, D.R. (1990). Less than Obvious: Statistical Treatment of Data Below the Detection Limit. *Environmental Science and Technology* **24**(12), 1766–1774.

Helsel, D.R., and T.A. Cohn. (1988). Estimation of Descriptive Statistics for Multiply Censored Water Quality Data. *Water Resources Research* **24**(12), 1997–2004.

Helsel, D.R., and R.J. Gilliom. (1986). Estimation of Distributional Parameters for Censored Trace Level Water Quality Data: 2. Verification and Applications. *Water Resources Research* **22**, 147–155.

Helsel, D.R., and R.M. Hirsch. (1988). Discussion of Applicability of the t-test for Detecting Trends in Water Quality Variables. *Water Resources Bulletin* **24**(1), 201–204.

Helsel, D.R., and R.M. Hirsch. (1992). *Statistical Methods in Water Resources Research.* Elsevier, New York, 522 pp.

Hettmansperger, T.P. (1984). *Statistical Inference Based on Ranks.* John Wiley & Sons, New York, 323 pp.

Heyde, C.C. (1963). On a Property of the Lognormal Distribution. *Journal of the Royal Statistical Society, Series B* **25**, 392–393.

Hill, B.M. (1963). The Three-Parameter Lognormal Distribution and Bayesian Analysis of a Point-Source Epidemic. *Journal of the American Statistical Association* **58**, 72–84.

Hinkley, D.V., and G. Runger. (1984). The Analysis of Transformed Data (with Discussion). *Journal of the American Statistical Association* **79**, 302–320.

Hirsch, R.M. (1988). Statistical Methods and Sampling Design for Estimating Step Trends in Surface-Water Quality. *Water Resources Bulletin* **24**(3), 493–503.

Hirsch, R.M., R.B. Alexander, and R.A. Smith. (1991). Selection of Methods for the Detection and Estimation of Trends in Water Quality. *Water Resources Research* **27**(5), 803–813.

Hirsch, R.M., and E.J. Gilroy. (1984). Methods of Fitting a Straight Line to Data: Examples in Water Resources. *Water Resources Bulletin* **20**(5), 705–711.

Hirsch, R.M., D.R. Helsel, T.A. Cohn, and E.J. Gilroy. (1993). Statistical Analysis of Hydrologic Data. In Maidment, D.R., ed. *Handbook of Hydrology.* McGraw-Hill, New York, Chapter 17.

Hirsch, R.M., and J.R. Slack. (1984). A Nonparametric Trend Test for Seasonal Data with Serial Dependence. *Water Resources Research* **20**(6), 727–732.

Hirsch, R.M., J.R. Slack, and R.A. Smith. (1982). Techniques of Trend Analysis for Monthly Water Quality Data. *Water Resources Research* **18**(1), 107–121.

Hirsch, R.M., and J.R. Stedinger. (1987). Plotting Positions for Historical Floods and Their Precision. *Water Resources Research* **23**(4), 715–727.

Hoaglin, D.C. (1988). Transformations in Everyday Experience. *Chance* **1**, 40–45.

Hoaglin, D.C., and D.F. Andrews. (1975). The Reporting of Computation-Based Results in Statistics. *The American Statistician* **29**(3), 122–126.

Hoaglin, D.C., F. Mosteller, and J.W. Tukey, eds. (1983). *Understanding Robust and Exploratory Data Analysis*. John Wiley & Sons, New York, 447 pp.

Hoeffding, W. (1948). A Class of Statistics with Asymptotically Normal Distribution. *Annals of Mathematical Statistics* **19**, 293–325.

Hoffman, F.O., and J.S. Hammonds. (1994). Propagation of Uncertainty in Risk Assessments: The Need to Distinguish between Uncertainty Due to Lack of Knowledge and Uncertainty Due to Variability. *Risk Analysis* **14**(5), 707–712.

Hollander, M., and D.A. Wolfe. (1999). *Nonparametric Statistical Methods*. Second Edition. John Wiley & Sons, New York, 787 pp.

Hoshi, K., J.R. Stedinger, and J. Burges. (1984). Estimation of Log-Normal Quantiles: Monte Carlo Results and First-Order Approximations. *Journal of Hydrology* **71**, 1–30.

Hosking, J.R.M. (1984). Testing Whether the Shape Parameter is Zero in the Generalized Extreme-Value Distribution. *Biometrika* **71**(2), 367–374.

Hosking, J.R.M. (1985). Algorithm AS 215: Maximum-Likelihood Estimation of the Parameters of the Generalized Extreme-Value Distribution. *Applied Statistics* **34**(3), 301–310.

Hosking, J.R.M. (1990). L-Moments: Analysis and Estimation of Distributions Using Linear Combinations of Order Statistics. *Journal of the Royal Statistical Society, Series B* **52**(1), 105–124.

Hosking, J.R.M., and J.R. Wallis (1995). A Comparison of Unbiased and Plotting-Position Estimators of L Moments. *Water Resources Research* **31**(8), 2019–2025.

Hosking, J.R.M., J.R. Wallis, and E.F. Wood. (1985). Estimation of the Generalized Extreme-Value Distribution by the Method of Probability-Weighted Moments. *Technometrics* **27**(3), 251–261.

Hubaux, A., and G. Vos. (1970). Decision and Detection Limits for Linear Calibration Curves. *Annals of Chemistry* **42**, 849–855.

Hughes, J.P., and S.P. Millard. (1988). A Tau-Like Test for Trend in the Presence of Multiple Censoring Points. *Water Resources Bulletin* **24**(3), 521–531.

Ibrekk, H., and M.G. Morgan. (1987). Graphical Communication of Uncertain Quantities to Nontechnical People. *Risk Analysis* **7**(4), 519–529.

Iman, R.L., and W.J. Conover. (1980). Small Sample Sensitivity Analysis Techniques for Computer Models, With an Application to Risk Assessment (with Comments). *Communications in Statistics—Volume A, Theory and Methods* **9**(17), 1749–1874.

Iman, R.L., and W.J. Conover. (1982). A Distribution-Free Approach to Inducing Rank Correlation Among Input Variables. *Communications in Statistics—Volume B, Simulation and Computation* **11**(3), 311–334.

Iman, R.L., and J.M. Davenport. (1982). Rank Correlation Plots For Use with Correlated Input Variables. *Communications in Statistics—Volume B, Simulation and Computation* **11**(3), 335–360.

Iman, R.L., and J.C. Helton. (1988). An Investigation of Uncertainty and Sensitivity Analysis Techniques for Computer Models. *Risk Analysis* **8**(1), 71–90.

Iman, R.L., and J.C. Helton. (1991). The Repeatability of Uncertainty and Sensitivity Analyses for Complex Probabilistic Risk Assessments. *Risk Analysis* **11**(4), 591–606.

Isaaks, E.H., and R.M. Srivastava. (1989). *An Introduction to Applied Geostatistics*. Oxford University Press, New York, 561 pp.

Israeli, M., and C.B. Nelson. (1992). Distribution and Expected Time of Residence for U.S. Households. *Risk Analysis* **12**(1), 65–72.

Jenkinson, A.F. (1955). The Frequency Distribution of the Annual Maximum (or Minimum) of Meteorological Events. *Quarterly Journal of the Royal Meteorological Society* **81**, 158–171.

Jenkinson, A.F. (1969). *Statistics of Extremes*. Technical Note 98, World Meteorological Office, Geneva.

Johnson, G.D., B.D. Nussbaum, G.P. Patil, and N.P. Ross. (1996). Designing Cost-Effective Environmental Sampling Using Concomitant Information. *Chance* **9**(1), 4–11.

Johnson, N.L., S. Kotz, and N. Balakrishnan. (1994). *Continuous Univariate Distributions, Volume 1*. Second Edition. John Wiley & Sons, New York, 756 pp.

Johnson, N.L., S. Kotz, and N. Balakrishnan. (1995). *Continuous Univariate Distributions, Volume 2.* Second Edition. John Wiley & Sons, New York, 719 pp.

Johnson, N.L., S. Kotz, and N. Balakrishnan. (1997). *Discrete Multivariate Distributions.* John Wiley & Sons, New York, 299 pp.

Johnson, N.L., S. Kotz, and A.W. Kemp. (1992). *Univariate Discrete Distributions.* Second Edition. John Wiley & Sons, New York, 565 pp.

Johnson, N.L., and B.L. Welch. (1940). Applications of the Non-Central t-Distribution. *Biometrika* **31**, 362–389.

Johnson, R.A., D.R. Gan, and P.M. Berthouex. (1995). Goodness-of-Fit Using Very Small but Related Samples with Application to Censored Data Estimation of PCB Contamination. *Environmetrics* **6**, 341–348.

Johnson, R.A., S. Verrill, and D.H. Moore. (1987). Two-Sample Rank Tests for Detecting Changes That Occur in a Small Proportion of the Treated Population. *Biometrics* **43**, 641–655.

Johnson, R.A., and D.W. Wichern. (1998). *Applied Multivariate Statistical Analysis.* Fourth Edition. Prentice-Hall, Englewood Cliffs, NJ, 816 pp.

Journel, A.G., and Huijbregts, C.J. (1978). *Mining Geostatistics.* Academic Press, London.

Judge, G.G., W.E. Griffiths, R.C. Hill, H. Lutkepohl, and T.C. Lee. (1985). Qualitative and Limited Dependent Variable Models: Chapter 18. In *The Theory and Practice of Econometrics.* John Wiley & Sons, New York, 1019 pp.

Kahn, H.D., W.A. Telliard, and C.E. White. (1998). Comment on "An Alternative Minimum Level Definition for Analytical Quantification" (with Response). *Environmental Science and Technology* **32**(5), 2346–2353.

Kahn, H.D., C.E. White, K. Stralka, and R. Kuznetsovski. (1997). Alternative Estimates of Detection. *Proceedings of the Twentieth Annual EPA Conference on Analysis of Pollutants in the Environment*, May 7–8, Norfolk, VA. U.S. Environmental Protection Agency, Washington, D.C.

Kaiser, H. (1965). Zum Problem der Nachweisgrenze. *Fresenius' Z. Anal. Chem.* **209**, 1.

Kalbfleisch, J.D., and R.L. Prentice. (1980). *The Statistical Analysis of Failure Time Data.* John Wiley & Sons, New York, 321 pp.

Kaluzny, S.P., S.C. Vega, T.P. Cardosa, and A.A. Shelly. (1998). *S+SPATIALSTATS User's Manual for Windows and UNIX.* Springer-Verlag, New York, 327 pp.

Kaplan, E.L., and P. Meier. (1958). Nonparametric Estimation from Incomplete Observations. *Journal of the American Statistical Association* **53**, 457–481.

Kapteyn, J.C. (1903). *Skew Frequency Curves in Biology and Statistics.* Astronomical Laboratory, Noordhoff, Groningen.

Keith, L.H. (1991). *Environmental Sampling and Analysis: A Practical Guide.* Lewis Publishers, Boca Raton, FL, 143 pp.

Keith, L.H., ed. (1996). *Principles of Environmental Sampling.* Second Edition. American Chemical Society, Washington, D.C., 848 pp. Distributed by Oxford University Press, New York.

Kempthorne, O., and T.E. Doerfler. (1969). The Behavior of Some Significance Tests under Experimental Randomization. *Biometrika* **56**(2), 231–248.

Kendall, M.G., and J.D. Gibbons. (1990). *Rank Correlation Methods.* Fifth Edition. Charles Griffin, London, 260 pp.

Kendall, M.G., and A. Stuart. (1991). *The Advanced Theory of Statistics, Volume 2: Inference and Relationship.* Fifth Edition. Oxford University Press, New York, 1323 pp.

Kennedy, W.J., and J.E. Gentle. (1980). *Statistical Computing.* Marcel Dekker, New York, 591 pp.

Kim, P.J., and R.I. Jennrich. (1973). Tables of the Exact Sampling Distribution of the Two Sample Kolmogorov-Smirnov Criterion. In Harter, H.L., and D.B. Owen, eds. *Selected Tables in Mathematical Statistics, Vol. 1.* American Mathematical Society, Providence, RI, pp. 79–170.

Kimbrough, D.E. (1997). Comment on "An Alternative Minimum Level Definition for Analytical Quantification" (with Response). *Environmental Science and Technology* **31**(12), 3727–3731.

Kitanidis, P.K. (1997). *Introduction to Geostatistics.* Cambridge University Press, London, 249 pp.

Kodell, R.L., H. Ahn, J.J. Chen, J.A. Springer, C.N. Barton, and R.C. Hertzberg. (1995). Upper Bound Risk Estimates for Mixtures of Carcinogens. *Toxicology* **105**, 199–208.

Kodell, R.L., and J.J. Chen. (1994). Reducing Conservatism in Risk Estimation for Mixtures of Carcinogens. *Risk Analysis* **14**(3), 327–332.

Kolmogorov, A.N. (1933). Sulla determinazione empirica di una legge di distribuzione. *Giornale dell' Istituto Italiano degle Attuari* **4**, 83–91.

Kotz, S., and N.L. Johnson, eds. (1985). *Encyclopedia of Statistics*. John Wiley & Sons, New York.

Kuchenhoff, H., and M. Thamerus. (1996). Extreme Value Analysis of Munich Air Pollution Data. *Environmental and Ecological Statistics* **3**, 127–141.

Kushner, E.J. (1976). On Determining the Statistical Parameters for Pollution Concentrations from a Truncated Data Set. *Atmospheric Environment* **10**, 975–979.

Lambert, D., B. Peterson, and I. Terpenning. (1991). Nondetects, Detection Limits, and the Probability of Detection. *Journal of the American Statistical Association* **86**(414), 266–277.

Lambert, J.H., N.C. Matalas, C.W. Ling, Y.Y. Haimes, and D. Li. (1994). Selection of Probability Distributions in Characterizing Risk of Extreme Events. *Risk Analysis* **14**(5), 731–742.

Land, C.E. (1971). Confidence Intervals for Linear Functions of the Normal Mean and Variance. *The Annals of Mathematical Statistics* **42**(4), 1187–1205.

Land, C.E. (1972). An Evaluation of Approximate Confidence Interval Estimation Methods for Lognormal Means. *Technometrics* **14**(1), 145–158.

Land, C.E. (1973). Standard Confidence Limits for Linear Functions of the Normal Mean and Variance. *Journal of the American Statistical Association* **68**(344), 960–963.

Land, C.E. (1975). Tables of confidence limits for linear functions of the normal mean and variance, in *Selected Tables in Mathematical Statistics, Vol. III*. American Mathematical Society, Providence, RI, pp. 385–419.

Landwehr, J.M., N.C. Matalas, and J.R. Wallis. (1979). Probability Weighted Moments Compared with Some Traditional Techniques in Estimating Gumbel Parameters and Quantiles. *Water Resources Research* **15**(5), 1055–1064.

Laudan, L. (1997). *Danger Ahead: The Risks You Really Face on Life's Highway*. John Wiley & Sons, New York, 203 pp.

Law, A.M., and W.D. Kelton. (1991). *Simulation Modeling & Analysis*. Second Edition. McGraw-Hill, Inc., New York, 759 pp.

Latta, R.B. (1981). A Monte Carlo Study of Some Two-Sample Rank Tests with Censored Data. *Journal of the American Statistical Association* **76**(375), 713–719.

Lee, E.T. (1980). *Statistical Methods for Survival Data Analysis.* Lifetime Learning Publications, Belmont, CA, 557 pp.

Lehmann, E.L. (1975). *Nonparametrics: Statistical Methods Based on Ranks.* Holden-Day, Oakland, CA, 457 pp.

Leidel, N.A., K.A. Busch, and J.R. Lynch. (1977). *Occupational Exposure Sampling Strategy Manual.* U.S. Department of Health, Education, and Welfare, Public Health Service, Center for Disease Control, National Institute for Occupational Safety and Health, Cincinnati, OH, January, 1977.

Lewis, H.W. (1990). *Technological Risk.* W.W. Norton & Company, New York, 353 pp.

Lettenmaier, D.P. (1976). Detection of Trends in Water Quality Data from Data with Dependent Observations. *Water Resources Research* **12**, 1037–1046.

Lettenmaier, D.P. (1988). Multivariate Non-Parametric Tests for Trend in Water Quality. *Water Resources Bulletin* **24**(3), 505–512.

Likes, J. (1980). Variance of the MVUE for Lognormal Variance. *Technometrics* **22**(2), 253–258.

Looney, S.W., and T.R. Gulledge. (1985). Use of the Correlation Coefficient with Normal Probability Plots. *The American Statistician* **39**(1), 75–79.

Lovison, G., S.D. Gore, and G.P. Patil. (1994). Design and Analysis of Composite Sampling Procedures: A Review. In Patil, G.P., and C.R. Rao, eds., *Handbook of Statistics, Vol. 12: Environmental Statistics.* North-Holland, Amsterdam, Chapter 4, 103–166.

Lu, L., and J.R. Stedinger. (1992). Variance of Two- and Three-Parameter GEV/PWM Quantile Estimators: Formulae, Confidence Intervals, and a Comparison. *Journal of Hydrology* **138**, 247–267.

Lucas, J.M. (1982). Combined Shewart-CUMSUM Quality Control Schemes. *Journal of Quality Technology* **14**, 51–59.

Lucas, J.M. (1985). Cumulative Sum (CUMSUM) Control Schemes. *Communications in Statistics A—Theory and Methods* **14**(11), 2689–2704.

Lundgren, R., and A. McMakin. (1998). *Risk Communication: A Handbook for Communicating Environmental, Safety, and Health Risks.* Battelle Press, Columbus, OH, 362 pp.

Macleod, A.J. (1989). Remark AS R76: A Remark on Algorithm AS 215: Maximum Likelihood Estimation of the Parameters of the Generalized Extreme-Value Distribution. *Applied Statistics* **38**(1), 198–199.

Maritz, J.S., and A.H. Munro. (1967). On the Use of the Generalized Extreme-Value Distribution in Estimating Extreme Percentiles. *Biometrics* **23**(1), 79–103.

Marsaglia, G. et al. (1973). *Random Number Package: "Super-Duper."* School of Computer Science, McGill University, Montreal.

Massart, D.L., B.G.M. Vandeginste, S.N. Deming, Y. Michotte, and L. Kaufman. (1988). *Chemometrics: A Textbook.* Elsevier, New York, Chapter 7.

McBean, E.A., and F.A. Rovers. (1992). Estimation of the Probability of Exceedance of Contaminant Concentrations. *Ground Water Monitoring Review* **Winter**, 115–119.

McBean, E.A., and R.A. Rovers. (1998). *Statistical Procedures for Analysis of Environmental Monitoring Data & Risk Assessment.* Prentice-Hall PTR, Upper Saddle River, NJ, 313 pp.

McCullagh, P., and J.A. Nelder. (1989). *Generalized Linear Models.* Second Edition. Chapman & Hall, New York, 511 pp.

McIntyre, G.A. (1952). A Method of Unbiased Selective Sampling, Using Ranked Sets. *Australian Journal of Agricultural Resources* **3**, 385–390.

McKay, M.D., R.J. Beckman., and W.J. Conover. (1979). A Comparison of Three Methods for Selecting Values of Input Variables in the Analysis of Output from a Computer Code. *Technometrics* **21**(2), 239–245.

McKean, J.W., and G.L. Sievers. (1989). Rank Scores Suitable for Analysis of Linear Models under Asymmetric Error Distributions. *Technometrics* **31**, 207–218.

McKone, T.E. (1994). Uncertainty and Variability in Human Exposures to Soil and Contaminants through Home-Grown Food: A Monte Carlo Assessment. *Risk Analysis* **14**(4), 449–463.

McKone, T.E., and K.T. Bogen. (1991). Predicting the Uncertainties in Risk Assessment. *Environmental Science and Technology* **25**(10), 1674–1681.

McLeod, A.I., K.W. Hipel, and B.A. Bodo. (1991). Trend Analysis Methodology for Water Quality Time Series. *Environmetrics* **2**(2), 169–200.

McNichols, R.J., and C.B. Davis. (1988). Statistical Issues and Problems in Ground Water Detection Monitoring at Hazardous Waste Facilities. *Ground Water Monitoring Review* **8**(4), 135–150.

Meeker, W.Q., and L.A. Escobar. (1998). *Statistical Methods for Reliability Data.* John Wiley & Sons, New York.

Meier, P.C., and R.E. Zund. (1993). *Statistical Methods in Analytical Chemistry*. John Wiley & Sons, New York, 321 pp.

Michael, J.R., and W.R. Schucany. (1986). Analysis of Data from Censored Samples. In D'Agostino, R.B., and M.A. Stephens, eds. *Goodness-of-Fit Techniques*. Marcel Dekker, New York, 560 pp., Chapter 11, 461–496.

Millard, S.P. (1987a). Environmental Monitoring, Statistics, and the Law: Room for Improvement (with Comment). *The American Statistician* **41**(4), 249–259.

Millard, S.P. (1987b). Proof of Safety vs. Proof of Hazard. *Biometrics* **43**, 719–725.

Millard, S.P. (1996). Estimating a Percent Reduction in Load. *Water Resources Research* **32**(6), 1761–1766.

Millard, S.P. (1998). *ENVIRONMENTALSTATS for S-PLUS User's Manual for Windows and UNIX*. Springer-Verlag, New York, 381 pp.

Millard, S.P., and S.J. Deverel. (1988). Nonparametric Statistical Methods for Comparing Two Sites Based on Data with Multiple Nondetect Limits. *Water Resources Research* **24**(12), 2087–2098.

Millard, S.P., and D.P. Lettenmaier. (1986). Optimal Design of Biological Sampling Programs Using the Analysis of Variance. *Estuarine, Coastal and Shelf Science* **22**, 637–656.

Millard, S.P., J.R. Yearsley, and D.P. Lettenmaier. (1985). Space-Time Correlation and Its Effects on Methods for Detecting Aquatic Ecological Change. *Canadian Journal of Fisheries and Aquatic Science*, **42**(8), 1391–1400. Correction: (1986), 43, 1680.

Miller, R.G. (1981a). *Simultaneous Statistical Inference*. Springer-Verlag, New York, 299 pp.

Miller, R.G. (1981b). *Survival Analysis*. John Wiley & Sons, New York, 238 pp.

Milliken, G.A., and D.E. Johnson. (1992). *Analysis of Messy Data, Volume I: Designed Experiments*. Chapman & Hall, New York, 473 pp.

Mode, N.A., L.L. Conquest, and D.A. Marker. (1999). Ranked Set Sampling for Ecological Research: Accounting for the Total Costs of Sampling. *Environmetrics* **10**, 179–194.

Molak, V., ed. (1997). *Fundamentals of Risk Analysis and Risk Management*. Lewis Publishers, Boca Raton, FL, 472 pp.

Montgomery, D.C. (1997). *Introduction to Statistical Quality Control.* Third Edition. John Wiley & Sons, New York.

Moore, D.S. (1986). Tests of Chi-Squared Type. In D'Agostino, R.B., and M.A. Stephens, eds. *Goodness-of-Fit Techniques.* Marcel Dekker, New York, pp. 63–95.

Morgan, M.G., and M. Henrion. (1990). *Uncertainty: A Guide to Dealing with Uncertainty in Quantitative Risk and Policy Analysis.* Cambridge University Press, New York, 332 pp.

Neely, W.B. (1994). *Introduction to Chemical Exposure and Risk Assessment.* Lewis Publishers, Boca Raton, FL, 190 pp.

Neerchal, N. K., and S. L. Brunenmeister. (1993). Estimation of Trend in Chesapeake Bay Water Quality Data. In Patil, G.P., and C.R. Rao, eds., *Handbook of Statistics, Vol. 6: Multivariate Environmental Statistics.* North-Holland, Amsterdam, Chapter 19, 407–422.

Nelson, W. (1969). Hazard Plotting for Incomplete Failure Data. *Journal of Quality Technology* **1**(1), 27–52.

Nelson, W. (1972). Theory and Applications of Hazard Plotting for Censored Failure Data. *Technometrics* **14**, 945–966.

Nelson, W. (1982). *Applied Life Data Analysis.* John Wiley & Sons, New York, 634 pp.

Nelson, W., and J. Schmee. (1979). Inference for (Log) Normal Life Distributions from Small Singly Censored Samples and BLUEs. *Technometrics* **21**, 43–54.

Nelson, W.R. (1970). Confidence Intervals for the Ratio of Two Poisson Means and Poisson Predictor Intervals. *IEEE Transactions on Reliability* **R-19**(2), 42–49.

Neptune, D., E.P. Brantly, M.J. Messner, and D.I. Michael. (1990). Quantitative Decision Making in Superfund: A Data Quality Objectives Case Study. *Hazardous Material Control* **May/June**, pp. 18–27.

Newman, M.C. (1995). *Quantitative Methods in Aquatic Ecotoxicology.* Lewis Publishers, Boca Raton, FL, 426 pp.

Newman, M.C., and P.M. Dixon. (1990). UNCENSOR: A Program to Estimate Means and Standard Deviations for Data Sets with Below Detection Limit Observations. *American Environmental Laboratory* **4(90)**, 26–30.

Newman, M.C., P.M. Dixon, B.B. Looney, and J.E. Pinder. (1989). Estimating Mean and Variance for Environmental Samples with Below Detection Limit Observations. *Water Resources Bulletin* **25**(4), 905–916.

Odeh, R.E., and D.B. Owen. (1980). *Tables for Normal Tolerance Limits, Sampling Plans, and Screening.* Marcel Dekker, New York, 316 pp.

Osborne, C. (1991). Statistical Calibration: A Review. *International Statistical Review* **59**(3), 309–336.

Ott, W.R. (1990). A Physical Explanation of the Lognormality of Pollutant Concentrations. *Journal of the Air and Waste Management Association* **40**, 1378–1383.

Ott, W.R. (1995). *Environmental Statistics and Data Analysis.* Lewis Publishers, Boca Raton, FL, 313 pp.

Owen, D.B. (1962). *Handbook of Statistical Tables.* Addison-Wesley, Reading, MA, 580 pp.

Owen, W., and T. DeRouen. (1980). Estimation of the Mean for Lognormal Data Containing Zeros and Left-Censored Values, with Applications to the Measurement of Worker Exposure to Air Contaminants. *Biometrics* **36**, 707–719.

Parkin, T.B., S.T. Chester, and J.A. Robinson. (1990). Calculating Confidence Intervals for the Mean of a Lognormally Distributed Variable. *Journal of the Soil Science Society of America* **54**, 321–326.

Parkin, T.B., J.J. Meisinger, S.T. Chester, J.L. Starr, and J.A. Robinson. (1988). Evaluation of Statistical Estimation Methods for Lognormally Distributed Variables. *Journal of the Soil Science Society of America* **52**, 323–329.

Parmar, M.K.B., and D. Machin. (1995). *Survival Analysis: A Practical Approach.* John Wiley & Sons, New York, 255 pp.

Patil, G.P., S.D. Gore, and A.K. Sinha. (1994). Environmental Chemistry, Statistical Modeling, and Observational Economy. In Cothern, C.R., and N.P. Ross, eds., *Environmental Statistics, Assessment, and Forecasting,* Lewis Publishers, Boca Raton, FL, Chapter 3, pp. 57–97.

Patil, G.P., and C.R. Rao, eds. (1994). *Handbook of Statistics, Vol. 12: Environmental Statistics.* North-Holland, Amsterdam, 927 pp.

Patil, G.P., A.K. Sinha, and C. Taillie. (1994). Ranked Set Sampling. In Patil, G.P., and C.R. Rao, eds., *Handbook of Statistics, Vol. 12: Environmental Statistics.* North-Holland, Amsterdam, Chapter 5, 167–200.

Pearson, E.S., and H.O. Hartley, eds. (1970). *Biometrika Tables for Statisticians, Volume 1*. Cambridge University Press, New York, 270 pp.

Persson, T. and H. Rootzen. (1977). Simple and Highly Efficient Estimators for a Type I Censored Normal Sample. *Biometrika* **64**, 123–128.

Peterson, B.P., S.P. Millard, E.F. Wood, and D.P. Lettenmaier. (1990). Design of a Soil Sampling Study to Determine the Habitability of the Emergency Declaration Area, Love Canal, New York. *Environmetrics* **1**(1), 89–119.

Pettitt, A.N. (1976). Cramér-von Mises Statistics for Testing Normality with Censored Samples. *Biometrika* **63**(3), 475–481.

Pettitt, A.N., and M.A. Stephens. (1976). Modified Cramér-von Mises Statistics for Censored Data. *Biometrika* **63**(2), 291–298.

Piegorsch, W.W., and A.J. Bailer. (1997). *Statistics for Environmental Biology and Toxicology*. Chapman & Hall, New York, 579 pp.

Pomeranz, J. (1973). Exact Cumulative Distribution of the Kolmogorov-Smirnov Statistic for Small Samples (Algorithm 487). *Collected Algorithms from CACM*.

Porter, P.S., R.C. Ward, and H.F. Bell. (1988). The Detection Limit. *Environmental Science and Technology* **22**(8), 856–861.

Prentice, R.L. (1978). Linear Rank Tests with Right Censored Data. *Biometrika* **65**, 167–179.

Prentice, R.L. (1985). Linear Rank Tests. In Kotz, S., and N.L. Johnson, eds. *Encyclopedia of Statistical Science*. John Wiley & Sons, New York. Volume 5, pp. 51–58.

Prentice, R.L., and P. Marek. (1979). A Qualitative Discrepancy between Censored Data Rank Tests. *Biometrics* **35**, 861–867.

Prescott, P., and A.T. Walden. (1980). Maximum Likelihood Estimation of the Parameters of the Generalized Extreme-Value Distribution. *Biometrika* **67**(3), 723–724.

Prescott, P., and A.T. Walden. (1983). Maximum Likelihood Estimation of the Three-Parameter Generalized Extreme-Value Distribution from Censored Samples. *Journal of Statistical Computing and Simulation* **16**, 241–250.

Press, W.H., S.A. Teukolsky, W.T. Vetterling, and B.P. Flannery. (1992). *Numerical Recipes in FORTRAN: The Art of Scientific Computing*. Second Edition. Cambridge University Press, New York, 963 pp.

Qian, S.S. (1997). Estimating the Area Affected by Phosphorus Runoff in an Everglades Wetland: A Comparison of Universal Kriging and Bayesian Kriging. *Environmental and Ecological Statistics* **4**, 1–29.

Ranasinghe, J.A., L.C. Scott, and R. Newport. (1992). *Long-term Benthic Monitoring and Assessment Program for the Maryland Portion of the Bay, Jul 1984-Dec 1991*. Report prepared for the Maryland Department of the Environment and the Maryland Department of Natural Resources by Versar, Inc., Columbia, MD.

Ripley, B.D. (1981). *Spatial Statistics*. John Wiley & Sons, New York.

Rocke, D.M., and S. Lorenzato. (1995). A Two-Component Model for Measurement Error in Analytical Chemistry. *Technometrics* **37**(2), 176–184.

Rodricks, J.V. (1992). *Calculated Risks: The Toxicity and Human Health Risks of Chemicals in Our Environment*. Cambridge University Press, New York, 256 pp.

Rogers, J. (1992). Assessing Attainment of Ground Water Cleanup Standards Using Modified Sequential t-Tests. *Environmetrics* **3**(3), 335–359.

Rosebury, A.M., and D.E. Burmaster. (1992). Lognormal Distributions for Water Intake by Children and Adults. *Risk Analysis* **12**(1), 99–104.

Rowe, W.D. (1994). Understanding Uncertainty. *Risk Analysis* **14**(5), 743–750.

Royston, J.P. (1982). Algorithm AS 177. Expected Normal Order Statistics (Exact and Approximate). *Applied Statistics* **31**, 161–165.

Royston, J.P. (1992a). Approximating the Shapiro-Wilk W-Test for Non-Normality. *Statistics and Computing* **2**, 117–119.

Royston, J.P. (1992b). Estimation, Reference Ranges and Goodness of Fit for the Three-Parameter Log-Normal Distribution. *Statistics in Medicine* **11**, 897–912.

Royston, J.P. (1992c). A Pocket-Calculator Algorithm for the Shapiro-Francia Test of Non-Normality: An Application to Medicine. *Statistics in Medicine* **12**, 181–184.

Royston, P. (1993). A Toolkit for Testing for Non-Normality in Complete and Censored Samples. *The Statistician* **42**, 37–43.

Rubinstein, R.Y. (1981). *Simulation and the Monte Carlo Method*. John Wiley & Sons, New York, 278 pp.

Ruffle, B., D.E. Burmaster, P.D. Anderson, and H.D. Gordon. (1994). Lognormal Distributions for Fish Consumption by the General U.S. Population. *Risk Analysis* **14**(4), 395–404.

Ryan, T.P. (1989). *Statistical Methods for Quality Improvement.* John Wiley & Sons, New York, 446 pp.

Saltelli, A., and J. Marivoet. (1990). Non-Parametric Statistics in Sensitivity Analysis for Model Output: A Comparison of Selected Techniques. *Reliability Engineering and System Safety* **28**, 229–253.

Sara, Martin N. (1994). *Standard Handbook for Solid and Hazardous Waste Facility Assessment.* Lewis Publishers, Boca Raton, FL. Chapter 11.

Sarhan, A.E., and B.G. Greenberg. (1956). Estimation of Location and Scale Parameters by Order Statistics for Singly and Doubly Censored Samples, Part I, The Normal Distribution up to Samples of Size 10. *Annals of Mathematical Statistics* **27**, 427–457.

Saw, J.G. (1961a). Estimation of the Normal Population Parameters Given a Type I Censored Sample. *Biometrika* **48**, 367–377.

Saw, J.G. (1961b). The Bias of the Maximum Likelihood Estimators of Location and Scale Parameters Given a Type II Censored Normal Sample. *Biometrika* **48**, 448–451.

Schaeffer, D.J., H.W. Kerster, and K.G. Janardan. (1980). Grab Versus Composite Sampling: A Primer for Managers and Engineers. *Journal of Environmental Management* **4**, 157–163.

Scheffé, H. (1959). *The Analysis of Variance.* John Wiley & Sons, New York, 477 pp.

Scheuer, E.M., and D.S. Stoller. (1962). On the Generation of Normal Random Vectors. *Technometrics* **4**(2), 278–281.

Schmee, J., D.Gladstein, and W. Nelson. (1985). Confidence Limits for Parameters of a Normal Distribution from Singly Censored Samples, Using Maximum Likelihood. *Technometrics* **27**(2), 119–128.

Schmoyer, R.L., J.J. Beauchamp, C.C. Brandt, and F.O. Hoffman, Jr. (1996). Difficulties with the Lognormal Model in Mean Estimation and Testing. *Environmental and Ecological Statistics* **3**, 81–97.

Schneider, H. (1986). *Truncated and Censored Samples from Normal Populations.* Marcel Dekker, New York, 273 pp.

Schneider, H., and L. Weisfeld. (1986). Inference Based on Type II Censored Samples. *Biometrics* **42**, 531–536.

Seiler, F.A. (1987). Error Propagation for Large Errors. *Risk Analysis* **7**(4), 509–518.

Seiler, F.A., and J.L. Alvarez. (1996). On the Selection of Distributions for Stochastic Variables. *Risk Analysis* **16**(1), 5–18.

Sengupta, S., and N. K. Neerchal. (1994). Classification of Grab Samples Using Composite Sampling: An Improved Strategy. *Calcutta Statistical Association Bulletin* **44**, 195–208.

Serfling, R.J. (1980). *Approximation Theorems of Mathematical Statistics.* John Wiley & Sons, New York, 371 pp.

Shaffner, G. (1999). *The Arithmetic of Life.* Ballantine Books, New York, 208 pp.

Shapiro, S.S., and C.W. Brian. (1981). A Review of Distributional Testing Procedures and Development of a Censored Sample Distributional Test. In Taillie, C., G.P. Patil, and B.A. Baldessari, eds. *Statistical Distributions in Scientific Work, Volume 5—Inferential Problems and Properties*, 1–24.

Shapiro, S.S., and R.S. Francia. (1972). An Approximate Analysis of Variance Test for Normality. *Journal of the American Statistical Association* **67**(337), 215–219.

Shapiro, S.S., and M.B. Wilk. (1965). An Analysis of Variance Test for Normality (Complete Samples). *Biometrika* **52**, 591–611.

Shea, B.L., and A.J. Scallon. (1988). Remark AS R72. A Remark on Algorithm AS 128. Approximating the Covariance Matrix of Normal Order Statistics. *Applied Statistics* **37**, 151–155.

Sheskin, D.J. (1997). *Handbook of Parametric and Nonparametric Statistical Procedures.* CRC Press, Boca Raton, FL, 719 pp.

Shlyakhter, A.I. (1994). An Improved Framework for Uncertainty Analysis: Accounting for Unsuspected Errors. *Risk Analysis* **14**(4), 441–447.

Shumway, R.H., A.S. Azari, and P. Johnson. (1989). Estimating Mean Concentrations under Transformations for Environmental Data with Detection Limits. *Technometrics* **31**(3), 347–356.

Silverman, B. W. (1986). *Density Estimation for Statistics and Data Analysis.* Chapman & Hall, New York.

Singh, A. (1993). Multivariate Decision and Detection Limits. *Analytica Chimica Acta* **277**, 205–214.

Singh, A.K., A. Singh, and M. Engelhardt. (1997). *The Lognormal Distribution in Environmental Applications.* EPA/600/R-97/006. Technology Support Center for Monitoring and Site Characterization, Technology Innovation Office, Office of Research and Development, Office of Solid Waste and Emergency Response, U.S. Environmental Protection Agency, Washington, D.C., December, 1997.

Size, W.B., ed. (1987). *Use and Abuse of Statistical Methods in the Earth Sciences.* Oxford University Press, New York, 169 pp.

Slob, W. (1994). Uncertainty Analysis in Multiplicative Models. *Risk Analysis* **14**(4), 571–576.

Smirnov, N.V. (1939). Estimate of Deviation between Empirical Distribution Functions in Two Independent Samples. *Bulletin Moscow University* **2**(2), 3–16.

Smirnov, N.V. (1948). Table for Estimating the Goodness of Fit of Empirical Distributions. *Annals of Mathematical Statistics* **19**, 279–281.

Smith, A.E., P.B. Ryan, and J.S. Evans. (1992). The Effect of Neglecting Correlations When Propagating Uncertainty and Estimating the Population Distribution of Risk. *Risk Analysis* **12**(4), 467–474.

Smith, E.P. (1994). Biological Monitoring: Statistical Issues and Models. In Patil, G.P., and C.R. Rao, eds., *Handbook of Statistics, Vol. 12: Environmental Statistics.* North-Holland, Amsterdam, Chapter 8, 243–261.

Smith, E.P., and K. Rose. (1991). Trend Detection in the Presence of Covariates: Stagewise Versus Multiple Regression. *Environmetrics* **2**(2), 153–168.

Smith, R.L. (1985). Maximum Likelihood Estimation in a Class of Nonregular Cases. *Biometrika* **72**(1), 67–90.

Smith, R.L. (1994). Use of Monte Carlo Simulation for Human Exposure Assessment at a Superfund Site. *Risk Analysis* **14**(4), 433–439.

Smith, R.M., and L.J. Bain. (1976). Correlation Type Goodness-of-Fit Statistics with Censored Sampling. *Communications in Statistics A—Theory and Methods* **5**(2), 119–132.

Snedecor, G.W., and W.G. Cochran. (1989). *Statistical Methods.* Eighth Edition. Iowa State University Press, Ames, IA, 503 pp.

Sokal, R.F., and F.J. Rolfe. (1981). *Biometry: The Principals and Practice of Statistics in Biological Research.* Second Edition. W.H. Freeman and Company, San Francisco, 859 pp.

Spiegelman, C.H. (1997). A Discussion of Issues Raised by Lloyd Currie and a Cross Disciplinary View of Detection Limits and Estimating Parameters That Are Often at or near Zero. *Chemometrics and Intelligent Laboratory Systems* **37**, 183–188.

Springer, M.D. (1979). *The Algebra of Random Variables*. John Wiley & Sons, New York, 470 pp.

Starks, T.H. (1988). *Evaluation of Control Chart Methodologies for RCRA Waste Sites*. Draft Report by Environmental Research Center, University of Nevada, Las Vegas, for Exposure Assessment Research Division, Environmental Monitoring Systems Laboratory—Las Vegas, Nevada. EPA Technical Report CR814342-01-3.

Stedinger, J. (1983). Confidence Intervals for Design Events. *Journal of Hydraulic Engineering* **109**(1), 13–27.

Stedinger, J.R. (1980). Fitting Lognormal Distributions to Hydrologic Data. *Water Resources Research* **16**(3), 481–490.

Stedinger, J.R., R.M. Vogel, and E. Foufoula-Georgiou. (1993). Frequency Analysis of Extreme Events. In Maidment, D.R., ed. *Handbook of Hydrology*. McGraw-Hill, New York, Chapter 18.

Stein, M. (1987). Large Sample Properties of Simulations Using Latin Hypercube Sampling. *Technometrics* **29**(2), 143–151.

Stephens, M.A. (1970). Use of the Kolmogorov-Smirnov, Cramér-von Mises and Related Statistics Without Extensive Tables. *Journal of the Royal Statistical Society, Series B* **32**, 115–122.

Stephens, M.A. (1986a). Tests Based on EDF Statistics. In D'Agostino, R. B., and M.A. Stevens, eds. *Goodness-of-Fit Techniques*. Marcel Dekker, New York, Chapter 4, pp. 97–193.

Stephens, M.A. (1986b). Tests Based on Regression and Correlation. In D'Agostino, R. B., and M.A. Stevens, eds. *Goodness-of-Fit Techniques*. Marcel Dekker, New York, Chapter 5, pp. 195–233.

Stokes, S.L. (1980). Estimation of Variance Using Judgment Ordered Ranked Set Samples. *Biometrics* **36**, 35–42.

Stolarski, R.S., A.J. Krueger, M.R. Schoeberl, R.D. McPeters, P.A. Newman, and J.C. Alpert. (1986). Nimbus 7 Satellite Measurements of the Springtime Antarctic Ozone Decrease. *Nature* **322**, 808–811.

Stoline, M.R. (1991). An Examination of the Lognormal and Box and Cox Family of Transformations in Fitting Environmental Data. *Environmetrics* **2**(1), 85–106.

Stoline, M.R. (1993). Comparison of Two Medians Using a Two-Sample Lognormal Model in Environmental Contexts. *Environmetrics* **4**(3), 323–339.

Suter, G.W. (1993). *Ecological Risk Assessment.* Lewis Publishers, Boca Raton, FL, 538 pp.

Taylor, J.K. (1987). *Quality Assurance of Chemical Measures.* Lewis Publishers, CRC Press, Boca Raton, FL, 328 pp.

Taylor, J.K. (1990). *Statistical Techniques for Data Analysis.* Lewis Publishers, Boca Raton, FL, 200 pp.

Thompson, K.M., and D.E. Burmaster. (1991). Parametric Distributions for Soil Ingestion by Children. *Risk Analysis* **11**(2), 339–342.

Thompson, K.M., D.E. Burmaster, and E.A.C. Crouch. (1992). Monte Carlo Techniques for Quantitative Uncertainty Analysis in Public Health Risk Assessments. *Risk Analysis* **12**(1), 53–63.

Tiago de Oliveira, J. (1963). Decision Results for the Parameters of the Extreme Value (Gumbel) Distribution Based on the Mean and Standard Deviation. *Trabajos de Estadistica* **14**, 61–81.

Tiago de Oliveira, J. (1983). Gumbel Distribution. In Kotz, S., and N. Johnson, eds. *Encyclopedia of Statistical Sciences.* John Wiley & Sons, New York. Volume 3, pp. 552–558.

Travis, C.C., and M.L. Land. (1990). Estimating the Mean of Data Sets with Nondetectable Values. *Environmental Science and Technology* **24**, 961–962.

Tsui, K.L., N.P. Jewell, and C.F.J. Wu. (1988). A Nonparametric Approach to the Truncated Regression Problem. *Journal of the American Statistical Association* **83**(403), 785–792.

USEPA. (1983). Standard for Remedial Actions at Inactive Uranium Processing Sites; Final Rule (40 CFR Part 192). *Federal Register* **48**(3), 590–604.

USEPA. (1987a). *Data Quality Objectives for Remedial Response Activities, Development Process.* EPA/540/G-87/003. U.S. Environmental Protection Agency, Washington, D.C.

USEPA. (1987b). *Data Quality Objectives for Remedial Response Activities, Example Scenario RI/FS Activities at a Site with Contaminated Soils and Ground Water.* EPA/540/G-87/004. U.S. Environmental Protection Agency, Washington, D.C.

USEPA. (1987c). List (Phase 1) of Hazardous Constituents for Ground-Water Monitoring; Final Rule. *Federal Register* **52**(131), 25942–25953 (July 9, 1987).

USEPA. (1988). Statistical Methods for Evaluating Ground-Water Monitoring from Hazardous Waste Facilities: Final Rule. *Federal Register* **53**, 39, 720–731.

USEPA. (1989a). *Methods for Evaluating the Attainment of Cleanup Standards, Volume 1: Soils and Solid Media.* EPA/230-02-89-042. Office of Policy, Planning, and Evaluation, U.S. Environmental Protection Agency, Washington, D.C.

USEPA. (1989b). *Statistical Analysis of Ground-Water Monitoring Data at RCRA Facilities, Interim Final Guidance.* EPA/530-SW-89-026. Office of Solid Waste, U.S. Environmental Protection Agency, Washington, D.C.

USEPA. (1989c). *Risk Assessment Guidance for Superfund, Volume 1: Human Health Evaluation Manual (Part A), Interim Final Guidance.* EPA/540/1-89/002. Office of Emergency and Remedial Response, U.S. Environmental Protection Agency, Washington, D.C.

USEPA. (1990). *Test Methods for Evaluating Solid Waste, 3rd Edition, Proposed Update I.* Office of Solid Waste and Emergency Response, U.S. Environmental Protection Agency, Washington, D.C.

USEPA. (1991a). *Data Quality Objectives Process for Planning Environmental Data Collection Activities, Draft.* Quality Assurance Management Staff, Office of Research and Development, U.S. Environmental Protection Agency, Washington, D.C.

USEPA. (1991b). *Risk Assessment Guidance for Superfund, Volume 1: Human Health Evaluation Manual, Supplemental Guidance/Standard Default Exposure Factors, Interim Final.* OSWER Directive 9285.6-03. Office of Emergency and Remedial Response, U.S. Environmental Protection Agency, Washington, D.C.

USEPA. (1991c). Solid Waste Disposal Facility Criteria: Final Rule. *Federal Register* **56**, 50978–51119.

USEPA. (1992a). *Characterizing Heterogeneous Wastes: Methods and Recommendations.* EPA/600/R-92/033. U.S. Environmental Protection Agency, Washington, D.C.

USEPA. (1992b). *Methods for Evaluating the Attainment of Cleanup Standards, Volume 2: Groundwater*. EPA/230-R-92-014. Office of Policy, Planning, and Evaluation, U.S. Environmental Protection Agency, Washington, D.C.

USEPA. (1992c). *Statistical Analysis of Ground-Water Monitoring Data at RCRA Facilities: Addendum to Interim Final Guidance*. Office of Solid Waste, U.S. Environmental Protection Agency, Washington, D.C. Currently available as part of: Statistical Training Course for Ground-Water Monitoring Data Analysis, EPA/530-R-93-003, which may be obtained through the RCRA Docket (202/260-9327).

USEPA. (1992d). *Supplemental Guidance to RAGS: Calculating the Concentration Term*. Publication 9285.7-081, May 1992. Intermittent Bulletin, Volume 1, Number 1. Office of Emergency and Remedial Response, Hazardous Site Evaluation Division, OS-230. Office of Solid Waste and Emergency Response, U.S. Environmental Protection Agency, Washington, D.C.

USEPA. (1994a). *Guidance for the Data Quality Objectives Process, EPA QA/G-4*. EPA/600/R-96/005. Office of Research and Development, U.S. Environmental Protection Agency, Washington, D.C.

USEPA. (1994b). *Statistical Methods for Evaluating the Attainment of Cleanup Standards, Volume 3: Reference-Based Standards for Soils and Solid Media*. EPA/230-R-94-004. Office of Policy, Planning, and Evaluation, U.S. Environmental Protection Agency, Washington, D.C.

USEPA. (1994c). *Use of Monte Carlo Simulation in Risk Assessments*. EPA/903-F-94-001. Hazardous Waste Management Division, U.S. Environmental Protection Agency, Region III, Philadelphia, PA.

USEPA. (1995a). *EPA Observational Economy Series, Volume 1: Composite Sampling*. EPA/230-R-95-005. Office of Policy, Planning, and Evaluation, U.S. Environmental Protection Agency, Washington, D.C.

USEPA. (1995b). *EPA Observational Economy Series, Volume 2: Ranked Set Sampling*. EPA/230-R-95-006. Office of Policy, Planning, and Evaluation, U.S. Environmental Protection Agency, Washington, D.C.

USEPA. (1995c). *EPA Risk Characterization Policy and Guidance*. Memorandum from Carol M. Browner, Administrator, U.S. Environmental Protection Agency, March 21, 1995.

USEPA. (1996a). *Guidance for Data Quality Assessment: Practical Methods for Data Analysis, EPA QA/G-9, QA96 Version*. EPA/600/R-96/084, July 1996. Office of Research and Development, U.S. Environmental Protection Agency, Washington, D.C.

USEPA. (1996b). *Soil Screening Guidance: User's Guide.* EPA/540/R-96/018, PB96963505. Office of Emergency and Remedial Response, U.S. Environmental Protection Agency, Washington, D.C., April, 1996.

USEPA. (1996c). *Soil Screening Guidance: Technical Background Document.* EPA/540/R-95/128, PB96963502. Office of Emergency and Remedial Response, U.S. Environmental Protection Agency, Washington, D.C., May, 1996.

USEPA. (1997a). *Guiding Principles for Monte Carlo Analysis.* EPA/630/R-97/001. Risk Assessment Forum, U.S. Environmental Protection Agency, Washington, D.C.

USEPA. (1997b). *Policy for Use of Monte Carlo Analysis in Risk Assessment, With Attachment (Draft).* Memorandum from William P. Wood, January 29, 1997. U.S. Environmental Protection Agency, Washington, D.C.

USEPA. (1998a). *Guidance for Data Quality Assessment: Practical Methods for Data Analysis, EPA QA/G-9, QA97 Version.* EPA/600/R-96/084, January 1998. Office of Research and Development, U.S. Environmental Protection Agency, Washington, D.C.

USEPA. (1998b). *Guidance for Quality Assurance Project Plans, EPA QA/G-5.* EPA/600/R-98/018. Office of Research and Development, U.S. Environmental Protection Agency, Washington, D.C.

van Belle, G., and J.P. Hughes. (1984). Nonparametric Tests for Trend in Water Quality. *Water Resources Research* **20**(1), 127–136.

van Belle, G., and D.C. Martin. (1993). Sample Size as a Function of Coefficient of Variation and Ratio of Means. *The American Statistician* **47**(3), 165–167.

Venables, W.N., and B.D. Ripley. (1999). *Modern Applied Statistics with S-PLUS.* Third Edition. Springer-Verlag, New York, 501 pp.

Verrill, S., and R.A. Johnson. (1987). The Asymptotic Equivalence of Some Modified Shapiro-Wilk Statistics—Complete and Censored Sample Cases. *The Annals of Statistics* **15**(1), 413–419.

Verrill, S., and R.A. Johnson. (1988). Tables and Large-Sample Distribution Theory for Censored-Data Correlation Statistics for Testing Normality. *Journal of the American Statistical Association* **83**, 1192–1197.

Vogel, R.M. (1986). The Probability Plot Correlation Coefficient Test for the Normal, Lognormal, and Gumbel Distributional Hypotheses. *Water Resources Research* **22**(4), 587–590. (Correction, *Water Resources Research* **23**(10), 2013, 1987.)

Vogel, R.M., and N.M. Fennessey. (1993). L Moment Diagrams Should Replace Product Moment Diagrams. *Water Resources Research* **29**(6), 1745–1752.

Vogel, R.M., and D.E. McMartin. (1991). Probability Plot Goodness-of-Fit and Skewness Estimation Procedures for the Pearson Type 3 Distribution. *Water Resources Research* **27**(12), 3149–3158.

Vose, D. (1996). Quantitative Risk Analysis: *A Guide to Monte Carlo Simulation Modelling.* John Wiley & Sons, New York, 328 pp.

Wald, A., and J. Wolfowitz. (1946). Tolerance Limits for a Normal Distribution. *Annals of Mathematical Statistics* **17**, 208–215.

Walsh, J. (1996). *True Odds: How Risk Affects Your Everyday Life.* Merritt Publishing, Santa Monica, CA, 401 pp.

Webster, R., and M.A. Oliver. (1990). *Statistical Methods in Soil and Land Resource Survey.* Oxford University Press, New York, 316 pp.

Weinstein, N.D., P.M. Sandman, and W.K. Hallman. (1994). Testing a Visual Display to Explain Small Probabilities. *Risk Analysis* **14**(6), 895–896.

Weisberg, S. (1985). *Applied Linear Regression.* Second Edition. John Wiley & Sons, New York, 324 pp.

Weisberg, S., and C. Bingham. (1975). An Approximate Analysis of Variance Test for Non-Normality Suitable for Machine Calculation. *Technometrics* **17**(1), 133–134.

Wicksell, S.D. (1917). On Logarithmic Correlation with an Application to the Distribution of Ages at First Marriage. *Medd. Lunds. Astr. Obs.* **84**, 1–21.

Wilk, M.B., and R. Gnanadesikan. (1968). Probability Plotting Methods for the Analysis of Data. *Biometrika* **55**, 1–17.

Wilk, M.B., and S.S. Shapiro. (1968). The Joint Assessment of Normality of Several Independent Samples. *Technometrics* **10**(4), 825–839.

Wilks, S.S. (1941). Determination of Sample Sizes for Setting Tolerance Limits. *Annals of Mathematical Statistics* **12**, 91–96.

WSDOE. (1992). *Statistical Guidance for Ecology Site Managers.* Washington State Department of Ecology, Olympia, WA.

WSDOE. (1993). *Statistical Guidance for Ecology Site Managers, Supplement S-6: Analyzing Site or Background Data with Below-Detection Limit or Below-PQL Values (Censored Data Sets)*. Washington State Department of Ecology, Olympia, WA.

Zacks, S. (1970). Uniformly Most Accurate Upper Tolerance Limits for Monotone Likelihood Ratio Families of Discrete Distributions. *Journal of the American Statistical Association* **65**, 307–316.

Zar, J.H. (1999). *Biostatistical Analysis*. Fourth Edition. Prentice-Hall, Upper Saddle River, NJ.

Zorn, M.E., R.D. Gibbons, and W.C. Sonzogni. (1997). Weighted Least-Squares Approach to Calculating Limits of Detection and Quantification by Modeling Variability as a Function of Concentration. *Analytical Chemistry* **69**, 3069–3075.

INDEX

A

Accuracy, 225, 226, 290
ACF. *See* Autocorrelation
ACL. *See* Groundwater monitoring
Alpha (α), 17, 296, 368
Alpha-level (α-level), 368
Alternative Concentration Limit. *See*
  Groundwater monitoring
Alternative hypothesis. *See* Hypothesis
  test
Analysis of variance, 328, 424, 450, 453,
  545
  in Levene's test, 464
  nonparametric, 459
  parametric, 454
  sample size and power, 502
ANOVA. *See* Analysis of variance
ArcView GIS, 11, 727
Assessment monitoring. *See* Groundwater
  monitoring
Autocorrelation, 651
  accounting for, 671
  AR(1) model, 652
  autocorrelation function (ACF), 654
  autoregressive (AR) model, 652, 671
  autoregressive-moving average
    (ARMA) model, 672
  confidence interval for, 659, 662
  correlogram, 654
  dealing with, 669
  dealing with missing values, 664
  effect on confidence intervals and
    hypothesis tests, 666
  eliminating, 669
  estimating, 655
  lag-1, 651
  lag-k, 651
  moving average (MA) model, 672
  random innovations, 652
  sample autocorrelation function
    (SACF), 656
  temporal, 327
  testing for, 659
  testing for trend in presence of, 676,
    686
Average. *See* Mean

B

Bar chart, 68, 73
Bar plot. *See* Bar chart
Bartlett's test, 462
Beta (β), 17, 296, 368
Beta distribution, 147
Bias, 212, 225, 226
Biased estimator, 212, 225
Binomial coefficient, 179
Binomial distribution, 147, 179
  comparing several proportions, 461
  comparing two proportions, 441
  confidence interval for proportion, 247
  likelihood function, 208
  mean and variance, 181
  method of moments estimator, 206
  minimum variance unbaised estimator,
    214
  normal approximation to, 197
  probability density function, 179
  test of proportion, 385
Blank, 563
Bonferroni method, 305, 321, 452
Bootstrap confidence interval, 258
  basing hypothesis test on, 390, 406,
    439
  bias-corrected and adjusted percentile
    method, 264
  bias-corrected percentile method, 261
  comparison of methods, 265
  empirical percentile method, 259
  for difference between means, 271
  for difference between medians, 271
  for mean, 266
  for mean of skewed distribution, 269
  standard normal method, 259
  vs. permutation test, 439
  when it works well, 265
Box-and-whisker plot. *See* Boxplot
Box-Cox transformations. *See* Data
  transformations
Boxplot, 68, 87
Brushing. *See* Scatterplot matrix
Bubble plot, 114, 121

C

M

Milton Keynes UK
Ingram Content Group UK Ltd.
UKHW021937071024
449327UK00022B/1837